PHYSICAL
CHEMISTRY
WITH
BIOLOGICAL
APPLICATIONS

THE BENJAMIN / CUMMINGS PUBLISHING CO., INC.

Menlo Park, California · Reading, Massachusetts
London · Amsterdam · Don Mills, Ontario · Sydney

PHYSICAL

CHEMISTRY

WITH

BIOLOGICAL

APPLICATIONS

KEITH J. LAIDLER

UNIVERSITY OF OTTAWA

Sponsoring Editor: Bruce Armbruster
Production Editor: Betsey Rhame
Book design: Design Office/Peter Martin
Cover design: Marjorie Spiegelman
Illustrator: Michael Fornalski

ISBN 0-8053-5680-0
ABCDEFGHIJK-HA-798

THE BENJAMIN/CUMMINGS PUBLISHING COMPANY, INC.
2727 SAND HILL ROAD
MENLO PARK, CALIFORNIA 94025

If anyone wishes to search out the
truth of things in serious earnest, he ought
not to select one special science; for
all the sciences are conjoined with
each other and interdependent

Descartes, *Rules for the Direction of the Mind*

PREFACE

Today it is not only students of chemistry who need to have some knowledge of the principles of physical chemistry; students of other sciences are finding it of increasing importance. In the life sciences, in particular, significant advances are now being made by researchers with a strong background in physical chemistry.

APPLICATIONS TO BIOLOGY: Most of the textbooks on physical chemistry include examples and applications that relate almost entirely to pure chemistry and chemical engineering. In teaching physical chemistry to my students, many of whom are majoring in the life sciences, I have felt the need for a book which, while developing the subject rigorously, includes applications not only to chemistry but also to biological systems. This book was written to fill that need.

SCOPE OF THE TEXT: The main aspects of physical chemistry—structure, thermodynamics, electrochemistry, and kinetics—have been dealt with, but I have deemphasized or omitted some topics which do not apply very much outside chemistry. This has allowed me to cover in greater depth certain topics such as transport properties in macromolecular systems, membrane equilibria, action potentials, and the use of isotopes in the study of reaction mechanisms. All of these are of great importance in biology as well as in chemistry.

LEVEL AND COURSE STRUCTURE: The book is intended for students who have taken mathematics and chemistry at the freshman level. They are assumed to have a sound, but not detailed, knowledge of the basic principles of mechanics and electricity, and of differential and integral calculus. The book has been written as a textbook for a course in physical chemistry given either to a class made up entirely of students of the life sciences or to a mixed class including biologists as well as chemists. In my own second-year course given to biologists (two lectures per week for one academic year) I cover much of what is in the book. However, since more advanced material is also included, the book can be used for a longer course or one at the third-year level.

ARRANGEMENT OF TOPICS: The topics have been arranged in what I find to be a convenient and logical sequence, but some instructors may well decide to follow a different order. For example, the first three chapters, which cover quantum mechanics, chemical bonding, molecular spectroscopy, and structure in biological systems, can be dealt with at a later stage. The last two chapters on equilibrium and transport in molecular systems and isotopes in biology stand somewhat apart and can be omitted if time does not permit their inclusion.

USE OF UNITS: The book employs mainly the units of the Système International d'Unités (SI), but I have used the atmosphere (101.325 kPa) and the calorie

(4.184 J), in view of the very wide use of these units in the chemical and biological literature. In many places I have worked with calories and joules side by side, so that the reader can acquire a feeling for their interrelationship during the present transition period.

MANY WORKED-OUT EXAMPLES AND PROBLEMS: A number of numerical examples are worked out in the text, and at the end of each chapter there are problems for the students to solve. Answers are given at the end of the book. Students are strongly urged to work these problems. In doing so they will gain a mastery of the quantitative aspects of physical chemistry. There is also a qualitative side to physical chemistry—indeed to all science—and a mastery of this is achieved by the skillful use of language. All of us are greatly helped in our understanding of a subject if we practice the art of setting down our thoughts in a logical and grammatical form. I have therefore included, at the ends of the chapters, a few essay questions, and my advice to students is to prepare careful answers to them. If the answers are put aside for a couple of weeks and then read, deficiencies will be apparent and can be remedied. There is no better advice to students than that given by Francis Bacon over three hundred years ago:

> *Reading maketh a full man, conference a ready man,*
> *and writing an exact man.*

ACKNOWLEDGMENTS: I am grateful to a number of people who have read all or part of the book and have made valuable suggestions: Drs. Francis Bonner, Peter S. Bunting, Brian E. Conway, A. Keith Dunker, John L. Holmes, R. Norman Jones, J. Gordin Kaplan, Barry A. Morrow, Charles P. Nash, L. Petersen, T. N. Solie, Charles G. Wade, and James S. Wright. My particular thanks go to Dr. John H. Meiser, who read the entire manuscript with great care and made many valuable suggestions.

Keith J. Laidler

January 1978

CONTENTS

CHAPTER 3 STRUCTURE IN BIOLOGICAL SYSTEMS 90

CHAPTER 4 THE FIRST LAW OF THERMODYNAMICS 144

CHAPTER 5 THE SECOND LAW AND CHEMICAL EQUILIBRIUM 182

CHAPTER 6   ELECTROLYTIC CONDUCTIVITY   259

CHAPTER 7   IONS IN AQUEOUS SOLUTION   290

CHAPTER 8   ELECTROCHEMICAL CELLS   332

CHAPTER 9   REACTION KINETICS   365

CHAPTER 10   ENZYME KINETICS   427

LIST OF SYMBOLS

For the most part the units, symbols, and terminology recommended by the International Union of Pure and Applied Chemistry and based on the Système Internationale (SI) have been used. A brief account of the SI is given in Appendix A (p. 556). Note that symbols for quantities are printed in *italics*, symbols for units in roman type. Non-SI units have been retained in a few cases—in particular calorie, atmosphere, and molarity—in view of the widespread use of them in chemical and biological work.

SYMBOL	QUANTITY OR UNIT
a	(linear decadic) absorption coefficient; activity; concentration of A
A	area of cross-section; frequency factor; Helmholtz energy (U-TS); (decadic) absorbance
A	area of cross-section; frequency factor; Helmholtz
Å	angstrom unit (non-SI unit of length $= 10^{-10}$ m)
b	Napierian absorption coefficient; concentration of B
B	Napierian absorbance; proportionality factor in activity-coefficient expression
Bq	becquerel (radioactivity unit $= s^{-1}$)
c	concentration; velocity of light (2.998×10^8 m s^{-1})
C	coulomb (SI derived unit of electric charge $=$ A s)
Ci	curie (radioactivity unit; superseded by Bq)
C_p, C_v	molar heat capacities
d	distance; diameter
D	dissociation energy; diffusion coefficient
D	debye unit (non-SI unit of dipole moment $= 10^{-18}$ esu cm)
e	electronic charge (1.602×10^{-19} C)
e	base of natural logarithms (2.718)
E	energy; energy of activation; electromotive force
f	activity coefficient used with mole fraction; molecular frictional coefficient
F	force; Faraday constant (96 487 C mol^{-1})
g	statistical weight

SYMBOL	QUANTITY OR UNIT
G	Gibbs energy (H-TS)
h	Planck's constant (6.626×10^{-34} J s)
H	enthalpy ($U + PV$)
$\hat{H}$	Hamiltonian operator
Hz	hertz (SI derived unit of frequency $= s^{-1}$)
i	degree of inhibition; square root of -1
I	intensity of radiation; electric current; ionic strength
J	flux (rate of flow through unit area)
J	joule (SI derived unit of energy $= $ kg m^2 s^{-2})
k	rate constant; decay constant
$\mathbf{k}$	Boltzmann constant (1.38×10^{-23} J K^{-1})
kg	kilogram (SI base unit of mass)
K	kelvin (SI base unit of temperature)
K	equilibrium constant
K_a	acid dissociation constant
K_b, K_f	boiling and freezing point constants
K_m	Michaelis constant
K_s	solubility product
l	azimuthal quantum number; thickness; path length
m	metre (SI base unit of length)
m	magnetic quantum number; mass; molality (mol kg^{-1})
mol	mole (SI base unit of amount of substance)
M	molecular weight
M	moles per liter (mol dm^{-3})
$[M]_\lambda^T$	molar optical rotation
n	principal quantum number; number of moles; number of electrons
N	newton (SI derived unit of force $= $ kg m s^{-2})
N_A	Avogadro number (6.022×10^{23} mol^{-1})
p	momentum; vapor pressure
P	pressure
Pa	pascal (SI derived unit of pressure $= $ N m^{-2} $= $ kg m^{-1} s^{-2})
q	heat added to a system
Q	quantity of electricity
r	distance; radius
R	radius; resistance; gas constant (8.314 J K^{-1} mol^{-1})

SYMBOL	QUANTITY OR UNIT
s	spin quantum number; solubility; sedimentation coefficient
s	second (SI base unit of time)
S	entropy
t	time; transport (transference) number
t_2	mean generation time
T	temperature in kelvins; transmittance
T	tesla (SI unit of magnetic flux density $= kg\ A^{-1}\ s^{-2}$ $= V\ m^{-2}\ s = 10^4$ gauss)
u	electric mobility
u_o	electrophoretic mobility
U	internal energy; potential energy
v	rate, velocity
$\overline{v}$	specific volume
V	volume; electric potential; limiting rate at high substrate concentration
V	volt (SI derived unit of electric potential difference $= kg\ m^2\ s^{-3}\ A^{-1} = J\ A^{-1}\ s^{-1}$)
w	work done on a system
x	distance; mole fraction; concentration of product of reaction
y	activity coefficient used with concentration
z	charge number of an ion
Z	atomic number
α (alpha)	degree of dissociation, Λ/Λ_o; alpha particle (helium nucleus, ^4_2He)
$[\alpha]$	optical rotation
β, β^- (beta)	beta particle (electron)
γ (gamma)	gamma photon; ratio C_p/C_v; activity coefficient used with molality
ε (epsilon)	absorption coefficient; relative permittivity (dielectric constant)
ε_o	permittivity of vacuum ($8.854 \times 10^{-12}\ C^2\ N^{-1}\ m^{-1}$)
ζ (zeta)	electrokinetic (or zeta) potential
η (eta)	coefficient of viscosity
θ (theta)	angle; fraction of surface covered
κ (kappa)	electrolytic conductivity; reciprocal of radius of ionic atmosphere
λ (lambda)	wavelength

SYMBOL	QUANTITY OR UNIT
Λ (lambda)	molar conductivity
μ (mu)	dipole moment
ν (nu)	frequency of radiation
$\bar{\nu}$	wavenumber; average number of bound molecules
π (pi)	osmotic pressure; ratio of circumference to radius of circle (3.1416)
ρ (rho)	density; charge density
σ (sigma)	molecular orbital; charge per unit area
τ (tau)	half-life
ϕ (phi)	angle; electric potential; wave function
ψ (psi)	wave function
ω (omega)	angular velocity
Ω (omega)	ohm (SI derived unit of electric resistance $= kg\ m^2\ s^{-3}\ A^{-2} = V\ A^{-1}$)

PHYSICAL CHEMISTRY WITH BIOLOGICAL APPLICATIONS

1

QUANTUM MECHANICS AND CHEMICAL BONDING

Physical chemistry is concerned with two main topics: the structure of matter, and the processes of chemical change. We must understand structure—i.e., matter at equilibrium—before we can deal with the more complicated problem of how chemical transformations occur. The first part of this book, therefore, will be more concerned with chemical and biological structure; the latter part will be concerned with processes.

An important characteristic of biological systems is that they involve large molecules. The structures of large molecules, and the transformations they undergo, are naturally more difficult to understand than the structures and transformations of small molecules. The subject of physical chemistry has been developed for the most part by chemists interested primarily in simple molecules, and much attention has been given to gaseous systems. By studying such relatively simple systems, the physical chemist has been able to develop a number of very important general relationships which also apply to the complex molecular systems in which the biologist is interested. Some of these general relationships (such as the concept of Gibbs energy, which is concerned with the direction of chemical change) are vital to a true understanding of biology. It is the object of this book to identify and explain a number of these important principles of physical chemistry and to show how these principles relate to living systems.

1.1 CHEMICAL STRUCTURE

In much of their work biologists need to have an understanding of the structures of molecules, and of molecular aggregates such as exist in the membranes and interiors of cells. A very general account of chemical structure is given in this chapter. Emphasis is more on the theoretical principles helpful in understanding structure and on the conclusions to which the theories lead, rather than on the mathematical operations, which are sometimes fairly complicated. It is assumed that the reader has had some introduction to these topics at the level of first-year university courses in physics and chemistry.

During the past few decades, there have been great advances in our understanding of chemical structure. A hundred years ago it had come to be realized that the atoms in a molecule are held together by chemical bonds, but little was then understood about the precise nature of such bonds. During the final years of the last century, scientists recognized that certain rules—the *valency* rules—apply to the number of bonds which can be formed by an atom; later, the concept of valency was interpreted in terms of the electronic configurations of the atoms. In particular, the American chemist Gilbert Newton Lewis (1875–1946) suggested that covalent bonds frequently involve the pairing of electrons, and that by sharing electrons atoms often attain the electronic configurations of the noble gases. The oxygen atom, for example, has two fewer electrons than the noble gas neon; thus oxygen has six, rather than eight, electrons in its valency shell. By combining with two atoms of hydrogen, each of which has one electron, oxygen thus achieves the noble-gas configuration. Lewis diagrams, in which only valency electrons are shown, are still widely used. The Lewis diagram for water, for example is

$$
\begin{array}{l}
\text{H} \\[4pt]
\overset{\cdot\cdot}{\underset{\cdot\cdot}{:\text{O}:}}\ \text{H}
\end{array}
$$

The development of theories of bonding naturally went hand in hand with the development of theories of the electronic structure of the atom. An important advance was made in 1911 by the British physicist Ernest Lord Rutherford (1871–1937). According to Rutherford, the atom is a miniature solar system in which the electrons move in orbits around the nucleus. This picture, however, involved a theoretical difficulty: electrons moving in orbits undergo an acceleration toward the nucleus and, according to Clerk Maxwell's laws of electromagnetics, should therefore emit radiation and lose energy. This dilemma was resolved in 1913 by the Danish physicist Niels Bohr (1855–1962). On the basis of quantum theory, Bohr postulated that the orbital angular momentum of the electron must be an integral multiple of a certain fundamental quantity, and that when this condition is satisfied the electron will not emit radiation. In other words, an electron can only move in certain "stationary" orbits. Radiation is emitted when an electron jumps from one stationary orbit to another one in which it has lower energy. The energy of the emitted radiation is equal to its frequency $v(\text{s}^{-1})$ multiplied by Planck's constant h (6.626×10^{-34} J s)†

$$E = hv \tag{1.1}$$

and is thus equal to the energy difference between the two orbits.

---

† See Appendix A for a discussion of units.

Bohr's original theory considered only circular orbits, and involved a quantum number n, known as the *principal* quantum number. In 1916 the German physicist Arnold Sommerfeld (1868–1951) suggested that the theory could be improved by taking elliptical orbits into account. This theory involved the introduction of a second quantum number. We need not pursue these theories in further detail, because they were replaced by a much more satisfactory theoretical treatment of atoms and molecules—one based on the concept that electrons have not only a particle character but also a wave character.

1.2   THE WAVE NATURE OF ELECTRONS

Several important investigations led to the conclusion that electrons have wave properties. The photoelectric phenomenon, and its interpretation by Albert Einstein (1879–1955), showed that in certain experiments light can behave as if it were a beam of particles, which are now known as *photons*. In a photoelectric experiment, a beam of light strikes the surface of a metal, and electrons are emitted. The kinetic energy of each electron emitted, $\frac{1}{2}mv^2$, is equal to the energy of the photon, hv, minus the minimum amount of energy required to remove the electron from the metal (the *work function* of the metal). Increasing the frequency v of the beam of light increases the kinetic energy of each electron emitted, but does not increase the number of electrons emitted. Increasing the intensity of the light increases the number of the electrons emitted but has no effect on the energy of each electron. The refraction of light and other effects show that light must have wave properties, but the wave theory cannot explain the photoelectric effect, which requires us to regard light as a beam of particles, each one of which can interact with an electron and remove it from the metal. In other words, we need a *dual theory* of radiation, which in some experiments shows wave properties and in other experiments shows particle properties.

This realization led the French physicist Louis Victor, Prince de Broglie, to suggest in 1924 the converse hypothesis, that particles such as electrons can also have wave properties. The wavelength of the wave associated with a particle of mass m moving with velocity v was obtained by de Broglie on the basis of the following reasoning: Einstein had suggested in his general theory of relativity that mass m and energy E are related by the equation

$$E = mc^2 \tag{1.2}$$

where c is the velocity of light. If this is combined with equation (1.1) the result is

$$hv = mc^2 \tag{1.3}$$

The wavelength λ is related to the frequency by

$$\lambda v = c \tag{1.4}$$

and elimination of v between equations (1.3) and (1.4) gives

$$\lambda = \frac{h}{mc} \tag{1.5}$$

This equation is for electromagnetic radiation, and it was suggested by de Broglie that a parallel expression would apply to a beam of particles of mass m and velocity v:

$$\lambda = \frac{h}{mv} = \frac{h}{p} \tag{1.6}$$

where p is the momentum of the particle. For example, an electron has a mass of 9.11×10^{-31} kg, and when accelerated by a potential of 100 volts it has a velocity of 5.90×10^6 metres† per second (m s^{-1}). The de Broglie wavelength is therefore

$$\lambda = \frac{6.626 \times 10^{-34} \text{ J s } (= \text{kg m}^2 \text{ s}^{-1})}{9.11 \times 10^{-31} \text{ kg} \times 5.90 \times 10^6 \text{ m s}^{-1}}$$

$$= 1.23 \times 10^{-10} \text{ m}$$

$$= 0.123 \text{ nanometres (nm)}$$

$$= 1.23 \text{ Å}$$

This wavelength is of a magnitude similar to that of the distance between neighboring atoms or ions in a crystal, and it was therefore suggested that a beam of electrons could be diffracted by using a crystal as a diffraction grating. This prediction was confirmed experimentally in 1927 by G. P. Thomson and A. Reid, and by C. Davisson and L. H. Germer. The diffraction of electrons is now commonly employed as a technique for investigating molecular structure.

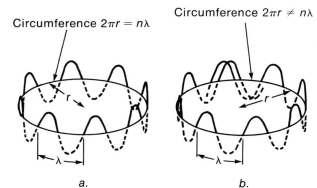

Circumference $2\pi r = n\lambda$

Circumference $2\pi r \neq n\lambda$

a.

b.

Figure 1.1

The de Broglie wave associated with an electron moving in an orbit.
a. Constructive interference; the wave fits into the orbit.
b. Destructive interference; the wave does not fit. If in this diagram we continued the waves indefinitely they would completely obliterate one another, the net amplitude becoming zero.

† Since the metre is an international (SI) unit (see Appendix A), the word should be spelled in this way. Many U. S. scientists now adopt this convention, which has the advantage that *metre* can be used for the unit, *meter* for the instrument.

The realization that electrons have wave properties leads at once to an interpretation of Bohr's hypothesis of stationary orbits. Figure 1.1 shows the de Broglie wave associated with an electron in a Bohr orbit of radius r. If the circumference $2\pi r$ is equal to an integer n multiplied by the wavelength λ, the wave will fit into the orbit an integral number of times, and *constructive interference* will occur (Figure 1.1a). The condition for this is

$$2\pi r = n\lambda \tag{1.7}$$

If, on the other hand, the circumference $2\pi r$ is not an integral multiplied by λ, the wave will not fit into the orbit, and there is *destructive interference* (Figure 1.1b); such an orbit is not a stationary one. The requirement that, for constructive interference, n must be an integer thus leads to a *quantum* condition. Substitution of the de Broglie equation (1.6) into (1.7) leads to

$$2\pi r = \frac{nh}{mv} \tag{1.8}$$

or

$$mvr = n\,\frac{h}{2\pi} \tag{1.9}$$

The quantity mvr is the orbital angular momentum of the electron, and equation (1.9) shows that it must be an integer multiplied by $h/2\pi$. This same relationship had been earlier suggested arbitrarily by Bohr, and it is of great interest that the wave concept provides a simple interpretation of this relationship.

An important consequence of the photoelectric effect is that it is impossible to make simultaneously accurate measurements of the position and momentum of a particle, such as an electron. The reason for this is that in order to make a measurement on an electron, we must cause light to interact with the electron, and observe what happens. To make an accurate determination of the position of an electron, we must use light of short wavelength; otherwise the position is not well defined. Light of short wavelength (high frequency), however, implies photons of high energy. When these interact with an electron, they change its momentum. We can avoid changing the momentum of the electron only by using light of long wavelength, but then the position of the electron will not be precisely determined. These conclusions are expressed in the uncertainty principle, proposed by the German physicist Werner Karl Heisenberg (1901–1976). According to this principle, the product of the uncertainty in the position of a particle, Δq, and the uncertainty in the momentum, Δp, is approximately equal to Planck's constant h divided by 4π:

$$\Delta q\,\Delta p \cong \frac{h}{4\pi} \tag{1.10}$$

Since $p = mv$,

$$\Delta q \, \Delta v \cong \frac{h}{4\pi m} \tag{1.11}$$

and the product of the uncertainties in the position and the velocity is therefore less for heavier particles. The uncertainty principle is thus of particular importance for light particles, such as electrons.

EXAMPLE

Consider a colloidal particle with a diameter of 10^3 nm and a mass of 6×10^{-16} kg. Suppose that we measure the position to within 1.0 nm, which is about the resolving power of an electron microscope (p. 109). Calculate the indeterminacy in the velocity.

SOLUTION

From equation (1.11), the product of the uncertainties in position and velocity is

$$\Delta q \, \Delta v \cong \frac{h}{4\pi m} = \frac{6.626 \times 10^{-34} \text{ J s} \ (= \text{kg m}^2 \text{ s}^{-1})}{4 \times 3.14 \times 6 \times 10^{-16} \text{ kg}}$$

$$\cong 10^{-19} \text{ m}^2 \text{ s}^{-1}$$

Therefore since

$$\Delta q = 10^{-9} \text{ m},$$

$$\Delta v = 10^{-10} \text{ m s}^{-1}$$

With this uncertainty in velocity, the position of the particle one second later would be uncertain to within 2×10^{-10} m, or 0.2 nm. This is only 0.02% of the diameter of the particle, and the uncertainty principle therefore does not present a serious problem for particles of this magnitude. For particles of molecular sizes, the uncertainty is much greater.

In the Bohr-Sommerfeld theories of the atom, the electrons are moving in orbits which are precisely specified, and the velocities are given exactly. Those theories are therefore concerned with properties which cannot be measured precisely. This difficulty is avoided, however, if one develops theories based on the wave properties of electrons; we have already seen, with reference to Figure 1.1, that such theories remove some of the arbitrariness inherent in the Bohr-Sommerfeld approach. Modern theories of atoms and molecules are, therefore, wave theories, which have led to a very considerable increase in our understanding. In the remainder of this chapter we will describe aspects of wave mechanics, or quantum mechanics, that will be of help to biologists in appreciating the nature of the molecular structures with which they are concerned.

1.3 QUANTUM MECHANICS
THE SCHRÖDINGER EQUATION

We have seen earlier in this chapter that, for the case of electron orbits, the fitting-together of de Broglie waves leads to quantum restrictions on the nature of the orbits. For this approach to be useful for a wide range of atomic and molecular problems, we must have a more general procedure. One is provided by the *wave equation* suggested in 1926 by the Austrian theoretical physicist Erwin Schrödinger (1887–1961). Another treatment, leading to essentially the same results, was suggested at about the same time by W. Heisenberg. Schrödinger's method is the more convenient for most atomic and molecular problems and is the one which will be presented here. Schrödinger did not derive his equation; he suggested it on the basis of analogy with equations for electromagnetic radiation.

In the study of an atom, the quantities of interest are the energies of the electrons and the positions of the electrons relative to the nucleus and to one another. The Schrödinger equation yields directly the electron energies, but does not yield precise information on the positions of the electrons; instead the equation leads to the *probability* that the electron is in a particular small region of space. That it does only this is consistent with the uncertainty principle. The Schrödinger wave equation gives rise to a function ψ of the coordinates representing the position of the electron (for example, the Cartesian coordinates x, y, and z). This function ψ is known as the *wavefunction* or the *eigenfunction*. Although ψ cannot be measured experimentally, we will see later that it is related to the probability that the electron is present in a particular small element of volume.

A very simple problem which can be treated on the basis of the Schrödinger equation is that of an electron of mass m which is able to move in one dimension only, say along the X axis. The total energy of the electron is the sum of its potential and kinetic energies. The potential energy depends on the environment and may be written in general as $U(x)$. In quantum-mechanical theory, the familiar $\frac{1}{2}mv^2$ term for kinetic energy is replaced by the differential operator

$$-\frac{h^2}{8\pi^2 m}\frac{\partial^2}{\partial x^2}$$

The Schrödinger equation then takes the form

$$-\frac{h^2}{8\pi^2 m}\frac{\partial^2\psi}{\partial x^2} + U(x)\psi = E\psi \tag{1.12}$$

where E is the total energy of the electron. For a particular system, the function $U(x)$ is known, and the problem is therefore to solve this differential equation and obtain ψ and E.

When this is attempted, solutions of the differential equation can only be obtained for certain values of the energy E. Such energy values are known as the *eigenvalues* of the equation. They are the permitted quantized energy levels for the system, and we see that the quantum restriction has appeared naturally, and not arbitrarily as in the Bohr treatment. Corresponding to each eigenvalue there is one or more eigen-function ψ.

Usually, of course, we are concerned with three-dimensional rather than one-dimensional problems. If we decide to work with Cartesian coordinates, the potential energy will be a function of x, y, and z, and can be written as $U(x,y,z)$. The Schrödinger equation now takes the form

$$-\frac{h^2}{8\pi^2 m}\left(\frac{\partial^2\psi}{\partial x^2} + \frac{\partial^2\psi}{\partial y^2} + \frac{\partial^2\psi}{\partial z^2}\right) + U(x,y,z)\psi = E\psi \qquad (1.13)$$

The wave function obtained, $\psi(x, y, z)$, depends on the three coordinates x, y, and z.

The solutions of the Schrödinger equation will be different for different potential functions $U(x,y,z)$. When the equation is applied to the hydrogen atom, or to hydrogen-like ions where there is a single electron, the function U takes a relatively simple form, especially when one uses polar rather than Cartesian coordinates. In such cases, it is not too difficult to obtain explicit solutions for the eigenvalues and eigenfunctions, so that one has a complete understanding of the permitted energy levels and electron probability distributions for such systems. For more complicated atomic systems, and for almost all molecules, exact solutions of the Schrödinger equation are impossible to obtain. One must then resort to approximate methods, one of which—the variation method—is considered later.

The three-dimensional Schrödinger equation (1.13) could have been written in the form

$$\left(-\frac{h^2}{8\pi^2 m}\nabla^2 + U\right)\psi = E\psi \qquad (1.14)$$

where ∇^2 (del squared) represents the operation of partial double differentiation with respect to x, y, and z:

$$\nabla^2 = \frac{\partial^2}{\partial x^2} + \frac{\partial^2}{\partial y^2} + \frac{\partial^2}{\partial z^2} \qquad (1.15)$$

The operator in the brackets in equation (1.14) is known as the Hamiltonian operator, since it is related to an expression for energy given by the Irish mathematician Sir William Rowan Hamilton (1805–1865); this operator is given the symbol $\hat{H}$:

$$\hat{H} = -\frac{h^2}{8\pi^2 m}\nabla^2 + U \qquad (1.16)$$

With this notation, the Schrödinger equation is very compactly written as

$$\hat{H}\psi = E\psi \qquad (1.17)$$

This is a very general form of the equation, and certain rules are employed to express the operator $\hat{H}$ in particular cases. The procedure is to express the total energy (kinetic plus potential) in Hamilton's form, and then to make certain transformations which convert the expression into an operator. The details are outside the scope of this book but can be found in most books on quantum mechanics.

NORMALIZATION

The eigenfunction ψ provides us with information about the probability that the electron is present at any given instant in a particular element of volume—in other words, about the *probability distribution* of the electron. We will now pursue this important matter in further detail. In certain cases, ψ is a *real*—as opposed to a complex—function, and in those cases the square of the wavefunction, ψ^2, is proportional to the probability that the electron is present in a certain small region of space. This is illustrated in Figure 1.2, which shows an element of volume $dx\,dy\,dz$ situated at a certain distance from the nucleus. If the value of the wavefunction within

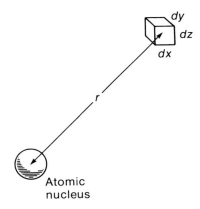

Figure 1.2

An infinitesimal cube of volume $dx\,dy\,dz$ is situated at a certain distance r from the nucleus of an atom. If ψ is the normalized eigenfunction (see the following discussion of normalization), the probability that the electron is in the cube is $\psi^2\,dx\,dy\,dz$.

the volume element is ψ, the probability that the electron is actually within the small cube is proportional to

$$\psi^2\,dx\,dy\,dz$$

If, on the other hand, the wavefunction is a *complex* function—that is, it involves i, the square root of minus one—the probability is instead proportional to

$$\psi\psi^*\,dx\,dy\,dz$$

where ψ^* is the complex conjugate of ψ (that is, the function ψ in which we have changed the sign of terms involving i).

We have said that the probabilities are *proportional* to $\psi\psi^*\,dx\,dy\,dz$, but it is possible to *normalize* wavefunctions so that the probability is *equal* to this function. First, we should note that there is an arbitrariness about a wavefunction obtained by solving the Schrödinger equation (1.17). Suppose that, on solving the equation for a particular system, we have obtained an eigenfunction ψ. Then, if we multiply ψ by any number a, the function $a\psi$ is still an eigenfunction, because

$$\hat{H}(a\psi) = E(a\psi) \tag{1.18}$$

Thus there is, for a given Schrödinger equation, an infinite set of eigenfunctions differing from each other by numerical factors. We want to know which particular eigenfunction is such that

$$\psi\psi^* \, dx \, dy \, dz$$

is *equal* to the probability that the electron is in the volume element of dimensions $dx \, dy \, dz$. To determine this, we make use of the fact that the total probability of finding the electron anywhere at all in space must be unity. The integral of $\psi\psi^* \, dx \, dy \, dz$, with x, y, and z running from infinity to minus infinity, must therefore be unity:

$$\int_{-\infty}^{\infty} \int_{-\infty}^{\infty} \int_{-\infty}^{\infty} \psi\psi^* \, dx \, dy \, dz = 1 \tag{1.19}$$

This integral is conveniently written as

$$\int \psi\psi^* \, d\tau = 1 \tag{1.20}$$

where $d\tau$ represents $dx \, dy \, dz$, and it is understood that the integration is done over all space.

Suppose that we have obtained an eigenfunction ϕ for the Schrödinger equation, and that when we carry out the integration we obtain not unity, but a number, b:

$$\int \phi\phi^* \, d\tau = b \tag{1.21}$$

The eigenfunction therefore requires adjustment to make the integral unity, and this is done by dividing it by $\sqrt{b}$, since

$$\int \frac{\phi}{\sqrt{b}} \frac{\phi^*}{\sqrt{b}} \, d\tau = 1 \tag{1.22}$$

The new function $\psi = \phi/\sqrt{b}$ is therefore such that $\psi\psi^* \, d\tau$ correctly represents the probability in any region. This process of adjusting an eigenfunction to satisfy equation (1.20) is known as *normalization*, and the resulting function a *normalized eigenfunction*.

1.4   THE STRUCTURE OF THE ATOM
HYDROGEN-LIKE ATOMS

It is not difficult to obtain solutions of the Schrödinger equation for systems containing a single electron, of charge $-e$, and a nucleus having a charge of $+Ze$, where Z is the atomic number. If $Z = 1$, this system is the hydrogen atom, while if $Z = 2$, it is the He$^+$ ion. In general such systems are referred to as "hydrogen-like." If the electron is at a distance r from the nucleus, its potential energy is $-Ze^2/r$ (the product of the charges divided by r), and the three-dimensional Schrödinger equation (1.14) becomes

$$-\frac{h^2}{8\pi^2 m} \nabla^2 \psi - \frac{Ze^2}{r} \psi = E\psi \tag{1.23}$$

The potential energy depends only on the distance r, and not on angles; that is, it is symmetrical about the nucleus. This suggests that it will be more convenient to work with polar coordinates r, θ, and ϕ, which are shown in Figure 1.3. These are related to the Cartesian coordinates in the manner shown in the figure, and the Schrödinger equation in polar coordinates is

$$-\frac{h^2}{8\pi^2 m} \frac{1}{r^2 \sin\theta} \left[\sin\theta \frac{\partial}{\partial r}\left(r^2 \frac{\partial\psi}{\partial r}\right) + \frac{\partial}{\partial\theta}\left(\sin\theta \frac{\partial\psi}{\partial\theta}\right) \right.$$
$$\left. + \frac{1}{\sin\theta}\frac{\partial^2\psi}{\partial\phi^2} \right] - \frac{Ze^2}{r} \psi = E\psi \tag{1.24}$$

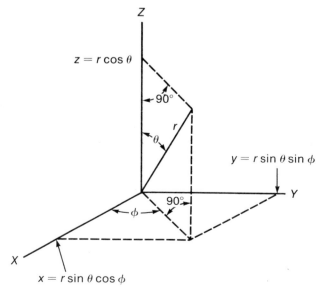

z = r cos θ

y = r sin θ sin φ

x = r sin θ cos φ

Figure 1.3

Polar coordinates and their relation to Cartesian coordinates.

It is outside the scope of this book to discuss the methods of solution of this equation, a matter dealt with in most texts on quantum mechanics. We shall be content here with presenting the form of the solutions.

The wavefunctions obtained as solutions of the equation are, in general, functions of r, θ, and ϕ, and can be written as $\psi(r,\theta,\phi)$. They can be expressed as the product of three functions: one is a function of r only, written as $R(r)$; one is a function of θ only, written as $\Theta(\theta)$; and the third $\Phi(\phi)$, is a function of ϕ only:

$$\psi(r,\theta,\phi) = R(r)\Theta(\theta)\Phi(\phi) \tag{1.25}$$

These individual functions R, Θ, and Φ give rise to the three orbital quantum numbers n, l, and m. We have seen that solutions of the Schrödinger equation are possible only for certain values of the total energy E. For hydrogen-like atoms, the permitted total energy values are given by the equation

$$E_n = -\frac{4\pi^2 me^2 Z^2}{h^2 n^2} \qquad n = 1, 2, 3, \ldots \tag{1.26}$$

The sign is negative because the energy relates to the completely separated nucleus and electron; since these attract each other, the energy decreases as they come together. We see that the energy depends on the quantum number n. The most negative energy occurs when $n = 1$, and this corresponds to the most stable state of the atom. It is known as the *ground state* and, for reasons which will be explained, as the $1s$ state. The next higher level is when $n = 2$, and so forth.

When $n = 1$, only one solution is possible for the wave equation. The normalized wavefunction obtained is

$$\psi_{1s} = \frac{1}{\sqrt{\pi}}\left(\frac{Z}{a_o}\right)^{3/2} e^{-Zr/a_o} \tag{1.27}$$

where a_o, which has the dimensions of a distance, is given by

$$a_o = \frac{h^2}{4\pi^2 me^2} \tag{1.28}$$

and is equal to 0.0529 nm = 0.529 Å. Note that this wavefunction is real and shows no dependence on the angles θ and ϕ, but only on the distance r. In other words, the wavefunction is *spherically symmetrical* about the nucleus.

We have seen that if an eigenfunction is a real function and is normalized, its square, known as the *probability density*, is a measure of the probability that the electron resides in a particular region of space. Plots of ψ_{1s} and ψ_{1s}^2 against r are shown for the hydrogen atom ($Z = 1$) in Figures 1.4a and 1.4b. The probability density ψ_{1s}^2 has a maximum value at $r = 0$ and falls toward zero at large r values. It is of greater interest, however, to ask about the probability that the electron is present at a distance between r and $r + dr$ from the nucleus—in other words, whether the electron is between two concentric spherical surfaces, one having a radius of r and the other of $r + dr$. The volume between such concentric surfaces is $4\pi r^2\ dr$, and the probability that the electron lies within the distances r and $r + dr$ is thus

$$4\pi r^2\psi^2\ dr$$

A plot of $4\pi r^2\psi^2$ against r is shown in Figure 1.4c, the curve passes through a maximum when $r = a_o = 0.0529$ nm $= 0.529$ Å. This distance is therefore the most probable distance between the nucleus and the electron, and it is of interest that this is the radius of the ground-state orbit given by the Bohr theory. The difference between the two theories is thus very clear: according to the Bohr theory the electron

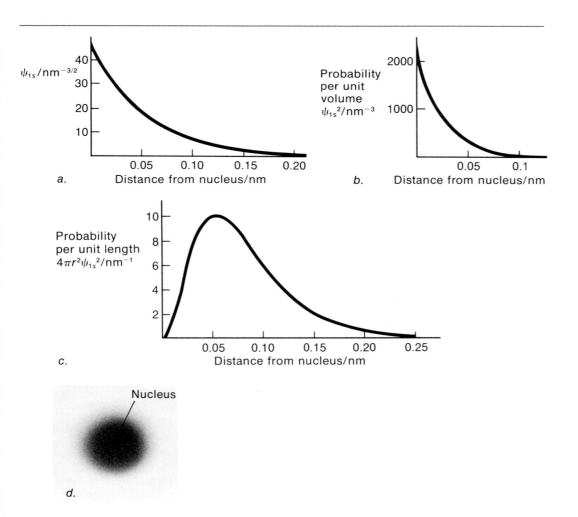

a. $\psi_{1s}/nm^{-3/2}$ Distance from nucleus/nm

b.   Probability per unit volume $\psi_{1s}{}^2/nm^{-3}$ Distance from nucleus/nm

c.   Probability per unit length $4\pi r^2\psi_{1s}{}^2/nm^{-1}$ Distance from nucleus/nm

d.   Nucleus

Figure 1.4

The ground state (1s) of the hydrogen atom: a. ψ_{1s} plotted against r.   b. $\psi_{1s}{}^2$ plotted against r. c. $4\pi r^2\ \psi_{1s}{}^2$ plotted against r.   d. Representation of the electron probability distribution.

in the ground state of the hydrogen atom is in a precise orbit of radius a_o, while according to quantum mechanics the *most probable* distance is a_o, the electron possibly being at other distances from the nucleus. In quantum mechanics, instead of referring to orbits, we speak of *orbitals*, and this term is used also for the wavefunction. Figure 1.4*d* shows a schematic representation of the electron "cloud" in the ground state of the hydrogen atom. The intensity of the shading gives some idea of the *electron density*. For this case of a single-electron atom, the electron-density distribution is the same as the probability distribution. However, if more electrons are present, the probability distributions for the individual orbitals must be added together to get the net electron-density distribution in the atom.

Table 1.1   The wavefunctions obtained for a hydrogen-like atom, when $n = 2$.

$R(r)$	$\Phi(\theta)$	$\Phi(\phi)$	l	m	Symbol
$\dfrac{1}{4\sqrt{2\pi}}\left(\dfrac{Z}{a_o}\right)^{3/2}\left(2 - \dfrac{Zr}{a_0}\right)e^{-Zr/2a_o}$	1	1	0	0	$2s$
$\dfrac{1}{4\sqrt{2\pi}}\left(\dfrac{Z}{a_o}\right)^{3/2}\dfrac{Zr}{a_o}e^{-Zr/2a_o}$	$\cos\theta$	1	1	0	$2p_z$
"	$\sin\theta$	$\cos\theta$	1	†	$2p_x$
"	$\sin\theta$	$\sin\theta$	1	†	$2p_y$

† A certain combination of p_x and p_y orbitals gives $m = 1$, and another combination gives $m = -1$.

Next we consider the eigenfunctions obtained when $n = 2$ (see equation 1.26). Four wavefunctions are now possible and are shown in Table 1.1, where they are split into three factors $R(r)$, $\Theta(\theta)$, and $\Phi(\phi)$. The first function listed, designated 2*s*, has no dependence on θ and ϕ, and is therefore *spherically symmetrical*. Plots of ψ_{2s} and ψ_{2s}^2 against r are shown in Figures 1.5*a* and 1.5*b*, and of $4\pi r^2\psi_{2s}^2$ against r in Figure 1.5*c*. This curve shows that the probability that the electron is at a distance between r and $r + dr$ from the nucleus goes through two maxima. This orbital is represented schematically in Figure 1.5*d*.

The other three functions in Table 1.1 form a set of three *p* orbitals which show angular dependence. The p_x orbital corresponds to a maximum electron density in both directions along the Z axis, since $\cos\theta$ has its maximum value of unity when $\theta = 0$ and a value of -1 when $\theta = 180°$. Figure 1.6*c* shows a plot of $4\pi r^2 R_{2p}^2$, i.e., $4\pi r^2\psi^2$ when $\theta = 0$ or $\theta = 180°$, and we now see a single maximum. This orbital is represented in Figure 1.6*d*. The orbital designated $2p_x$ is equivalent to $2p_z$ except in spatial orientation; it has maximum value when $\theta = 90°$ ($\sin\theta = 1$) and when $\phi = 0$ or $180°$ ($\cos\phi = \pm 1$). For this orbital shown in Figure 1.6*f* there is, therefore, a maximum electron density along the X axis (see Figure 1.3). The orbital designated $2p_y$ is shown in Fig. 1.6*e*; it is equivalent to $2p_z$ and $2p_x$, but the maximum electron density is now along the Y axis.

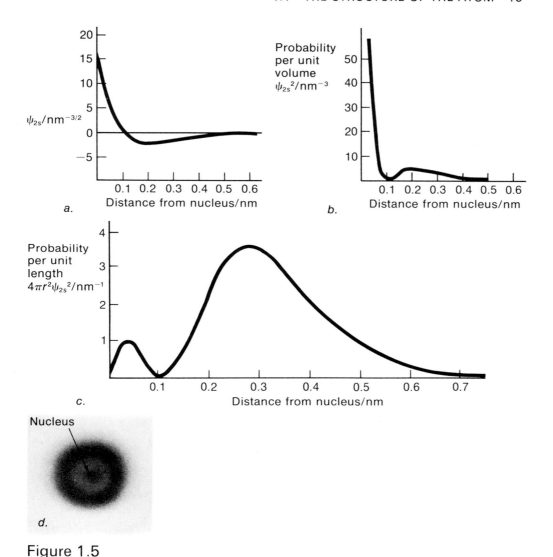

Figure 1.5

The 2s state of the hydrogen atom: *a.* ψ_{2s} plotted against *r*.   *b.* ψ_{2s}^2 plotted against *r*.
c. $4\pi r^2 \psi_{2s}^2$ plotted against *r*.   *d.* Schematic representation of the orbital.

Besides the principal quantum number *n*, there are two other orbital quantum numbers. One of these is a measure of the orbital angular momentum of the electron; it is called the *azimuthal* quantum number and is given the symbol *l*. Its permitted values are related to the value of *n* as follows:

$$l = 0, 1, 2, 3, 4, \ldots, n - 1$$

$$s \quad p \quad d \quad f \quad g$$

Letters are assigned to the various values of *l*, as shown. If, for example, the electron has a principal quantum number *n* equal to 3, the possible *l* values are 0, 1, and 2.

An electron having $n = 3$ and $l = 1$ is called a $3p$ electron. These symbols were employed in Table 1.1. The general shapes of the orbitals are determined by the l value. For an s electron ($l = 0$), there is spherical symmetry; for a p electron ($l = 1$), the shapes are as shown in Figure 1.6d–f.

The quantum-mechanical treatment also leads to a third quantum number, the *magnetic quantum number*, which is given the symbol m. This quantum number is related to the *orientation in space* of the orbital. Its permitted values are related to the value of l and are

$$l, l - 1, \ldots, 0, \ldots, -l$$

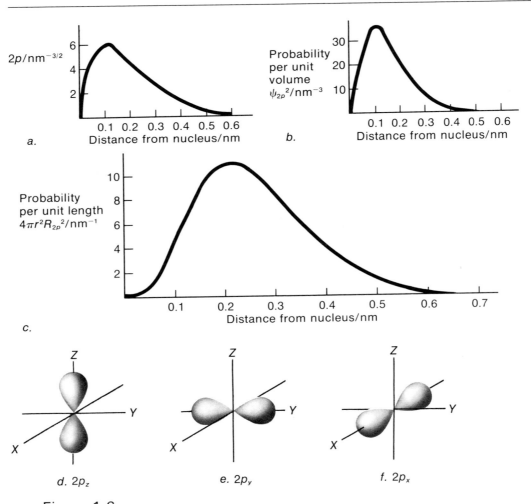

d. $2p_z$ e. $2p_y$ f. $2p_x$

Figure 1.6

The $2p$ state of the hydrogen atom: *a.* R_{2p} plotted against r.   *b.* R_{2p}^2 plotted against r.
c. $4\pi r^2 R_{2p}^2$ plotted against r.   *d.*–*f.* Shapes of the $2p$ orbitals.

We saw in Table 1.1 the eigenfunctions for a p orbital, for which the permitted m values are 1, 0, and -1, and Figure 1.6 shows their orientation in space. The case of d orbitals is shown in Figure 1.7. The magnetic properties of atoms are related to the magnetic quantum number.

For hydrogen-like atoms, which have one electron, the energy depends only on the principal quantum number n; the energies of the 2s and 2p orbitals are the same. For atoms having more than one electron, however, the energy also depends on the value of l.

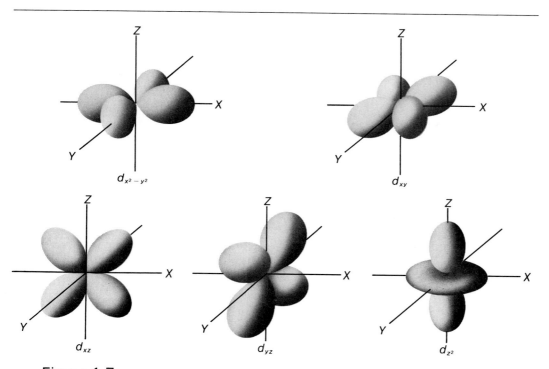

Figure 1.7
The shapes of d orbitals.

SPIN QUANTUM NUMBERS

The three orbital quantum numbers n, l, and m appear naturally when the Schrödinger equation is solved. There is also a spin quantum number, s, the value of which can be visualized as related to the spin of the electron on its own axis. This quantum number does not appear when the wave equation is solved, but certain quantum-mechanical treatments lead to the conclusion that an elementary particle such as an electron can have two spin angular momentum values, which are specified by the spin quantum numbers $+\frac{1}{2}$ and $-\frac{1}{2}$. We can conveniently think of these two quantum numbers as representing a spin either in one direction (clockwise) or in the other direction (counterclockwise).

Experimental evidence for electron spin was first obtained in 1921 by the German-American physicist Otto Stern, working with W. Gerlach. Their experiment involved a beam of neutral silver atoms, each atom having an odd electron. As the beam passed between the poles of a magnet especially designed to produce an inhomogeneous magnetic field, it was split into two separate beams. The experiment showed that if half of the unpaired electrons are spinning in one direction with respect to the magnetic field, and half are spinning in the opposite direction, half of the atoms are deflected to one side and half to the other.

MANY-ELECTRON ATOMS

We have seen that exact solutions of the Schrödinger equation are possible for atoms or ions containing only one electron. For more complicated systems, however, exact mathematical solutions are not possible, and it is necessary to resort to approximate methods. These have led to the conclusion that it is permissible, as a useful first approximation, to deal with the electrons of a many-electron atom one at a time — each electron being subject to the effect of the nucleus and of the other electrons averaged over the positions they occupy. This is sometimes done by assigning to the nucleus an effective charge $+Z_{eff}e$, which is less than its true charge $+Ze$.

When procedures such as this are adopted, it is still possible to assign to each orbital four quantum numbers n, l, m, and s, the latter being for spin. However, whereas for hydrogen-like atoms the energy of the electron depends only on the principal quantum number n, for atoms containing more than one electron the energy depends also on l. This is illustrated in Figure 1.8, which shows schematically the relative

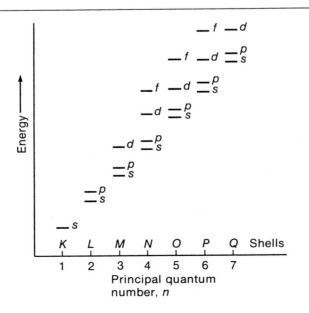

Figure 1.8

Relative energies for different atomic orbitals.

energy levels. The level depends to a considerable extent on the principal quantum number n. However, note that, for example, the $4s$ level is actually lower than the $3d$, in spite of the fact that the principal quantum number is higher. As shown in Figure 1.8, it is convenient to employ the term *shell* to include all electrons having a given principal quantum number. The term *subshell* can be used to include electrons having the same values of both n and l.

The way in which the electrons in a given atom are distributed among the orbitals involves two important principles. First, in a normal atom the electrons will be arranged as far apart as possible so that the energy level is at a minimum. If this were the only factor involved, all of the electrons in an atom would be in the lowest orbital, which is the $1s$ orbital ($n = 1, l = 0$). However, the number of electrons in a given orbital is limited by a principle proposed by the Austrian-American physicist Wolfgang Pauli (1900–1958). Pauli's *exclusion principle* can be stated as follows:

In an atom no two electrons can have all four quantum numbers (n, l, m, and s) the same.

Another formulation is:

If, in an atom, two electrons have the same three orbital quantum numbers (n, l, and m), their spins must be coupled (i.e., the spin quantum number of one must be $+\frac{1}{2}$, the other $-\frac{1}{2}$).

It follows at once from the Pauli principle that, in a given shell in an atom, the maximum number of electrons that can be in the various subshells is as follows:

$$s \quad 2$$

$$p \quad 6$$

$$d \quad 10$$

$$f \quad 14 \quad \text{(etc.)}$$

In general, the maximum number is $2(2l + 1)$. A very convenient device for representing the energy levels and the possible number of electrons in each subshell is the "box" diagram shown in Figure 1.9. The electrons can be represented as arrows pointing up (↑) or down (↓), corresponding to the two spin quantum numbers. Figure 1.9 shows the 6 electrons in the carbon atom.

ELECTRONIC CONFIGURATIONS AND THE PERIODIC TABLE

The reader will already be familiar with the structure of the periodic table and with how it is interpreted in terms of adding electrons successively into the various orbitals. The carbon diagram was shown in Figure 1.9. When a total of 10 electrons has been added, the $1s$, $2s$, and $2p$ orbitals are completely filled, and we have the inert gas structure of neon. An additional 8 electrons, making a total of 18 electrons, gives the inert gas structure of argon. Later, $3d$ and $4f$ levels are filled. The reader is referred to textbooks of general chemistry for further details.

An important question arises with certain atoms, such as the carbon atom (see Figure 1.9), where alternative arrangements are possible for the electrons in corresponding orbitals. In the case of carbon, the two $2p$ electrons could have been in orbitals having the same m values, and the spin quantum numbers of the electrons would then have to be different:

(a) 2p

Alternatively, the magnetic quantum numbers could be different, and then the spins could either be the same or opposed:

(b) 2p [↑|↑|]   or   (c) 2p [↑|↓|]
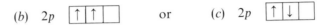

An important rule, formulated by the German physicist Friedrich Hund, enables us to decide which of these possibilities exists. According to Hund's rule, the favored configuration is the one in which the electrons have different magnetic quantum numbers and have the same spin quantum numbers; thus (b) is the favored configuration for carbon. The other arrangements are allowed but correspond to slightly higher energies. When electrons have different magnetic quantum numbers, they tend to be further apart, and they therefore repel each other less than they do if they have the same quantum number. Moreover, if the spins are opposed, there is somewhat less repulsion than if the spins are parallel.

Experimental support for the conclusions reached on the basis of Hund's rule (e.g., for the configuration shown in Figure 1.9 for carbon) is provided from spectroscopic measurements of the magnetic properties of atoms. The magnetism of an atom is

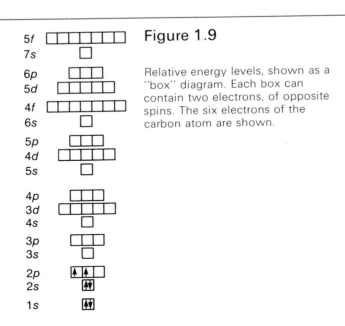

Figure 1.9

Relative energy levels, shown as a "box" diagram. Each box can contain two electrons, of opposite spins. The six electrons of the carbon atom are shown.

related to the number of unpaired electrons in orbitals where there is orbital angular momentum (i.e., p and d orbitals, but not s orbitals). If carbon has two unpaired p electrons, as predicted by Hund's rule, the magnetism of the carbon atom should be roughly twice that of the boron atom, which has one p electron. In fact, this is the case. If the two $2p$ electrons in the carbon atom were paired with each other [i.e., had opposite spins as in (a) and (c) above], the atom would exhibit zero magnetism.

We will see later in this chapter that the precise configurations of the electrons in atoms have an important bearing on the chemical bonds formed by the atom.

SIZES OF ATOMS AND IONS

The eigenfunction for an electron in an atom decreases exponentially toward zero as the distance from the nucleus increases, and therefore the probability of finding the electron also falls exponentially. Thus, an atom has no sharp boundary, and strictly speaking we cannot assign a definite radius or volume to an atom. At a certain distance from the nucleus, however, $\psi\psi^*$ has become negligibly small, so that as a result it is possible, as an approximation, to assign a radius to an atom or an ion—the electron density being negligible at greater distances.

For the hydrogen atom, for example, the electron density is negligible at a distance beyond about 1.2×10^{-10} m (1.2 Å or 0.12 nm) from the nucleus. The atom can therefore be regarded as a sphere with this radius. This radius is referred to as the *van der Waals radius*, after the Dutch physicist J. D. van der Waals (1837–1923) who related such radii and the corresponding atomic and molecular volumes to the pressure-volume behavior of gases.

Another radius that can be assigned to an atom is the *covalent* radius, which relates to the sizes of chemical bonds formed by the atom. For example, in the hydrogen molecule, H_2, the atoms are 0.074 nm (0.74 Å) apart; each hydrogen atom is said to have a covalent radius of $\frac{1}{2} \times 0.074 = 0.037$ nm. The van der Waals and covalent radii are shown in Figure 1.10 for the hydrogen and chlorine molecules. With the use of the two radii, the effective volumes of the molecules can be calculated easily. By analyzing data for a variety of molecules, it is possible to assign van der Waals and covalent radii to atoms—the assumption being that the covalent radii are

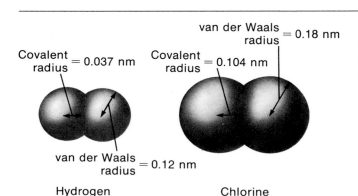

Covalent radius = 0.037 nm

van der Waals radius = 0.12 nm

Hydrogen

Covalent radius = 0.104 nm

van der Waals radius = 0.18 nm

Chlorine

Figure 1.10

The van der Waals and covalent radii in the hydrogen and chlorine molecules.

approximately additive—i.e., the lengths of chemical bonds can be obtained by adding covalent radii. Tables of radii are of considerable value in providing us with approximate values for the volumes occupied by various kinds of molecules.

1.5 TYPES OF CHEMICAL BONDS

Atoms are attracted to each other in various ways, and Table 1.2 gives information about some of the more important types of chemical bonds. The ionic and covalent bonds are the strongest, but ion-dipole and dipole-dipole forces also play very important roles in chemical and biological structures. The best known bond formed largely as a result of dipole-dipole attractions is the hydrogen bond, which is so very important in the structure of liquid water. The hydrophobic bonds, which are considered in more detail later (p. 249), are of an indirect kind; they occur when nonpolar groups are present in aqueous solution, and result from the fact that such groups have an effect on the hydrogen-bonded structure of water.

Table 1.2 The Main Types of Chemical Bonds

Type of force or chemical bond	Example	Equilibrium separation, nm	Dissociation energy[†] kJ mol^{-1}
Ionic bond (ion-ion force)	$Na^+ \ldots F^-$	0.23	670.0
Covalent bond	$H-H$	0.074	435.0
Ion-dipole force	$Na^+ \ldots O \overset{H}{\underset{H}{<}}$	0.24	84.0
Hydrogen bond (dipole-dipole force)	$O \ldots H-O$ with H groups	0.28	20.0
Hydrophobic bond	$CH_2 \ldots H_2C$	≈ 0.30	≈ 4.0
Van der Waals (dispersion) force	$Ne \ldots Ne$	≈ 0.33	≈ 0.25

† This is the energy that would be required per mole to dissociate the species into its units, e.g., $H-H$ into $H + H$, or $Na^+ \ldots OH_2$ into $Na^+ + H_2O$, in a vacuum.

Ionic interactions between species present in aqueous solutions are considered in later chapters. The remainder of this chapter will be concerned with the application of quantum mechanics to the covalent bond.

1.6 THE VARIATION PRINCIPLE

There are many systems for which it is impossible to obtain explicit mathematical solutions of the Schrödinger equation. For many atoms and molecules it is therefore necessary to employ approximate methods. Exact solutions for molecular systems can be obtained only for the hydrogen molecule ion, H_2^+, which has a single electron. Approximate methods are required even for the hydrogen molecule, H_2.

The most useful and commonly used approximate method is the *variation method*. A derivation of the variation principle is outside the scope of this book, but it is important to have some understanding of how the method is applied, since some of its applications shed considerable light on the nature of the chemical bond.

Suppose, first, that the Schrödinger equation can be solved exactly and that we have obtained an eigenfunction ψ, which need not be normalized. Then if $\hat{H}$ is the Hamiltonian operator, this eigenfunction satisfies the equation

$$\hat{H}\psi = E\psi \tag{1.29}$$

Suppose that we multiply both sides by the complex conjugate ψ^* of the eigenfunction:

$$\psi^*\hat{H}\psi = \psi^*E\psi = E\psi\psi^* \tag{1.30}$$

The transformation of $\psi^*E\psi$ into $E\psi\psi^*$ is permissible because E is a number giving the energy, and ψ and ψ^* are functions of the coordinates; as a result ψ, ψ^*, and E all commute.† However, it is not permissible to interchange $\hat{H}$ with ψ or ψ^*, because the operator $\hat{H}$, which involves a differentiation, does not commute with these functions.

Integration of both sides of equation (1.30) over all space gives

$$\int \psi^*\hat{H}\psi \, d\tau = E \int \psi\psi^* \, dr \tag{1.31}$$

and hence

$$E = \frac{\int \psi^*\hat{H}\psi \, d\tau}{\int \psi\psi^* \, d\tau} \tag{1.32}$$

The denominator of this equation is, of course, unity if the eigenfunction has been normalized.

On the other hand, suppose that ψ was not an eigenfunction of the operator $\hat{H}$ but was instead a trial function obtained in some way (examples will be given later). We could then calculate the values of the integrals on the right-hand side of equation (1.32) and calculate an energy. (If an integral cannot be evaluated explicitly, it can

---

† *A* and *B* are said to commute if $AB = BA$.

always be evaluated by numerical methods.) The variation principle now tells us that the energy calculated from equation (1.32), with *any* trial function, *cannot be below* the true energy for the ground state of the system. If the function chosen happened to be the true eigenfunction, the exact energy would be obtained. Otherwise, the calculated energy is *higher* than the true energy.

The usefulness of this method is that we can calculate energies from a variety of functions, and we know that the lowest energy obtained is closest to the truth. More systematically, we can express a trial eigenfunction as a function of certain variables and calculate the energy from equation (1.32) as a function of these variables. The energy can be minimized with respect to these variables, and we will then have obtained the best energy value from this type of function.

We will now see how this method is applied to some simple molecular systems.

1.7   THE HYDROGEN MOLECULE-ION, H_2^+

The simplest molecular system, H_2^+, consisting of two protons and a single electron, does not commonly occur, although it is observed in electrical discharges and some of its characteristics have been studied by spectroscopic methods. The experimental potential-energy curve is shown in Figure 1.11a. The minimum energy occurs at an

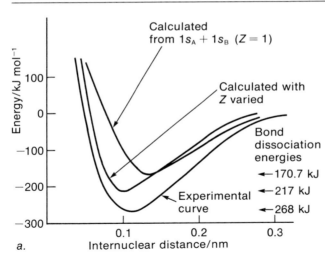

Figure 1.11

The hydrogen molecule ion, H_2^+:
a. Experimental and theoretical potential-energy curves.
b. The system of two protons and an electron.

internuclear distance of 0.106 nm (1.06 Å), and corresponds to an energy 268.1 kJ mol^{-1} (= 64.1 kcal mol^{-1} = 2.78 eV) below that of H + H$^+$. With the use of elliptical coordinates, it is possible to obtain an exact quantum-mechanical solution for this species, but we will here consider a much simpler solution based on the variation principle.

Figure 1.11b shows the system and the coordinates that will be used. The electron is being attracted by the two nuclei, and its potential energy is therefore

$$U = -\frac{e^2}{r_A} - \frac{e^2}{r_B} + \frac{e^2}{r_{AB}} \tag{1.33}$$

The Hamiltonian operator of equation (1.16) is thus

$$\hat{H} = -\frac{h^2}{8\pi^2 m}\nabla^2 - \frac{e^2}{r_A} - \frac{e^2}{r_B} + \frac{e^2}{r_{AB}} \tag{1.34}$$

and the Schrödinger equation to be solved is

$$\left(-\frac{h^2}{8\pi^2 m}\nabla^2 - \frac{e^2}{r_A} - \frac{e^2}{r_B} + \frac{e^2}{r_{AB}}\right)\psi = E\psi \tag{1.35}$$

To apply the variation method, we choose a trial eigenfunction ϕ and calculate an energy using equation (1.32) and the Hamiltonian operator in (1.34)—knowing that the resulting energy cannot be below the experimental one. In fact, we will be interested in calculating a number of energies for various internuclear distances r_{AB} in order to obtain a potential-energy curve to compare with the experimental one.

A procedure that is commonly used in such cases is to construct trial functions for molecules from the exact eigenfunctions that apply to the atoms from which the molecules are formed. The eigenfunction for the 1s state of the hydrogen atom is given by equation (1.27) with $Z = 1$, and we may therefore write the eigenfunction for the electron bound to proton A as

$$1s_A = \frac{1}{\sqrt{\pi}}\left(\frac{1}{a_o}\right)^{3/2} e^{-r_A/a_o} \tag{1.36}$$

The corresponding wave function for the electron associated with nucleus B is

$$1s_B = \frac{1}{\sqrt{\pi}}\left(\frac{1}{a_o}\right)^{3/2} e^{-r_B/a_o} \tag{1.37}$$

In the variation method, we can use any trial function we like, and a reasonable procedure is to use the sum of the above functions.

$$\phi = 1s_A + 1s_B \tag{1.38}$$

The fact that this sum is not normalized does not matter, since this is taken care of in the denominator of the expression in equation (1.32).

When the integrals in equation (1.32) are evaluated and the energies calculated for various internuclear distances, the results are as shown in Figure 1.11a. As expected, the calculated energies are all higher than the experimental energies. The theory certainly gives a curve of the right form, but the calculated energies are not very accurate. For example, compare the calculated dissociation energy, 170.7 kJ mol^{-1}, with the experimental value of 268.1 kJ mol^{-1}. A simple procedure, which leads to considerable improvement, is to vary the nuclear charge in order to get the lowest possible energy at each internuclear separation. In other words, the trial eigenfunction used is

$$\phi = \frac{1}{\sqrt{\pi}}\left(\frac{Z}{a_o}\right)^{3/2} e^{-Zr/a_o} + \frac{1}{\sqrt{\pi}}\left(\frac{Z}{a_o}\right)^{3/2} e^{-Zr/a_o} \tag{1.39}$$

instead of equation (1.38), and Z is varied after the energy is calculated from equation (1.32). The best value for Z is found to be 1.23, and with this value the calculated potential-energy curve is much closer to the experimental curve, as shown in Figure 1.11a. Now the calculated dissociation energy is 217.0 kJ mol^{-1}, much closer to the experimental value of 268.1 kJ mol^{-1}.

In the procedure outlined we have used a *linear combination of atomic orbitals* as the trial wave function. This procedure is known as the LCAO method or as the LCAO-MO method, the latter designation meaning that we have constructed a molecular orbital (MO) as a linear combination of atomic orbitals. The LCAO method is frequently employed, but unless additional terms are added the agreement with experiment is never very close.

1.8   THE HYDROGEN MOLECULE

The hydrogen molecule H_2 is the simplest molecule which forms an electron-pair bond. Many calculations have been made for this molecule, which is a prototype for many other chemical bonds. One of the two basic quantum-mechanical treatments of the hydrogen molecule involves constructing a molecular orbital for the bond from a *linear combination* of atomic orbitals (LCAO method). The other involves constructing the molecular orbital as the *product* of wave functions for each of the two electrons forming the bond. Both of these methods will be outlined.

LCAO METHOD

The experimental potential-energy curve for the H_2 molecule is shown in Figure 1.12a, and the molecule is shown in Figure 1.12b. The potential energy is

$$U = \frac{e^2}{r_{AB}} + \frac{e^2}{r_{12}} - \frac{e^2}{r_{A1}} - \frac{e^2}{r_{A2}} - \frac{e^2}{r_{B1}} - \frac{e^2}{r_{B2}} \tag{1.40}$$

and thus the Hamiltonian operator is

$$\hat{H} = -\frac{h^2}{8\pi^2 m}\nabla^2 + e^2\left(\frac{1}{r_{AB}} + \frac{1}{r_{12}} - \frac{1}{r_{A1}} - \frac{1}{r_{A2}} - \frac{1}{r_{B1}} - \frac{1}{r_{B2}}\right) \qquad (1.41)$$

As a trial function we might consider

$$\phi_1 = 1s_A(1)1s_B(2) \qquad\qquad (1.42)$$

where $1s_A(1)$ is the $1s$ wave function for electron 1 associated with nucleus A, and $1s_B(2)$ is the wave function for electron 2 associated with nucleus B. However, the energies calculated from this function are in poor agreement with the experimental results, as shown in Figure 1.12a. The fault with this trial function was recognized by the German physicists Walter Heitler and F. London, who carried out in 1927 (just after Schrödinger's equation appeared in 1926) the first calculations of molecular energies. What is wrong with the function is that, although electrons are indistinguishable particles, equation (1.42) implies that one electron can be designated as electron 1

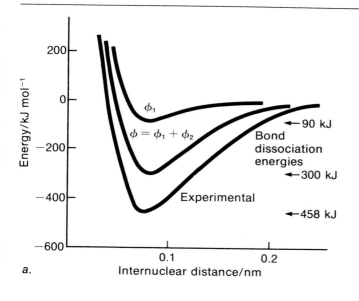

a.

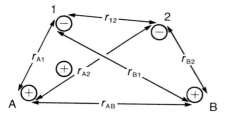

b.

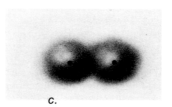

c.

Figure 1.12

The hydrogen molecule, H_2:
a. Experimental and theoretical potential-energy curves.
b. The system of two protons and two electrons.
c. The piling-up of electron density between the nuclei.

and is particularly associated with a nucleus designated nucleus A. An alternative function is

$$\phi_2 = 1s_A(2)1s_B(1) \tag{1.43}$$

but this has the fault of associating electron 2 with nucleus A and electron 1 with nucleus B. Heitler and London realized that these difficulties are avoided if one uses the sum of these wave functions as the trial function†

$$\phi = 1s_A(1)1s_B(2) + 1s_A(2)1s_B(1) \tag{1.44}$$

In this expression, although the electrons and the nuclei have been labeled, identical roles are assigned to the two electrons and the two nuclei.

When this function is used and the energy calculated from the variation equation (1.32), the agreement with experiment is more satisfactory, as seen in Figure 1.12*a*. There is still much room for improvement, however; the calculated dissociation energy is only 66% of the experimental value.

When the integrals in equation (1.32) are evaluated, the calculated energy is the sum of two terms

$$E = J + K \tag{1.45}$$

where J is known as the *coulombic* energy and K as the *exchange* energy. At the normal internuclear separation both of these terms are negative in value (with respect to the energy of the separated atoms). The coulombic energy accounts for only about 10% of the binding; the remaining 90% is exchange energy. The coulombic energy is approximately the energy that would be calculated on the basis of electrostatic effects in a purely classical treatment, and we see that such a treatment is inadequate. The exchange energy, which accounts for most of the binding, is a purely quantum-mechanical contribution, arising from the interchange of electrons allowed for in the Heitler-London wave function.

The Heitler-London treatment can be improved in various ways. For example, we can add terms corresponding to ionic states. If both electrons are associated with nucleus A, we have the function

$$\phi_3 = 1s_A(1)1s_A(2) \tag{1.46}$$

but if both are associated with B, we have

$$\phi_4 = 1s_B(1)1s_B(2) \tag{1.47}$$

Therefore, a reasonable trial function is

$$\phi = c_1(\phi_1 + \phi_2) + c_2(\phi_3 + \phi_4) \tag{1.48}$$

---

† Here and elsewhere in this chapter we will give eigenfunctions in unnormalized forms, since such functions are satisfactory in the variation treatment of equation (1.32).

where c_1 and c_2 are numbers which can be varied, after the integrals in equation (1.32) have been evaluated, to get the lowest energy. We have multiplied ϕ_1 and ϕ_2 by the same number, c_1, because these two functions are equivalent. Similarly, ϕ_3 and ϕ_4 have both been multiplied by c_2, because in the symmetrical H_2 molecule one ionic state cannot be favored over the other. After the variation method is applied, the function (1.48) leads to better energies than those obtained when only the first two terms are used.

All of the successful wave functions for the H_2 molecule lead to the conclusion that there is a piling-up of electron density between the two nuclei. This is illustrated in Figure 1.12c. As a result of the high probability of finding an electron between the nuclei, the electron cloud forming the bond screens the nuclei from each other, thus reducing the repulsion between the nuclei. This high electron density between the nuclei is characteristic of the covalent bond.

MOLECULAR ORBITALS

Instead of constructing the molecular orbital for the electron-pair bond by taking a linear combination of atomic orbitals, we can take the product of two molecular orbitals—one for electron 1 and the other for electron 2. The function for electron 1 can be the sum of two atomic orbitals

$$\sigma(1) = 1s_A(1) + 1s_B(1) \tag{1.49}$$

and that for electron 2

$$\sigma(2) = 1s_A(2) + 1s_B(2) \tag{1.50}$$

The trial molecular orbital obtained from these two functions is thus

$$\sigma = \sigma(1)\sigma(2) = \left[1s_A(1) + 1s_B(1)\right]\left[1s_A(2) + 1s_B(2)\right] \tag{1.51}$$

$$= 1s_A(1)1s_B(2) + 1s_B(1)1s_A(2) + 1s_A(1)1s_A(2) + 1s_B(1)1s_B(2) \tag{1.52}$$

The first two terms correspond to the Heitler-London wave function; the other two are for ionic states. The function (1.52) is, in fact, the same as (1.48) with $c_1 = c_2$, i.e., with equal weighting of the ionic and covalent states. Equation (1.52) gives fairly good energy values, but not as good as those given by equation (1.48) after variation of c_1 and c_2, because equation (1.52) gives too great a contribution from ionic states.

The difference between this approach and the Heitler-London method is that here we are considering the molecule as a whole, rather than just the bond. We are constructing for the molecule the orbital

$$\sigma = 1s_A + 1s_B \tag{1.53}$$

into which two electrons can be placed. When one electron is in this orbital, we have the H_2^+ ion, and there is bonding between the nuclei—the energy of the H_2^+ being below that of $H + H^+$. Therefore this orbital is said to be a *bonding* orbital, and it corresponds to a high electron density between the nuclei. Two electrons with opposite spins can go into this orbital, and there is now more bonding. The electron-pair bond in H_2 has a *bond order* of one; the single-electron bond in H_2^+ has a bond order of 0.5.

Alternatively, use can be made of the orbital

$$\sigma^* = 1s_A - 1s_B \tag{1.54}$$

This corresponds to a low electron density between the nuclei. When one electron goes into this orbital, the energy of the H_2^+ ion is greater than that of $H + H^+$. This orbital is referred to as an *antibonding* orbital. Figure 1.13 is an energy diagram showing the separated atoms, which have electrons in $1s$ states, and the bonding and antibonding orbitals, designated σ and σ^* respectively. In the ground state of H_2, both electrons are in the bonding orbital and have opposite spins. If one or both electrons are in the antibonding orbital, there is repulsion and the molecule would at once dissociate into $H + H$.

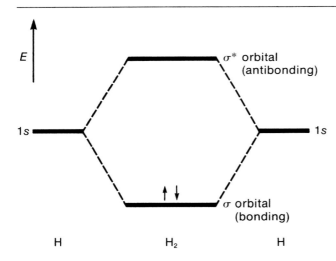

Figure 1.13

Bonding and antibonding orbitals for the H_2 molecule. Two electrons, with opposite spins, are shown in the bonding orbital.

1.9 VALENCE-BOND THEORY

The procedure of taking a linear combination of atomic orbitals, which we have considered with respect to the H_2 molecule, is very fruitful when applied to other covalent bonds. Consider, for example, the hydrogen fluoride molecule, HF, formed from a hydrogen atom with one electron in the $1s$ state and a fluorine atom with an electron configuration of $1s^2 2s^2 2p^5$. Fluorine has an unpaired $2p$ electron, and we can form a wave function of the Heitler-London type by making use of the atomic orbitals for this $2p$ electron and for the $1s$ electron in the hydrogen atom:

$$\phi = c_1 1s(1)2p(2) + c_2 1s(2)2p(1) \tag{1.55}$$

This wave function could be improved by adding the contributions for ionic states:

$$\phi = c_1 1s(1)2p(2) + c_2 1s(2)2p(1) + c_3 1s(1)1s(2) + c_4 2p(1)2p(2) \qquad (1.56)$$

Calculation of the bond energy on the basis of the variation treatment gives a reasonably good approximation to the experimental value. Better results are obtained if account is taken of the other electrons in the fluorine atom. Similar treatments can be applied to other covalent bonds.

ELECTRONEGATIVITY

As with the hydrogen molecule, the calculations for hydrogen fluoride lead to the conclusion that there is a piling-up of electron density between the nuclei. The H_2 molecule is symmetrical, so that the electron cloud lies symmetrically between the nuclei. The quantum-mechanical calculations for hydrogen fluoride, on the other hand, lead to the result that the electron cloud lies more toward the fluorine atom. In other words, the last term in the wave function (1.56), in which both electrons are related to the fluorine atom, is more important than the third term, in which both are associated with the hydrogen atom (i.e., $c_4 > c_3$).

The consequence of this asymmetry is that the hydrogen fluoride molecule has a dipole moment,† the fluorine atom being more negative than the hydrogen atom. In an electric field, the molecule turns (orients)—the fluorine end moving toward the positive pole and the hydrogen end toward the negative pole. The dipole moment of a diatomic molecule is equal to the effective charge q at the positive and negative ends multiplied by the distance between them:

$$\mu = qd \qquad\qquad\qquad (1.57)$$

This is shown in Figure 1.14. It is common to express the charge in electrostatic units (esu), and the distance in angstroms. If an electronic charge of 4.8×10^{-10} esu were separated by a distance of 1 Å (10^{-8} cm or 0.1 nm) from an equal charge of

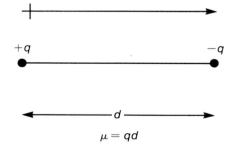

Figure 1.14

Two charges $+q$ and $-q$, separated by a distance d; the dipole moment is qd. The direction of the moment is often represented by an arrow $\leftrightarrow$, as shown.

† Dipole moments are further considered in Chapter 3 (p. 113).

opposite sign, the dipole moment would be 4.8×10^{-18} esu cm. This quantity is usually written as 4.8 debyes, 1 debye (D) being 10^{-18} esu cm.

If we know the distance d between two atoms, we can calculate a theoretical dipole moment on the assumption that the atoms bear a full single charge; this value is $\mu_{\text{theor}} = 4.8 \times 10^{-10}d = 4.8d$ debyes. The percent ionic character of the bond can then be calculated as

$$\% \text{ ionic character} = \frac{\mu_{\text{exp}}}{\mu_{\text{theor}}} \times 100 \qquad (1.58)$$

where μ_{exp} is the experimental dipole moment. The percent ionic character can also be calculated from valence bond theory. Suppose that a wavefunction is constructed as a linear combination of covalent and ionic wavefunctions,

$$\psi = \psi_{\text{covalent}} + \lambda\psi_{\text{ionic}} \qquad (1.59)$$

and suppose that to give the lowest energy the ionic function has to be weighted by a particular value λ, as compared with unity for the covalent function. When the energy is calculated using the variation equation (1.32), the integral

$$\int \psi^* \hat{H} \psi \, d\tau$$

is employed. As a result, the ratio of the contribution to the energy of the ionic wavefunction as compared to the covalent wavefunction is $\lambda^2 : 1$. The theoretical percent ionic character is therefore

$$\% \text{ ionic character} = \frac{\lambda^2}{1 + \lambda^2} \times 100 \qquad (1.60)$$

The American chemist Linus Pauling has made important contributions to our understanding of electronegativity and the ionic character of bonds by using both quantum-mechanical theory and experimental results. He considered a reaction such as

$$AA + BB \rightarrow 2AB$$

in which two homonuclear molecules form two heteronuclear molecules. He regarded the molecules AA and BB as purely covalent in the sense that the molecules are symmetrical and cannot have dipole moments. However, the unsymmetrical molecule AB can have a dipole moment, and there will be an ionic contribution to its energy, which will make the molecule more stable (i.e., will cause it to have a higher dissociation energy D than if the bond were purely covalent). Pauling concluded empirically that the purely covalent bond dissociation energy of AB would be the geometric mean† of the values for AA and BB:

---

† He also tried the arithmetic mean, but found that the geometric mean gave more satisfactory results.

$$E_{\text{covalent}} = [D(AA)D(BB)]^{\frac{1}{2}} \tag{1.61}$$

The measured dissociation energy, $D(AB)$, will in general be greater than this, and the difference is taken to be the ionic energy of the bond:

$$E_{\text{ionic}} = D(AB) - [D(AA)D(BB)]^{\frac{1}{2}} \tag{1.62}$$

This quantity, therefore, can be calculated from the dissociation energies, which are usually known.

Pauling found empirically that the square roots of these ionic energies, $(E_{\text{ionic}})^{\frac{1}{2}}$, were additive with respect to the atoms A and B. In other words, the $(E_{\text{ionic}})^{\frac{1}{2}}$ values were proportional to the difference between certain numbers χ assigned to each atom:

$$(E_{\text{ionic}})^{\frac{1}{2}} = K|\chi_A - \chi_B| \dagger \tag{1.63}$$

Pauling chose his proportionality constant K in such a way that the difference $\chi_A - \chi_B$ also gave a reliable estimate of the dipole moment of AB measured in debyes. If the energies are in kcal, $K \cong 5$; for energies in kJ, $K \cong 10$. In this way, he was able to construct a table of χ values, or *electronegativities*. A few such values are given in Table 1.3. These values are very useful for making rough estimates of dipole moments. For hydrogen fluoride, for example, we see that hydrogen has an electronegativity of 2.1 and fluorine of 4.1; the estimated dipole moment is thus $4.1 - 2.1 = 2.0$ debyes—the fluorine atom being at the negative end of the dipole.

Table 1.3   Atomic Electronegativities

H						
2.1						
Li	Be	B	C	N	O	F
1.0	1.5	2.0	2.5	3.1	3.5	4.1
Na	Mg	Al	Si	P	S	Cl
1.0	1.23	1.5	1.7	2.1	2.4	2.8

ORBITAL HYBRIDIZATION

In the examples of H_2 and hydrogen fluoride, HF, we have constructed molecular orbitals by taking a linear combination of atomic orbitals—one atomic orbital being used for each atom. In H_2, for example, we combined the $1s$ orbital for each atom; for HF we used the $1s$ orbital of hydrogen and the $2p$ orbital of fluorine. However, sometimes this is not satisfactory and we must use two or more orbitals from a given atom.

A very simple example is provided by the molecule of methane, CH_4. The electronic

---

† The two vertical lines indicate the *absolute value*, i.e., the positive value. Thus, we always subtract the smaller value from the larger.

configuration of the carbon atom is shown in Figure 1.9, and we see that there are two unpaired $2p$ electrons with the same spin. It might appear that carbon would form only two single bonds involving these two electrons. Divalent compounds of carbon are indeed known, the simplest being the methylene radical CH_2. This, however, is unstable, and carbon more commonly shows a valency of 4. To obtain 4 unpaired electrons, we can imagine one of the two $2s$ electrons being "promoted" to the $2p$ state to give the following configuration for the excited (C^*) carbon atom

$$2p \quad \boxed{\uparrow \mid \uparrow \mid \uparrow}$$

$$2s \qquad \boxed{\uparrow} \qquad C^*,\ 1s^2\ 2s\ 2p^3$$

$$1s \qquad \boxed{\downarrow\uparrow}$$

Initially it would appear that, of the four bonds formed by a carbon atom, one would involve the $2s$ orbital and the other three the three $2p$ orbitals. This, however, implies that one bond is different from the other three, whereas experimentally the CH_4 molecule is perfectly symmetrical—all four bonds being identical. The solution to this dilemma was given by Pauling who suggested that we should use a linear combination of orbitals instead of the pure $2s$ and $2p$ orbitals of the carbon atom. On the basis of a simple quantum-mechanical treatment, he concluded that from the one $2s$ and three $2p$ orbitals one can construct four *hybridized* orbitals as follows:

$$\psi_1 = 2s + 2p_x + 2p_y + 2p_z \tag{1.64}$$

$$\psi_2 = 2s + 2p_x - 2p_y - 2p_z \tag{1.65}$$

$$\psi_3 = 2s - 2p_x + 2p_y - 2p_z \tag{1.66}$$

$$\psi_4 = 2s - 2p_x - 2p_y + 2p_z \tag{1.67}$$

These four hybrid orbitals are equivalent to each other and project out from the nucleus toward the corners of a regular tetrahedron, as shown in Figure 1.15c. The molecular orbital for each bond is formed by taking a linear combination of one of these hybrid orbitals and the $1s$ orbital for the hydrogen atom. This can be visualized as the overlapping of the electron clouds, as shown in Figure 1.15c. Obviously the maximum overlapping is obtained when the hydrogen atoms are located along the axes of the orbitals, a conclusion that is expressed in Pauling's principle of maximum overlapping. This principle is of great value in predicting the shapes of molecules, which follow at once from the shapes of the orbitals.

A similar situation arises with the bonding in H_2O, but in this case matters are not as clear. The ground-state oxygen atom configuration is

$$2p \quad \boxed{\uparrow\downarrow \mid \uparrow \mid \uparrow}$$

$$2s \qquad \boxed{\uparrow\downarrow} \qquad O,\ 1s^2 2s^2 2p^4$$

$$1s \qquad \boxed{\uparrow\downarrow}$$

We note that, in accordance with Hund's rule, there are two unpaired $2p$ electrons in orbitals at right angles to each other. We can imagine the H_2O molecule to be

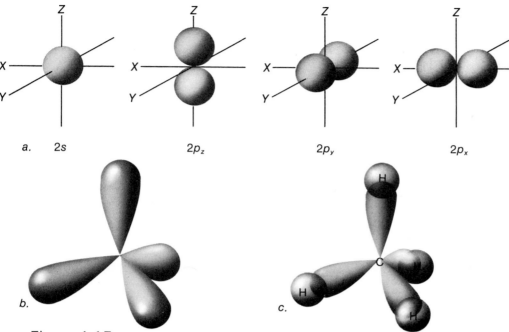

Figure 1.15

Orbitals for the carbon atom.
a. The unhybridized orbitals $2s$, $2p_x$, $2p_y$ and $2p_z$.
b. The hybrid orbitals.
c. The bonding in CH_4, involving overlapping orbitals.

made up directly from these $2p$ orbitals. The principle of maximum overlapping predicts, in the first instance, an angle of $90°$ shown in Figure 1.16a. Experimentally, however, the angle is about $104.5°$.

Alternatively, we can first hybridize the one $2s$ and the three $2p$ orbitals to obtain the tetrahedral sp^3 arrangement, in which the angle between the orbitals is $109.5°$, as shown in Figure 1.16b. Thus the molecule will have this angle, which is rather larger than the experimental angle. Detailed quantum-mechanical treatments suggest that the truth lies somewhere between these two models, there being only partial hybridization of the orbitals.

MULTIPLE BONDS

Carbon also forms molecules having double and triple bonds. Simple examples are

ethylene   formaldehyde   acetylene

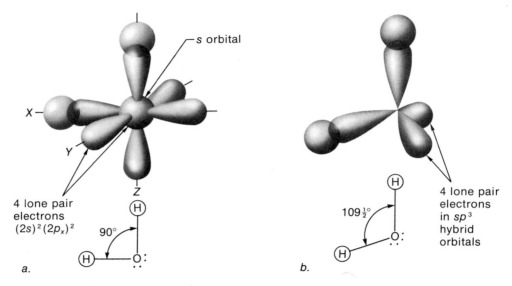

Figure 1.16

Two formulations for the H_2O molecule:
a. Bonds formed from pure *p* orbitals.
b. Bonds formed from *sp*3 hybrid orbitals.

Such molecules are explained in valence-bond theory in terms of two different kinds of hybridization, sp^2 and sp, which are illustrated in Figures 1.17*b* and *c*. In sp^2 hybridization, the bonding orbitals involve a linear combination of an *s* orbital and two *p* orbitals (e.g., p_x and p_y) as follows:

$$\psi_1 = s + \sqrt{2}p_x \tag{1.68}$$

$$\psi_2 = s - \frac{1}{\sqrt{2}}p_x + \frac{\sqrt{3}}{\sqrt{2}}p_y \tag{1.69}$$

$$\psi_3 = s - \frac{1}{\sqrt{2}}p_x - \frac{\sqrt{3}}{\sqrt{2}}p_y \tag{1.70}$$

As shown in Figure 1.17*b*, these orbitals lie symmetrically in the XY plane, the angle between the orbitals being 120°. In *sp* hybridization, we combine the *s* orbitals with one *p* orbital (e.g., p_x) as follows:

$$\psi_1 = s + p_x \tag{1.71}$$

$$\psi_2 = s - p_x \tag{1.72}$$

If a carbon atom is attached to three other atoms by two single bonds and a double bond, as in ethylene, there will be sp^2 hybridization. The *p* orbital, which is not

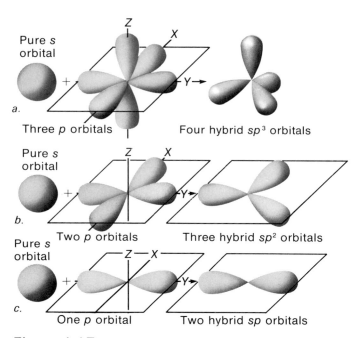

Figure 1.17

The three sets of hybrids formed from s and p orbitals.

involved in the hybridization, lies above and below the plane of the hybrid bonds, as shown in Figure 1.18. Thus the three atoms attached to the carbon atom will lie in the same plane with angles of 120°. This accounts for one of the two bonds between the two carbon atoms in ethylene. This bond, in which the two hybrid orbitals point toward each other, is known as a σ (sigma) bond. The second bond, which is not as

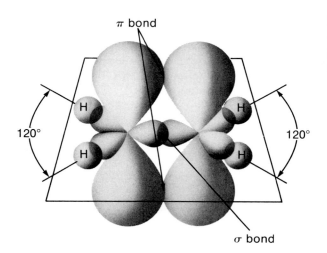

Figure 1.18

The ethylene molecule, showing the σ and π bonds between the carbon atoms.

strong as the σ bond, is formed by the sideways overlapping of the unhybridized p orbitals. Such a bond is known as a π bond and is also shown in Figure 1.18.
This bond holds the entire molecule in a planar configuration. If the two CH_2 residues were twisted with respect to each other, the sideways overlapping would be reduced, and the π bond would be weakened. We can therefore conclude that it requires energy to distort the molecule from the planar configuration, which is the most stable form.

When a carbon atom is attached to two other atoms, as in acetylene or carbon dioxide, there is sp hybridization and the bonds lie in a straight line. The electronic structure of carbon dioxide is shown in Figure 1.19. Two π bonds are now formed in addition to the σ bonds. In the figure, the unhybridized p orbitals above and below the plane overlap with the p orbitals of the right-hand oxygen atom, and the unhybridized p orbitals in the plane overlap with the p orbitals of the left-hand oxygen atom.

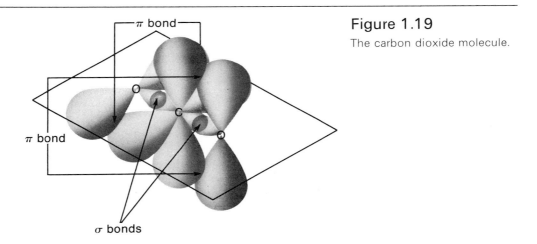

Figure 1.19

The carbon dioxide molecule.

This valence-bond method, which takes account of the possibility of orbital hybridization, provides us with a useful procedure for predicting the shapes of molecules. The method has been used here to predict the shapes of molecules containing carbon, but the method can also be applied to other types of molecules. The procedure helps us to predict the angles between the bonds emanating from each atom, and thus allows us to visualize the entire structure of the molecule.

DELOCALIZED ORBITALS

In the molecules considered so far, the electrons forming the σ and π bonds remain in the region between the atomic nuclei. Such electrons are called *localized* electrons. In some molecules, however, some of the electrons are *delocalized* and do not remain between a given pair of nuclei.

A simple example of delocalization occurs with benzene, for which the structure

was originally written as

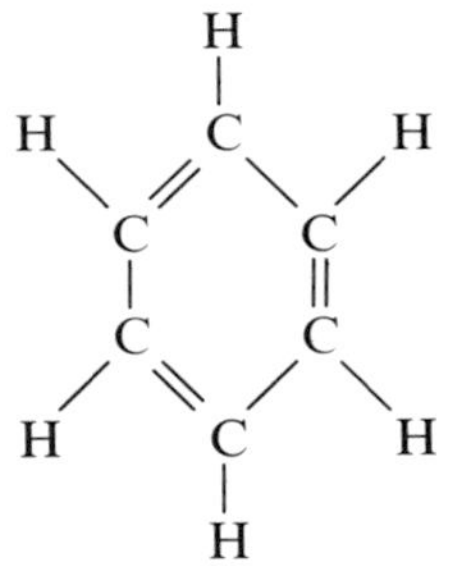

The above structure is not satisfactory, because the properties of benzene are not the properties of molecules containing double bonds. For example, ethylene reacts rapidly with bromine to form the addition compound CH_2BrCH_2Br, whereas benzene reacts slowly and in a different manner with bromine. The fact that benzene and other molecules containing alternate single and double bonds do not exhibit double-bonded behavior has been explained by the hypothesis that the molecule is in a *hybrid* or *resonance state* between the two following structures:

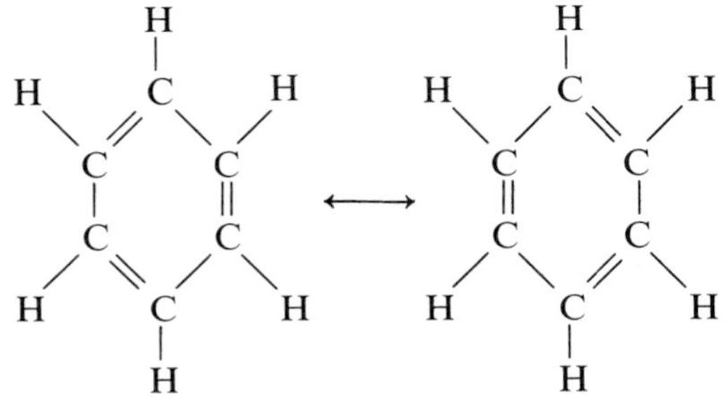

The symbol $\longleftrightarrow$ is used to denote resonance. It is important to realize that we are *not* implying that benzene exists as an equilibrium mixture of the two forms. In that case, the symbol $\rightleftharpoons$ would have been used. Instead, each molecule is a hybrid, each carbon-carbon bond being halfway between a single and a double bond; the C—C bond has a bond order of 1.5. The C—C bonds are all the same length, 0.140 nm, which is intermediate between the distance for a single bond (bond order = 1), 0.153 nm, and the distance for a double bond (bond order = 2), 0.133 nm.

Instead of employing the above notation, a convenient way of expressing the resonance in benzene is by writing the molecule as

When such resonance occurs, the energy level of the resulting hybrid structure is substantially lower than that of the single- and double-bonded structure. This means

that the molecule is more stable than if it had the simple single- and double-bonded structure. The reduction in the energy is known as the *resonance energy*.

The use of the valence-bond method sheds an interesting light on the situation. Applying this method leads to the electronic structure shown in Figure 1.20. Since

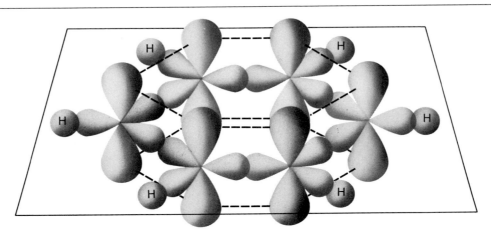

Figure 1.20

The benzene molecule. The dashed lines represent the overlapping of p orbitals.

each carbon atom is attached to three atoms, there is sp^2 hybridization, leading to the planar arrangement. Between each pair of neighboring carbon atoms there is a σ bond, formed by the overlapping of sp^2 hybrid orbitals; the result is the formation of a regular hexagon, with all of the carbon and hydrogen atoms lying in a plane. The remaining unhybridized p orbitals project at right angles above and below the plane of the molecule, as shown in Figure 1.20. These p orbitals overlap all around the ring, forming π bonds. This way of envisioning the molecule does not lead to alternating single and double bonds. Instead, all of the bonds are identical and are halfway between a single and a double bond, as postulated in the resonance theory.

1.10   MOLECULAR ORBITALS

We saw in our discussion of the hydrogen molecule that there are two procedures for constructing the orbitals. One is to consider the two electrons in the bond and to form a wave function for the pair of electrons by taking a linear combination of atomic orbitals. This is the valence-bond treatment, and we considered its application to other bonds in the preceding section. The other procedure was a little different, in that instead of focusing attention on the electron-pair bond, we considered the molecular orbitals for the molecule as a whole, and then placed the electrons appropriately in those orbitals. One way of carrying out this second treatment is to bring the nuclei together to form a "united atom," and then to consider the orbitals for this united atom. In other words, the first treatment starts with the separated atoms and brings them together; the second treatment is related to the united atom. This second treatment is usually referred to as *molecular-orbital* (MO) theory. It offers some advantages over the valence-bond treatment because the MO treatment generally leads to better energies when detailed quantum-mechanical calculations are carried out. Also, MO theory is very helpful in leading to conclusions about the numbers of unpaired electrons in molecules. The treatment has the disadvantage of not leading to simple predictions about molecular shapes. In general, application of both procedures is recommended, since in this way a deeper understanding of the structures of molecules is reached.

The simplest systems to consider in terms of MO theory are those in which the orbitals are constructed from two $1s$ orbitals. We saw for the H_2 molecule that combining two $1s$ atomic orbitals leads to a bonding σ orbital ($1s_A + 1s_B$) and an antibonding σ^* orbital ($1s_A - 1s_B$). Four species can be considered in terms of these orbitals, namely H_2^+, H_2, He_2^+, and He_2, which contain 1, 2, 3, and 4 electrons respectively. Each σ or σ^* orbital can contain two electrons. Figure 1.21 shows the electronic arrangements for these four species. In H_2^+ there is one bonding electron

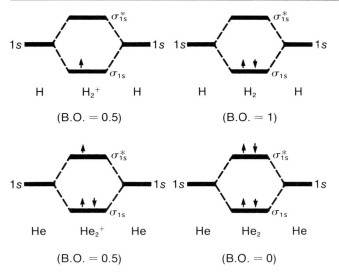

Figure 1.21

The assignment of electrons to the molecular orbitals for the species H_2^+, H_2, He_2^+, He_2, with the bond orders (B.O.) shown. They are defined as $\frac{1}{2}$ (net number of bonding electrons).

(bond order = 0.5), and the ion is stable. In H_2 there are two bonding electrons, and we have a stable species with a bond order of 1. In He_2^+ one electron has to go into the antibonding σ^* orbital; there are thus two bonding electrons and one antibonding electron, and a net attraction. The bond order, defined as one-half the net number of bonding electrons, is 0.5 in this instance. In He_2, however, there are two bonding and two antibonding electrons, and thus no attraction. Experimentally, H_2^+, H_2, and He_2^+ have been observed, but He_2 has not. This is a good example of how MO theory can lead to predictions about molecular stabilities, predictions which cannot be made as easily using valence-bond theory.

The species lithium hydride (LiH) and beryllium hydride (BeH) are treated a little differently. The electronic configuration of lithium is $1s^2 2s$ and, as a first approximation, we can regard the two $1s$ electrons as remaining in the lithium atomic orbital even when the molecule is formed. We construct MOs from the $2s$ orbital of the lithium and the $1s$ orbital of the hydrogen, as shown in Figure 1.22a. There are two electrons to go into these orbitals, and both will go into the σ orbitals. In BeH (see Figure 1.22b) there will be two electrons in the $1s$ orbital of Be; of the three remaining electrons, two will go into the bonding σ orbital and one into the antibonding σ^* orbital.

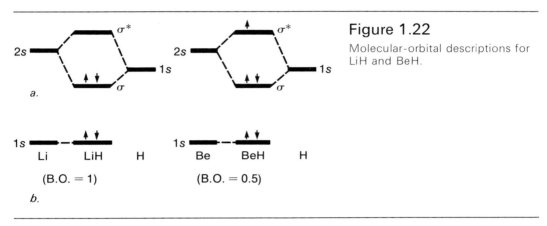

Figure 1.22

Molecular-orbital descriptions for LiH and BeH.

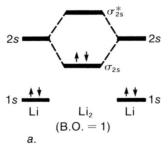

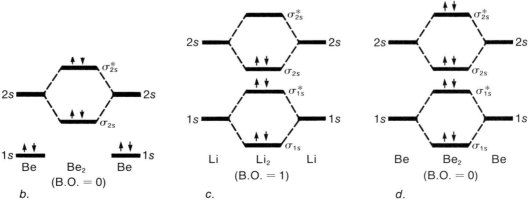

Figure 1.23

Molecular-orbital descriptions for Li_2 and Be_2. In (a) and (b) the $1s$ electrons are regarded as remaining in atomic orbitals, while in (c) and (d) they are in σ and σ^* molecular orbitals.

Thus there is net bonding for both LiH and BeH, the bonding being less for the latter because of the antibonding electron.

The molecular orbital descriptions for Li_2 and Be_2 are shown in Figure 1.23a–d. Again, we can regard the 1s electrons as remaining in their atomic orbitals, as shown in Figure 1.23a and b, or we can distribute them among molecular orbitals formed from 1s orbitals, as in Figure 1.23c and d. How we proceed makes little difference, because net bonding does not occur in either case. In Li_2 the two electrons coming from the 2s orbitals of the lithium atoms will be in the σ orbital, giving bonding. In Be_2 there are four electrons coming from atomic 2s orbitals. Two will go into bonding σ orbitals and two into antibonding σ^* orbitals, and there is therefore no net bonding.

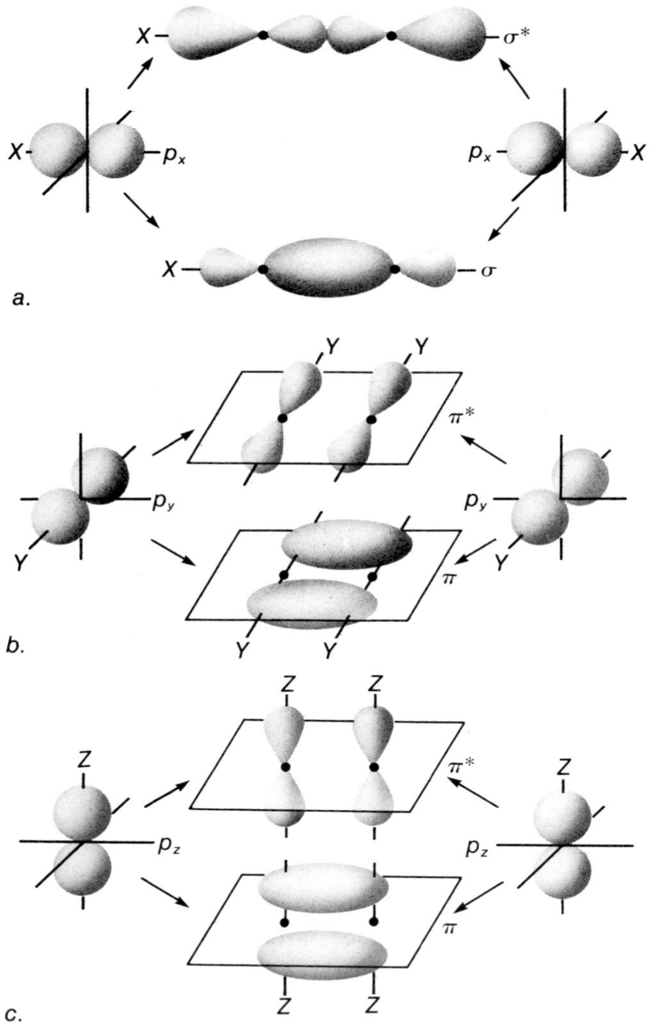

a.

b.

c.

Figure 1.24

The combination of *p* orbitals to give molecular orbitals.

a. σ orbitals formed from two p_x orbitals.

b. π orbitals formed from p_y orbitals.

c. π orbitals formed from p_z orbitals.

We have seen that in Be_2 all of the molecular orbitals formed from the $1s$ and $2s$ orbitals are filled. To deal with diatomic molecules containing more electrons, such as C_2, O_2, and N_2, we must therefore construct molecular orbitals from p orbitals. Two situations now arise. If we start with p orbitals that lie along the axis of the diatomic molecule, these combine in two ways according to whether we add or subtract the functions. Orbitals lying along the axis of the molecule are usually designated p_x, and the two molecular wavefunctions are thus

$$\sigma = p_x(A) + p_x(B) \tag{1.73}$$

$$\sigma^* = p_x(A) - p_x(B) \tag{1.74}$$

The former is a bonding orbital and the latter an antibonding orbital, as shown in Figure 1.24a.

Atomic p orbitals that lie perpendicular to the molecular axis (i.e., the p_y and p_z orbitals) also combine to give bonding and antibonding MOs, depending on whether we add or subtract them. These are referred to as π or π^* orbitals. Note the analogy with the π bonds formed in valence-bond theory by the sideways overlapping of p orbitals. The π and π^* molecular orbitals formed from the p_y and p_z orbitals are equivalent, and the π^* levels are also the same. When two energy levels fall in the same position, they are said to be *degenerate*.

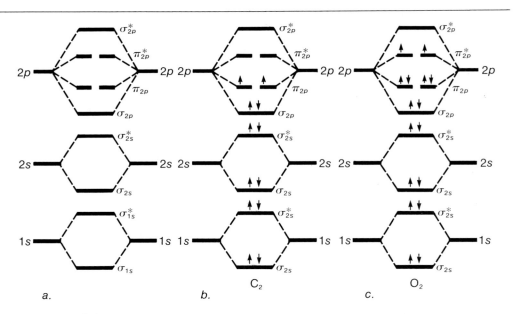

Figure 1.25

a. Typical order for the M.O. energy levels in a diatomic molecule. The π and π^* levels are shown as two lines, the levels being degenerate and capable of holding four electrons.
b. Electron distribution in C_2 (12 electrons).
c. Electron distribution in O_2 (16 electrons).

In order to know into which levels the electrons will go, we must construct an energy diagram showing the order of the levels. Detailed quantum-mechanical calculations reveal that the order varies somewhat from molecule to molecule. A typical arrangement is shown in Figure 1.25a. The distribution of electrons in a diatomic molecule can then be determined by feeding electrons into the successive levels, two electrons going into the σ and σ^* levels and four into the π and π^* levels, since these are degenerate.

For example, the electron distribution of the fourteen electrons in C_2 is shown in Figure 1.25b. We can write the arrangement as

$$C_2, (\sigma_{1s})^2(\sigma_{1s}^*)^2(\sigma_{2s})^2(\sigma_{2s}^*)^2(\sigma_{2p})^2(\pi_{2p})^2$$

There is no net bonding from $(\sigma_{1s})^2(\sigma_{1s}^*)^2$ or from $(\sigma_{2s})^2(\sigma_{2s}^*)^2$, because the bonding and antibonding effects cancel out. The higher $(\sigma_{2p})^2$ arrangement gives us the equivalent of one bond and $(\pi_{2p})^2$ of the other bond, and the molecule therefore has two bonds (i.e., the bond order is 2).

The situation in O_2 is shown in Figure 1.25c. The two antibonding π_{2p}^* electrons cancel out two of the bonding π_{2p} electrons. We are thus left with one $(\sigma_{2p})^2$ bond and one $(\pi_{2p})^2$ bond, and the bond order is 2. An interesting result appears if we apply Hund's rule (p. 20) to the two π_{2p}^* electrons. These electrons can go into two different π_{2p}^* orbitals, one arising from the atomic p_y orbitals and the other from the p_z orbitals, and Hund's rule tells us that the electrons will do so and will have the same spin. Two important consequences arise from the fact that O_2 has two unpaired π_{2p}^* electrons. One is that the molecule is magnetic, and the other is that the molecule is chemically quite reactive, having some of the properties of a free radical. Molecules such as oxygen, which have two electrons with the same spin not paired with each other, are often referred to as *diradicals*.

1.11   VALENCE-SHELL ELECTRON-PAIR REPULSIONS AND MOLECULAR GEOMETRY

A theory which is particularly useful in making predictions about the shapes of molecules is the *valence-shell electron-pair repulsion* (VSEPR) theory. The ideas behind this theory were first suggested in 1940 by the British chemists N. V. Sidgwick (1873–1952) and H. E. Powell. These concepts were later developed further by Sir Ronald Nyholm (1917–1971) and, more particularly, by the Canadian chemist Ronald S. Gillespie.

The basis of the theory is the following fundamental rule:

The pairs of electrons in a valence shell adopt that arrangement which maximizes their distance apart; i.e., the electron pairs behave as if they repel each other.

A very simple example is mercuric chloride, $HgCl_2$. Mercury has two valence electrons, which form two covalent bonds with electrons from two chlorine atoms. If we write only the electrons in the valence shell of the mercury atom, the result is

Cl : Hg : Cl

We deduce at once that the molecule is linear, because of the repulsion between the two pairs of electrons. Thus molecules of the type A : B : A, which have only two pairs of valence electrons associated with the central atom, are invariably linear.

By the same theory the boron trifluoride molecule (BF_3) is triangular and planar:

This arrangement, with 120° bond angles, gives the greatest separation between the electron pairs.

In the methane molecule, CH_4, the carbon atom has four pairs of electrons in its valence shell—all involved in bonding. The electrostatic repulsions between these four pairs leads to the tetrahedral structure shown in Figure 1.26a; the four bonds are equidistant from each other, the angles being 109.5°. The $NH_4{}^+$ ion is isoelectronic† with CH_4, and therefore has exactly the same structure. The phosphate ion, $PO_4{}^{3-}$, is also tetrahedral for the same reasons.

In the molecules considered so far, the electrons in the valence shell are all bonding electrons. In ammonia, NH_3, on the other hand, one pair of electrons is a *lone pair*

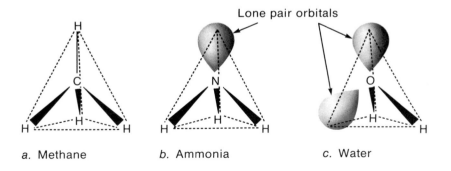

a. Methane   b. Ammonia   c. Water

Figure 1.26

The geometries of methane, ammonia, and water.

---

† That is, it contains exactly the same numbers of atoms and electrons, which are therefore in the same electronic configurations.

not involved in bonding. Nitrogen has five valence electrons and the Lewis structure of ammonia is

$$
\begin{array}{c}
H \\
\overset{\cdot\cdot}{H : N : H} \\
\cdot\cdot
\end{array}
$$

Again, the four electron pairs will arrange themselves as far apart as possible because of electrostatic repulsion, and the molecular structure will be like that of ammonia shown in Figure 1.26b, the lone pair of electrons being in an orbital directed toward one corner of the tetrahedron. According to the VSEPR theory, lone-pair electrons are less localized, i.e., occupy more volume than the bond pairs. A lone pair therefore exerts a greater repulsion on the bond pairs than the bond pairs do on each other. As a result, the three bonds of the NH_3 molecule are forced slightly closer together than in the tetrahedral arrangement, the $H : N : H$ angles being 107.3° rather than the tetrahedral angle of about 109.5°.

The water molecule has two pairs of bonding electrons and two lone pairs:

$$
\begin{array}{c}
\overset{\cdot\cdot}{: O : H} \\
\overset{\cdot\cdot}{} \\
H
\end{array}
$$

Again, the four pairs are arranged in an approximately tetrahedral manner as shown in Figure 1.26c. The lone-pair orbitals point toward two corners of the tetrahedron, the bonds toward two other corners. Because of the greater repulsions of the lone-pair electrons, the angle $H : O : H$ is less than the tetrahedral angle; the experimental value is 104.5°.

The central atoms of some molecules have six electron pairs, and the geometries are then related to the octahedron. A simple example is sulfur hexafluoride (SF_6). The sulfur atom has six valence electrons, and therefore there are six bond pairs in the molecule. In order for these pairs to be as far from one another as possible, the molecule has the shape of a regular octahedron, as shown in Figure 1.27a. In this structure the bonds are all equivalent, and the bond angle is exactly 90°. This type of arrangement is found in many molecules of biological importance, such as the molecule of hemoglobin, which is a complex compound of the ferrous ion, Fe^{2+}.

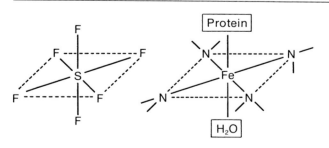

a. Sulfur hexafluoride, SF_6 b. Hemoglobin

Figure 1.27

The octahedral arrangement found when the central atom has six electron pairs, none of them lone pairs: (a) SF_6 and (b) hemoglobin. In the hemoglobin structure the four N atoms form part of the heme molecule.

The Fe^{2+} is surrounded by four nitrogen atoms, which lie in a plane at the corners of a square. A fifth bond, at right angles to the plane, goes to a protein molecule (globin). The sixth bond, also at right angles to the plane, binds a water molecule. The arrangement is shown schematically in Figure 1.27b. In oxyhemoglobin the water molecule is replaced by an oxygen molecule.

The octahedral geometry is also the basis of certain structures which have lone pairs. The iodine pentafluoride (IF_5) molecule, for example, has five bond pairs and one lone pair, since iodine has seven valence electrons. The structure of the molecule is shown in Figure 1.28a; the square pyramid is slightly distorted because of the greater repulsion of the lone pair.

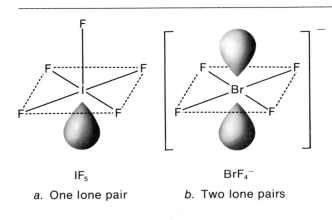

IF$_5$

a. One lone pair

BrF$_4^-$

b. Two lone pairs

Figure 1.28

Geometries of molecules in which the central atoms have six electron pairs: (a) one lone pair, (b) two lone pairs.

In the bromine tetrafluoride BrF_4^- ion there are four bonds and two lone pairs, and the structure is shown in Figure 1.28b with the fluorine atoms at the corners of a square.

1.12   COORDINATION COMPOUNDS

The basic principles of valency were developed during the latter years of the last century. It was recognized then that there are certain compounds to which the usual rules did not apply. The German-Swiss chemist Alfred Werner (1866–1919) suggested that in substances such as salt hydrates, metal amines, and double salts, an element might display a secondary valency in addition to its normal valency. These compounds are known as *complex compounds* or *coordination compounds*. Many compounds of biological importance are of this class. Examples are hemoglobin, chlorophyl, vitamins of the B_{12} group, the cytochromes, and a number of enzymes containing metal atoms.

Werner recognized that the secondary valence bonds in his theory were directed in space, with the result that geometrical and optical isomersion is possible. For example, there are nine isomers with the empirical formula $Co(NH_3)_3(NO_2)_3$, corresponding to different spatial arrangements of the NH_3 and NO_2 group. In 1916 G. N. Lewis included Werner's complex compounds in his electronic theory of valency and suggested that each bond consists of a pair of electrons, *both* of which are donated to the central metal ion by the group attached to the ion. The atom donating the two electrons is called the *donor*, and the atom or ion accepting them the *acceptor*. Lewis called the bond formed in this way a *dative bond*, but it is now usually called a *coordinate bond*. A donor molecule which forms such a bond with a metal or other atom is called a *ligand* (from the Latin *ligare*, to bind).

A simple molecule having a coordinate bond is formed by the addition of ammonia, NH_3, to boron trifluoride, BF_3. The Lewis structures of these molecules are:

$$
\begin{array}{ccc}
\text{H} & & \text{F} \\
\overset{\cdot\cdot}{\text{H}:\text{N}:} & \text{and} & \overset{\cdot\cdot}{\text{B}:\text{F}} \\
\underset{\cdot\cdot}{\text{H}} & & \underset{\cdot\cdot}{\text{F}}
\end{array}
$$

The ammonia molecule has a lone pair of nonbonding electrons, and the BF_3 has a vacant orbital—there being only six electrons in the vacant shell of the boron atom. An additive compound, or *adduct*, is therefore formed:

$$
\begin{array}{cccc}
\text{H} & \text{F} & & \text{H F} \\
\overset{\cdot\cdot}{\text{H}:\text{N}:} + & \overset{\cdot\cdot}{\text{B}:\text{F}} & \rightarrow & \overset{\cdot\cdot\ \cdot\cdot}{\text{H}:\text{N}:\text{B}:\text{F}} \\
\underset{\cdot\cdot}{\text{H}} & \underset{\cdot\cdot}{\text{F}} & & \underset{\cdot\cdot\ \cdot\cdot}{\text{H F}}
\end{array}
$$

Once the molecule has been formed, the dative bond is of the same general character as any other covalent bond. We could, in fact, have regarded the molecule as being formed by the combination of NH_3^+ and BF_3^-:

$$
\begin{bmatrix} \text{H} \\ \overset{\cdot\cdot}{\text{H}:\text{N}\cdot} \\ \underset{\cdot\cdot}{\text{H}} \end{bmatrix}^+ + \begin{bmatrix} \text{F} \\ \overset{\cdot\cdot}{.\,\text{B}:\text{F}} \\ \underset{\cdot\cdot}{\text{F}} \end{bmatrix}^- \rightarrow \overset{\cdot\cdot\ \cdot\cdot}{\underset{\cdot\cdot\ \cdot\cdot}{\text{H}:\text{N}:\text{B}:\text{F}}}
$$

This last equation emphasizes the fact that in the resulting molecule the nitrogen atom has an effective positive charge and the boron atom an effective negative charge. It is convenient to represent a coordinate bond either by an arrow pointing from the donor atom to the acceptor atom:

$$H_3N \rightarrow BF_3$$

or as a covalent bond between ions:

$$H_3N^+{-}B^-F_3$$

Both methods of representation remind us that the bond usually has a fairly high dipole moment in the direction indicated by the arrow. The second representation is useful in allowing us to draw a conclusion immediately about the shape of the molecule.

The ions N^+ and B^- are both isoelectronic with carbon. Therefore, in both cases the bonds emanating from these ions will be directed toward the corners of a regular tetrahedron.

The reason that transition-metal ions frequently enter into complex compounds as acceptors is that these ions have high electron affinities as a result of their having vacant d orbitals. As an example, consider the formation of the anion $[Ni(CN)_4]^{2-}$. The electronic configurations of Ni and Ni^{2+} are as follows:

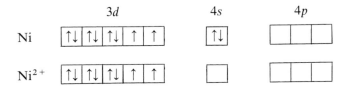

The Lewis structure of the cyanide ion is

$$[:C \vdots N:]^-$$

and the pair of unshared electrons in the carbon atom can be donated to the Ni^{2+}. We can imagine the two unpaired electrons in Ni^{2+} pairing up, and a pair of electrons being donated by four CN^- ions, which go into a d, an s, and two p orbitals. Hybridization of these four orbitals, dsp^2 hybridization, leads to a square-planar arrangement:

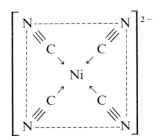

An example of a compound having six coordinated groups is $[Pt Cl_6]^{2-}$. The platinum (IV) ion has the structure

$$Pt^{4+}$$

The six Cl^- ions donate six pairs of electrons, which go into two d, one s, and three p orbitals. The d^2sp^3 hybridization gives an octahedral arrangement:

$$\begin{bmatrix} & & \text{Cl} & & \\ & \text{Cl} & | & & \\ & & \downarrow & \nwarrow & \text{Cl} \\ & & \text{Pt} & & \\ \text{Cl} & & \uparrow & & \text{Cl} \\ & & | & & \\ & & \text{Cl} & & \end{bmatrix}^{2-}$$

This arrangement is frequently found with ions in water; an example is the Fe^{2+} ion, which is surrounded octahedrally by six water molecules:

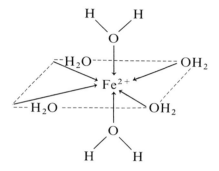

In Figure 1.27a and b, note this octahedral structure in SF_6 and in hemoglobin.

The number of ligands that can surround a metal ion is determined to some extent by geometry. Thus the ions Cu^{2+} and Zn^{2+}, being small, can be surrounded by a maximum of four ligands only; Fe^{2+} can have six. The number of ligands in a coordination compound is known as the *coordination number*. The maximum number of ligands that an atom or ion in a coordination compound can have is called the *maximum coordination number*. The following ions have a maximum coordination number of 4:

$$Au^+, Cu^+, Zn^{2+}, \text{ and } Au^{3+}$$

These ions have a maximum coordination number of 6:

$$Fe^{2+}, Co^{2+}, Ni^{2+}, Cu^{2+}, Mg^{2+}, Fe^{3+}, \text{ and } Co^{3+}$$

Table 1.4 shows the shapes associated with certain types of hybridization, and gives a few examples of complexes having those shapes.

Of particular interest in biology are the coordination compounds formed between metal ions and ligand groups on amino acids. A number of enzymes are active only when they have formed a complex with a metal ion. The best understood metallo-enzyme is carboxypeptidase; this occurs in the digestive system, and its structure has been determined by X-ray diffraction (p. 107). The usual form of the enzyme contains one Zn^{2+} ion about which the arrangement is approximately tetrahedral. Three of the ligands are provided by the protein and the fourth ligand is a water molecule. When the enzyme forms an addition compound with a substrate molecule prior to breaking it down, this molecule replaces the water molecule as the fourth ligand. The complexing in this enzyme is shown in a simplified manner in Figure 1.29.

During recent years the properties of complexes have been interpreted on the basis of a theory which is concerned with the electrostatic field created by the ligands. This theory was developed by the German-American physicist Hans Bethe and applied first to crystal structures. The original theory was known as *crystal field theory*.

Table 1.4 Hybrid Orbitals and Spatial Arrangements

Coordination Number	Hybridization	Spatial Arrangement	Examples
2	sp	linear	$[Ag(CN)_2]^-$
4	dsp^2	square planar	$[PtCl_4]^{2-}$
			$[Cu(NH_3)_4]^{2+}$
4	sp^3	tetrahedral	carboxypeptidase† (Figure 1.29)
6	d^2sp^3	octahedral	$[Co(CN)_6]^{3-}$,
			SiF_6^{2-},
			hemoglobin (Figure 1.27*b*)
8	d^4sp^3	dodecahedral	$[Mo(CN)_8]^{4-}$

† The arrangement around the zinc atom in carboxypeptidase is in between the sp^3 tetrahedral configuration and the dsp^2 square-planar configuration.

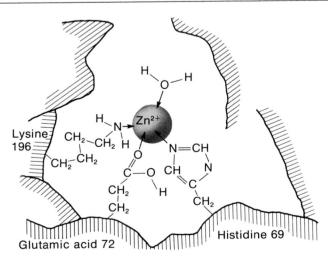

Figure 1.29

Schematic representation of the molecule of carboxypeptidase, a metalloenzyme containing a Zn^{2+} ion.

The modified theory which applies to complex compounds is known as *ligand field theory*. An important aspect of the theory is the recognition that the action of the electrostatic field of the ligands on the *d* orbitals of the central ion creates an additional bonding effect called the *ligand field stabilization energy*.

Ligand field theory is of chief relevance in the calculation of the energy levels of coordination compounds, a matter of importance when one is interpreting the spectroscopic and magnetic properties of these compounds. Only a very brief account of ligand field theory will be given here, with reference to the two octahedral complexes $FeF_6{}^{3-}$ and $[Fe(CN)_6]^{3-}$. These involve the five $3d$ orbitals of Fe^{3+}, which, in the absence of an electrostatic field due to the ligands, would all be equal in energy (i.e., having fivefold degeneracy). However, when ligands are added, these orbitals are then in an electrostatic ligand field which is not spherically symmetrical. The energies of the five $3d$ orbitals are no longer exactly equal; the energies of two of the orbitals are raised and those of the other three are lowered. This is shown in Figure 1.30. The extent to which the energy levels are displaced depends on the magnitude of the field; we can distinguish between weak fields, exemplified by the F^- ligands, and strong fields, exemplified by CN^-.

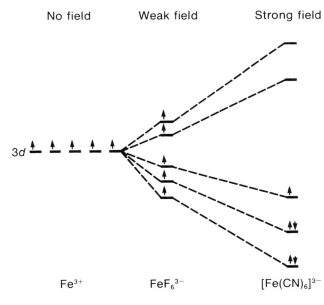

No field   Weak field   Strong field

Figure 1.30

The splitting of *d* orbitals by a weak octahedral ligand field and by a strong field.

Fe^{3+} $FeF_6{}^{3-}$ $[Fe(CN)_6]^{3-}$

The extent of the splitting has an important bearing on the electron distribution. We have seen (p. 20) that, according to Hund's rule, electrons tend to occupy orbitals in such a way that the electron spins can be the same, with a consequent lowering of the electrostatic repulsion energy. In Fe^{3+} there are five $3d$ electrons, all with the same spin (see Figure 1.30). In $FeF_6{}^{3-}$ there is not much difference between the energy levels of the *d* orbitals. Again one electron will go into each orbital, and all will have the same spin. In $[Fe(CN)_6]^{3-}$, on the other hand, the splitting of the *d* levels by the ligand field is large, and it requires too much energy to put electrons into the higher d orbitals. Instead, electrons will go into the three lower orbitals, as shown in Figure 1.30, and four of the spins will be paired. Since magnetism depends on the number of unpaired electrons, $FeF_6{}^{3-}$ exhibits much more magnetism than $[Fe(CN)_6]^{3-}$.

PROBLEMS

1.1 Calculate the wavelength of the de Brogie waves associated with:

(a) An electron (mass $= 9.11 \times 10^{-31}$ kg) moving with a velocity of 6.0×10^7 m s^{-1} (this is the approximate velocity produced by a voltage of 10 000 V).

(b) An oxygen molecule moving with a velocity of 425 m s^{-1} (this is the average kinetic-theory velocity at $0°$ C).

(c) An α particle (He nucleus) emitted by radium, moving at a velocity of 1.5×10^7 m s^{-1}.

(d) A car weighing 1000 kg moving with a velocity of 100 km per hour.

1.2 Calculate the wavelength of (a) an electron, and (b) a neutron (1.007 atomic mass units), each having kinetic energy of 1 eV. $[1 \text{ eV} = 1.602 \times 10^{-19} \text{ J}]$.

1.3 An electron accelerated by a potential of 100 volts has a velocity of 5.90×10^6 m s^{-1}. If the velocity of such an electron can be measured with an uncertainty of 10%, what is the uncertainty in the position of the electron?

1.4 In a photoelectric experiment with metallic sodium, light of 450 nm wavelength ejects electrons with a maximum energy of 3.36×10^{-19} J. Calculate the maximum wavelength of light that will eject electrons from sodium. [Velocity of light $= 3.00 \times 10^8$ m s^{-1}].

1.5 The attractive electrostatic energy between two univalent ions M^+ and A^- separated by a distance d in nanometres is $137.2/d$ J mol^{-1}. Suppose that there is also a repulsive energy term given by $0.0975/d^6$ J mol^{-1}, with d also in nanometres. On the same graph, plotting energy against d, sketch the attractive and repulsive curves, and the resultant. Then, by differentiating the energy equation, calculate the equilibrium interionic distance and the net energy at that distance.

1.6 The following are bond dissociation energies:

$$D(Li_2) = 113 \text{ J mol}^{-1}$$

$$D(H_2) = 435 \text{ J mol}^{-1}$$

$$D(LiH) = 243 \text{ J mol}^{-1}$$

(a) Estimate the electronegativity difference between the lithium and hydrogen atoms, making use of Pauling's relationship that

$$(E_{ionic})^{\frac{1}{2}} = 10 |\chi_A - \chi_B|$$

(b) Estimate the percent ionic character of the Li—H bond, given the following covalent radii:

$$Li : 0.126 \text{ nm}$$

$$H : 0.036 \text{ nm}$$

and with $e = 4.8 \times 10^{-10}$ esu.

1.7 Write down Lewis structures for the following molecules, and deduce the molecule shapes using VSEPR theory:

$$BeCl_2, SF_6, H_3O^+, NH_4^+, PCl_6^-, AlF_6^{3-}, PO_4^{3-}.$$

1.8 Suggest an explanation, on the basis of VSEPR theory, for the experimental fact that in ethylene the H—C—C angles are greater than the H—C—H angles.

1.9 Construct molecular orbital diagrams for the following molecules:

$$B_2, CO, NO, CN, BN, BN^{2-}, BF$$

Deduce the bond order in each case.

1.10 Two functions ψ_1 and ψ_2 are said to be *orthogonal* if

$$\int \psi_1 \psi_2^* \, d\tau = 0$$

Suppose that there are four real functions ψ_1, ψ_2, ψ_3, and ψ_4, which are normalized and mutually orthogonal. Normalize the following functions

(a) $\psi_1 + \psi_2$ (see equation (1.38))

(b) $\psi_1 - \psi_2$ (see equation 1.54))

(c) $\psi_1 + \psi_2 + \psi_3$

(d) $\psi_1 - \psi_2 + \psi_3 - \psi_4$ (see equation (1.64))

(e) $\psi_1 - \dfrac{1}{\sqrt{2}} \psi_2 + \dfrac{\sqrt{3}}{\sqrt{2}} \psi_3$ (see equation (1.66))

ESSAY QUESTIONS

1.11 Give a brief account of the Valence-Shell Electron-Pair Repulsion (VSEPR) theory of the shapes of molecules.

1.12 Give a brief account of the main principles underlying the Variation Method in quantum mechanics.

1.13 Explain precisely what is meant by a *normalized* eigenfunction.

1.14 Explain the principles underlying the construction of trial wave functions:
 (a) in the valence-bond method
 (b) in the molecular-orbital method

1.15 Give an account of orbital hybridization with special reference to sp, sp^2, sp^3, and d^2sp^3 hybridization.

SUGGESTED READING

Barrow, G. M. *The Structure of Molecules.* New York: Benjamin, 1963.

Cartmell, E., and Fowles, G. W. A. *Valency and Molecular Structure.* London: Butterworth, 1961.

Coulson, C. A. *Valence.* 2nd ed. Fair Lawn, New Jersey: Oxford University Press, 1961.

Gillespie, R. J. *Molecular Geometry.* London: Van Nostrand Reinhold, 1972.

Gray, H. B. *Electrons and Chemical Bonding.* New York: Benjamin, 1964.

Heitler, W. *Elementary Wave Mechanics.* Fair Lawn, New Jersey: Oxford University Press, 1945.

Karplus, M., and Porter, R. N. *Atoms and Molecules: An Introduction for Students of Physical Chemistry.* New York: Benjamin, 1970.

Linnett, J. W. *Wave Mechanics and Valency.* London: Methuen, 1960.

Linnett, J. W. *The Electronic Structure of Molecules: A New Approach.* London: Methuen, 1964.

Moore, W. J. *Physical Chemistry.* 4th ed. Chapters 13, 14, and 15. Englewood Cliffs, New Jersey: Prentice-Hall, 1972.

Murrell, J. N., Kettle, S. F. A., and Tedder, J. M. *Valence Theory.* New York: John Wiley, 1965.

Pauling, L. *The Nature of the Chemical Bond.* 3rd ed. Ithaca. New York: Cornell University Press, 1960.

2

MOLECULAR SPECTROSCOPY

Very detailed information about structure is obtained from investigations in which electromagnetic radiation interacts with matter. An important area of study, known as spectroscopy, is concerned mainly with the extent to which substances absorb radiation at various wavelengths. The information obtained through spectroscopy has contributed greatly to our understanding of chemical structure and is particularly important in biology.

Besides providing basic information about structure, investigations with electro-magnetic radiation are also valuable in determining concentrations of substances. The amount of radiation absorbed depends upon the amount of the absorbing substance. Optical techniques for determining concentrations by light absorption are extremely important in biology, and are used wherever possible. These techniques have various advantages over other methods. Radiation often does not bring about any destruction of the material being studied, and the techniques can determine much smaller amounts of material than is possible with other procedures. Also, optical techniques measure concentrations instantaneously, and the results can be recorded immediately. The techniques can therefore be employed to follow the concentration changes of substances involved in very rapid reactions. These analytical aspects of light absorption are referred to as *optical analysis* or *photometric analysis*, and will be dealt with briefly in this chapter.

2.1 THE SPECTRUM OF ELECTROMAGNETIC RADIATION

There are two types of radiation—electromagnetic radiation and particle radiation. Examples of particle radiation are α rays, which are beams of helium nuclei, and β rays, which are beams of electrons. These are subatomic particles and can move with various velocities. Beams of electrons can be produced in different ways, as in a cathode-ray tube, and the speeds of the beams can be controlled by electrical and magnetic devices.

   Electromagnetic radiation is different from particle radiation in various ways. The most important difference is that the particles in electromagnetic radiation (photons) have zero mass. In contrast, the particles (e.g., electrons and atomic nuclei) in particle radiation have a nonzero mass. A second essential difference between electromagnetic and particle radiation is that a beam of particles can be moving at any velocity, whereas electromagnetic radiation always travels with a fixed velocity in a given medium. This fixed velocity, the velocity of light, has a value of 3.00×10^8 m s^{-1} in a vacuum. Electromagnetic radiation can be characterized by either a frequency v or a wavelength λ, the product of these being c, the velocity of light. Various units are used to express frequency and wavelength. The basic unit of frequency is reciprocal seconds (s^{-1}) or hertz (Hz), which is the number of vibrations per second. However, for practical applications of spectroscopy this unit is somewhat inconvenient, because the numbers involved are often large. A common practice is to use the reciprocal of the wavelength instead of the frequency, i.e., $1/\lambda$ instead of c/λ. The term *wavenumber* is then employed, and the usual symbol is $\bar{v}$. If λ is expressed in metres (m), $\bar{v} = 1/\lambda$ will be in reciprocal metres (m^{-1}), and multiplication by the velocity of light in m s^{-1} will give v. Wavenumbers are more often expressed in the units of reciprocal centimetres; 1 cm^{-1} = 100 m^{-1}. Multiplication of $\bar{v}$(cm^{-1}) by 10^2 c(m s^{-1}) will then give the frequency v.

   Various units have been used for wavelength. The most important of these are given below and are related to the metre:

$$1 \text{ angstrom (Å)} = 10^{-8} \text{ cm} = 10^{-10} \text{ m}$$

$$1 \text{ micron } (\mu) = 10^{-4} \text{ cm} = 10^{-6} \text{ m}$$

$$1 \text{ millimicron (m}\mu) = 10^{-7} \text{ cm} = 10^{-9} \text{ m}$$

The micron is not accepted in SI; it should be called a *micrometre* and given the symbol μm. Similarly, the millimicron should be called the *nanometre* and given the symbol nm. Use of the angstrom is acceptable in SI but not recommended. However, spectroscopists and crystallographers find this unit very convenient, and it seems likely that it will remain in use for some time. In particular, the angstrom is convenient for expressing bond lengths, most of which lie between 1 and 2 Å. When the nanometre is used, bond lengths lie between 0.1 and 0.2 nm. Bond lengths can also be expressed in picometres (10^{-9} m), in which case they are in the range 100–200 pm.

   In 1665–66 Sir Isaac Newton (1642–1727) demonstrated that when white light is passed through a prism, it is split into a spectrum of colors ranging from red to violet. The red end of the spectrum corresponds to the longer wavelengths ($\approx$ 700 nm or 7000 Å) and lower frequencies. The violet end of the spectrum is the short-wavelength high-frequency end; the wavelengths are about 400 nm or 4000 Å. Since the energies of the photons are equal to the frequency multiplied by Planck's constant (6.626×10^{-34} J s), the violet end corresponds to higher photon energies. Violet light is therefore, in general, more effective than red light in bringing about chemical and biological change.

The colors visible to humans are a very small part of the electromagnetic spectrum, as shown in Figure 2.1. Beyond the violet end of the visible spectrum is a region known as *ultraviolet*, where the frequencies (and therefore the photon energies) are higher and the wavelengths shorter than in the violet region. Beyond the ultraviolet are the X rays and the γ rays. Below the red end of the visible spectrum are the *infrared* and radio waves. These are of lower frequencies (and lower energies) and have longer wavelengths than the red.

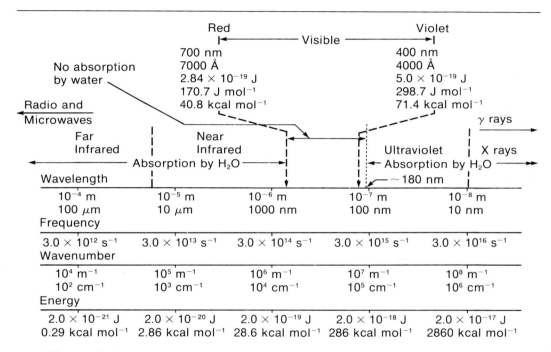

Figure 2.1

The spectrum of electromagnetic radiation.

2.2   MOLECULAR ENERGY LEVELS

Energy can reside in molecules in a number of forms, of which the most important are (a) translational energy, (b) electronic energy, (c) vibrational energy, and (d) rotational energy. Translational energy, the kinetic energy of a molecule associated with the movement of its center of gravity, is not of importance in spectroscopy, because light cannot interact directly with a molecule and cause it to change its translational energy. However, the interaction of electromagnetic radiation with molecules can bring about a change in electronic, vibrational, and rotational energy. Such interactions provide us with very valuable information about molecular structure.

The relationship between the type of energy and the spectroscopic region where there is absorption is summarized in Table 2.1.

Table 2.1 Types of Optical Spectra

Spectroscopic Region	Approximate Range			Types of Molecular Energy	Information Obtained
	Frequency	Wavenumber	Energy		
	(s^{-1})	(cm^{-1})	$(kJ\ mol^{-1})$		
Microwave and far infrared	$10^9–10^{12}$	0.03–30	4×10^{-4}–0.4	Rotation	Interatomic distances
Infrared	$10^{12}–10^{14}$	30–3000	0.4–40	Vibration and rotation	Interatomic distances and force constants of bonds
Visible and ultraviolet	$10^{14}–10^{16}$	3×10^3–3×10^5	40–4000	Electronic, vibration and rotation	Electronic energy levels, bond dissociation energies, force constants of bonds, and interatomic distances

ROTATIONAL ENERGY

We see in Table 2.1 that the smallest energies are associated with rotation. The manner in which a molecule rotates must be considered separately for a linear and for a nonlinear molecule. Figure 2.2a shows a linear molecule with the axis X the axis of the molecule, and with two axes Y and Z drawn at right angles to the axis of the molecule and passing through the center of gravity. Rotation about the X axis does not cause any displacement of the atoms, and therefore does not involve any energy. Rotations about the Y and Z axes, on the other hand, do involve energy. Any rotation that the molecule undergoes can be treated as a rotation about the two axes Y and Z, and we say that there are *two degrees of rotational freedom* for a linear molecule.

The case of a nonlinear molecule is illustrated by H_2O in Figure 2.2b. We can draw three Cartesian axes through the center of gravity, and rotation about any of these three axes now involves a displacement of atoms and therefore requires energy. There are three degrees of rotational freedom for any nonlinear molecule.

VIBRATIONAL ENERGY

Higher energies are associated with the vibrations of molecules. Again we must distinguish between linear and nonlinear molecules. Analysis of the vibrational motion

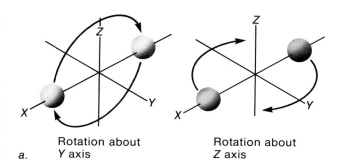

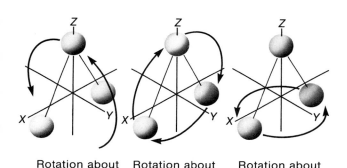

Figure 2.2

Molecular rotations in (*a*) a linear molecule, and (*b*) a nonlinear molecule.

Rotation about
Y axis

Rotation about
Z axis

a.

Rotation about   Rotation about   Rotation about
X axis   Y axis   Z axis

b.

of a linear molecule such as carbon dioxide (CO_2) shows that the vibrations can be treated as a combination of four basic vibrational motions. These are known as *normal modes of vibration*, and in each mode every atom in the molecule vibrates with the same frequency. The four modes for carbon dioxide are illustrated in Figure 2.3*a*; two of the modes correspond to stretching of bonds and two to bending — i.e., to changes in bond angles. The two latter, which correspond to bending, have the same frequency and are said to be *degenerate*.

The normal modes of vibration of the nonlinear water molecule are shown in Figure 2.3*b*. Modes 1, 2, and 3 are equivalent to modes 1, 2, and 3 for the carbon dioxide molecule. However, mode 4 for carbon dioxide has no counterpart in water; a motion equivalent to mode 4 in carbon dioxide would, for the bent water molecule, be a *rotation* and not a vibration. In water there is thus one more rotational mode and one fewer vibrational mode, as compared with carbon dioxide.

The number of normal modes of vibration (i.e., of vibrational degrees of freedom) can be calculated easily for any molecule. The total number of degrees of freedom must be three times the number of atoms in the molecule, since each atom has three components of motion along the X, Y, and Z axes. Thus for any molecule there will be three degrees of translational freedom, because the center of gravity of the molecule has three components of motion along the three Cartesian axes. For any linear molecule, there are two degrees of rotational freedom; for any nonlinear molecule there are three degrees of rotational freedom. For a molecule having n number of atoms, the various numbers of degrees of freedom are:

	Linear	Nonlinear
Translational	3	3
Rotational	2	3
Vibrational	$3n - 5$	$3n - 6$
Total	$3n$	$3n$

For benzene, a nonlinear molecule having 12 atoms, the number of normal modes of vibration is 30, and the vibration of the molecule is very complex. Associated with each normal mode is a characteristic frequency of vibration; some of the normal modes of vibration have identical vibrational frequencies, and we then say that the vibrations are *degenerate*, e.g., modes 3 and 4 in Figure 2.3a.

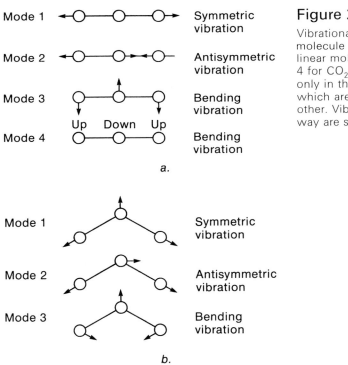

Mode 1 — Symmetric vibration

Mode 2 — Antisymmetric vibration

Mode 3 — Bending vibration

Up   Down   Up

Mode 4 — Bending vibration

a.

Mode 1 — Symmetric vibration

Mode 2 — Antisymmetric vibration

Mode 3 — Bending vibration

b.

Figure 2.3

Vibrational motions in (*a*) the linear molecule CO_2, and (*b*) the non-linear molecule H_2O. Modes 3 and 4 for CO_2 are equivalent, differing only in that they occur in planes which are perpendicular to each other. Vibrations equivalent in this way are said to be *degenerate*.

The relationship between vibrational and rotational energy levels is illustrated for a diatomic molecule in Figure 2.4, which shows a potential-energy curve for a molecule AB. There is now one normal mode of vibration $[(3 \times 2) - 3 - 2]$. The figure shows the quantized vibrational levels. The lowest vibrational level, with the vibrational quantum number 0, corresponds to the vibration with the minimum energy possible, the *zero-point* energy. The next energy level is given the vibrational quantum number 1, the next 2, and so on.

The energy difference between successive rotational levels is much smaller than the energy difference between successive vibrational levels. A molecule in the zero-point vibrational level can be in various rotational levels, and these are indicated in the figure. Similarly, there is a series of rotational levels associated with each of the other vibrational levels.

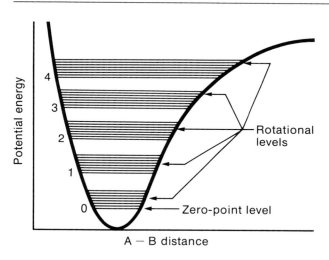

Figure 2.4

A potential-energy curve for a diatomic molecule, showing vibrational and rotational levels. For polyatomic molecules a multidimensional surface is needed to represent the levels, but the principles are the same as for the diatomic case.

ELECTRONIC ENERGY

A molecule can also exist in a number of electronic states. In Chapter 1 we considered the most stable states, or the ground states, of molecules. In these ground states the electrons are present in the lowest molecular orbitals consistent with the Pauli principle. We saw, for example, in Figure 1.25c the ground-state configuration for the oxygen molecule. The molecular-orbital diagram for this state is reproduced in Figure 2.5a. In the ground state there are two unpaired electrons, which have the same spin quantum number. A state in which there are two unpaired electrons with the same spin is known as a *triplet* state; this particular state in O_2 is designated the $^3\Sigma$ state. The reason such states are called "triplet states" is that an applied electric or magnetic field splits the energy into three energy states; thus the state is said to have a *multiplicity* of 3. There also exists an excited state of O_2 in which the electrons in the $2\pi^*$ state are paired; this is shown in Figure 2.5b and is known as a Δ state. Application of an electric or magnetic field to an O_2 molecule in this state produces no splitting. The multiplicity is 1, and the state is called a *singlet* state.

The energy level of the Δ state of O_2 lies 92 kJ mol^{-1} ($= 22$ kcal mol^{-1}) above that of the ground ($^3\Sigma$) state. This energy difference is rather larger than that usually associated with vibrations, and much larger than the energy associated with rotations.

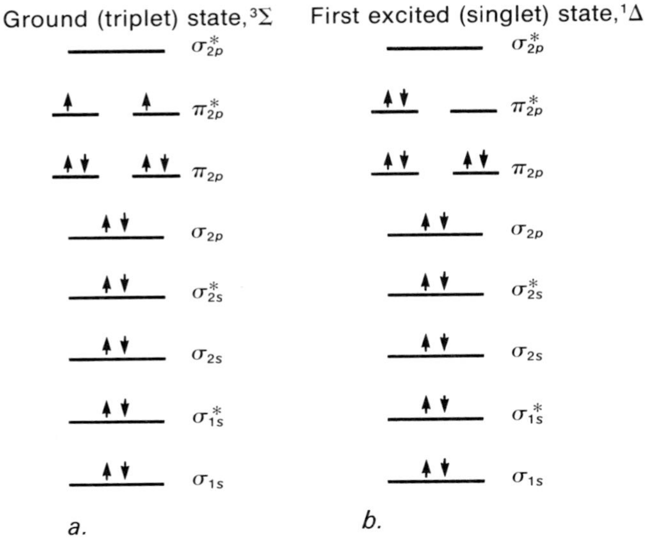

Ground (triplet) state, $^3\Sigma$ First excited (singlet) state, $^1\Delta$

Ground (triplet) state, $^3\Sigma$	First excited (singlet) state, $^1\Delta$
σ_{2p}^*	σ_{2p}^*
↑ ↑ π_{2p}^*	↑↓ ___ π_{2p}^*
↑↓ ↑↓ π_{2p}	↑↓ ↑↓ π_{2p}
↑↓ σ_{2p}	↑↓ σ_{2p}
↑↓ σ_{2s}^*	↑↓ σ_{2s}^*
↑↓ σ_{2s}	↑↓ σ_{2s}
↑↓ σ_{1s}^*	↑↓ σ_{1s}^*
↑↓ σ_{1s}	↑↓ σ_{1s}
a.	b.

Figure 2.5

Molecular-orbital diagrams for the ground state and the first excited state of the oxygen molecule.

2.3   THE ABSORPTION OF RADIATION

Spectra can be either *emission* spectra or *absorption* spectra. In order for the *emission* spectrum of a substance to be observed, the substance must be excited in some way. This may be done by introducing the substance into a flame or by passing an electric discharge through it. If the emitted light is observed through a spectrometer, the characteristic emission spectrum of the substance will be seen. For example, if a compound of sodium is heated in a flame and the yellow light emitted is observed through a spectrometer, most of the radiation has a wavelength of 589 nm (5890 Å).

The procedure for observing an *absorption* spectrum is quite different. An absorption spectrum can be seen by passing continuous radiation, such as white light, through the substance and observing the spectrum with a spectrometer. When this is done, certain wavelengths are found to be missing from the spectrum. For example, if a continuous radiation of white light is used, its normal spectrum consists of colors from red to violet, which blend smoothly into one another. After the white light has passed through the substance, however, black lines may be superimposed on the continuous spectrum. These black lines occur because the substance through which the light passed has removed light corresponding to certain wavelengths. The energy corresponding to a transition is the frequency of the line multiplied by Planck's constant.

In biological work one is rarely concerned with the far infrared section of the spectrum, where the energies are small and the only changes occurring are in the rotational levels of the molecules. Work in the near infrared is of greater importance in biology. Here the molecule can change its vibrational state and, at the same time, may also experience a change in its rotational state. Infrared spectra thus consist of series of *bands*, each band corresponding to a change in the vibrational state of

the molecule and each line in the band corresponding to a change in the rotational
state. If the material under study is in the liquid state or in solution, there is no
free rotation of the molecules, and the structure of the bands is considerably blurred.

The basis of electronic absorption, i.e., absorption in the visible and ultraviolet
regions, is illustrated in Figure 2.6, which shows potential-energy curves for a molecule
in its ground electronic state and in an excited electronic state. There are a number

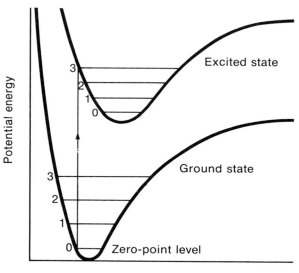

Figure 2.6

Potential-energy curves for a
molecule in its ground state and
its excited state, showing
vibrational levels. The arrow
represents a transition from the
ground state at $v = 0$ to the excited
state at $v = 3$.

of vibrational states associated with the ground electronic state. These are represented
by horizontal lines in the diagram. Similarly, the excited electronic state has a number
of vibrational levels. At ordinary temperatures molecules are almost all in their ground
vibrational levels, which correspond to the zero-point vibrational energy. When light
of a suitable wavelength passes through the substance, a molecule may absorb a
photon and be converted into the upper electronic state. At the same time there
may be a change in vibrational level. In Figure 2.6 a transition is shown from the
lowest vibrational level of the original molecule to the third vibrational level ($v = 3$)
of the electronically-excited state. The Franck-Condon principle states that the most
likely transitions will be the most "vertical" ones, in which there is little change in
the distance between the atoms when the transition occurs.

Since an electronically-excited molecule can make a transition to several different
vibrational states, and since, in addition, changes in rotational energy occur, the
electronic spectrum of a molecule will involve a series of bands. If the electronic
spectrum is observed in the vapor phase, the bands will consist of a series of sharp
lines, which result from the closely-spaced rotational levels (see Figure 2.4). These
sharp lines are known as the *fine structure* of the bands. In the liquid state or in
solution the molecules cannot rotate freely. In these states the fine structure is
considerably blurred.

SELECTION RULES

When electromagnetic radiation interacts with a molecule, certain transitions between states can occur, and others cannot. There are *selection rules* which tell us which transitions are possible and which are not. The detailed theory underlying these selection rules is fairly complicated, but some important general conclusions can be stated.

Pure rotational transitions, which give spectra in the far infrared and radio regions, will be considered first. A molecule can only absorb electromagnetic radiation if it can interact with the oscillating electric field associated with the radiation. If a molecule has a permanent dipole moment, this dipole oscillates as rotation occurs, and a pure rotational spectrum is obtained. This is the case with molecules such as carbon monoxide (CO) and hydrogen chloride (HCl), which have permanent dipole moments. Molecules such as H_2 and N_2, which do not have permanent dipole moments, do not have pure rotational spectra.

Another selection rule which applies to both rotational and vibrational transitions is that the quantum number can only change by unity; in other words, there can be a transition only to a neighboring rotational or vibrational level.

The selection rule for vibration is in other respects different from that for rotation. For a transition to be possible, the dipole moment must oscillate as vibration occurs. This rule at once excludes *diatomic* molecules with zero dipole moments, which remain zero as the molecule vibrates (e.g., H_2, O_2). However, any molecule with a dipole moment (e.g., HCl and H_2O) will give a vibrational spectrum, since the dipole moment will change somewhat as the molecule vibrates. Furthermore, a molecule containing more than two atoms can have a vibrational spectrum even if it has no dipole moment, because some of the vibrational modes will give rise to a change in dipole moment. Carbon dioxide, the vibrations of which are shown in Figure 2.3a, provides a good example. This molecule, being symmetrical, has no dipole moment. If it vibrates in the symmetric mode (Mode 1), the dipole moment remains zero and there is no infrared spectrum. However, vibration in the antisymmetric mode (Mode 2) produces an oscillating dipole moment, and a vibrational (infrared) spectrum will result from this type of vibration. Modes 3 and 4 also produce an oscillating dipole moment, and give rise to an infrared spectrum.

We will see later (p. 76) that the selection rules for Raman spectra are different from those for vibrational-rotational spectra. Thus a molecule giving no infrared spectrum may have a Raman spectrum and vice versa.

Selection rules also apply to visible and ultraviolet spectra, which involve a transition from one electronic state to another. An important selection rule is that transitions are more likely to occur if there is no change of multiplicity. For example, a singlet state is more likely to be converted into another singlet state than into a triplet state. We have seen that in a singlet state all of the electrons are paired. A transition into a triplet state would involve the uncoupling of a pair of electrons, one of which would have to change its spin quantum number, and such changes occur with difficulty. Violations of this selection rule are found with molecules containing heavy atoms but only rarely with ordinary compounds.

2.4   THE LAWS OF LAMBERT AND BEER

When spectroscopic studies of substances are combined with measurements of light absorbed and transmitted at various wavelengths, the term *spectrophotometry* is employed. This type of investigation is based on laws propounded by the German mathematician Johann Heinrich Lambert (1728–1777) and the German astronomer Wilhelm Beer (1797–1850). These laws are concerned with the intensities of light absorbed or transmitted when incident light is passed through some material.

LAMBERT'S LAW

Lambert's law states that the proportion of radiation absorbed by a substance is independent of the intensity of the incident radiation. This means that each successive layer of thickness dx of the medium absorbs an equal fraction $-dI/I$ of the radiation of intensity I incident upon it. In other words,

$$-\frac{dI}{I} = bdx \tag{2.1}$$

The proportionality constant b, which is characteristic of the absorbing medium, is known as the *Napierian absorption coefficient*, or the *Napierian extinction coefficient*.

Integration of equation (2.1), for passage of light through a thickness l, proceeds as follows:

$$\int \frac{dI}{I} = -b \int_0^l dx \tag{2.2}$$

$$\ln I = -bl + d \tag{2.3}$$

The constant of integration d can be evaluated using the boundary condition that when $l = 0$, $I = I_0$, where I_0 is the intensity of the radiation before passage through the medium. Thus the constant d is equal to $\ln I_0$, and equation (2.3) thus becomes

$$\ln I - \ln I_0 = -bl \tag{2.4}$$

or

$$I = I_0 e^{-bl} \tag{2.5}$$

The product bl may be written as B, which is known as the *Napierian absorbance* or *Napierian extinction*:

$$I = I_0 e^{-B} \tag{2.6}$$

or

$$B = \ln \frac{I_0}{I} \tag{2.7}$$

For many purposes it is convenient to use common instead of natural logarithms. Now

$$\ln \frac{I_0}{I} = 2.303 \log_{10} \frac{I_0}{I} \tag{2.8}$$

so that equation (2.6) becomes

$$\log_{10} \frac{I_0}{I} = \frac{B}{2.303} = \frac{bl}{2.303} \tag{2.9}$$

The quantity $B/2.303$ can be written as A and is known as the *decadic absorbance*, or simply as the *absorbance*. This is the term recommended in the SI (see Appendix A). However, the terms *decadic extinction* and *optical density* are still commonly used, and the student should be familiar with them. The quantity $b/2.303$ (i.e., the absorbance divided by the light path) is called the *linear decadic absorption coefficient* and usually is given the symbol a. Thus equation (2.9) can be written as

$$\log_{10} \frac{I_0}{I} = A = al \tag{2.10}$$

Another quantity that is frequently used is the *transmittance, T*. This is the ratio of the intensities of transmitted to incident light:

$$\frac{I}{I_0} = T \tag{2.11}$$

It follows from equations (2.10) and (2.11) that

$$\log_{10} \frac{1}{T} = A \tag{2.12}$$

The percentage light transmittance, $T\%$, is given by

$$T\% = \frac{100I}{I_0} = 100T \tag{2.13}$$

so that, from equation (2.13),

$$\log_{10} T\% = \log_{10} 100 + \log_{10} T \tag{2.14}$$
$$= 2 - A \tag{2.15}$$

BEER'S LAW

Beer studied the influence of the concentration of a substance in solution on the absorbance, and he found the same linear relationship between absorbance and concentration as Lambert had found between absorbance and thickness (see equation 2.10). Thus, for a substance in solution of concentration c, Beer's law states that

$$\log_{10} \frac{I_0}{I} = A = \text{const.} \times c \tag{2.16}$$

THE LAMBERT-BEER LAW

The two equations (2.10) and (2.16) can be combined in the single equation

$$A = \log_{10} \frac{I_0}{I} = \varepsilon c l \tag{2.17}$$

where ε is a constant, known as the *absorption coefficient*, and l is the light path. The absorption coefficient is usually written as

$$\varepsilon_l^c(\lambda)$$

in order to specify the concentration units, light path and wavelength. If, for example, the concentration units are moles per cubic decimeter,† the light path 1 cm, and the wavelength 430 nm, the coefficient would be written as

$$\varepsilon_{1\,cm}^{1\,M}\,(430\ nm)$$

and would then be called the *molar absorption coefficient*. Since $\varepsilon c l$ is dimensionless, the molar absorption coefficient has the units $dm^3\ mol^{-1}\ cm^{-1}$. The International Union of Pure and Applied Chemistry recommends that molar absorption coefficients be expressed in these units. However, since 1 liter $(dm^3) = 10^3\ cm^3$, another unit which can be used (and is commonly used by biochemists) is $cm^2\ mol^{-1}$, where

$$10^3\ cm^2\ mol^{-1} = 1\ dm^3\ mol^{-1}\ cm^{-1}$$

Biologists and biochemists also frequently express molar absorption coefficients as $cm^2\ mmol^{-1}$; this unit is numerically the same as $dm^3\ mol^{-1}\ cm^{-1}$. In the case of substances for which the molecular weight is not well known (e.g., proteins), it is customary to use a 1% (weight/volume) solution as standard, again with a 1 cm light path. The absorption coefficient is then known as the *specific absorption coefficient* and is denoted by the symbol $\varepsilon_{1\,cm}^{1\%}(\lambda)$. However, this coefficient is sometimes employed with reference to a volume/volume concentration.

---

† In SI the liter is replaced by the cubic decimetre (dm^3). The symbol M (molar) is used for moles per cubic decimetre $(mol\ dm^{-3})$.

EXAMPLE

An aqueous solution of iridine-5'-triphosphate, at a concentration of 57.8 mg per dm^3 of the trisodium dihydrate (molecular weight 586), gave an absorbance of 1.014 with a light path of 1 cm. Calculate the molar absorption coefficient. What would be the absorbance of a 10 μM solution, and what would be its percentage of light transmittance?

SOLUTION

From equation (2.17)

$$A = 1.014 = \varepsilon c l$$

and

$$c = \frac{57.8 \times 10^{-3}}{586} = 9.86 \times 10^{-5} \text{ M}$$

Thus

$$\varepsilon = \frac{1.014}{9.86 \times 10^{-5} \text{ mol dm}^{-3} \times 1 \text{ cm}}$$

$$= 1.028 \times 10^{-4} \text{ dm}^3 \text{ mol}^{-1} \text{ cm}^{-1}$$

For a 10 μM solution,

$$A = 1.028 \times 10^4 \times 10^{-5}$$

$$= 0.1028$$

The percentage light transmittance $T\%$ is given by equation (2.15):

$$\log_{10} T\% = 2 - A = 1.8972$$

and thus

$$T\% = 78.9\%$$

The Lambert-Beer equation is always considered to be obeyed exactly. However, apparent deviations are sometimes observed, and it is then necessary to find an explanation in terms of extraneous effects such as dissociation and complex formation. Thus, when certain substances are present in solution, there are shifts in equilibria, and consequent changes in the relative proportions of the various molecular components. If the absorption of light by one of these components is being observed, there will be an apparent deviation from Beer's law, equation (2.16). Also, solutions that scatter light will not obey the law, and it is therefore important to exclude dust particles and large aggregated molecules in photometric experiments.

2.5   SPECTROPHOTOMETRY

The most commonly used instrument for measuring the absorbance of samples is the spectrophotometer. It uses monochromatic light (i.e., light covering a narrow band of wavelengths) and allows the absorbance to be measured at different wavelengths. Figure 2.7 shows a commercial instrument, which consists of a suitable light source, a prism or diffraction-grating monochromator to select a narrow range of wavelengths, a cuvette to hold the sample, and a photoelectric detector to determine the intensity of light that passes through the sample. For work in the visible region of the spectrum, the source of light is a tungsten filament lamp; for work in the ultraviolet, a hydrogen-discharge lamp is used.

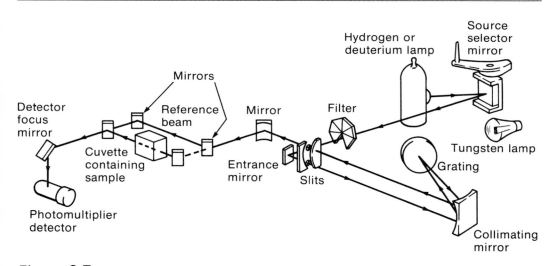

Figure 2.7

Schematic diagram of a spectrophotometer suitable for work in the visible and ultraviolet. This drawing is based on a diagram kindly provided by the Beckman Instruments Company.

The spectrophotometer measures absorbance [i.e., $\log_{10}(I_0/I)$, see equation (2.10)] by comparing the light transmitted by a blank with that of the test solution; the blank might, for example, be the pure solvent. With the absorbance A measured for a known concentration of solution, the molar absorption coefficient ε can be calculated. Many modern spectrophotometers are also recording instruments which can scan automatically a range of wavelengths and, at the same time, record the absorbance of the solution on a moving sheet. Thus these instruments can provide a complete spectrum in which absorbance (or transmittance) is plotted against wavelength or wavenumber, or these instruments may be used at a fixed wavelength to measure absorbance. If absorbance has been determined for a known amount of a given substance, the concentration of another solution can be calculated from its absorbance at the same wavelength.

Whenever concentrations are determined by spectrophotometry, it is best to prepare a *calibration curve* covering a range of concentrations. This is done by preparing solutions having various concentrations and determining the absorbance in each case. The absorbance is then plotted against concentration. If Beer's law applies, the plot will be a straight line through the origin. There are sometimes deviations from linearity, which might be caused by shifts in equilibria. In either case, concentrations can be determined in a reliable fashion by interpolation from such calibration curves. If there is good linearity, it is also justifiable to extrapolate beyond the range of the calibration curve.

EXAMPLE

An aqueous solution of sodium fumarate, at a concentration of 0.454 mM, gave an absorbance of 0.65 at 250 nm with the light path 1 cm. Calculate the molar absorption coefficient. Another solution under the same conditions gave a percentage light transmittance of 19.23%. Calculate the concentration of this solution.

SOLUTION

From equation (2.17)

$$1.65 = \varepsilon \times 0.454 \times 10^{-3} \times 1$$

and thus

$$\varepsilon_{1\ \mathrm{cm}}^{1\ \mathrm{M}}\ (250\ \mathrm{nm}) = 1.432 \times 10^3\ \mathrm{dm^3\ mol^{-1}\ cm^{-1}}$$

The relation between percentage light transmittance and absorbance is given by equation (2.15):

$$A = 2 - \log_{10} T\%$$
$$= 2 - \log_{10} 19.23$$
$$= 0.716$$

The concentration of the second solution is

$$\frac{A}{\varepsilon l} = \frac{0.716}{1.432 \times 10^3} = 5 \times 10^{-4}\ \mathrm{M}$$
$$= 500\ \mu\mathrm{M}$$

2.6   OPTICAL TECHNIQUES IN BIOLOGY

Spectroscopic measurements have been very valuable in the investigation of biological systems. They have led to much important information about structure; they have been invaluable in the identification of substances; and they are used routinely for the measurement of concentrations of known substances.

Aside from purely spectroscopic techniques, other optical methods have been employed to study substances of interest to biologists. Examples are the study of optical rotation and of optical rotatory dispersion. These latter types of investigation will be considered in the next chapter with special application to biological problems.

INFRARED SPECTROSCOPY

We have seen that, in the infrared region of the spectrum, there are spectral bands associated with both vibrational and rotational transitions. Pure rotational transitions are found in the microwave region of the spectrum, which is beyond the infrared, but studies of these are not of great value in the investigation of large molecules. Studies of the vibrational-rotational spectra in the infrared have, however, made very important contributions to biological research.

A disadvantage of studies in the infrared is that it is not usually possible to make measurements on substances in aqueous solution. Water absorbs intensely throughout much of the infrared. By contrast, it is transparent throughout the entire visible region of the spectrum and in the ultraviolet down to a wavelength of about 180 nm. As a result of this strong absorption of light by water, it is not possible to study the infrared spectrum of a substance in aqueous solution because of complete masking by the water spectrum.

Measurements in the infrared therefore must be made with the substance present in a material that does not absorb. Certain organic solvents are used frequently for this purpose. Alternatively, the solvent is eliminated completely. A common modern technique is to disperse the sample in a suitable inorganic salt, usually potassium bromide. The sample is mixed with the powdered crystalline salt, which is then pressed into a transparent disk measuring 0.5 mm in thickness and 10 mm in diameter. The disk is then mounted in a holder which is supported in the beam of the infrared instrument. There are some experimental difficulties which can be overcome by a skillful investigator. Since aqueous systems cannot be used in such experiments, infrared spectroscopy has no direct value in the study of biological systems, which are always aqueous. The usefulness of spectroscopy to the biologist is in the study of substances that have been extracted from biological systems.

Let us suppose that, using the techniques indicated above, we have the infrared spectrum of a protein. What information can be obtained from it? For a molecule of such magnitude a complete analysis of the spectrum is impossible because there will be an enormous number of different modes of vibration. However, a great simplification arises from the fact that individual functional groups in molecules absorb in characteristic regions of the infrared spectrum. For example, a molecule containing the alcohol group, —O—H, will always show an absorption in the 3100–3500 cm^{-1} region of the spectrum. Absorption in that region is associated with the stretching of the O—H bond. On the other hand, a molecule containing the double-bonded carbonyl group, $\diagdown\!\!\!\!\!\!\!\diagup$ C=O, will absorb in the 1600–1800 cm^{-1} region. Thus, if we were studying a molecule containing a single oxygen atom and found absorption in the 3100–3500 cm^{-1} region, we would know at once that the oxygen atom belonged to a hydroxyl group. Table 2.2 gives a few characteristic wavenumbers for different kinds of bonds as well as the corresponding wavelengths in nanometres.

Table 2.2 Characteristic Wavenumbers and Wavelengths for Bonds

Bond	Wavenumber, $\bar{\nu}$ (cm^{-1})	Wavelength, λ (nm)
O—H, N—H	3100–3500	3200–2850
C—H	2800–3100	3600–3250
C—O, C—N	800–1300	12500–7700
C=C, C=O	1600–1800	6250–5550
P—O	1200–1300	8300–7700

Besides the usefulness of infrared spectra in the identification of individual bonds, the technique is of great value in leading to an absolute identification of a pure compound. Because of the large numbers of vibrational modes in a molecule, the complete infrared spectrum of a given molecule will be very complicated and quite

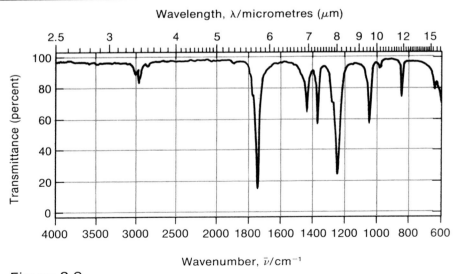

Figure 2.8

The infrared spectrum of methyl acetate.

unique to the molecule. Thus, if one suspects that an unknown substance is the same as a known sample, confirmation is obtained when the infrared spectra are identical. This technique is often referred to as "obtaining the fingerprints" of the molecule. Figure 2.8 shows a typical infrared spectrum; the precise details of this spectrum will not be reproduced by any other molecule.

Infrared spectra provide important information about *hydrogen bonding*. Figure 2.9 is an example of the kind of spectral shift that takes place when hydrogen bonding

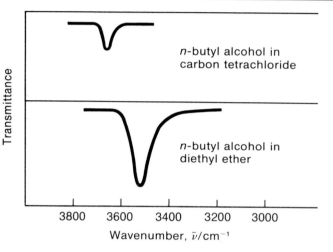

Figure 2.9

The infrared spectra of *n*-butyl alcohol in carbon tetrachloride (no hydrogen bonding) and in diethyl ether. The hydrogen bonding with the latter solvent leads to a shift to lower frequencies and an intensification of the band.

occurs. The upper diagram is for n-butyl alcohol dissolved in carbon tetrachloride. There is no hydrogen bonding, and we see the usual —O—H bond absorption. When the alcohol is dissolved in diethyl ether, however, there is a shift in absorption to lower frequencies as a result of the formation of hydrogen bonds:

Similar studies have been carried out with water dissolved in various solvents.

RAMAN SPECTROSCOPY

When a beam of light passes through a medium, a certain amount of light is scattered and can be detected by making observations perpendicular to the incident beam. Most of the light is scattered without change in wavelength, and the term employed for this effect is *Rayleigh scattering*. There is, in addition, a small amount of scattered light, the wavelength of which is altered. If the incident light is monochromatic, which means that only a narrow range of wavelengths is represented, the scattered spectrum exhibits a number of lines displaced from the original wavelength. Some are of higher and others of lower wavelength. This effect was first observed in 1928 by the Indian physicist Sir Chandrasekhara Venkata Raman and his coworker K. S. Krishnan, and is known as the *Raman effect*.

The Raman effect is a result of the interaction that occurs between a molecule and a photon of light. The photon can give energy to the molecule or accept energy from it. In the former case, the frequency of the light is reduced and its wavelength is increased, and in the latter case, the frequency is increased. The energy the photon accepts or gives up corresponds to the difference between two quantized energy states in the molecule. If the molecule undergoes a simple change in rotational energy, the resulting spectrum is a pure rotational Raman spectrum. The differences between the frequency of the incident light and the frequencies of the other lines then correspond to rotational energy transitions.

An important advantage of Raman spectroscopy is that one does not have to work in the infrared and microwave regions of the spectrum. The region used is determined by the choice of frequency of the incident light and can be the visible region. Also, with Raman spectra it is substantially less difficult to work with aqueous systems.

Another advantage of Raman spectroscopy is that certain lines appear in the Raman spectrum that do not appear in the infrared spectrum. This happens because of different selection rules. To be active in the infrared, a vibration must produce an oscillating dipole; to be active in the Raman spectrum, a molecule must produce an oscillating *polarizability*. The polarizability of a molecule is a measure of the effectiveness of an applied field in disturbing the electron clouds. A molecule with loosely-bound electrons is more polarizable than one in which the electrons are tightly bound to the nucleus. In general, a bond is more polarizable when it is lengthened than when it is shortened, because in the extended bond the electrons are less under the control of the nuclei. The vibration of a homonuclear molecule such as H_2 and N_2 therefore gives rise to an oscillating polarizability. Consequently, such a molecule, which has no infrared spectrum, does have a Raman spectrum.

We saw earlier (p. 66) that the symmetric vibration of carbon dioxide does not give rise to an infrared spectrum because there is no change in dipole moment, although the antisymmetric mode does give an infrared spectrum as a result of the oscillating dipole moment. The situation with Raman spectra is the converse. The symmetric vibration stretches and compresses both bonds, and the polarizability therefore varies. As a result, this vibration gives rise to Raman lines. The antisymmetric vibration, on the other hand, does not lead to a net polarizability variation because as one bond lengthens the other bond shortens, and the two effects cancel. Thus there are no Raman lines associated with the antisymmetric vibration. To summarize the situation for CO_2:

Symmetric vibrations are infrared inactive and Raman active.
Antisymmetric vibrations are infrared active and Raman inactive.

In view of these differences, it is advantageous to make a parallel study of both infrared and Raman spectra.

VISIBLE-ULTRAVIOLET SPECTROSCOPY

As has been explained, transitions in which there is a change from one electronic state to another involve considerably higher energies than those in which there is a vibrational change only. An electronic transition gives rise to absorption in the visible and ultraviolet regions of the spectrum. For several reasons investigations in these regions are much more frequent in studies of biological systems than are investigations in the infrared. First, the techniques are somewhat easier to apply. Second, there are many commercial spectrophotometers which permit measurements to be made easily in the visible and ultraviolet regions. Lastly, water does not absorb at all in the visible or in the near ultraviolet; it absorbs only in the ultraviolet at wavelengths of less than about 180 nm. Consequently, it is possible to study substances in aqueous solution over a wide spectral range in the visible and ultraviolet regions. Since biological materials normally exist in water, this is a considerable advantage.

Generally, electronic spectra are recorded as the logarithm of the absorbance plotted against wavelength, in contrast to infrared spectra which are usually given as transmittance against wavenumber. A typical electronic absorption spectrum is shown in Figure 2.10. Compared with the infrared spectrum of Figure 2.8, it is seen to be much smoother in outline. As with infrared spectra, electronic spectra have

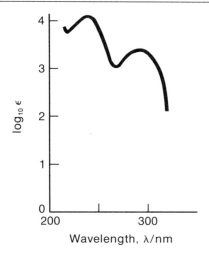

Figure 2.10

Electronic (ultraviolet) absorption spectrum of L-tyrosine, at pH 12.55; the logarithm of the absorbance, ε, is plotted against wavelength.

characteristic shapes, and substances may be identified by comparison with a known sample. Also, the concentration of a substance can be determined by studying the absorption at a certain wavelength, as explained previously.

The curve corresponding to an electronic transition from the ground electronic state to a particular excited electronic state is referred to as a *band*. Most molecules will exhibit more than one absorption band, corresponding to transitions between different electronic states. For example, substances, such as acetone, which contain the

carbonyl group, $C=O$, all exhibit a weak absorption band having a maximum at about 285 nm and a stronger absorption band near 200 nm. The absorption coefficient ε for the band at 285 nm is usually around 10 dm^3 mol^{-1} cm^{-1}; at 200 nm the absorption coefficient is usually between 10^3 and 10^4 dm^3 mol^{-1} cm^{-1}. These two absorption bands are understood readily with reference to the nature of the carbonyl bond. The localized part of the molecule which gives rise to the absorption band is known as the *chromophore*.

Some of the electrons relating to the carbonyl group are nonbonding and are given the symbol n. Others are in the π bond, if we use valence-bond theory; in terms of molecular-orbital theory, these electrons are in a π orbital (see Figure 2.5 for the O$_2$ molecule). Figure 2.11 shows the types of transitions that can occur. A nonbonding (n) electron can be promoted into an antibonding π orbital, and the

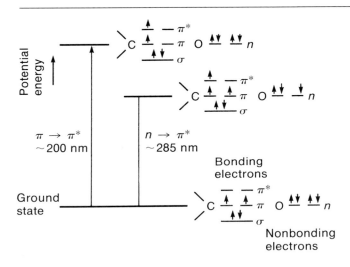

Figure 2.11

Two types of electronic transitions which can occur with a chromophore such as the carbonyl group.

transition is known as an $n \rightarrow \pi^*$ transition. It is this transition which leads to the absorption at approximately 285 nm. The more intense absorption at about 200 nm is caused by the promotion of a π electron into the antibonding π^* orbital in the electronically excited state. This is known as a $\pi \rightarrow \pi^*$ transition. The majority of the absorption bands commonly observed are of these two types. Other transitions involve higher energies and therefore give absorption at such low wavelengths (<180 nm) that there is interference by the absorption of the water molecules. The contribution of a particular chromophore to the absorption spectrum is considerably altered by conjugation with other chromophores. Thus, isolated carbon-carbon double bonds, C=C, exhibit a $\pi \rightarrow \pi^*$ transition at about 180 nm. However, if there is conjugation, i.e., alternating double and single bonds, the absorption is shifted to much higher wavelengths of around 450–500 nm.

Numerous electronic-spectrum assay procedures for determining concentrations have been devised. For example, the reaction of ninhydrin with most amino acids yields a purple product having a maximum absorption at 570 nm. This provides a very convenient procedure for amino acid analysis. In a biological study we frequently have to work with materials present in such minute quantities that ordinary methods of chemical analysis are impractical. If, however, the substance contains a chromophore which is strongly absorbing in the visible or ultraviolet regions, it is possible to assay even very minute quantities. Usually the absorption coefficients for chromophoric groups are sufficiently high that one can often make reliable assays even when concentrations are in micromoles per dm^3 or less.† An additional advantage of the spectrophotometric technique is that there is usually no destruction of the material being assayed.

Spectrophotometric techniques are employed frequently in kinetic studies, in which concentrations must be measured as a function of time. The technique provides an instantaneous record of the concentration, and if the reaction is not too fast a pen-and-ink recorder can be employed to follow the reaction. If the recorder paper is moving at constant speed, a record of concentration against time will be obtained, and this is precisely what is needed in a kinetic experiment. For more rapid reactions, recording can be done by use of an oscilloscope with a rapid sweep. As an example of a kinetic investigation, consider the reaction between L-lactate and nicotinamide adenine dinucleotide (NAD^+) to give pyruvate and NADH:

$$CH_3 \, CHOH \, COO^- + NAD^+ \rightleftharpoons CH_3 \, CO \, COO^- + NADH + H^+$$

As shown in Figure 2.12, NAD^+ exhibits little absorption above 300 nm, but NADH has a strong absorption band at 340 nm. Thus the reaction can be studied easily in either direction by measuring the change in absorbance at 340 nm. Many similar procedures are available for biological reactions.

The case just considered is a particularly simple one, because there are wavelengths

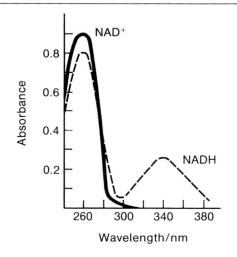

Figure 2.12

The absorption spectra of nicotinamide adenine dinucleotide (NAD^+) and its reduced form (NADH).

---

† For example, ninhydrin, a reagent for amino acids, has a molar absorption coefficient $\varepsilon_{1 \, cm}^{1 \, M}$ of $\approx 10^{11} \, dm^3 \, mol^{-1} \, cm^{-1}$, which means that amino acids in amounts of the order of 10 nanomoles can be estimated.

at which only one component of the mixture absorbs. Sometimes, however, there are no such wavelength regions. An example of this is shown in Figure 2.13 for 2,4-dinitrophenylhydrazones prepared from pyruvic and α-oxoglutaric acids. For the analysis of a mixture of these two substances, it would be necessary to work at two wavelengths, λ_1 and λ_2. Suppose that one of the substances has a molar absorption

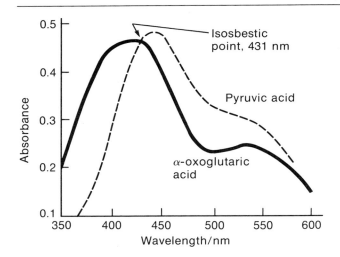

Figure 2.13

The absorption spectra of equal concentrations of the 2,4-dinitrophenylhydrazones of pyruvic acid and α-oxoglutaric acids, showing an isosbestic point at 431 nm.

coefficient of $\varepsilon_1(\lambda_1)$ at wavelength λ_1 and of $\varepsilon_1(\lambda_2)$ at wavelength λ_2, and suppose that the corresponding values for the second substance are $\varepsilon_2(\lambda_1)$ and $\varepsilon_2(\lambda_2)$. At concentrations c_1 and c_2 the absorbances at the two wavelengths would be

$$A(\lambda_1) = c_1\varepsilon_1(\lambda_1) + c_2\varepsilon_2(\lambda_1) \tag{2.18}$$

$$A(\lambda_2) = c_1\varepsilon_1(\lambda_2) + c_2\varepsilon_2(\lambda_2) \tag{2.19}$$

If the four molar absorption coefficients are known and absorption is measured at the two wavelengths, the concentrations c_1 and c_2 then can be determined by solution of the simultaneous equations (2.18) and (2.19). If three components are present, it is necessary to make absorption measurements at three wavelengths, and three simultaneous equations are solved. The number of wavelengths used must be at least as many as the number of components.

In certain cases, as in the example shown in Figure 2.13, the absorption spectra for two compounds at equal concentrations may intersect. The point of intersection is known as the *isosbestic*[†] point. For the case shown in Figure 2.13 the isosbestic point is at 431 nm, and it follows that at this particular wavelength a calibration curve relating absorption to molar concentration is the same for each compound.

[†] This word comes from the Greek prefix iso-, meaning *the same*, and sbestos, meaning *quench*. The word is sometimes written "isobestic," but this is etymologically incorrect.

This can be made the basis of a method for determining concentrations of the compounds in a mixture of the two. The absorption is measured at the isosbestic point, and this gives the total molar concentration of the mixture. A measurement can also be made at another wavelength, such as 500 nm. The ratio of the two absorptions at 500 nm and 431 nm, A_{500}/A_{431}, varies linearly with the proportion of either of the components; the ratio varies from the value characteristic of one of the pure compounds to that of the other. In this way the relative proportions of the two compounds can be obtained. Since the total molar concentration is known from the A_{431} measurement, the absolute amounts of each component can be calculated.

LASER SPECTROSCOPY

During recent years, the techniques of spectroscopy have been greatly enhanced by the introduction of lasers. The word *laser* is an acronym for "Light Amplification by Stimulated Emission of Radiation." The development of this field began in 1953 with the introduction by the American physicist Charles H. Townes of the *maser*, which stands for Microwave Amplification by Stimulated Emission of Radiation. Work in the microwave region was soon extended to other regions of the spectrum, including the visible and ultraviolet. Today, laser spectroscopy is carried out in many laboratories and a variety of applications to biological systems have been developed.

In a laser beam, excitation energy is supplied to a substance in such a way as to produce a *population inversion*, in which more atoms are in a specific excited level than in the ground-state level. Normally there are more molecules in ground states than in any of the excited states, but if energy is supplied, this situation can be reversed. Such an excited state can release energy spontaneously, but the unique feature of a laser is that the release may be accomplished by a process known as *stimulated emission*. During this process, a photon released by one atom or molecule will interact with an excited species having a population inversion and will stimulate the species to release a photon of the same wavelength and traveling in the same direction. Thus, as a beam progresses through the excited laser medium, the intensity is greatly increased. A common device is to reflect the initial beam by means of a mirror, with the result that it passes again through the medium and is further amplified. When the beam reaches a partially reflecting mirror, a portion of the beam escapes, and this is the active emission from the laser.

A commonly used laser is the ruby laser, in which the medium is a rod of ruby; this is sapphire, Al_2O_3, in which one out of every 10^2 or 10^3 Al^{3+} ions has been replaced by a Cr^{3+} ion—giving the resulting crystal its characteristic red color. The population inversion is usually brought about by means of xenon arc flashlamps which are placed in a highly reflecting housing in order to focus the lamp emission on the laser rod. Some lasers operate continuously, but a common technique is to pulse the exciting light for short periods of time (e.g., a few milliseconds). The power developed in a pulsed laser may be as high as a gigawatt (10^9 W). The ruby laser produces light of 694.3 nm wavelength, which lies toward the red end of the visible spectrum. Of particular importance is the fact that the light is highly monochromatic (i.e., it covers a very narrow range of wavelengths). Also, if the laser medium (e.g., the ruby) has a suitable geometry, the beam divergence can be extremely small, perhaps of the order of a milliradian.

Because of these special features, laser investigations have many advantages over conventional spectroscopic studies. The fact that very narrow beams can be produced allows measurements to be made on exceedingly small samples, such as individual biological cells. Lasers are now being used in surgery, including eye surgery. Analytical determinations on chemical and biological material are carried out routinely using laser beams. For many years Raman spectroscopy presented serious experimental difficulties, but laser Raman spectroscopy has many practical advantages.

COLORIMETRY

For substances absorbing in the visible region of the spectrum, simple instruments called *colorimeters* are often used in place of spectrophotometers for measuring concentrations. In the earliest form of this instrument, the Duboscq colorimeter, the observer visually matched the unknown and standard solutions. These visual instruments have now been superseded largely by photoelectric devices, but the Duboscq colorimeter illustrates well the principle involved, and its use will be described briefly.

A schematic view of the instrument is shown in Figure 2.14. A solution of known concentration, c, is placed in one of the compartments and the solution of unknown

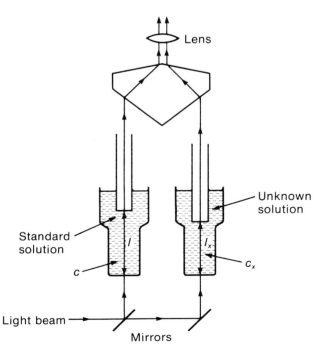

Figure 2.14

The Duboscq colorimeter.

concentration in the other. By an optical device the observer can see through both solutions at once, the standard solution filling half of the field of vision and the unknown filling the other half. The observer adjusts the heights l and l_x until the intensities of the two solutions appear the same. According to the Lambert-Beer equation (2.17), the absorbance is proportional to the product of the concentration and the path length. Thus, when the absorbances of the two solutions are the same,

$$cl = c_x l_x \qquad (2.20)$$

where c_x is the unknown concentration. This concentration is therefore given by

$$c_x = \frac{cl}{l_x}$$

The light source in this simple colorimeter is, of course, ordinary white light; complications arise if the solutions are of different colors.

EXAMPLE

A solution containing 2.5 mg of ammonia per dm^3 was treated with Nessler's reagent, which produces a yellow color. An unknown solution was treated similarly. When the two solutions were matched in the Duboscq colorimeter, the depths were 12.5 mm for the standard solution and 17.0 mm for the unknown. Calculate the concentration of the unknown solution.

SOLUTION

Since a longer light path was required for the unknown, it was less concentrated than the standard, the ratio being 12.5/17.0. Thus the concentration of the unknown is

$$c_x = \frac{12.5 \times 2.5}{17.0} = 1.84 \text{ mg dm}^{-3}$$

Modern chemical and biological laboratories routinely make use of photoelectric colorimeters. Typically, the light used covers a narrow range of wavelengths, accomplished by means of filters. The light passes through the solution under investigation and then impinges on a photocell. The current generated by the photocell is fed into a milliammeter which is usually calibrated to give absorbance readings directly. The standard and blank solutions are introduced separately. The ratio of the individual absorbance readings gives the ratio of the solution concentrations.

2.7   ELECTRON SPIN RESONANCE SPECTROSCOPY

We have seen in the last chapter that electrons have a spin angular momentum corresponding to the quantum numbers $+\frac{1}{2}$ and $-\frac{1}{2}$. In most molecules the electron spins are paired, but some molecules contain an odd number of electrons; there is therefore at least one unpaired electron spin. An example of an ordinary molecule with an unpaired electron is nitric oxide (NO), which has an odd number (15) of electrons. Of particular interest are the free radicals, such as the methyl radical (CH_3). This radical is produced if a hydrogen atom is abstracted from a methane (CH_4) molecule, and its Lewis structure is

$$
\begin{array}{c}
H \\[2pt]
\overset{\displaystyle \cdot\cdot}{H : C} \cdot \\[-2pt]
\overset{\displaystyle \cdot\cdot}{} \\
H
\end{array}
$$

We will see in Chapter 8 that free radicals of this kind play important roles as intermediates in chemical reactions.

When a species having an odd electron is placed in a magnetic field, the spin angular momentum of the electron is quantized along the direction of the magnetic field. The components of the spin angular momentum along the field are

$$+\frac{1}{2}\left(\frac{h}{2\pi}\right) \quad \text{and} \quad -\frac{1}{2}\left(\frac{h}{2\pi}\right)$$

The result is a splitting of the energy of the molecule into two levels. If a magnetic field of 0.3 tesla (T)† is used, the energy difference between the two levels is about 7.0×10^{-24} J. This corresponds to a wavelength of about 3 cm, which is in the microwave region.

This behavior has led to the development of the important experimental technique of *electron spin resonance spectroscopy* (ESR), also known as *electron paramagnetic resonance spectroscopy* (EPR). The sample is placed in a magnetic field of about 3000 G, and microwaves are passed through it. If the energy of these is equal to the energy of splitting, a resonance effect occurs. This phenomenon is analogous to a sound wave setting up a resonance vibration in an object that has the same natural frequency as the sound wave. The occurrence of resonance provides information about the energy splitting.

Electron spin resonance spectroscopy is a valuable technique for the study of chemical species having unpaired electrons, such as the free-radical intermediates in many reactions of interest to the biologist. It allows us to determine the concentrations of such species. It also provides information about the distribution of the

---

† The tesla (T) is the SI unit of magnetic field or magnetic flux density.
The old unit, the gauss (G), is equal to 10^{-4} tesla; 0.3 T = 3000 G.

odd electron within the free radical. In some investigations the method of *spin labeling* has been employed. For example, a molecule having an unpaired electron can be attached to a biological membrane; analysis of the electron spin resonance then provides important information about the properties and functioning of the membrane.

2.8   NUCLEAR MAGNETIC RESONANCE SPECTROSCOPY

ESR spectroscopy cannot be used with most molecules, because they do not have an unpaired electron. However, many atomic nuclei have spin angular momenta, which will also be quantized along the direction of a magnetic field. The magnetic moment for nuclear spin is only a very small fraction of that for electron spin, and the consequent splitting of the energy levels is much smaller. In order to observe resonance when radiation is passed through the system, it is necessary either to use much higher magnetic fields (which is difficult) or to use radiation of lower energy. Radio waves having a frequency of 60 megahertz (6×10^7 Hz), corresponding to a wavelength of 5 metres, are employed frequently for this purpose. Usually, the radio frequency is maintained constant, and the magnetic field is varied. Resonance is then found at certain magnetic fields. Figure 2.15 shows an experimental arrangement for nuclear magnetic resonance spectroscopy.

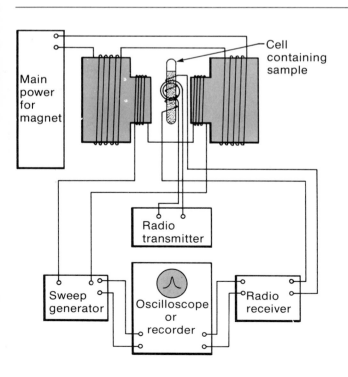

Cell containing sample

Main power for magnet

Radio transmitter

Sweep generator

Oscilloscope or recorder

Radio receiver

Figure 2.15

The basic experimental arrangement in a nuclear magnetic resonance experiment. The sample is placed in a tube between the poles of a powerful electromagnet and the magnetic field is varied by the application of a "sweep field." At certain magnetic fields the record from the radio receiver indicates that there is resonance, which means that the magnetic energy difference corresponds to the energy of the magnetic field.

Work of this kind has been of enormous value in structural determinations—particularly with organic molecules—and in providing more detailed information about the nature of chemical bonds. The principle underlying such studies is that the magnetic field which penetrates to the nucleus is highly dependent on the nature of the electron clouds surrounding the nucleus. Therefore, the circumstances in which nuclear magnetic resonance occurs depend on the nature of the neighboring atoms, on the chemical bonds, and on the nucleus. A modification brought about by the chemical environment is referred to as a *chemical shift*. For example, consider the nuclear magnetic resonance spectrum of ethanol (CH_3CH_2OH):

$$
\begin{array}{c}
\quad\; H \;\; H \\
\quad\; | \quad\; | \\
H-C-C-O \\
\quad\; | \quad\; | \quad\quad \backslash \\
\quad\; H \;\; H \quad\quad H
\end{array}
$$

which is shown in Figure 2.16. Figure 2.16*a* shows the results obtained with a low-resolution apparatus, and *b* shows the results with a high-resolution apparatus capable of splitting the main peaks into a number of smaller peaks.

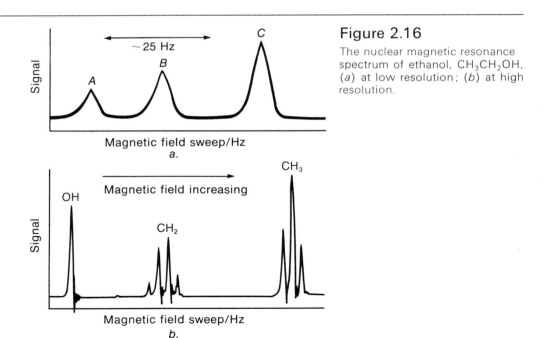

Figure 2.16

The nuclear magnetic resonance spectrum of ethanol, CH_3CH_2OH, (*a*) at low resolution; (*b*) at high resolution.

The characteristic feature of the record in 2.16*a* is that there are three peaks, *A*, *B*, and *C*. The areas below the respective peaks are in the ratio 1:2:3. These peaks correspond to the three different "kinds" of proton in the molecule—more exactly, to three groups of protons in three different environments. One group consists of the single proton bonded to the oxygen atom, the second group consists of the two protons bonded to the carbon atom next to the oxygen atom, and the third group consists of the three protons bonded to the carbon atom next but one to the oxygen atom. The areas under the peaks are proportional to the numbers of protons in each group. Obviously, a record of this kind is valuable in giving information about molecular structure.

When the experiments are carried out at higher resolution, as in Figure 2.16b, the peak for CH_2 is split into four peaks, and the peak for CH_3 into three peaks. The extent of this splitting (as opposed to the splitting into the three main peaks) does not depend on the strength of the magnetic field, and therefore cannot be a chemical shift due to shielding of the nuclei by the surrounding electrons. The effect is caused by the interaction of the nuclear spins of one set of protons with those of another and is known as *spin-spin coupling*. An analysis of spin-spin coupling in various structures has provided very valuable information about the detailed nature of chemical bonds in molecules and in crystals.

PROBLEMS

2.1 The wavelength of the sodium D_1 line is 589.0 nm. What is the wavelength in angstroms? What is the wavenumber in reciprocal centimetres and the frequency in reciprocal seconds?

2.2 Calculate the energy changes in joules per atom and in kcal per mole associated with the electronic transitions which give rise to the following spectral lines:

    (a) The sodium D_1 line of wavelength 589.0 nm.
    (b) The mercury resonance line of wavelength 253.7 nm.

In what region of the spectrum would you look for these lines?

2.3 The wavenumber of the O—H stretching in CH_3OH is 3300 cm^{-1}. Calculate the frequency in s^{-1}. In what region of the spectrum would you look for a line of wavenumber 3300 cm^{-1}?

2.4 A spectrophotometer has a meter which gives a reading directly proportional to the amount of light reaching the detector. When the light source is off, the reading is zero. With pure solvent in the light path, the meter reading is 78; with a solution of a solute in the same solvent, the reading is 55. Calculate the absorbance, the transmittance, and the percentage transmittance.

2.5 An aqueous solution containing 0.95 g of oxygenated myoglobin (mol. wt. 18 800) in 100 cm^3 gave a transmittance of 0.87 at 580 nm, with a path length of 10.0 cm. What is the absorption coefficient $\varepsilon_{1\,cm}^{1\,M}$ (580 nm)?

2.6 The molar absorption coefficient $\varepsilon_{1\,cm}^{1\,M}$ (430 nm) of human hemoglobin (mol. wt. 64 000) is 532 dm^3 mol^{-1} cm^{-1}. A solution of hemoglobin was found to have an absorbance of 0.155 at 430 nm with a light path of 1.00 cm. Calculate the concentration of the hemoglobin in mol dm^{-3} and in g dm^{-3}.

2.7 The ground state and lowest electronic state for a molecule in the vapor state are separated by 120.0 kcal mol^{-1}. On solution in water the energy of the ground state decreases by 4.0 kcal mol^{-1} and the energy of the excited state by 6.0 kcal mol^{-1}. Calculate the wavelength corresponding to this electronic transition (a) in the vapor state and (b) in water. In what region of the spectrum do these wavelengths lie?

2.8 The absorption coefficient $\varepsilon_1^{1\,M}{}_{cm}$ (250 nm) of sodium fumarate is 1.432×10^3 dm^3 mol^{-1} cm^{-1}. A solution in a cuvette with 0.5 cm light path gave an absorption of 0.225 at this wavelength. What is the concentration of the solution?

2.9 The molar absorption coefficient of reduced nicotinamide-adenine dinucleotide phosphate at 340 nm is 6.22×10^3 dm^3 mol^{-1} cm^{-1}. A solution was placed in a cuvette of 1.05 cm light path, and the percentage light transmittance was found to be 42.7%. Calculate the concentration of the solution.

2.10 A solution of partially reduced nicotinamide-adenine dinucleotide, containing NAD$^+$ and NADH, was placed in a 1 cm cuvette, and the absorptions were measured at 340 nm and 260 nm. The values obtained were 0.215 and 0.850 respectively. Calculate the concentrations of the two forms, given the following molar absorption coefficients:

$$\begin{aligned}
\text{NAD}^+: \quad & \varepsilon_1^{1\,M}{}_{cm}\,(260\ \text{nm}) = 1.8 \times 10^4\ \text{dm}^3\ \text{mol}^{-1}\ \text{cm}^{-1} \\
& \varepsilon_1^{1\,M}{}_{cm}\,(340\ \text{nm}) = 0 \\
\text{NADH}: \quad & \varepsilon_1^{1\,M}{}_{cm}\,(260\ \text{nm}) = 1.8 \times 10^4\ \text{dm}^3\ \text{mol}^{-1}\ \text{cm}^{-1} \\
& \varepsilon_1^{1\,M}{}_{cm}\,(340\ \text{nm}) = 6.22 \times 10^3\ \text{dm}^3\ \text{mol}^{-1}\ \text{cm}^{-1}
\end{aligned}$$

2.11 The molar absorption coefficient of adenine (mol. wt. 135) in hydrochloric acid solution, $\varepsilon_1^{1\,M}{}_{cm}$ (262.5 nm), has been given as 1.34×10^4 dm^3 mol^{-1} cm^{-1}. Calculate the absorption and the percent transmittance of a solution containing 7.5 μg cm^{-3}, if the path length is 0.5 cm.

2.12 A colored solution of concentration 0.50 mol dm^{-3} was placed in a Duboscq colorimeter and compared with an unknown solution of the same substance. The observed depths were 14.5 mm for the standard and 9.6 mm for the unknown. Calculate the concentration of the unknown.

ESSAY QUESTIONS

2.13 Give a brief account of the fundamental origins of ultraviolet, infrared, and nuclear magnetic resonance spectra.

2.14 State the laws of Lambert and Beer, and write down an equation comprising the two laws.

2.15 Explain clearly what is meant by absorbance and transmittance, and derive a relationship between them.

2.16 Explain the selection rules as they apply to infrared and Raman spectra.

SUGGESTED READING

Barrow, G. M. *Introduction to Molecular Spectroscopy.* New York: McGraw-Hill, 1962.

Hecht, H. G. *Magnetic Resonance Spectroscopy.* New York: John Wiley, 1967.

Kasha, M., and Khan, A. U. "The Physics, Chemistry and Biology of Singlet Molecular Oxygen," *Annals of the New York Academy of Science*, 171, 5 (1970).

Leach, S. J., ed. *Physical Principles and Techniques of Protein Chemistry.* New York: Academic Press, 1969.

Moore, W. J. *Physical Chemistry.* 4th ed. Englewood Cliffs, N.J.: Prentice-Hall, 1972, Chapter 17.

Roberts, J. D. *Nuclear Magnetic Resonance.* New York: McGraw-Hill, 1959.

Whiffen, D. H. *Spectroscopy.* New York: John Wiley, 1966.

See *Scientific American*, Vol. 219, No. 3. September, 1969, for a series of articles on light.

3

STRUCTURE IN BIOLOGICAL SYSTEMS

The term structure, as used in chemistry and biology, relates both to molecules and to molecular aggregates. A simple example is provided by water. We can deal with the structure of the H_2O molecules and also with the way in which these molecules are associated in solid ice and in liquid water. A more complicated example is a biological membrane, where we face the problem of the molecular structures of protein, lipid, and other molecules, and also of the way these molecules are aggregated in the membrane. In this chapter we are concerned with both of these problems, but more attention is given to the structures of individual molecules. We deal with some of the experimental methods used for investigating structure and with some of the structural information which has been accumulated.

The first stage in the investigation of molecular structure is the determination of the chemical composition—the proportions of the various elements. The reader will be familiar with such procedures and with the methods for obtaining the empirical formula of a molecule. The next stage is the determination of the general size and shape of the molecule. With smaller molecules this can be done precisely by a molecular-weight determination followed by a structural study from which the bond lengths and bond angles are determined. In cases involving larger molecules, precise information of this kind is difficult to obtain, and we must be content with information of a more general nature.

For many years chemists have had available to them a number of methods for determining molecular weights. Many of these methods are based on the *colligative* properties of solutions—vapor-pressure lowering, elevation of the boiling point, depression of the freezing point, and osmotic pressure. These methods are particularly useful for analyzing substances of fairly low molecular weight. For proteins and other large molecules, i.e., *macromolecules*,† these methods are difficult to use or cannot be used at all. Consequently, it was necessary to develop a number of other techniques, some of which will be described in this chapter. One group of methods requires that the substance be present in pure form and is useful mainly when we are working with small molecules; this group of methods comprises studies of chemical composition. Another group, including diffusion, viscosity, sedimentation, flow birefringence, and light scattering, does not require chemical purity, and these methods are more widely applicable to biological materials. Some of the latter methods also provide information about molecular shapes.

3.1   CHEMICAL COMPOSITION

When a material is present in pure form, it is possible to obtain tentative information about its molecular weight by employing various chemical analytical procedures, primarily (a) elemental analysis, (b) determination of combining weights, and (c) end-group analysis.

ELEMENTAL ANALYSIS

Elemental analysis refers to the determination of the proportions of the various elements in a compound. A compound must contain in each molecule at least one atom of every element present, and the mass of the compound that contains one mole of a given element is thus the minimum possible value of its molecular weight. Therefore, if the percentage of the element present in the compound is known, the minimal molecular weight is given by the expression

$$M_{min} = \frac{\text{atomic weight of element} \times 100}{\text{percentage of element in compound}} \qquad (3.1)$$

If the molecule contains n atoms of the element, the true molecular weight is

$$M_{true} = nM_{min} \qquad (3.2)$$

To obtain the value of n, which must be a whole number, one can obtain an approximate value of the molecular weight by measuring one of the colligative properties. However, as the value of n increases, its evaluation becomes more difficult.

---

† The prefix comes from the Greek *makros*, large, in contrast to *mikros*, small.

EXAMPLE

The amino acid lysine is found, on elemental analysis, to contain
19.17% nitrogen (at. wt. 14.01). A molecular weight determination
by the osmotic-pressure method yields an approximate value of
150. Calculate a more accurate value of the molecular weight.

SOLUTION

The minimal molecular weight is

$$M_{true} = \frac{14.01 \text{ g mol}^{-1} \; 100}{19.17} = 73.1 \text{ g mol}^{-1}$$

This is about half the approximate molecular weight of 150.
Therefore, $n = 2$, i.e., the molecule of lysine contains two atoms
of nitrogen. Thus a more accurate molecular weight is

$$M_{true} = nM_{min} = 2 \times 73.1 = 146.2 \text{ g mol}^{-1}$$

In determining a minimal molecular weight, it is obviously better to choose an
element in which only a few atoms occur in each molecule. For biological molecules
it would be of little value to work out an M_{min} from the percentage of carbon,
because most biological molecules contain a considerable number of carbon atoms,
and M_{true} would be many times larger than M_{min}. In the case of proteins, sulfur and
metals are good elements to choose because large numbers of these atoms are not
expected to be present.

EXAMPLE

The following are the percentages of iron and sulfur (other than
disulfide sulfur) found in horse hemoglobin:

Fe: 0.335%

S: 0.390%

Calculate the minimal molecular weight.

SOLUTION

From the iron data (at. wt. of Fe = 55.85)

$$M_{min} = \frac{55.85 \text{ g mol}^{-1} \; 100}{0.335} = 16\,670 \text{ g mol}^{-1}$$

From the sulfur data (at. wt. of S = 32.06)

$$M_{min} = \frac{32.06 \times 100}{0.390} = 8220 \text{ g mol}^{-1}$$

The latter value is about half of the former; thus if we combine the two, we obtain a minimum of $\approx 16\,500$ g mol^{-1} for the molecular weight. This value would correspond to one atom of iron and two atoms of sulfur per molecule. Of course the true molecular weight may be any integral multiple of $\approx 16\,500$.

Some proteins do not contain any one element in a sufficiently small proportion to enable this method to be used in a reliable manner. A variation of the method is to determine the percentage composition of a particular amino acid. This can be done by hydrolyzing the protein into its component amino acids. It is best to use an amino acid present in a very small percentage. Tyrosine, tryptophan, and cystine are often convenient to use because they are usually present in small amounts and can be determined accurately.

EXAMPLE
The wheat protein glutenin is found to contain 4.5% tyrosine. Calculate the minimal molecular weight. What is the true molecular weight if an approximate osmotic pressure determination leads to a value of $\approx 36\,000$?

SOLUTION
The molecular weight of tyrosine is 181.18, and the minimal molecular weight of the protein is therefore

$$M_{min} = \frac{181.18 \times 100}{4.50} = 4030 \text{ g mol}^{-1}$$

This is about one-ninth of the approximate molecular weight, and a more accurate molecular weight for glutenin is thus

$$M_{true} = n \times M_{min} = 9 \times 4030 = 36\,300 \text{ g mol}^{-1}$$

COMBINING WEIGHTS

A minimal molecular weight may be determined by the weight of the compound which combines or reacts with one mole of a suitable reagent. Protein molecules contain a number of free carboxyl and amino groups, and these may be titrated with base and acid, respectively. If, for example, there were n carboxyl groups per molecule, then one mole of protein would react with n equivalents of base, and the true molecular weight would be n times the minimal molecular weight obtained in this way. Again, n may be determined if an approximate molecular weight can be obtained by other methods. A disadvantage of this method is that with proteins the values of n are rather large.

In the case of the respiratory proteins, combination with oxygen has been used to determine minimal molecular weights. A method used for the hormone insulin was to cause it to combine with the reagent 1-fluoro-2, 4-dinitrobenzene. There was a complication in that both monosubstituted and disubstituted products were obtained, but the monosubstituted material could be separated and analyzed for the dinitrophenyl group. The result was that 6000 g of protein contained 1 mol of the dinitrophenyl group; the minimum molecular weight of insulin is therefore 6000 g mol^{-1}.

EXAMPLE
One gram of hemoglobin is found to combine with a maximum of 1.34 cm^3 of oxygen gas at standard temperature and pressure (0° C and 1 atm). Calculate the minimal molecular weight of hemoglobin.

SOLUTION
22.4 dm^3 of oxygen at S.T.P. is one mole, and therefore 1.34 cm^3 is

$$\frac{1.34 \text{ cm}^3}{22.4 \times 10^3 \text{ cm}^3 \text{ mol}^{-1}} = 5.98 \times 10^{-5} \text{ mol}$$

Therefore one mole of oxygen combines with

$$\frac{1}{5.98 \times 10^{-5}} = 16\ 700 \text{ g}$$

which is therefore the minimal molecular weight of hemoglobin.

END-GROUP ANALYSIS

This method can be used only when one has certain information about the structure of the molecule under investigation. A good example of the use of the method is the estimation by the British organic chemist Walter Norman Haworth (1883–1950) of the chain length of cellulose, which consists of chains of glucose molecules. The hydroxyl groups can be completely methylated, and then gentle hydrolysis can be carried out in such a way that the methylated glucose units are separated from each other without removal of the methyl group. Most of the glucose units formed are 2,3,6-trimethyl glucose, but the units at the ends of the chains are 2,3,4,6-tetramethyl glucose molecules. It is thus possible to determine the lengths of the glucose chains in the original cellulose molecule. Complications arise from the branching of chains and from the breaking of chains during the methylation procedure.

3.2 TYPES OF MOLECULAR WEIGHTS

Biological material is often inhomogeneous in the molecular sense; the molecules, even of a substance such as a specific protein, may not all be of the same size. The experimental procedures used to determine molecular weight therefore will yield some kind of an average value. The methods just described involve, essentially, the counting of molecules, as do the various methods based on the colligative properties. Thus these methods provide the *number-average molecular weight*, which we will denote by the symbol N_n. If the numbers of molecules of the i-th kind is N_1, the *number fraction*, or *mole fraction*, is given by

$$x_i = \frac{N_i}{\sum_i N_i} \tag{3.3}$$

where the summation is taken over all types of molecules present. The number-average molecular weight is then

$$M_n = \sum_i x_i M_i = \frac{\sum_i N_i M_i}{\sum_i N_i} \tag{3.4}$$

EXAMPLE
Suppose that a protein sample consists of 30% of molecules of mol. wt. 20 000, 40% of molecules of mol. wt. 30 000, and 30% of molecules of mol. wt. 60 000. Calculate the number-average molecular weight.

SOLUTION
The mole fractions are 0.3, 0.4, and 0.3, respectively. Therefore, from equation (3.4),

$$M_n = (0.3 \times 20\,000) + (0.4 \times 30\,000) + (0.3 \times 60\,000)$$

$$= 36\,000$$

Other methods of molecular-weight determination, such as diffusion, sedimentation, flow birefringence, and light scattering (all of which are considered later in this chapter) involve terms related to the size and shape of the molecules and give a molecular weight known as the *weight-average molecular weight*, M_w. If W_i is the weight of the i-th species, the *weight fraction*, w_i, is

$$w_i = \frac{W_i}{\sum_i W_i} \tag{3.5}$$

and the weight-average molecular weight is defined as

$$M_w = \sum_i w_i M_i = \frac{\sum_i W_i M_i}{\sum_i W_i} \tag{3.6}$$

Since $W_i = N_i M_i$, the weight-average molecular weight can be expressed as

$$M_w = \frac{\sum_i N_i M_i^2}{\sum_i N_i M_i} \tag{3.7}$$

EXAMPLE
Using the data given in the previous example, calculate the weight-average molecular weight.

SOLUTION
We may insert percentages into equation (3.7) instead of the numbers of molecules, N_i, the numbers being proportional to the percentages and the proportionality factor cancelling out. Thus

$$M_w = \frac{(30)(20\,000)^2 + (40)(30\,000)^2 + (30)(60\,000)^2}{(30)(20\,000) + (40)(30\,000) + (30)(60\,000)} = 43\,300$$

Note that in this example M_w is greater than M_n. It cannot, in fact, be less than M_n. The two molecular weights would have been equal if the substance had consisted of molecular units all of the same size; such a system is said to be *monodisperse*. If the substance consists of molecules of different sizes, as in the example given, it is said to be *polydisperse*. The M_w value is then greater than M_n, because in M_w the heavier units have been weighted more strongly than in M_n.

Other molecular weights are occasionally used. For example, there is the so-called Z-average molecular weight defined by

$$M_z = \frac{\sum_i N_i M_i^3}{\sum_i N_i M_i^2} \tag{3.8}$$

Again this is larger than M_w, because the heavier components are weighted even more strongly.

3.3 COLLIGATIVE PROPERTIES

The word *colligative* means "bound together" (Latin *colligare*, to bind together) and serves as a class name for several properties of solutions, namely:

1. Lowering of the vapor pressure.

2. Elevation of the boiling point.

3. Depression of the freezing point.

4. Osmotic pressure.

The essential feature of these four properties is that they depend only on the *number* of solute molecules present—not on the chemical and physical characteristics. Consequently, there is an exact relationship between each of the colligative properties and any other one; if one property is measured, the others can be calculated. From the measurement of a colligative property we can calculate the number of moles of solute present if we also know the mass of the substance.

It is assumed that the reader has already been introduced to colligative properties; here we will simply state the basic relationships, which will be derived in Chapter 5 from thermodynamics.

Raoult's law is concerned with the partial vapor pressure p exerted by a solvent of mole fraction x_1. The equation is

$$p = p_0 x_1 \tag{3.9}$$

where p_0 is the vapor pressure of the pure solvent. Alternatively, the *lowering* of the vapor pressure is given by

$$\Delta p = p_0 - p = p_0 x_2 \tag{3.10}$$

where x_2 is the mole fraction of the solute.

The elevation of the boiling point is given by

$$\Delta T_b = K_b m \tag{3.11}$$

where K_b is a constant characteristic of the solvent and m is the *molality* (number of moles of solute per 1000 g of *pure solvent*).

The depression of the freezing point is given by a similar equation:

$$\Delta T_f = K_f m \tag{3.12}$$

where K_f is a constant characteristic of the solvent.

The equation for osmotic pressure π is

$$\pi V = nRT \tag{3.13}$$

or

$$\pi = cRT \tag{3.14}$$

where n is the number of moles, V the volume, $c\,(=n/V)$ the molar concentration, R the gas constant ($8.314\ \mathrm{J\ K^{-1}\ mol^{-1}} = 0.0820\ \mathrm{atm\ dm^3\ mol^{-1}\ K^{-1}}$), and T the Kelvin temperature. Note that the equations given above for the four colligative properties are approximations, and that significant deviations can occur.

The principle involved in a molecular-weight determination by a colligative method is that one determines the mole fraction, molarity, or molality. If the concentration of the solution is known in grams per liter, the molecular weight is easily calculated.

The colligative property that one chooses to measure in a molecular-weight determination depends upon the nature of the substance under investigation. It is easiest to measure the boiling-point elevations and freezing-point depressions, and these properties are used frequently. However, if the substance is of very high molecular weight, it will be present only in rather low molar concentrations. The boiling-point elevations and freezing-point depressions are then so small that accurate measurements cannot be made. In that case, the methods are not suitable.

The vapor-pressure and osmotic-pressure methods are more useful for determinations of substances of high molecular weight. The reason for this is that it is more accurate to measure directly a small difference between two quantities than to measure the quantities separately and subtract one from the other. In using the freezing-point and boiling-point methods, a measurement must be made in one experiment for the solvent and a measurement made for the solution in a separate experiment. In the vapor-pressure lowering method, however, it is possible to allow the solvent and solution to come to equilibrium with their vapors, then measure the difference between the vapor pressures directly. Special techniques have been developed for magnifying vapor-pressure differences in such a way that they can be measured accurately. By the use of such methods, molecular weights have been measured up to about 10 000.

The measurement of an osmotic pressure can also be carried out more accurately than can the measurement of a boiling-point elevation or a freezing-point depression. One difficulty in measuring very small osmotic pressures is the long time required for the system to reach equilibrium. This difficulty is sometimes overcome by imposing a pressure on the solution side of the membrane and observing how the rate of flow of liquid varies over time. The osmotic pressure can be calculated from this variation. Molecular weights of up to 3 000 000 have been measured by the use of such techniques.

One problem with all colligative-property measurements is that a small amount of an impurity of low molecular weight will have a disproportionate effect on the results obtained on a substance of high molecular weight. The molar concentration of the low-molecular-weight substance may be, relatively, very large. Suppose, for example, that a substance under investigation has a molecular weight of 1 000 000, but there is an impurity of molecular weight 100 present at 0.1%. If 1 g of the substance were present in 1 dm^3 of solution, its molar concentration would be 10^{-6} M. Also, there would be present in the solution 10^{-3} g of the impurity, which has a molar concentration of 10^{-5} M. Therefore, the effect of the impurity on any of the colligative properties would be ten times that of the substance under investigation, and the molecular weight calculated from the experimental results would be seriously in error. When osmotic-pressure measurements are made, a related difficulty arises from the Donnan effect (see page 326); certain complications arise when ions are present and result in highly spurious molecular weights. Because of these problems, studies of colligative properties must be carried out with extremely pure materials— particularly if the molecular weight is high.

For molecular weights of the order of 10^6 or higher, none of the colligative properties can be measured accurately enough to provide reliable molecular weights. In such cases, alternative methods must be used.

3.4 METHODS BASED ON TRANSPORT PROPERTIES

Information about the size of molecules is also provided by the results of experiments on the rates of movement of the molecules in solution. Properties which depend on rates of movement are referred to as *transport properties*. If the motion occurs in aqueous solution, we can also speak of *hydrodynamic properties*.† The most important transport properties are diffusion, viscosity, and sedimentation. The theory of transport properties will be dealt with in Chapter 11, after the treatment of chemical kinetics. In the present section we will see, in a very general way, how measurements of transport properties lead to estimates of molecular sizes and shapes.

DIFFUSION

When a solution is at equilibrium, the distribution of solute species is uniform. If a solution is placed in contact with pure solvent, the solute molecules tend to move from the high-concentration region into the solvent until equilibrium is established. This molecular motion, known as *diffusion*, occurs as a result of the thermal energy of the molecules. At any given temperature, all particles in the solution have the same average kinetic energy. Macromolecules thus have a lower velocity than the smaller molecules and therefore distribute themselves more sluggishly throughout the solution. From a knowledge of diffusion velocities, we can obtain information about the sizes of molecules in a solution and can estimate the various molecular weights.

The fundamental law of translational diffusion was formulated in 1855 by the German physiologist Adolf Eugen Fick (1829–1901) and is known as Fick's first law.

---

† This word (from the Greek *hydor*, water; *dynamis*, power) is sometimes used inappropriately for liquids other than water, when *fluid dynamics* would be a better term.

Fick noted that the rate at which solute diffuses across a plane is proportional to the area, A, of the plane. The rate will also depend on the variation of the concentration, c, with the distance, x, across that plane, and will be larger the greater the *decrease* in c with x. Fick concluded that the rate of diffusion is proportional to $-dc/dx$ as well as to A. His law for the rate of diffusion dn/dt, where dn is the number of moles of solute crossing the plane in time dt, is therefore

$$\frac{dn}{dt} = -DA\frac{dc}{dx} \tag{3.15}$$

The proportionality factor, D, is known as the *diffusion coefficient*. Its usual units are square centimetres per second ($cm^2\ s^{-1}$), but $cm^2\ min^{-1}$ and even $cm^2\ day^{-1}$ are sometimes used, particularly in biological work. For smaller molecules, such as amino acids in water at room temperature, D is of the order of $10^{-5}\ cm^2\ s^{-1}$, while for proteins values of 10^{-6} or $10^{-7}\ cm^2\ s^{-1}$ are more common.

We will see in Chapter 11 that in 1850 the British physicist Sir George Gabriel Stokes (1819–1903) proposed a very simple relationship between the diffusion coefficient D and the radius r of the diffusing molecule, on the assumption that it is spherical. Stokes's law is

$$D = \frac{\mathbf{k}T}{6\pi\eta r} \tag{3.16}$$

where $\mathbf{k}$ is the *Boltzmann constant* (the gas constant R per molecule, i.e., $\mathbf{k} = R/N_A$), T is the temperature in kelvins, and η is the viscosity† of the solvent. It therefore follows that if D and η are measured, we can estimate the radius of the molecule, on the assumption that it is spherical. From the radius we can calculate the volume, and if the density of the substance is also known the mass of the molecule can be calculated.

The assumption that the molecule is spherical is, of course, an unsatisfactory one. If we have an alternative way of measuring the molecular weight we can make use of Stokes's law to obtain information about the actual shape of the molecule. From the molecular weight and the density we can calculate the radius *on the assumption that the molecule is spherical*. Insertion of this radius into equation (3.16) then gives a hypothetical diffusion constant, which we can call D_o. If the directly determined diffusion constant D is the same as D_o we have evidence that the molecule is in fact spherical. Often, however, D is found to be smaller than the D_o value calculated from equation (3.16). A nonspherical molecule will diffuse more slowly than a spherical molecule of the same volume.

The ratio D_o/D is known as the *dissymmetry constant* or the *frictional ratio*, and is a measure of the extent to which the molecule deviates from a perfect sphere. The enzyme ribonuclease has a dissymmetry constant of 1.04; thus the molecule is nearly spherical. Tobacco mosaic virus, on the other hand, has a D_o/D value of 3.12, and is cigar-shaped. If a molecule is assumed to have the shape of an ellipsoid, the ratio of the major and minor axes can be calculated from the dissymmetry constant.

---

† Viscosity is further discussed on pp. 102 and 497.

SEDIMENTATION

Particles suspended in a liquid tend to gravitate towards the bottom of the vessel. The rate with which they fall depends on a number of factors, including the density of the liquid and its viscosity. The more viscous the liquid, the more slowly the particle moves. The rate of movement also depends on the volume of the particle. The volume of a molecule, and hence its molecular weight, thus can be estimated by measuring its rate of sedimentation.

In 1908 the French physicist Jean Baptiste Perrin (1870–1942) studied with a microscope the rate of sedimentation of particles of the pigment gamboge and related their sizes to their rates of movement. In the Earth's gravitational field the forces are too small for a reasonable rate of sedimentation except for rather large particles, such as those studied by Perrin. This difficulty is overcome by artificially creating the equivalent of a large gravitational field by spinning the solution in a centrifuge. The development of centrifugal techniques—particularly ultracentrifuges with very high gravitational fields—is largely due to the Swedish scientist Theodor Svedberg. Beginning his work about 1923, Svedberg has designed ultracentrifuges able to produce effective gravitational fields up to 300 000 times the Earth's field and has used them to study molecular weights and the properties of macromolecules such as proteins.

The tendency of a molecule to sediment is countered by its tendency to diffuse back from the more concentrated to the less concentrated regions. At relatively low spinning speeds an equilibrium is reached in which sedimentation is exactly opposed by diffusion and there is no change in the distribution of macromolecules in the solution. This is called *sedimentation equilibrium*, and an analysis of the molecular distribution at this equilibrium allows molecule sizes to be determined. Alternatively, if high centrifugal forces are employed, the rate of sedimentation is much greater than the rate of the diffusion, which can be neglected. The *sedimentation velocity* method is based on making measurements under these conditions. A third technique involves studying the *approach to sedimentation equilibrium*.

The *sedimentation velocity* method is the most widely used technique, and its theory is the simplest. At high centrifugal forces the solute molecules form a fairly sharp boundary between the solution and the pure solvent; the movement of this boundary can be followed by refractometry. When the solute molecules are moving at a constant velocity, the centrifugal force is balanced by the frictional resistance. The rate of sedimentation is usually expressed in terms of the *sedimentation coefficient, s*, which is expressed in units of time and is the velocity for a unit centrifugal force. In order to calculate the molecular weight M in sedimentation velocity experiments, s is determined by analyzing the rate of movement of the boundary, and the diffusion coefficient D is also determined. The molecular weight is given by an equation derived by Svedberg:

$$M = \frac{RTs}{D(1 - \bar{v}\rho)} \tag{3.17}$$

where $\bar{v}$ is the specific volume of the macromolecule and ρ is the density of the solvent.

In the *sedimentation equilibrium* technique the centrifuge is operated at a lower speed—e.g., about 8000 rpm for a molecular weight of 60 000. When the centrifuge has been operated for a sufficient length of time there is no further change in the concentration of the solute as a function of the distance from the axis of rotation. The distribution that is established at the rate of rotation can then be analyzed to give the molecular weight.

It often takes an inconveniently long time for sedimentation equilibrium to be established. In 1947 the Canadian physicist William J. Archibald developed a method based on the *approach to sedimentation equilibrium*. The so-called Archibald method now is used commonly by biologists and biochemists. For details of the various methods used for studying sedimentation the student is referred to Chapter 11 and to texts listed at the end of this chapter. The interested reader is also referred to a useful new technique in sedimentation studies, *density gradient centrifugation* (see p. 497).

VISCOSITY

Valuable information about the sizes and shapes of large molecules is provided by measurements of the viscosities of their solutions. The viscosity of a fluid is a measure of the frictional resistance it offers to an applied shearing force. If a fluid is flowing past a surface, the layer of fluid adjacent to the surface is stagnant; successive layers have increasingly higher velocities. Figure 3.1 shows two parallel planes in a fluid,

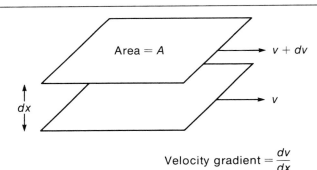

Figure 3.1

Two parallel moving layers in a fluid.

Area = A

$v + dv$

dx

v

Velocity gradient $= \dfrac{dv}{dx}$

separated by a distance dx and having velocities of flow differing by dv. According to Newton's law of viscous flow, the frictional force F_f resisting the relative motion of the two adjacent layers is proportional to the area A and to the velocity gradient dv/dx:

$$F_f = \eta A \frac{dv}{dx} \tag{3.18}$$

The proportionality constant η is known as the *coefficient of viscosity*. Its dimensions are (mass) (length)$^{-1}$ (time^{-1}) and the SI unit is kg m^{-1} s^{-1}. However, it is still common to use the *poise* (g cm^{-1} s^{-1}) as the unit. One poise is equal to one-tenth of the SI unit. The way in which viscosity coefficients are measured experimentally is discussed in Chapter 11.

The addition of a solute to a solvent affects its viscosity—sometimes very markedly. The magnitude of a viscosity change depends upon the concentration of the solution and on the size and shape of the solute molecules. Molecules which are long and thin, such as some protein molecules, may form a tangled network in solution and cause a high viscosity. More spherical molecules have a lesser effect on viscosity. By studying the variation of viscosity with concentration, we can gain some idea of the general size and shape of the solute molecules.

The theory of the viscosity of a solution of spherical solute molecules was first worked out in 1906 by Albert Einstein (1879–1955). The equation he derived was

$$\lim_{\phi \to o} \left[\frac{1}{\phi} \frac{\eta - \eta_o}{\eta_o} \right] = 2.5 \tag{3.19}$$

where η is the coefficient of viscosity of the solution and η_o that of the pure solvent; ϕ is the volume fraction of the solution occupied by solute particles, and we note that the quantity in brackets must be extrapolated to zero value of ϕ. The fraction η/η_o is known as the *viscosity ratio*, and $(\eta/\eta_o) - 1$ is the fractional increase in viscosity brought about by the dissolved particles. This equation was later extended by the American physical chemist Robert Simha to ellipsoidal particles; the factor 2.5 in the Einstein equation is now replaced by a factor which depends on the ratio of the two axes of the ellipsoid.

Since it is difficult to measure ϕ directly, the expression on the left-hand side of equation 3.19 is often replaced by

$$[\eta] = \lim_{c \to o} \left[\frac{1}{c} \frac{(\eta - \eta_o)}{\eta_o} \right] \tag{3.20}$$

where c is the concentration in mass per unit volume. The quantity $[\eta]$ is known as the *limiting viscosity number* or the *intrinsic viscosity*.

In practice, molecular sizes have usually been determined from viscosities on the basis of formulae which are semiempirical (i.e., based partly on theory and partly on experiment). A particularly useful relationship was suggested by Herman F. Mark and R. Houwink:

$$[\eta] = \beta M^\alpha \tag{3.21}$$

where M is the molecular weight and α and β are constants. Thus,

$$\log_{10}[\eta] = \log_{10}\beta + \alpha \log_{10} M \tag{3.22}$$

A plot of $\log_{10}[\eta]$ against $\log_{10} M$ should thus give a straight line of slope α. For spherical solute molecules, such as globular proteins, the value of α is zero; in other words, all globular proteins, irrespective of size, have essentially the same limiting viscosity number. Therefore molecular weights of globular molecules cannot be determined in this way. On the other hand, for molecules which exist in the form of flexible randon coils, the value of α is close to 2. In these cases, there is a strong dependence of η on M, and the molecular weights of such molecules can be determined from a measurement of η.

GEL FILTRATION

Gel filtration is a technique which allows macromolecules of different sizes to be separated and provides approximate values of molecular weights by making use of gels having different pore sizes. A commercially-available material, Sephadex, is available in a range of pore sizes and is commonly employed. If a solution of macromolecular material is passed through a column containing such a gel, the various components appear at different times. A gel containing small pores will only accept material of low molecular weight; large molecules will be unable to diffuse through it. The molecular weights obtained by the use of this method are very approximate and anomalies arise in the case of highly asymmetric molecules. The method does, however, have the advantage of being rapid, and it permits the molecular weights of different molecules in the same solution to be determined in one experiment.

3.5   OPTICAL METHODS OF DETERMINING MOLECULAR SIZES AND SHAPES

LIGHT SCATTERING

When light passes through a medium containing suspended particles, some of it is scattered and the incident beam passes through with weakened intensity. The first investigation of this phenomenon was made in 1871 by the British physicist John Tyndall (1820–1893). The type of apparatus he used, and which is still used in modern investigations, is shown schematically in Figure 3.2. The detector of the scattered light is mounted in such a way that the intensity of scattering can be measured at various positions. Analysis of the scattering as a function of the angle of scattering gives valuable information about the shapes and sizes of molecules.

The theory of light scattering is very complicated and the details are beyond the scope of this book; the student is referred to books listed at the end of this chapter. The following account will outline the main ideas.

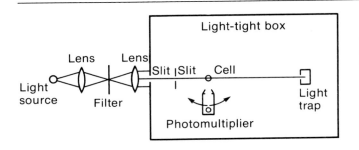

Figure 3.2

Schematic diagram of the type of apparatus used in light-scattering experiments.

The proportion of incident light that is scattered increases with an increase in the number and size of the particles in a solution. If the intensity of the incident radiation is I_o, and l is the length of the light path through the scattering solution, the intensity of the transmitted light is given by

$$I = I_o\, e^{-\tau l} \tag{3.23}$$

where τ is *turbidity*. In 1871 the British physicist John William Strutt, Lord Rayleigh (1843–1919), deduced that, for spherical particles having dimensions much smaller than the wavelength λ of the radiation, the turbidity is given by

$$\tau = \frac{32\pi^3 n_o^2 [(n - n_o)/c]^2 cM}{3\lambda^4 N_A} \tag{3.24}$$

where n and n_o are the refractive indices of the solution and solvent respectively, c is the concentration of the scattering particles, N_A is the Avogadro number, and M is the molecular weight. Thus, simply by measuring turbidity and assuming the particles to be spherical, we can determine the molecular weight.

When the particle dimensions are not small compared to λ, and the particles are not spherical, the theory is much more complicated. Important contributions to the theory have been made by Gustav Mie in 1908, Peter J. W. Debye (1884–1966) in 1947, Bruno H. Zimm in 1948, and Paul Doty in recent years. By applying their equations to the analysis of the intensity of scattering at various angles it is possible to obtain reliable information about the shapes and sizes of macromolecules.

FLOW BIREFRINGENCE

Certain solid substances have different indices of refraction along different axes, and are said to be *birefringent* or to exhibit *double refraction*. However, when solutions of such substances are examined under a polarizing microscope, no birefringence is observed; the molecules are oriented at random. If the molecules are caused to orient, birefringence is observed. This can be done by causing the solution to flow. Figure 3.3 shows the alignment in a schematic manner. In practice, a flow birefringence experiment is usually conducted by enclosing the solution between two concentric cylindrical vessels, the outer one of which is rotated. Analysis of the results indicates the molecular weight and the lengths of the axes of the ellipsoidal molecule.

Liquid at rest

Flowing liquid

Figure 3.3

The orientation of ellipsoidal molecules in a flowing liquid.

3.6   COMPARISON OF METHODS

Table 3.1 summarizes the various methods described in this chapter for the experimental determination of sizes and shapes of molecules, indicating which average molecular weight is determined when there is a mixture of macromolecules. The range of molecular weights satisfactorily covered by the different types of investigation is also stated.

Table 3.1   Methods for Determining Molecular Sizes and Shapes

Method	Type of Molecular Weight	Remarks
Chemical analysis	M_n	Requires an atom or group present in small proportions; best for low molecular weights.
Osmotic pressure	M_n	Applicable to molecular weights of up to 10^6.
Vapor-pressure lowering	M_n	Applicable to molecular weights of up to 10^4.
Diffusion	M_w	For molecular weights up to 10^6. Gives minimal molecular weights based on spherical shape.
Sedimentation	M_w	For molecular weights up to 5×10^7. Result is independent of shape.
Viscosity	M_v†	Best used in association with other methods.
Light scattering	M_w	Applicable to molecular weights up to 10^4.
Flow birefringence	M_w	
Gel filtration	M_w	Gives approximate molecular weights only

† This is a special viscosity-average value; it is much closer to M_w than to M_n.

Figure 3.4 shows the sizes and shapes of some protein molecules, as determined by various methods.

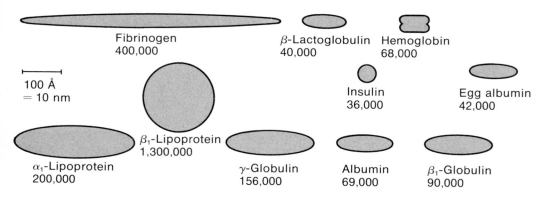

Figure 3.4

The sizes and shapes of various protein molecules, as deduced on the basis of diffusion, viscosity and light-scattering experiments. The approximate molecular weights are indicated.

3.7   BOND LENGTHS AND BOND ANGLES

Thus far we have considered methods for determining molecular weights, or overall sizes and shapes of molecules. There are also techniques for the measurement of bond lengths and bond angles, which provide much more detailed information about molecular structure. The most powerful and widely-used method is *X-ray diffraction,* which has provided a great deal of information about the detailed structures of proteins and other molecules of biological importance. Valuable information is also provided by use of *X-ray absorption* and the *electron microscope.*

X-RAY DIFFRACTION

The X-ray diffraction method is a rather specialized technique, being used only in a limited number of laboratories. Only a brief account will be given here, sufficient to indicate the type of information that can be obtained.

   If light strikes a very finely ruled grating in which the distance between the lines is close to the wavelength of the light, a characteristic diffraction pattern is observed. In an ionic crystal or a covalent molecule the distance between neighboring ions or atoms is of the order of 0.1 nm, and in order to obtain diffraction it is necessary to employ radiation having such wavelengths. X rays have wavelengths of the right magnitude (see Figure 2.1) and produce diffraction patterns when they strike matter in the solid, liquid, or gaseous state.

   The X-ray diffraction method is most easily applied to crystalline solids. The pioneering work in this field was done in 1912 and 1913 by the British physicist Sir William H. Bragg (1862–1942) and his son Sir Lawrence Bragg. Figure 3.5 shows a schematic diagram of the Bragg X-ray spectrometer. The same principle is used in most modern instruments, which are widely used to study crystals and other solid materials. The technique can be applied either to single crystals or to powders. From the analysis of the X-ray diffraction pattern it is possible to calculate the distance between the successive planes of atoms or ions in the crystal, and hence the internuclear distances.

The X-ray diffraction method has been applied to many macromolecules, including a number of proteins. The analysis of an X-ray diffraction pattern is very difficult for large molecules, because of the many interatomic distances involved. It is the electrons within the molecules which scatter the X rays, and the image calculated from the diffraction pattern thus reveals the distribution of electrons within the molecule. The usual procedure is to use high-speed computers to calculate the electron density at a regular array of points and to make the image visible by drawing contour lines through points of equal electron density. These contour maps can be drawn on clear plastic sheets and a three-dimensional image can be obtained by

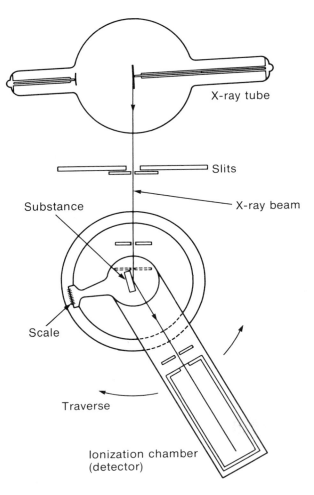

Figure 3.5

Schematic representation of a Bragg X-ray spectrometer.

X-ray tube

Slits

Substance

X-ray beam

Scale

Traverse

Ionization chamber (detector)

stacking the maps one above the other. The amount of detail that can be seen depends upon the resolving power of the instrument; if the resolving power is sufficiently good the atoms appear as individual peaks in the image map. At lower resolutions there are groups of unresolved atoms, which can be frequently recognized by their characteristic shapes.

An important contribution to the analysis of X-ray patterns for complex molecules was made by the British molecular physicist Max Ferdinand Perutz in 1953. His method, known as *isomorphous replacement*, involves the preparation and study of crystals into which heavy atoms, such as atoms of uranium, have been introduced without altering the crystal structure. This technique led rapidly to the detailed analysis of the structures of a number of protein molecules.

X-RAY ABSORPTION

A completely different X-ray approach to the problem of molecular structure was proposed in 1971 by the American physicists Dale E. Sayers, Edward A. Stern, and Farrel W. Lytle. This method is based not on the diffraction of X rays by an array of atoms or ions, but on the absorption of X rays by individual atoms or ions. This new technique measures how the absorption is affected by the atoms in the immediate neighborhood of the atom which is absorbing the X rays, and yields a knowledge of the atomic environment of each type of atom in a molecule. This technique promises to provide valuable information about biological structures and the reactions they undergo.

ELECTRON MICROSCOPY

An electron (mass = 9.1×10^{-31} kg) accelerated by 10 000 volts has a velocity of 5.9×10^7 m s^{-1}, and the de Broglie equation (1.6) leads to a wavelength of 1.2×10^{-11} m, or 1.2×10^{-2} nm. This is somewhat less than the usual distances between neighboring atoms in a molecule, and therefore it is possible to diffract an electron beam in the same way as X rays, using a crystal as a diffraction grating.

Electron beams have one advantage over X rays: because of their negative charge, electrons can be brought into sharp focus by appropriate arrangements of electric and magnetic fields. This has led to the development of *electron microscopes*, which are capable of resolving images as small as 0.5 nm in diameter. This is not quite fine enough to resolve interatomic distances, but valuable information about tissues and other biological structures has been provided by use of this technique.

3.8 MOLECULAR SYMMETRY

Important information about the symmetry properties of molecules is provided by measuring their *optical rotation* and *optical rotatory dispersion*.

OPTICAL ROTATION

Ordinary light vibrates in all directions perpendicular to its path. When light is passed through certain crystals, such as calcite and tourmaline, it is caused to vibrate in only one plane and is said to be *plane-polarized*. If plane-polarized light is passed through certain substances, the plane of polarization is rotated in a clockwise or anticlockwise direction. In order for a substance to rotate the plane of polarization of light, the molecule must have neither a center of symmetry nor a plane of symmetry. The simplest type of molecule which satisfies this requirement is one in which a carbon atom is attached to four different groups, as is shown in Figure 3.6 for lactic acid. Such a carbon atom is known as an *asymmetric carbon atom*. There are two possible forms for such a molecule; one is the mirror image of the other, and the two cannot be superimposed.

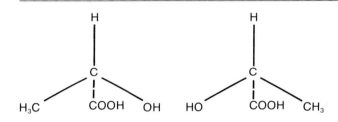

D-(—)-lactic acid   L-(+)-lactic acid

Figure 3.6

The two enantiomeric forms of lactic acid. The + and − signs indicate optical rotation to the right and to the left. The letters D and L indicate the relationship to D- and L-glyceraldehyde (see p. 112).

Molecules having this type of symmetry are said to exhibit "handedness" or *chirality* (Greek *cheiros*, hand); a pair of hands exhibits the same kind of relationship as do the two molecules of lactic acid shown in Figure 3.6. Two molecules related to each other in this way are known as *enantiomorphs* (from the Greek *enantios*, opposite; *morphe*, form) or *enantiomers* (Greek *meros*, part). Enantiomorphs have many identical chemical and physical properties, but they differ in rotating the plane of polarized light in opposite directions. Compounds which rotate the plane of polarized light are said to be *optically active*, and enantiomorphs are also called *optical isomers*. As far as chemical properties are concerned, two enantiomers behave identically except when they undergo reaction with other asymmetric molecules, in which case there can be a significant difference in reaction rates and equilibrium constants. This has an important bearing on reactions catalyzed by enzymes, the biological catalysts, and will be referred to again in Chapter 10.

If an asymmetric substance such as lactic acid is separated into its two enantiomers, one will rotate the plane of polarization to the right and the other to the left. The instrument used to measure optical activity is known as a *polarimeter*. A solution of the substance is placed in a tube, usually one decimetre (ten centimetres) long, each end of which is transparent. A beam of plane-polarized light, produced by passing ordinary light through a suitable crystal, is passed through the tube, and the direction of polarization of the emerging beam is determined with another crystal of the same kind; the direction can be found by rotating this crystal until no light passes through it. The rotation of the plane of polarized light is proportional to the concentration of the solution and to the path length. The observed rotation for a path length of one decimetre of a solution containing one gram per cubic centimetre (1 g cm^{-3}) is a constant for any substance and is referred to as its *specific rotation*. Thus the observed rotation $[\alpha]$ is given by

$$[\alpha] = [\alpha]_\lambda^T lc \qquad (3.25)$$

where $[\alpha]_\lambda^T$ ($\text{deg cm}^3 \text{ dm}^{-1} \text{ g}^{-1}$)† is the specific rotation at a temperature T and wavelength λ, l is the length of the tube in decimetres and c the concentration in grams per cubic centimetre (g cm^{-3}). The specific rotation is a characteristic property of a substance. The *molar rotation* $[M]_\lambda^T$ ($\text{deg cm}^3 \text{ dm}^{-1} \text{ mol}^{-1}$) is the product of the specific rotation and the molecular weight M:

$$[M]_\lambda^T = [\alpha]_\lambda^T M \qquad (3.26)$$

The light used in polarimetric work is usually the yellow Sodium-D line, produced by a special lamp, and the specific rotation is then recorded as $[\alpha]_D^{T(^\circ C)}$, e.g., $[\alpha]_D^{25^\circ}$ at 25° C.

EXAMPLE
An optical rotation of $+5.68°$ was observed with a solution of D-glucose (mol. wt. 180) in a 1 dm cell at 20° C with sodium light. If the specific rotation $[\alpha]_D^{20°}$ is $+52.7°$, what was the concentration of the solution?

SOLUTION
From equation (3.26), with $l = 1$,

$$5.68 = 52.7 \times 1 \times c$$

thus

$$c = 0.108 \text{ g cm}^{-3}$$

As the molecular weight is 180, the molar concentration (mol dm^{-3}, M) is

$$\frac{0.108 \times 1000}{180} = 0.60 \text{ M}.$$

---

† The SI unit, $\text{deg m}^2 \text{ kg}^{-1}$, is one-tenth of this unit.

For many years it was not known which of a pair of enantiomorphs rotated the plane of polarized light to the right, and which to the left. The problem was solved in 1951 by a crystal-structure determination of glyceraldehyde, using X-ray methods. It was found in this way that the enantiomorph of glyceraldehyde that rotates the plane of polarized light in a positive, or clockwise, direction has the following structure:

$$
\begin{array}{c}
H \\
| \\
C \\
\end{array}
$$

HOCH$_2$ CHO OH

This particular enantiomorph is known as D-(+)-glyceraldehyde. The + sign tells us that the rotation of the plane of polarized light is clockwise; the D refers to the particular structure shown above. Once the absolute configuration† of D-(+)-glyceraldehyde had been determined it became possible to assign absolute configurations to other molecules by preparing them from glyceraldehyde. For example, it is possible to convert glyceraldehyde into lactic acid by processes in which there is no disturbance of the bonds to the asymmetric carbon atom. When this is done with D-(+)-glyceraldehyde as the starting point, the product will also have the same absolute configuration of the groups. This is indicated by calling the product D-lactic acid; its configuration is shown in Figure 3.6. This form, unlike D-(+)-glyceraldehyde, rotates the plane of polarized light in the counterclockwise direction and is called D-(−)-lactic acid.

Molecules having an asymmetric carbon atom are not the only ones to rotate the plane of polarization. A molecule having a helical (spiral) structure also exists as two enantiomorphs, one resembling a right-handed screw and the other a left-handed screw. These two enantiomorphs also rotate the plane of polarization of light in opposite directions.

OPTICAL ROTATORY DISPERSION

Studies of optical rotation at a single wavelength do not permit us to distinguish between effects due to asymmetric carbon atoms and effects due to helical or other asymmetric conformations. Such a distinction can be made by studying the dependence of optical rotation on the wavelength of the light employed.

A plot of the specific rotation against the wavelength may show maxima and minima, with the specific rotation actually changing sign, as shown in Figure 3.7. This phenomenon, discovered by the American chemist Frank Albert Cotton, is known as the *Cotton effect*. Recent workers, in particular the American chemist

---

† The term *configuration* refers to different spatial arrangements of a given group of atoms in a molecular structure, interconversion between the structures requiring the breaking of primary (e.g., covalent) bonds. Configuration is contrasted with *conformation*, where interconversion does not require the breaking of primary bonds.

Carl Djerassi, have studied this phenomenon in detail. Various theories have been developed which relate these optical rotatory dispersion curves for solutions of macromolecules to their conformations. For example, synthetic long-chain molecules present in solution in a helical form will behave very differently from the molecules present in a random arrangement (Figure 3.7). On the basis of a detailed analysis of the Cotton effect it has been possible to determine what proportion of a protein molecule occurs in the helical form.

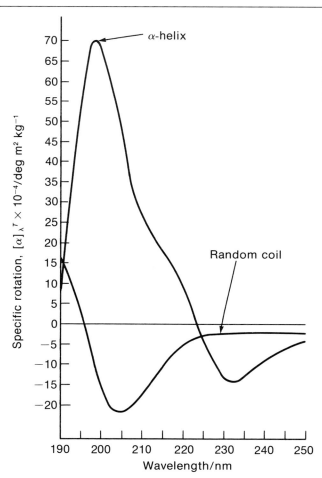

Figure 3.7

Optical rotatory dispersion curves for poly-L-lysine, in helical and random-coil conformations. The helical form is the α-helix (see p. 133).

3.9   DIPOLE MOMENTS

We saw in Chapter 1 that a chemical bond can have a dipole moment, resulting from the unsymmetrical distribution of the electron cloud. The dipole moment is a vector quantity; thus, the resultant dipole moment can be calculated from individual moments just as a resultant force can be calculated from individual forces. A common procedure is to estimate the dipole moment of a molecule from the dipole moments of the individual bonds and the bond angles. Conversely, if the overall dipole moment of a molecule is known together with the shape of the molecule, the bond moments can sometimes be estimated. The application of this procedure to the water molecule is illustrated in Figure 3.8. The observed dipole moment of the molecule is 1.85 debyes and the bond angle is about 104°; use of the parallelogram method then leads to a bond moment of 1.60 debyes. Unfortunately, however, this procedure is somewhat unreliable, since it neglects the contributions from nonbonding electrons. In the water molecule, for example, the nonbonding lone-pair electrons also make an important contribution to the overall dipole moment of the molecule.

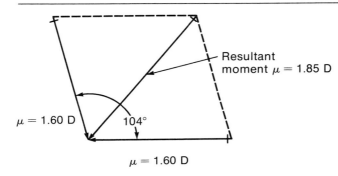

Figure 3.8

The resultant dipole moment of the water molecule, as related to that of two individual bonds. If the bond moments are represented by the lengths and directions of the arrows directed along the bonds, the dipole moment of the whole molecule is represented by the arrow lying along the diagonal of the parallelogram.

The dipole moment of a substance is measured experimentally by indirectly determining the extent to which the molecules orient in an electric field. The dielectric constant of a solution of the substance is obtained by measuring the electrical capacitance of a capacitor in which the solution is between its plates. It is most convenient to use as a solvent a liquid such as benzene, which is little affected by an electric field. If pure benzene, C_6H_6, is placed between the plates of a capacitor and an electric field is applied, there will be a certain tendency for the electron clouds on the benzene molecules to move towards the positive plate. This effect is known as *polarization*. By measuring the capacitance, the extent of this polarization can be estimated. Since benzene is a perfectly symmetrical molecule having no dipole moment, there is no tendency for the molecules to orient in the electric field.

Suppose that we dissolve in the benzene a substance such as hydrogen chloride, HCl, which has a dipole moment. When the electric field is applied, there again will be a tendency for the electron cloud to move towards the positive electrode; this effect is referred to as *atom* and *molecular polarization*. In addition, because the hydrogen chloride molecule has a permanent dipole moment, there will be some *orientation polarization*; that is, the molecules will tend to orient in such a way that the negative chlorine ends will be towards the positive electrode and the positive hydrogen ends will be towards the negative electrode. This orientation polarization makes an additional contribution to the capacitance.

The theory underlying the usual method of determining dipole moments was formulated in 1912 by the Dutch-American physical chemist Peter J. W. Debye

(1884–1966). According to his theory the total molar polarizability, P_m, is related to the dielectric constant ε by

$$P_m = \frac{\varepsilon - 1}{\varepsilon + 2} \frac{M}{\rho} \tag{3.27}$$

where M is the molecular weight and ρ the density of the substance. The total molar polarizability thus can be easily calculated from the dielectric constant. This total molar polarizability arises from the atom, molecular, and orientation polarizations. The atom and molecular polarizations do not depend upon the temperature, but the orientation polarization does. The tendency of an electric field to orient a molecule is proportional to μ^2—the square of the dipole moment— but this effect is resisted by the thermal motions of the molecules, which tend to cause the molecules to be oriented at random rather than in the direction of the electric field. Since the tendency of the orientations to be random is proportional to the absolute temperature T, the net tendency of the molecules to be oriented in the direction of the applied field is proportional to μ^2/T. By measuring the polarization at various temperatures it is therefore possible to separate the temperature-dependent orientation polarization from the temperature-independent atom and molecular polarizations, and hence to calculate the dipole moment.

Table 3.2 gives some dipole moments for individual bonds, determined by methods similar to that used for the water molecule (see Figure 3.8). We saw in Chapter 1 how individual moments can be estimated from the atomic electronegativities, some of which were listed in Table 1.3.

Table 3.2 Dipole Moments of Individual Bonds

Bond	Dipole moment (debyes)
O—H	1.6
N—H	1.3
C—H	0.4
C—F	1.4
C—Cl	1.5
C—Br	1.4
P—H	0.4
S—H	0.7

Dipole moments provide two kinds of information about molecular structure. In the first place, they tell us the extent to which a bond is polarized, a matter of importance in connection with intermolecular forces. Later we will see many examples of how interactions between dipoles and interactions between dipoles and ions contribute to an understanding of biological systems. Secondly, dipole moments provide valuable clues to the structures of simple molecules. For example, the fact that the water molecule has an appreciable dipole moment shows that the molecule is not linear but bent. Conversely, the zero dipole moment of carbon dioxide, CO_2, suggests that this molecule is linear.

3.10   MOLECULAR MAGNETISM

Another property which provides valuable information about molecular structure is molecular magnetism. The theory of the magnetism of molecules is similar to that of the electric dipole moment. As discussed in Section 3.9, an electric field can induce a moment in a molecule, and some molecules have permanent electric dipole moments. Similarly, a magnetic field induces a magnetic moment in a molecule; the molecule also may have a permanent magnetic moment. The extent to which a magnetic field induces a temporary magnetic moment in a molecule is expressed by the *diamagnetic susceptibility* of the molecule. The measure of the permanent magnetism of a molecule is its *paramagnetic susceptibility*. Diamagnetic susceptibilities are not of great chemical interest, but paramagnetic susceptibilities provide useful information about molecular structure.

Magnetic susceptibilities, both diamagnetic and paramagnetic, are usually measured by the use of a Gouy magnetic balance. The substance is suspended in a glass container in a very sensitive balance between the poles of a powerful magnet. The magnet is designed to produce a magnetic field which varies considerably from one region to another, i.e., it is an *inhomogeneous* field. A diamagnetic molecule, having no permanent magnetic moment, is less permeable to magnetic lines of force than is a vacuum and thus tends to move from a stronger to a weaker part of an inhomogeneous magnetic field. A paramagnetic substance, on the other hand, has a permanent magnetic dipole moment, and shows a tendency to move from a weaker to a stronger magnetic field. Paramagnetic behavior almost always overshadows the diamagnetic effect.

We saw in Section 3.9 that the temporary and permanent electric dipole effects can be separated by utilizing the temperature dependence of the latter. In a similar way, diamagnetic and paramagnetic susceptibilities are separated by studying the effect of temperature on paramagnetic behavior, and absence of temperature effect on diamagnetic behavior. The tendency of the paramagnetic substance to move in the magnetic field is proportional to the square of the magnetic moment μ, and this tendency is countered by the thermal energy, which is proportional to the absolute temperature T. The net tendency of the substance to move as a result of its permanent moment is proportional to μ^2/T. Consequently, μ can be measured by studying the behavior over a range of temperatures.

The paramagnetism of a molecule is caused by the orbital motion of the electrons and their spin angular momenta. However, the contributions due to spin are much more important than those due to orbital motion, which may be neglected. This is the case because in a molecule there are strong internal electric fields directed along the chemical bonds which hold the orbits of the electrons in fixed orientations. Thus the orbits cannot easily line up with an external magnetic field. As a result they make little contribution to paramagnetism. The electron spin, on the other hand, is not influenced by the internal field, and the paramagnetism is therefore a measure of electron spin. Since the spins of paired electrons cancel each other, the observed paramagnetism is a measure of the number of unpaired electrons in the molecule.

The various experimental techniques which provide information about the structures of molecules and of molecular aggregates are summarized in Table 3.3.

Table 3.3 Techniques for the Study of Molecular Structure

Property	Technique	Remarks
Molecular rotations	Microwave spectroscopy	Not suitable for large molecules
Molecular vibrations and rotations	Infrared spectroscopy	Not suitable for aqueous systems
	Raman spectroscopy	More suitable for aqueous systems
	Visible-ultraviolet spectroscopy	Used for assay as well as for molecular structure (as is colorimetry)
Bond lengths and bond angles	X-ray diffraction	Of great importance for macromolecules
	Electron microscope	Of great importance for macromolecules
Molecular symmetry	Optical rotation Optical rotatory dispersion	Valuable for the study of protein conformations
Dipole moments	Measurement of dielectric constant	Particularly useful for small molecules
Molecular magnetism	Gouy magnetic balance	Used for determining the number of unpaired electrons

3.11   WATER

Water is an abundant and important ingredient in all living systems, and it is appropriate that we should consider some aspects of its structure before going on to the organic compounds.

Water has some remarkable properties. One unusual feature is that its boiling point is very much higher than would be expected on the basis of its molecular weight. For example, water boils at 100° C, which is very much higher than the boiling point of its heavier homolog hydrogen sulfide, H_2S, whose boiling point is $-61.8°$ C. Water also has an unusually high heat of vaporization. These and other properties indicate that there are large attractive forces between neighboring water molecules.

The main contribution to these attractive forces is *hydrogen bonding*. A simple example of hydrogen bonding is found in the vapor of hydrogen fluoride, in which species such as $(HF)_2$, $(HF)_3$, etc., are present. We saw in Chapter 1 that HF has a substantial dipole moment, so that a string of HF molecules can be held together by dipole-dipole attractions:

$$\overset{\leftrightarrow}{H-F} \ldots \overset{\leftrightarrow}{H-F} \ldots \overset{\leftrightarrow}{H-F} \ldots$$

The theoretical treatment of hydrogen bonding leads to the conclusion that electro-static attractions play the most important role in holding the structure together—although purely quantum-mechanical effects are also significant. The small radius of the hydrogen atom allows the molecules to come close together and the attractive forces are therefore large. Hydrogen bonds are thus the strongest bonds that result from electrostatic attractions between dipoles.

The structure of the water molecule is such that a three-dimensional hydrogen-bonded structure can be formed very readily. The best that HF can do is to form strings of molecules. By contrast, because of its electronic and dipolar character, the water molecule tends to form structures of the type shown in Figure 3.9. The oxygen atom of one water molecule is attached by hydrogen bonds to *two* hydrogen atoms in two neighboring water molecules. When a water molecule is hydrogen-bonded through its oxygen atom to two other water molecules, the oxygen atom is surrounded tetrahedrally by four hydrogen atoms. Two hydrogen atoms are attached to the oxygen atom by covalent bonds, the other two by hydrogen bonds. The structure is essentially three-dimensional and may be compared with the structure of diamond, in which each carbon atom is also surrounded tetrahedrally by four other carbon atoms.

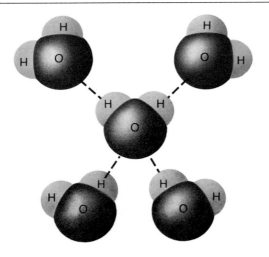

Figure 3.9

A water molecule surrounded by four other water molecules held to it by hydrogen bonding.

In liquid water and ice each H_2O molecule tends to be surrounded tetrahedrally by four other H_2O molecules. However, the arrangement of water molecules in the liquid state is not completely orderly, and the tetrahedral structure is constantly changing. The high boiling point of water is directly related to this type of structure. In order for a water molecule to pass from the liquid to the vapor state, several hydrogen bonds must be broken and, as a result, the heat of vaporization has the high value of about 10 kcal mol^{-1}. It is believed that, on the average, about three hydrogen bonds are broken for each water molecule vaporized; the heat of vaporization is thus consistent with the fact that each hydrogen bond has a strength equivalent to about 3 kcal mol^{-1}.

A very important property of liquid water is its high *dielectric constant*. The force of electrostatic attraction or repulsion between two charged bodies is inversely proportional to the dielectric constant of the medium. Water has a dielectric constant ε of about 78 at ordinary temperatures. This means that, for example, a Na^+ and a Cl^- ion are attracted to one another only 1/78th as strongly in water as compared with a vacuum (where $\varepsilon = 1$). Consequently, if we introduce solid sodium chloride into water the attractive forces between the Na^+ and Cl^- ions—which hold the crystal together—are greatly reduced and the salt goes into solution.

The very high dielectric constant of liquid water results from its peculiar structure. The structure shown in Figure 3.9 is not a completely rigid one, since individual water molecules can rotate fairly freely on their axes. The structure thus has a high polarizability, in that it easily accommodates itself to an electric field. This high polarizability results in the high dielectric constant.

3.12 AQUEOUS SOLUTIONS

A number of different kinds of interactions must be taken into account in considering aqueous solutions. Several of these are depicted in Figure 3.10, which is an example of a protein molecule present in water. The various attractive forces shown in the figure all play important roles in connection with the conformations of proteins and other macromolecules.

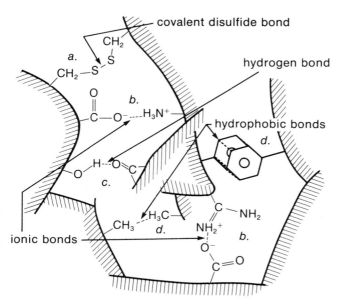

Figure 3.10

Schematic representation of a protein molecule, showing different kinds of bonds which contribute to the folded structure:
a. Covalent disulphide bond.
b. Ionic bonds (salt bridges).
c. Hydrogen bond.
d. Hydrophobic bonds.

IONIC BONDS (SALT BRIDGES)

Figure 3.10 includes two examples of "salt bridges," which involve attractions between ions of opposite sign. We have seen that when small ions such as Na^+ and Cl^- are present in aqueous solution there is little attraction between them, since the electrostatic force has been reduced by a factor of about 80. In addition, the ions are stabilized by *hydration*, an effect considered further in Chapter 7. The situation is considerably different, however, when the charged groups are surrounded mostly by protein or other macromolecular material. The effective dielectric constant is then reduced, and the attractive forces are therefore much stronger. As a result, salt bridges play an important structural role in aqueous solutions of large molecules.

HYDROGEN BONDS

Hydrogen bonds, which are so important in determining the structure of liquid water, also play a significant role in determining the conformations of proteins and other molecules. The most important hydrogen bonds found with proteins are between

$$\ce{>C=O} \text{ and } \ce{H-N<} \text{ groups:}$$

$$\ce{>\overset{\leftrightarrow}{C}=O \cdots H-\overset{\leftrightarrow}{N}<}$$

This matter is further considered later in this chapter. In addition, there is considerable hydrogen bonding between water molecules and the alcohol groups and carboxyl groups present in solute molecules:

alcohol $\ce{-O-H \cdots O<^{H}_{H}}$

carboxyl

Molecules containing a large proportion of such groups are soluble in water, whereas those containing mainly nonpolar groups are much less soluble.

HYDROPHOBIC BONDS

In aqueous systems there is also an indirect type of interaction between nonpolar groups, which is related to the hydrogen-bonded structure of water itself.

Consider first the behavior of isolated molecules of methane, CH_4. There is not much difference in electronegativity between the hydrogen atom (2.1) and the carbon atom (2.5) so that the C—H bond has only a very small dipole moment. In contrast to water molecules, isolated methane molecules have very little tendency to stick together. Because of their tetrahedral geometry the approach of two methane molecules would involve the approach of two slightly positive hydrogen atoms, which leads to some repulsion. There is a small net attraction between methane molecules resulting from the Van der Waals or dispersion forces, and the smallness of this attraction is reflected in the very low boiling point of methane.

For the same reason, methane molecules have little attraction to water molecules, so that the solubility of methane in water is very low. Similarly, any nonpolar group, such as the methyl group, —CH_3, and the phenyl group, —C_6H_5, will exert very little attraction on neighboring water molecules. Such groups will, in fact, interfere with the hydrogen-bonded structure in liquid water. The evidence suggests that the water molecules which are next to nonpolar groups, being unable to form hydrogen bonds with these groups, tend to orient in a particular way so as to form hydrogen bonds with neighboring water molecules. The thermodynamic implications of this are discussed in Chapter 5 (p. 249), where it will be seen that there is a tendency for nonpolar groups either to escape completely from aqueous solution or at least to stick together so as to cause minimum disturbance to the water structure.

There is thus an *apparent* attraction between nonpolar groups when they are present in water, and we speak of *hydrophobic* (Greek *hydor*, water; *phobos*, fear). It is important to realize that hydrophobic bonding does not result from a *direct* attraction between nonpolar groups, since the Van der Waals attraction is too weak to explain the observed effects (see Table 1.2).

3.13   LIPIDS

Lipids are fatty substances which are important structural components of all living systems. One of their functions is to lubricate the tissues. Lipids also occur in the cell walls and play an essential role in controlling the flow of substances into and out of cells. They are also of great importance as fuels; that is, as substances which provide the necessary energy to living systems. They are very effective in this regard because on complete oxidation they provide more than twice as much energy as an equal weight of carbohydrate or protein.

The *simple lipids* are esters of the alcohol glycerol with a variety of fatty acids. Glycerol is an alcohol which contains three hydroxyl groups, its structure being

CH_2OH
|
$CHOH$
|
CH_2OH

Each of the three OH groups can become esterified with a carboxylic acid as follows:

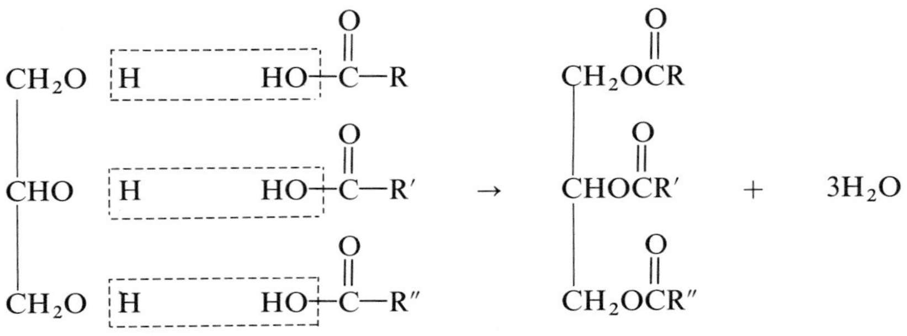

$$
\begin{array}{ccc}
CH_2O{-}H & HO{-}\overset{\displaystyle O}{\overset{\|}{C}}{-}R & CH_2OCR \\[2mm]
CHO{-}H & HO{-}\overset{\displaystyle O}{\overset{\|}{C}}{-}R' \rightarrow & CHOCR' \quad + \quad 3H_2O \\[2mm]
CH_2O{-}H & HO{-}\overset{\displaystyle O}{\overset{\|}{C}}{-}R'' & CH_2OCR''
\end{array}
$$

glycerol   carboxylic acids   triglyceride

In the naturally-occurring fats the acids, known as fatty acids, are long-chain molecules. Fats containing more than one double bond are usually liquids at ordinary temperatures, and they are then known as oils.

In addition to these triglycerides, other types of compounds are classed as lipids. One important group consists of the phospholipids, which are derived from α-glycerophosphoric acid:

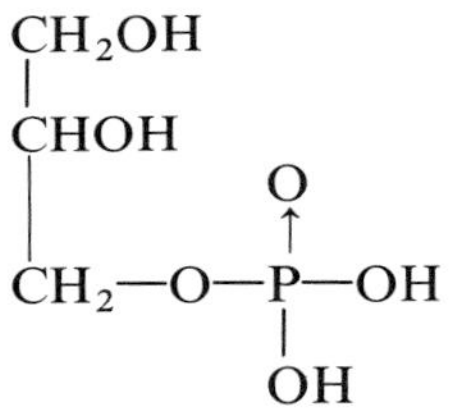

$$
\begin{array}{l}
CH_2OH \\
CHOH \\
CH_2{-}O{-}\overset{\displaystyle O}{\overset{\uparrow}{P}}{-}OH \\
\qquad\quad OH
\end{array}
$$

The remaining two —OH groups are esterified with a long-chain fatty acid. Another important group of lipids comprises the steroids, some of which show some relationship to the aromatic hydrocarbon phenanthrene:

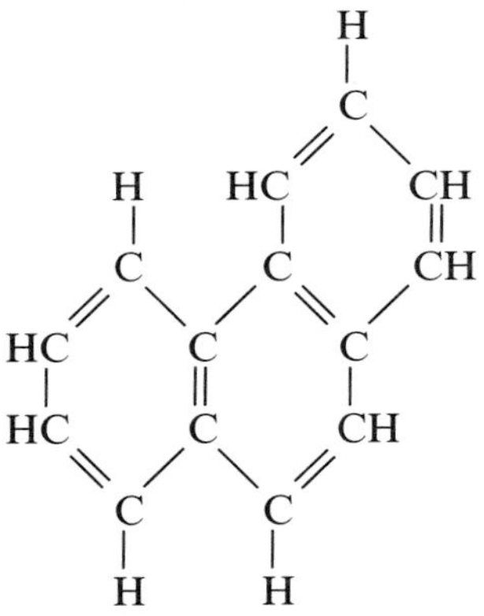

The most important member of this group is cholesterol :†

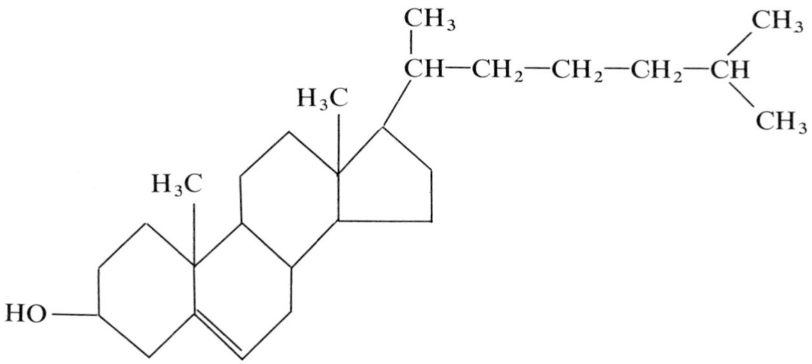

Lipids are only very slightly soluble in water. This is to be expected since, in a lipid molecule, most of the chemical bonds are C—C and C—H bonds, which are of the nonpolar variety; the more polar C—O bonds represent only a small fraction of the bonds. We have seen that compounds containing a high proportion of nonpolar bonds tend to be insoluble in water; they are much more soluble in solvents such as acetone and ether.

3.14 CARBOHYDRATES

The carbohydrates, which include sugars, starches, and cellulose, are the most abundant organic compounds in nature. A considerable variety of them exists, and only a few examples can be given here. One of the commonest—and the most important— carbohydrate is glucose, the structure of a common form of which is

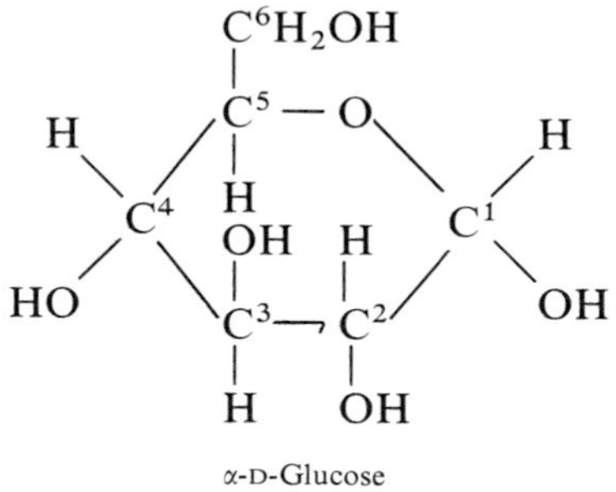

α-D-Glucose

---

† For simplicity, the ring carbon atoms in this structure have not been written out; it is to be understood that there is a carbon atom at each corner of the rings and that the appropriate numbers of hydrogen atoms are attached to them.

This is known as α-D-glucose. There is another form of glucose in which the arrangement around atom number 1 is

This form is known as β-D-glucose. In the crystalline form, glucose exists predominantly as α-D-glucose; in solution, an equilibrium is established between the two forms, the β form preponderating.

Examination of the structure shown above for α-D-glucose reveals that five of the carbon atoms in the molecule are *asymmetric* carbon atoms (see p. 110). Since there are two alternative arrangements of the groups around each of the five carbon atoms, the total number of structures of this kind that can exist is $2 \times 2 \times 2 \times 2 \times 2 = 32$. We have already seen that two of them, in which there are different arrangements around C^1 but in which everything else is the same, are designated α-D-glucose and β-D-glucose. The exact mirror images of these structures are called α-L-glucose and β-L-glucose. The remaining 28 members of the group are given different names altogether, such as α-D-galactose and β-D-galactose.

The above structures are known as *aldoses*, because they may be regarded as formed from aldehydic structures by the closing of a ring:

In the case of D-glucose, this process involves the —OH group on carbon atom number 5 and leads to the formation of a six-membered ring. Besides the aldoses, there is also an important class of carbohydrates known as the ketoses, in which there is a ketonic rather than an aldehydic group; in other words, the $\diagdown$ $C{=}O$ $\diagup$

group is not at the end of the molecule. An example is D-fructose, which is a ketohexose (having six carbon atoms):

The structure above shows a five-membered ring, in contrast to glucose which is six-membered. Fructose in the crystalline form exists mainly as a six-membered ring, but the five-membered structure is important in solution and in most of the derivatives of fructose, including sucrose.

The carbohydrates considered thus far are known as *monosaccharides*. Since all of them contain a high proportion of —OH groups they are very soluble in water.

MORE COMPLEX CARBOHYDRATES

Besides the monosaccharides, there are more complicated carbohydrate structures such as disaccharides and polysaccharides. These may be regarded as being formed as a result of a reaction in which a molecule of water is eliminated between two hydroxyl groups (a type of reaction known as a *condensation* reaction):

$$\text{w}O{-}H \qquad H{-}O\text{w} \quad \rightarrow \quad H_2O \quad + \quad \text{w}O\text{w}$$

For example, the sugar maltose, which is known as a *disaccharide*, can be regarded as resulting from the condensation of two glucose molecules.

α-D-Glucose + α-D-Glucose → Maltose

Similarly, ordinary cane sugar, known scientifically as sucrose, can be regarded as being formed by the condensation of a molecule of glucose with one of fructose:

α-D-Glucose   α-D-Fructose

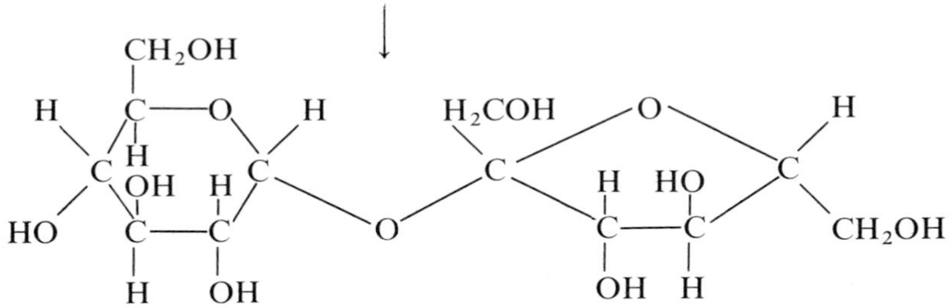

$$\text{CH}_2\text{OH} \qquad \downarrow$$

$$\text{Sucrose}$$

Since in these disaccharides there still remains a large proportion of free —OH groups, there is a high possibility of hydrogen bonding with water molecules, so that the solubility in water is large.

A number of much larger carbohydrate molecules, known as *polysaccharides*, are also of great biological importance. Starch, which is present extensively in plants, is a polymer of α-D-glucose, containing structures of the following type:[†]

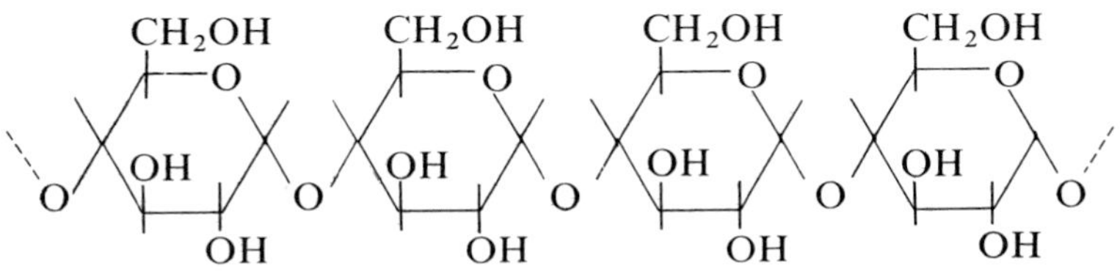

The orientation of the bonds at the oxygen atoms connecting the glucose units is such as to give the molecule a helical (spiral) structure. In addition to this feature, the starch molecule has a considerable amount of branching, with structures of the following type being found:

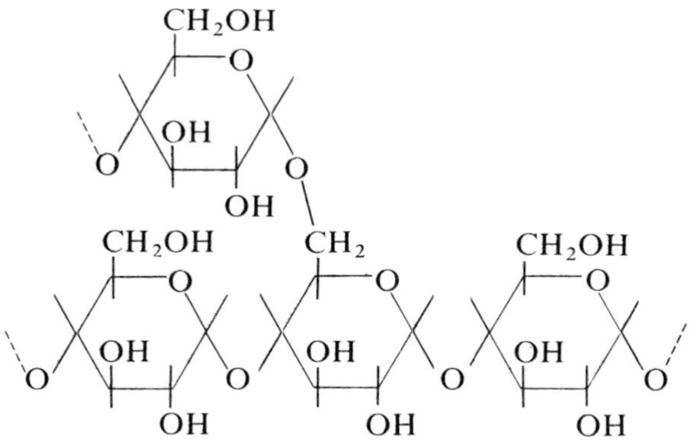

Glycogen, or animal starch, is very similar to plant starch, since glucose residues are condensed together in the same manner as in maltose. There is even more branching

---

[†] For simplicity, an abbreviated notation is used for these ring structures.

in glycogen, however. Figure 3.11 gives a schematic representation of the glycogen molecule.

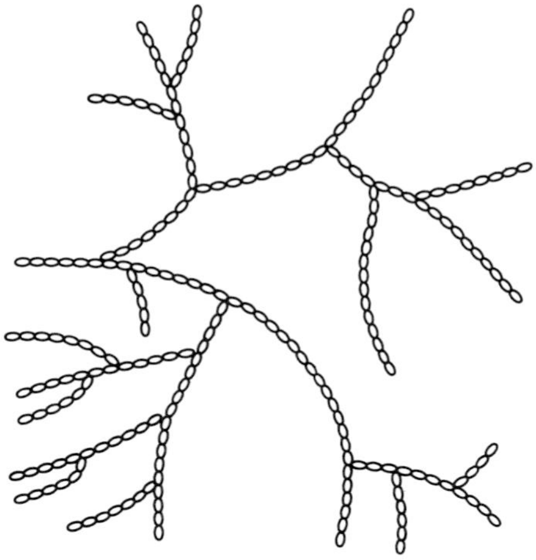

Figure 3.11

Schematic representation of the structure of a glycogen molecule. Each ellipse represents a glucose residue. There is considerable branching of chains.

In cellulose, on the other hand, the glucose molecules are in the configuration:

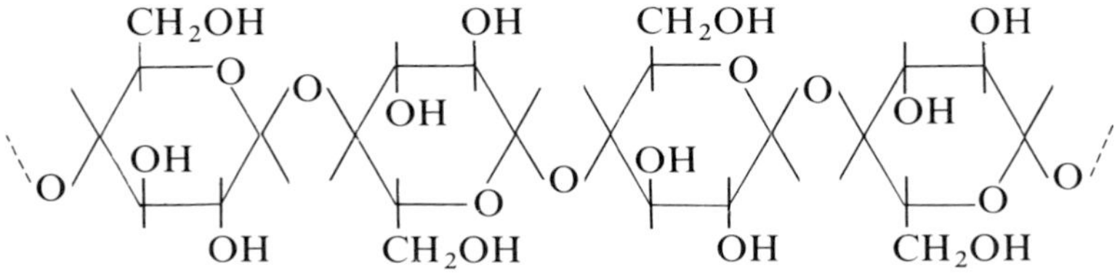

The polymer is now linear, as is appropriate for the fibrous materials in plants.

In these polysaccharides there are free —OH groups. However, there is considerable *intramolecular* hydrogen bonding, between —OH groups on one part of a molecule and those on another. This has the effect of diminishing the solubility in water, since these groups are not readily available for hydrogen bonding with water molecules. In addition, there can be considerable *intermolecular* hydrogen bonding between —OH groups of different molecules. For example, in cellulose, which is a linear polymer, the molecules form bundles in which they are held together by hydrogen bonding. Since many of the —OH groups are not readily available for hydrogen bonding with water molecules, cellulose is not very soluble in water.

3.15 PROTEINS

The proteins are of biological importance for two main reasons: they are the principal structural elements of the cells, and some of them are catalysts for the chemical reactions that occur in living systems.

The components of the proteins are the α-amino acids, of which about twenty are important. The α-amino acids possess both an amino group and a carboxylic acid group attached to the same carbon atom:

This carbon atom also is attached to a hydrogen atom and to a group R, which is different for each amino acid; examples are

Glycine $R = H$ Valine $R = CH(CH_3)_2$

Histidine $R = C$

All amino acids except glycine contain an asymmetric carbon atom, which is connected to the four different groups H, R, NH_2, and COOH. They therefore exist in two optically active mirror-image forms. Valine, for example, occurs as D-valine and L-valine, which rotate the plane of polarized light to the same extent but in opposite directions. Their configurations are shown in Figure 3.12. The amino acids occurring in nature are nearly all of the L variety; a few microorganisms contain the D-amino acids, but the proteins that occur in the tissues of animals contain exclusively the L forms.

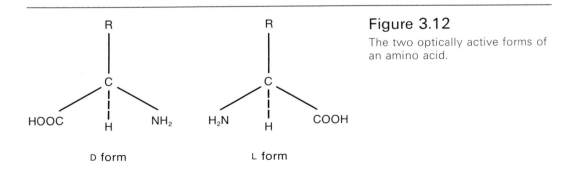

D form   L form

Figure 3.12

The two optically active forms of an amino acid.

The two amino-acid molecules can be condensed together to form a molecule known as a dipeptide; thus in the case of glycine,

$$H_2N-\underset{\underset{H}{|}}{\overset{\overset{H}{|}}{C}}-C\underset{\overset{\cdot}{O-H}}{\overset{\overset{O}{\|}}{\diagdown}} \qquad \underset{\overset{H}{H}}{\overset{H}{\diagdown}}N-\underset{\underset{H}{|}}{\overset{\overset{H}{|}}{C}}-C\underset{OH}{\overset{O}{\diagup}}$$

$$\downarrow$$

$$\underset{\overset{H}{|}}{\overset{H}{\diagdown}}N-\underset{\underset{H}{|}}{\overset{\overset{H}{|}}{C}}-\overset{\overset{O}{\|}}{C}-\underset{\underset{H}{|}}{\overset{\overset{H}{|}}{N}}-\underset{\underset{H}{|}}{\overset{\overset{H}{|}}{C}}-C\underset{OH}{\overset{O}{\diagup}} \quad + H_2O$$

The product is known as diglycine, and since it still contains an amino group and a carboxylic acid group, it can condense with additional molecules of an amino acid:

$$\underset{\overset{H}{|}}{\overset{H}{\diagdown}}N-\underset{\underset{H}{|}}{\overset{\overset{H}{|}}{C}}-C\underset{OH}{\overset{O}{\diagdown}} \quad \underset{\overset{H}{|}}{\overset{H}{\diagdown}}N-\underset{\underset{H}{|}}{\overset{\overset{H}{|}}{C}}-\overset{\overset{O}{\|}}{C}-\underset{\underset{H}{|}}{\overset{\overset{H}{|}}{N}}-\underset{\underset{H}{|}}{\overset{\overset{H}{|}}{C}}-C\underset{OH}{\overset{O}{\diagdown}} \quad \underset{\overset{H}{|}}{\overset{H}{\diagdown}}N-\underset{\underset{H}{|}}{\overset{\overset{H}{|}}{C}}-C\underset{OH}{\overset{O}{\diagdown}}$$

A very large molecule can be produced in this way, and this type of *condensation polymerization* is possible because the amino acids contain two functional groups capable of undergoing condensation reactions.

The $\overset{\overset{O}{\|}}{\underset{\underset{H}{\diagdown}}{C}-N}\diagup$ linkage that occurs between each of the amino-acid residues

in the proteins is known as a peptide linkage. The naturally occurring proteins are made up of about twenty *different* amino acids, and there may be many hundreds of amino-acid residues in each protein molecule, with the twenty different acids being used several times over.

THE SEQUENCE OF AMINO ACIDS IN PROTEINS

Living systems consists of many different proteins. Some of them are enzymes, which catalyze chemical reactions; others are structural units in skin or hair; still others are hormones, which perform a regulatory function. Each individual protein has a specific role to play in the living organism; indeed *specificity* is one of the most important characteristics of the proteins.

During recent years it has become increasingly clear that specificity is imparted not only by the number and nature of the amino acids in the protein molecule, but also by the sequence in which they occur and by the conformation of the molecule. Some protein molecules, such as those in muscle and hair, exist as a more or less linear polymer.† Others, like the molecules of hemoglobin, are wound up somewhat like a ball of wool. These different shapes are determined by the nature of the amino acids and their sequences. All of these factors have a profound bearing on biological behavior.

Various chemical techniques have been used to determine the sequence of amino acids in proteins. Much painstaking work is involved, and sequences have been established for several of the smaller protein molecules, such as insulin (mol. wt. 6000), myoglobin (mol. wt. 17 000), ribonuclease (mol. wt. 12 600), lysozyme (mol. wt. 14 000), and chymotrypsin (mol. wt. 25 000). At one time it was suspected that the amino acids in proteins are arranged in a regular and repeating pattern. However, now that the sequence has been determined in a number of cases, it has become apparent that there is no evidence for such a conclusion. It is also clear that the sequence is the same for all samples of protein obtained from a given source. Thus, all hemoglobin molecules obtained from normal human beings appear to be identical in amino-acid sequence. All hemoglobin samples from horse blood appear to be identical, but show small differences from hemoglobin obtained from humans.

PROTEIN CONFORMATIONS

The properties of a protein cannot be understood entirely on the basis of the sequence of amino acids; they also depend on the three-dimensional structure of the molecule. This three-dimensional structure, or molecular *conformation*, determines the way in which the various side groups on the molecule are brought into close proximity with one another; this has an important effect on the chemical and physical properties of the protein. In particular, if the protein is an enzyme, the relative positions of certain groups have a large effect on the enzyme's catalytic action.

Certain physical methods have been used for many years to gain a general idea of the overall shapes of protein molecules. These include measurements of the viscosity and light scattering of solutions. These properties are very different for long, thin molecules than for molecules that have a more or less spherical form. By the use of such physical methods, proteins have been grouped into two main classes: the *fibrous* proteins, in which the molecules are fairly extended, and the *globular* proteins, which are roughly spherical in shape.

More detailed information about the structures of proteins is provided by the technique of X-ray diffraction. The analysis of an X-ray pattern is difficult for a molecule as large as a protein because of the large number of interatomic distances involved. We have already referred on p. 109 to Perutz's method of isomorphous replacement, which has been of great help in the determination of protein structures.

Figure 3.13 shows schematically the kind of structure that has been found in the insulin molecule. Each of the two amino-acid chains, known as the A and B chains, occurs as a special type of helix, about which more will be said later. The B chain

---

† Actually, as we shall see, they have a spiral or helical structure.

is entirely a right-handed helix. The A chain, however, is partly right-handed and partly left-handed.

X-ray structures also have been determined for a number of larger proteins, such as myoglobin (mol. wt. 17 000), hemoglobin (mol. wt. 65 000), lysozyme (mol. wt. 14 000), and chymotrypsin (mol. wt. 25 000). In myoglobin and hemoglobin there is a considerable amount of helical structure – in myoglobin about 75% of the protein chain is in this form. In lysozyme and chymotrypsin there is considerably less helical structure. Figure 3.14 shows the conformation of the main chain in the chymotrypsin molecule, which was studied by the British X-ray crystallographer David M. Blow. Some helical structure is to be noted at the left of the diagram, at the end of the C chain.

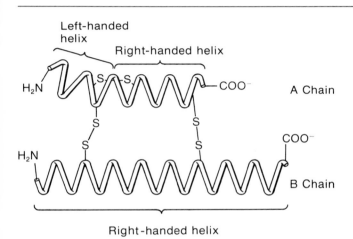

Figure 3.13

Schematic representation of the type of structure found in the insulin molecule. The amino acids are indicated by numbers.

PRINCIPLES OF PROTEIN STRUCTURE

Before any detailed X-ray study of the structure of proteins had been done, a number of important suggestions had been made about protein conformations, many of which proved to be of great value in the development of the subject. One suggestion, made by the British physical chemist Sir Eric K. Rideal (1890–1974) and the American physical chemist Irving Langmuir (1881–1957), is that a protein molecule is "an oil drop with a polar coat." We have already seen that nonpolar groups, such as alkyl groups, tend to stick together in an aqueous environment because of the formation of hydrophobic bonds. Some of the side groups on the amino acids which form the polypeptide chains are nonpolar groups, while polar groups such as —OH, —COO⁻, and —NH₃⁺, tend to form hydrogen bonds with water. The essence of Rideal and Langmuir's suggestion is that the polypeptide chains in proteins become folded in such a way that the nonpolar groups come into contact with each other as much as possible in the interior of the molecule, while polar groups are as far as possible on the exterior, where they can form hydrogen bonds with the surrounding water molecules.

Globular proteins, such as chymotrypsin and myoglobin, as a rule have a larger proportion of nonpolar groups than fibrous proteins. The fact that proteins with a larger proportion of nonpolar groups tend to take a globular form is almost certainly due to the formation of numerous hydrophobic bonds in the interior of the molecules.

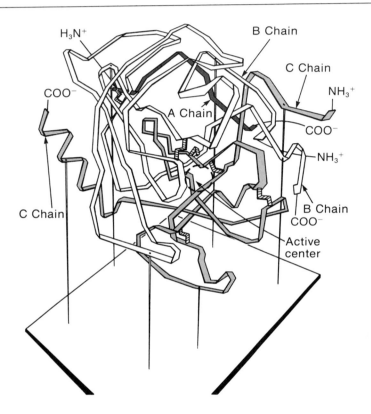

Figure 3.14

A schematic drawing showing the general conformation of the polypeptide chains in chymotrypsin. There are three polypeptide chains, A, B, and C, held together by S—S linkages, hydrogen bonds, and hydrophobic bonds. The dotted circle shows the active center—the portion of the molecule mainly concerned with the catalytic action.

On the other hand, the proteins of wool, hair, and muscle are fibrous and have a considerable amount of helical structure. In 1951 the American chemists Linus Pauling and Elias James Corey pointed out that certain helical structures formed by polypeptide chains allow a considerable amount of hydrogen bonding between the carboxyl group on one part of the chain and the amino group on another:

$$\begin{array}{c}\diagdown\\ /\end{array}C=O\cdots H-N\begin{array}{c}/\\ \diagdown\end{array}$$

Pauling and Corey considered in detail the known bond lengths and angles in the flexible chains and concluded that two different helices provide the maximum numbers of hydrogen bonds (see Figure 3.15). X-ray work has shown that one of these helices, the so-called α helix, is quite common in protein structures, particularly in the fibrous proteins. In this arrangement, each N—H group is hydrogen bonded to the

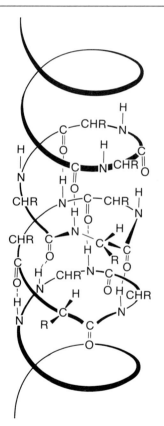

Figure 3.15

The α-helix structure of Pauling and Corey.

third C=O group beyond it along the helix, the result being that there are about 3.6 amino-acid groups per turn of the helix. Such a helix may be regarded as a spiral staircase in which the amino-acid residues form the steps; the height of each step is 0.15 nm. Just as a spiral staircase or a screw can be right-handed or left-handed, so can a protein helix. It appears that the right-handed helix occurs more commonly in protein structures. However, we have already noted with insulin (Figure 3.13) that a left-handed helix sometimes occurs.

There are three important forces at work in producing protein conformations:

1. The tendency of nonpolar groups to form hydrophobic bonds and to remain in the interior of the molecule.

2. The tendency of polar groups to remain at the exterior of the molecule so that they can form hydrogen bonds with water molecules.

3. The tendency of C=O and N—H groups to form hydrogen bonds, with the formation of a helical structure.

The actual structure of a given protein is determined in a very subtle way by the sometimes conflicting demands of these different effects, and thus depends to a considerable extent on the nature and positions of the various amino-acid side groups.

3.16   PURINES, PYRIMIDINES, AND NUCLEIC ACIDS

An important aspect of living systems is their ability to reproduce themselves. The compounds responsible for this are the nucleic acids, which tell the living system what kinds of protein to make.

In some ways the nucleic acids are similar in structure to the proteins. They are immense molecules, often having molecular weights of several million. Like the proteins, they are polymers composed of a relatively small number of units replicated many times over. In fact, the variety of constituent units is even less in a nucleic acid than in a protein. The constituent units are two closely related sugar molecules, an inorganic phosphate group, and five organic compounds containing nitrogen. The individuality of a nucleic acid results from differences in modes of combination and in variations in the order of the units in the chain.

The two sugar molecules involved in nucleic acid structure are ribose

$$
\begin{array}{ccccc}
H & H & H & H & O \\
| & | & | & | & \diagup\!\!\diagup \\
H\!-\!C\!-\!C\!-\!C\!-\!C\!-\!C & & & & \\
| & | & | & | & \diagdown \\
OH & OH & OH & OH & H
\end{array}
$$

and 2-deoxyribose

$$
\begin{array}{ccccc}
H & H & H & H & O \\
| & | & | & | & \diagup\!\!\diagup \\
H\!-\!C\!-\!C\!-\!C\!-\!C\!-\!C & & & & \\
| & | & | & | & \diagdown \\
OH & OH & OH & H & H
\end{array}
$$

The nucleic acids are of two types, depending on whether they contain ribose or 2-deoxyribose. If they contain ribose they are called ribonucleic acids (RNA). If they contain 2-deoxyribose, they are called deoxyribonucleic acids (DNA).

The organic compounds containing nitrogen, or nitrogenous bases, are of five types:

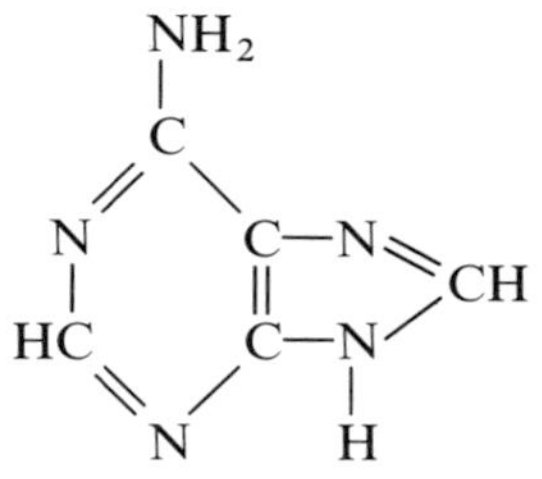

Adenine

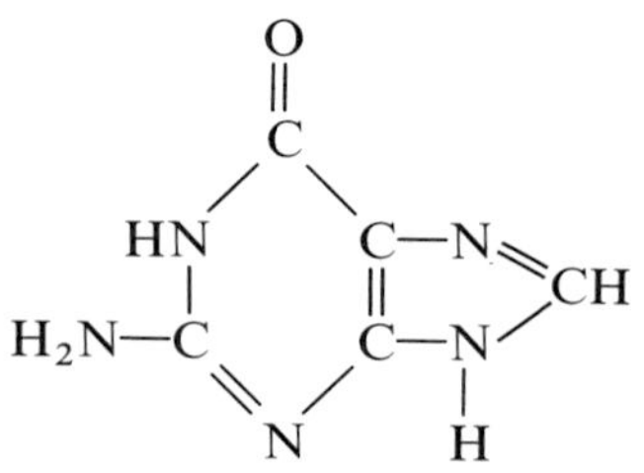

Guanine

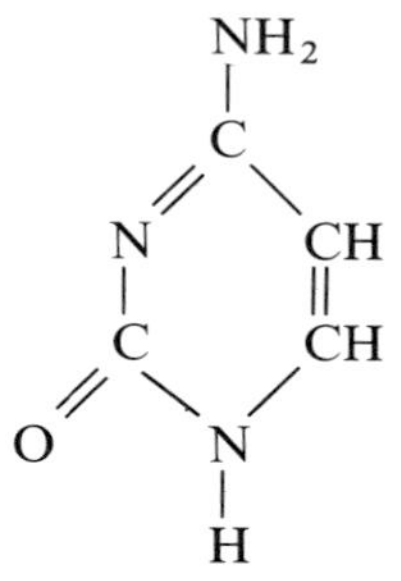

Cytosine

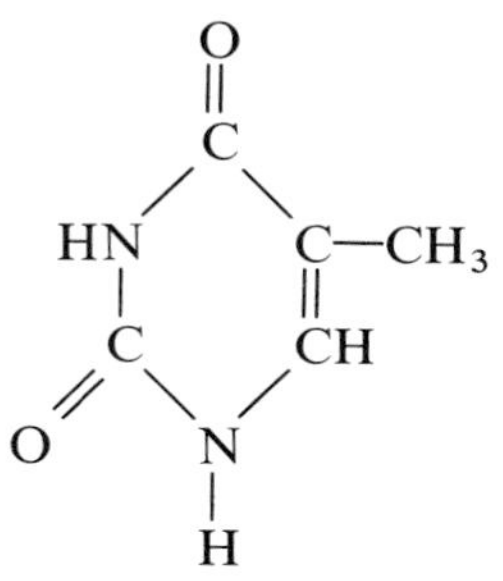

Thymine

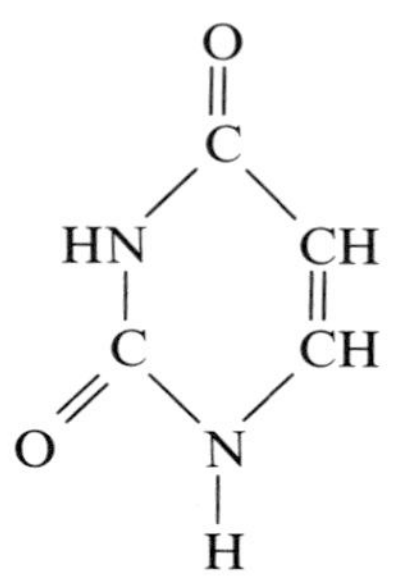

Uracil

The double-ring compounds, adenine and guanine, are known as *purines*. The single-ring compounds cytosine, thymine, and uracil are known as *pyrimidines*. They belong to the class of heterocyclic compounds, having atoms other than carbon atoms in the rings.

RNA contains ribose, phosphate, adenine, guanine, cytosine, and uracil. DNA has a very similar structure except that 2-deoxyribose replaces ribose and thymine replaces uracil. Thus adenine, guanine, and cytosine are found in combination with either sugar, but uracil is associated only with ribose (in RNA) and thymine is associated only with 2-deoxyribose (in DNA).

A molecule in which one of the organic bases is combined with a sugar is called a *nucleoside*. An example is cytosine riboside:

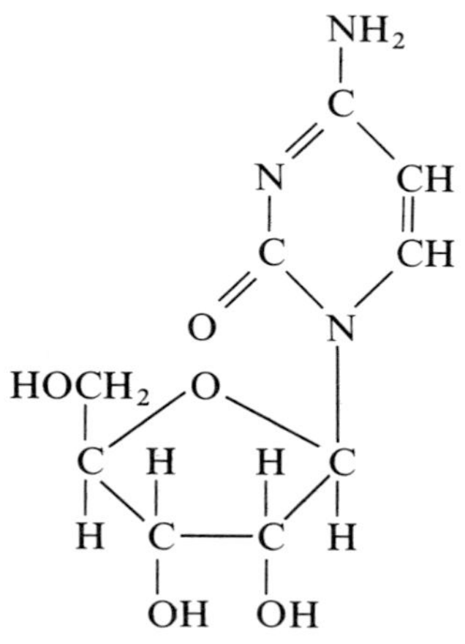

When a phosphate group is also attached, the compound is known as a *nucleotide*. An example is cytosine ribotide:

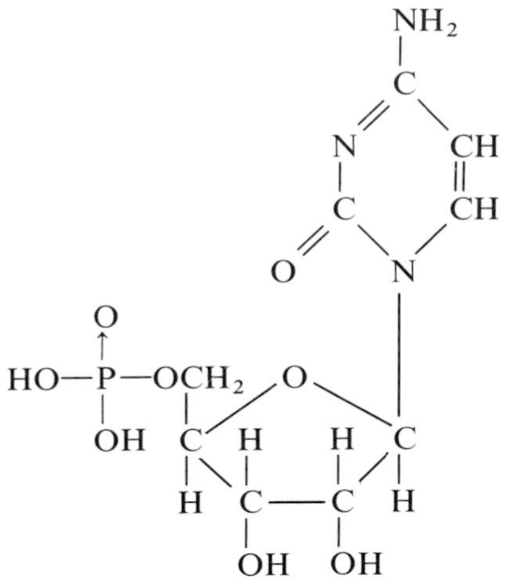

Both RNA and DNA consist of many nucleotides joined by phosphate bridges between carbon atoms three and five of the adjacent sugar molecules:

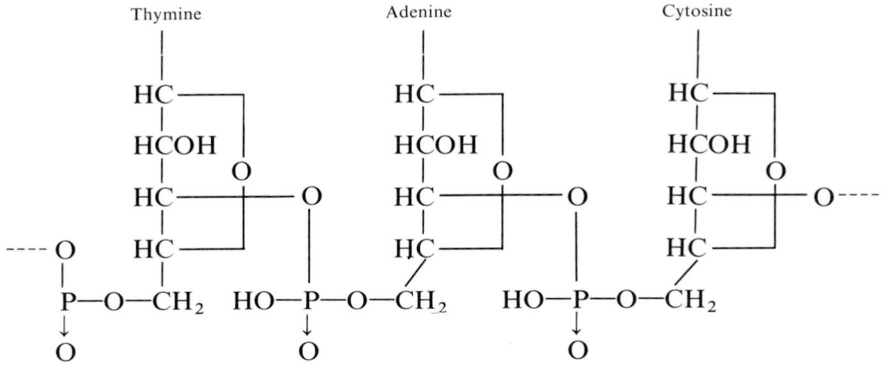

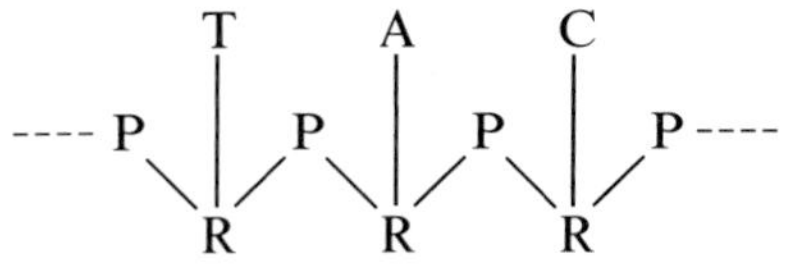

This structure is represented more conveniently by a shorthand notation such as where P represents the phosphate group, R ribose, T thymine, A adenine, and C cytosine. Nucleic acid molecules occurring in nature are composed of many nucleotides in a single extended strand.

The exact three-dimensional shape of molecules is a matter of great importance. The American molecular biologist James Dewey Watson and the British molecular biologist Francis H. C. Crick suggested in 1953 that the structure of DNA consists of two strands of nucleotides coiled about one another in helical fashion, as shown in Figure 3.16. The nitrogenous bases are in the center, facing one another, and the phosphate and sugar molecules are outside.

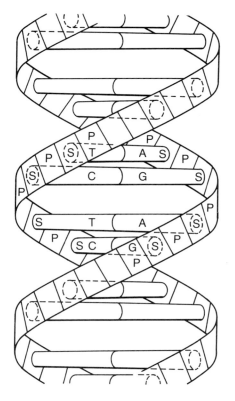

Figure 3.16

The Watson and Crick model of DNA. S = 2-deoxyribose; P = phosphate; T = thymine; A = adenine; C = cytosine; and G = guanine. This structure was based on the experimental work of a number of people, including the British physicists Maurice H. F. Wilkins and Rosalind Franklin.

3.17 TWO SUBSTANCES OF SPECIAL BIOLOGICAL IMPORTANCE

Of particular importance in biology are adenosine triphosphate (ATP) and nicotinamide adenine dinucleotide (NAD^+). The structure of ATP is

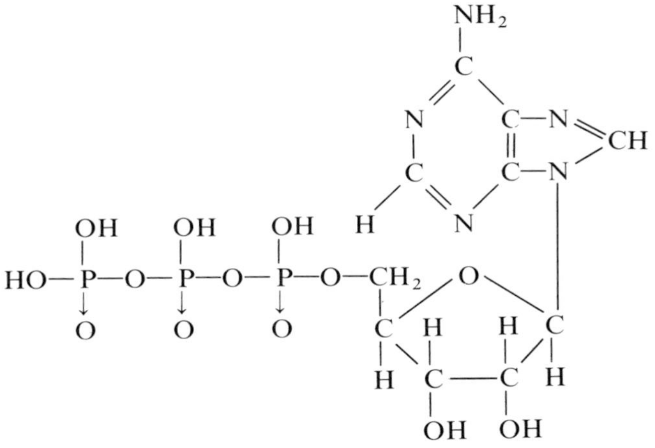

On hydrolysis, particularly under the action of the enzyme adenosine triphosphatase (ATPase), ATP is converted into adenosine diphosphate (ADP) and phosphate. This process is of great importance in metabolism (see p. 246). Under ordinary physiological conditions ATP bears four negative charges, as shown above.

The structure of NAD^+ is

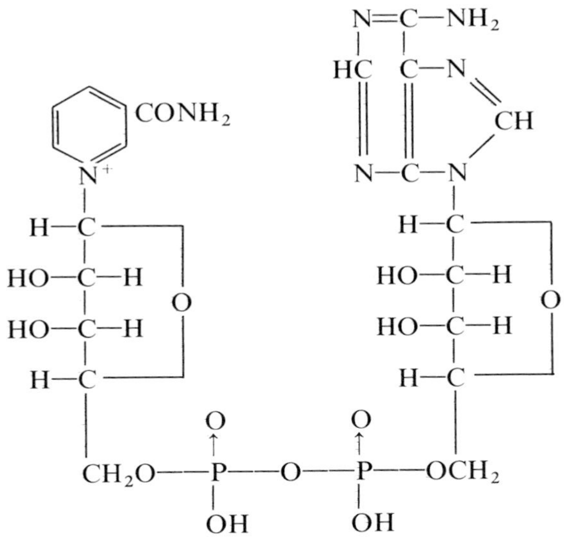

The importance of this substance is that under the action of an appropriate enzyme it can undergo reduction. If we write NAD^+ simply as the reaction with lactate ions is

$$CH_3CHOHCOO^- + NAD^+ \rightleftharpoons CH_3COCOO^- + H^+ +$$

Lactate   Pyruvate

(NADH)

The enzyme which brings about this particular interconversion is known as lactate dehydrogenase, and is normally found associated with NAD^+, which is said to be its *coenzyme*. This reaction is further discussed on p. 433.

3.18 BIOLOGICAL MEMBRANES

It was originally assumed that the composition of a cell, such as an erythrocyte (red blood cell), was more or less uniform throughout — its surface having the same character as its interior. However, a number of investigations made it clear that such cells are surrounded by a membrane which has distinctly different properties from the interior of the cell. At first, it was thought that this membrane was composed entirely of lipid material, but then it was discovered that protein is present also.

Investigations on the nature of the cell membrane have been made by the use of various techniques, including electron microscopy and X-ray diffraction. In addition, a considerable number of studies have been made of the rates of diffusion of various materials such as water, lipids, and ions through biological membranes. These studies have all led to the conclusion that the structure of membranes is to a large extent as represented in Figure 3.17. This model, which was suggested in 1952 by the British physiologists Hugh Davson and James Frederick Danielli, involves protein molecules at the inner and outer surfaces, and two layers of phospholipid within. The phospholipid molecules are arranged with the nonpolar groups held together by hydrophobic bonds; the polar phosphate groups lie near the protein molecules. In 1972 this model was modified by the American biologists S. J. Singer and G. L. Nicholson, who suggested that protein in globular form is interpolated into the lipid bilayers. Their model is illustrated in Figure 3.18.

While these models have been very valuable and must be close to the truth, they need further modification in detail to explain some of the behavior observed. For example, water molecules and ions pass through membranes much more rapidly than would be possible if the structure were exactly as represented in these models. It is necessary to assume in addition that membranes contain water-filled pores along which

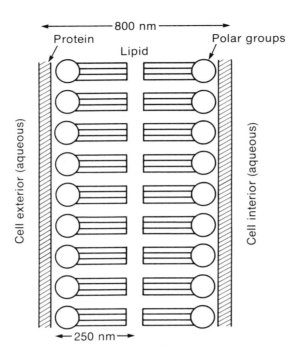

Figure 3.17

The cell membrane model proposed by Davson and Danielli.

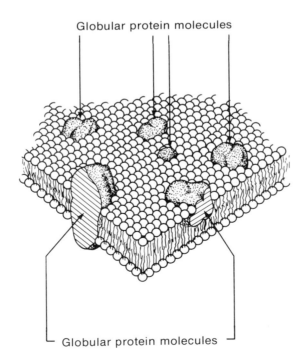

Figure 3.18

The cell membrane model of Singer and Nicholson.

water molecules and ions can travel. Also, as will be mentioned in Chapter 11, some of the diffusion results indicate that certain membranes contain *carrier molecules*, which can transport molecules from one surface to the other.

PROBLEMS

3.1 The amino acid serine contains 13.33 percent of nitrogen (at. wt. 14.01), and the osmotic pressure method gives an approximate molecular weight of 100. Obtain a more precise value for the molecular weight.

3.2 The following are the percentages of iron and sulfur (other than disulfide sulfur) found in pig hemoglobin:

Fe (at. wt. 55.85): 0.40%
S (at. wt. 32.06): 0.48%

What is the minimum molecular weight consistent with these two values?

3.3 The following results have been obtained with ox hemoglobin:

Fe (at. wt. 55.85): 0.336%
S (at. wt. 32.06): 0.48%
Arginine content (mol. wt. 174.20): 4.24%

What is the minimum molecular weight consistent with these three values?

3.4 A sample of gelatin has a maximum acid-combining capacity of 9.6×10^{-4} equivalents per gram. Calculate the minimum molecular weight.

3.5 Serum albumin has been reported to possess maximal acid- and base-combining capacities of 72 and 70×10^{-5} moles per gram respectively. The molecular weight obtained by other methods is 67 100. Calculate values for the approximate numbers of acidic and basic groups per protein molecule.

3.6 A polydisperse protein consists of 10% of protein of mol. wt. 10 000, 80% of protein of mol. wt. 20 000 and 10% of protein of mol. wt. 40 000. Calculate the number-average and weight-average molecular weights.

3.7 The composition of a protein corresponds to 5 moles of molecular weight 30 000 and 10 moles of molecular weight 60 000. Calculate M_n and M_w.

3.8 A sample of blood has a freezing point of $-0.560°$ C. What is the effective molal concentration of the solutes present in the sample, if $K_f = 1.86°$ C for water?

3.9 The fluid in certain protoplasts has an osmotic pressure of 5 atm at 37° C. What must be the molar concentration of an aqueous sucrose solution which has the same osmotic pressure as (i.e., is *iso-osmotic* with) the fluid of these cells at 37° C?

3.10 The osmotic pressure of a solution of poly-L-proline in water, containing 11.12 g dm^{-3}, was found to be 9.22 cm H$_2$O at 30° C. Calculate the molecular weight of that sample of polymer. (Density of $Hg = 13.59$ g cm^{-3}; 76 cm $Hg = 1$ atm; $R = 0.0820$ atm dm^3 mol^{-1} K^{-1}.)

3.11 A solution containing 0.5 g of urea per 100 cm^3 exerted an osmotic pressure of 2.037 atm at 25°C. Calculate the molecular weight of urea.

3.12 A solution of serum albumin at concentration of 7.8 g dm^{-3} gave an osmotic pressure of 2.39 cm H_2O at 0°C. Calculate the molecular weight.

3.13 An optical rotation of $+9.49°$ was observed with 1 M solution of D-glucose (mol. wt. 180) in a 10 cm cell at 20° with sodium-D light. Calculate the specific rotation, $[\alpha]_D^{20°}$, for glucose.

3.14 Which of the following compounds occur in enantiomeric forms?

(a) 1-chlorobutane
(b) 2-chlorobutane
(c) 1,2-dichlorobutane
(d) glycine, H_2N CH_2 $COOH$
(e) alanine, H_2N $CH(CH_3)COOH$
(f) 3-phosphoglyceric acid,

$$
\begin{array}{c}
COOH \\
| \\
HCOH \quad\quad O \\
| \quad\quad\quad \uparrow \\
CH_2-O-P-H \\
| \\
OH
\end{array}
$$

ESSAY QUESTIONS

3.15 Explain what is meant by optical rotation and optical rotatory dispersion.

3.16 Explain clearly why water has a higher boiling point than hydrogen sulfide, methane, and hydrogen fluoride.

3.17 What predictions would you make as to the relative solubilities in water of the following pairs of compounds? Give reasons in each case.

(a) Ethanol, C_2H_5OH, and dimethyl ether, CH_3OCH_3,
(b) Glycine, H_2NCH_2COOH, and phenylalanine,

$$
\begin{array}{c}
\text{(phenyl ring)} \\
| \\
CH_2 \\
| \\
H_2NCH_2COOH
\end{array}
$$

(c) 3-Phosphoglyceric acid (see question 3.14) and tristearin

$$
\begin{array}{c}
H_2COCOC_{17}H_{35} \\
/ \\
HCOCOC_{17}H_{35} \\
\backslash \\
H_2COCOC_{17}H_{35}
\end{array}
$$

3.18 Give an account of the more important factors that determine whether a protein will be in a fibrous or globular form.

3.19 Describe briefly three physical methods for determining the molecular weight of a soluble protein, and comment on their relative merits.

3.20 Suggest a method for determining the degree of hydration of a protein in aqueous solution.

3.21 What information can be derived from light-scattering experiments on macromolecules in aqueous solution?

3.22 Give an account of the colligative properties, and comment on their relative merits for the determination of molecular weights of proteins.

SUGGESTED READING

Barrow, G. M. *The Structure of Molecules.* New York: Benjamin, 1963.

Bull, H. B. *An Introduction to Physical Biochemistry*, 2nd ed. Philadelphia: F. A. Davis, 1971.

Dawes, E. A. *Quantitative Problems in Biochemistry.* Edinburgh and London: Churchill Livingstone, 1972.

Dickerson, R. E. and Geis, I. *The Structure and Action of Proteins.* New York: Harper & Row, 1969.

Djerassi, C. *Optical Rotatory Dispersion.* New York: McGraw-Hill, 1960.

Flory, P. J. *Principles of Polymer Chemistry.* Ithaca, N.Y.: Cornell University Press, 1953.

Leach, S. J., ed. *Physical Principles and Techniques of Protein Chemistry.* New York: Academic Press, 1969.

Phillips, D. C. "The Three-Dimensional Structure of an Enzyme Molecule." *Scientific American*, November 1966.

Stern, E. A. "The Analysis of Materials by X-Ray Absorption." *Scientific American*, April 1976.

Tanford, C. *The Physical Chemistry of Macromolecules.* New York: John Wiley and Sons, 1961.

Van Holde, K. E. *Physical Biochemistry.* Englewood Cliffs, New Jersey: Prentice-Hall, 1971.

Wentworth, W. E. and Ladner, S. J. *Fundamentals of Physical Chemistry.* Belmont, California: Wadsworth Publishing Co., 1972.

Wheatley, P. J. *The Determination of Molecular Structure.* Fair Lawn, N.J.: Oxford University Press, 1959.

Williams, V. R. and Williams, H. B. *Basic Physical Chemistry for the Life Sciences*, 2nd ed. San Francisco: W. H. Freeman, 1973.

4

THE FIRST LAW
OF
THERMODYNAMICS

Up to now we have been concerned with the structures of molecules and with properties such as light absorption and diffusion, which relate to individual molecules. These are known as *microscopic* properties. More familiar to us in everyday life are *macroscopic* properties, which relate not to individual molecules but to large assemblies of molecules; examples are pressure, volume, and temperature. The general relationships which exist between macroscopic properties are dealt with by a branch of science known as *thermodynamics*.

Strictly speaking, thermodynamics pays no attention to the existence of atoms and molecules. Indeed, some texts on the subject make no reference to them at all. However, many of us find it helpful to clarify our understanding of thermodynamics by using our knowledge of the behavior of molecules. In the branch of science known as *statistical thermodynamics*, the knowledge of molecules and the formal principles of thermodynamics are blended together; this subject has made important contributions to our understanding of the behavior of matter.

At first sight it might be thought that a study of thermodynamics without a consideration of microscopic behavior would not be very fruitful. However, it has been possible to develop some very far-reaching conclusions on the basis of purely thermodynamic arguments. These conclusions are all the more convincing because they do not depend on the truth or falsity of any particular theory of atomic or molecular structure. Pure thermodynamics starts with a small number of assumptions which are based on well-established experimental results and makes logical deductions from them, finally arriving at a set of relationships which must be true if the original premises are true.

Thermodynamics, then, is concerned with the relationships between the various macroscopic properties. One of the most important conclusions from thermodynamics is that there are two properties which are particularly important in explaining the behavior of matter. These properties are *energy* and *entropy*. The first law of thermodynamics is conveniently summarized in the statement:

The energy of the universe remains constant.

The second law of thermodynamics, which will be dealt with in the following chapter, can be summarized as:

The entropy of the universe is constantly increasing.

Pursued to their logical conclusions, these statements provide us with an understanding of the behavior of chemical systems at equilibrium and of the circumstances under which chemical change will occur. For example, they allow us to calculate the equilibrium constants of reactions that have never been carried out and show how these equilibrium constants vary with the temperature.

It may be well to emphasize at the outset that all available evidence indicates that the laws of thermodynamics apply to biological systems. There has occasionally been a misunderstanding on this point. Earlier workers in biology thought that living systems operated through a so-called "vital force," and that the laws of nature that had been discovered for nonliving systems did not necessarily apply to living systems. Even the great American physical chemist Gilbert Newton Lewis (1875–1946) once accused living systems of being "cheats in the game of entropy." However, the careful thermodynamic studies that have more recently been made have shown that biological systems do comply with the laws of thermodynamics—and indeed to the other general laws of nature. The very great complexity of living systems often makes it hard for us to understand how the systems work.

It is very important for the biologist to have a sound appreciation of thermodynamics, because thermodynamics makes such a profound contribution to the understanding of living systems. Unfortunately, all of us find thermodynamics a difficult subject when we study it for the first time. There is an amusing anecdote told of the distinguished German physicist Arnold Sommerfeld (1868–1951), who wrote lucid books on every subject in physics except thermodynamics. When asked why he did not write on that field he replied somewhat as follows:

Thermodynamics is a funny subject. The first time you go through the subject, you don't understand it at all. The second time you go through it, you think you understand it—except for one or two small points. The third time you go through it, you know *you don't understand it, but by that time you are so used to the subject that it doesn't bother you any more.*

Students who initially find the subject difficult should be heartened by the knowledge that this happens to all of us, even to brilliant people like Sommerfeld. If the student of biology will persist with his study of thermodynamics until he is "so used to the subject that it doesn't bother him any more" he will find that he is amply rewarded by acquiring a much deeper understanding of the functioning of living systems.

4.1   THERMODYNAMIC SYSTEMS

In thermodynamics, it is important to distinguish between a *system* and its *surroundings*. In experimental work we are accustomed to confining materials in order to make measurements on them. For example, a solution of chemicals may be confined in a sealed glass vessel. In this case there is a clear boundary between the solution, which is the *system*, and the vessel, which is part of the *surroundings*. Heat, but not matter, can pass between the system and the surroundings, and the system is said to be a *closed* system. Another example of a system is a muscle which has been removed from an animal and which is being investigated while being bathed in a solution. Since matter can now pass between the muscle and the surrounding solution we speak of an *open* system. In both closed and open systems there can be transfer of heat between the system and the environment. If a system is surrounded by an insulating container so that there can be transfer of neither matter nor heat, the system is said to be *isolated* and the processes which occur are described as *adiabatic*.

The distinction between the system and the surroundings is very important, since we are constantly concerned with transfer of heat between the system and the surroundings. We are also concerned with the work done by the system on its surroundings or by the surroundings on the system. In all cases the system must be carefully defined.

4.2   STATES AND STATE FUNCTIONS

Certain of the macroscopic properties have fixed values for a particular *state* of the system, others do not. For example, suppose that we maintain 1 gram of water in a vessel at $0°$ C and 1 atmosphere pressure; it will have a volume of 1 cm^3. These quantities specify the *state* of the system. Any time we satisfy these conditions we have the water in the same state and as long as it is in that state it will have these particular specifications. The macroscopic properties which we have mentioned—mass, pressure, temperature, and volume—are known as *state functions* or *state variables*.

One important characteristic of a state function is that once we have specified the state of a system by giving the values of *some* of the state functions, the values of all other state functions are fixed. Thus, in the example just given, once we have specified the mass, temperature, and pressure of the water, the volume is fixed. So, too, is the total energy of the system, and energy is therefore another state function. When the concept of entropy is discussed in the next chapter we shall see that entropy is also a state function.

Another important characteristic of a state function is that when the state of a system is altered, the change in any state function depends only on the initial and final states of the system, and not on the path followed in making the change. For example, if we heat our sample of water from $0°$ C to $25°$ C, the change in temperature is equal to the difference between the initial and final temperatures:

$$\Delta T = T_{\text{final}} - T_{\text{initial}} = 25° \text{ C} \qquad (4.1)$$

The way in which the temperature change is brought about has no effect on this result.

This example may seem trivial, but it should be emphasized that not all functions have this characteristic. For example, the amount of heat supplied to a system in order to change it from one state to another varies with the way in which the change is brought about. Similarly, the work done by a system when it passes from one state to another is not a fixed quantity, but depends upon the path. This will be discussed in greater detail later.

It is useful at this point to consider an analogy. Suppose that there is a point A on the earth's surface which is 1000 feet above sea level, and another point B which is 4000 feet above sea level. The difference, 3000 feet, is clearly the height of B with respect to A. In other words, the difference in height Δh can be expressed as

$$\Delta h = h_B - h_A \qquad (4.2)$$

where h_A and h_B are the heights of A and B above sea level. Height above sea level is thus a state function, the difference Δh being in no way dependent on the path chosen. However the distance that one has to travel in order to get from A to B *is* dependent on the path; one can go by the shortest route, or take a longer route. Therefore, distance traveled is not a state function.

4.3 EQUILIBRIUM STATES AND REVERSIBILITY

Thermodynamics is only concerned with *equilibrium states*, in which the state functions have constant values throughout the system. It provides us with information about the circumstances under which nonequilibrium states will move towards equilibrium, but it tells us nothing directly about the nonequilibrium states.

Suppose, for example, that we have a gas confined in a cylinder that has a frictionless movable piston (Figure 4.1). If the piston is motionless, the state of the gas can be specified by giving the values of pressure, volume, and temperature. However, if the gas is compressed very rapidly it passes through states in which pressure and temperature cannot be specified, since these properties vary throughout the gas; the gas near the piston is compressed and heated and the gas at the far end of the cylinder is not. The gas then would be said to be in a nonequilibrium state, and pure thermodynamics could not deal with such a state, although it could tell us what kind of a change would spontaneously occur in order for equilibrium to be attained.

The criteria for equilibrium are very important, and may be summarized as follows:

1. The mechanical properties must be uniform throughout the system and constant in time.

2. The chemical composition of the system must be uniform, with no net chemical change taking place.

3. The temperature of the system must be uniform and must be the same as the temperature of the surroundings.

The first of these criteria means that the force acting on the system must be exactly balanced by the force exerted by the system; otherwise the volume will change. If, for example, we consider the system illustrated in Figure 4.1, we see that for the system

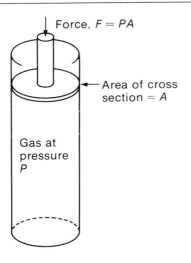

Force, $F = PA$

Area of cross section $= A$

Gas at pressure P

Figure 4.1

A gas at pressure P maintained at equilibrium by an external force F, equal to PA, where A is the area of cross section of the piston.

to be at equilibrium the force F exerted on the piston must exactly balance the pressure P of the gas; if A is the area of the piston,

$$PA = F \tag{4.3}$$

If we increase the force, the gas will be compressed; if we decrease it, the gas will expand.

Suppose that we increase the force F by an *infinitesimal* amount dF. The pressure we exert on the gas will now be infinitesimally greater than the pressure of the gas; i.e., it will be $P + dP$. The gas will therefore be compressed infinitely slowly. We can make dP as small as we like, and at all stages during the infinitely slow compression we therefore maintain the gas in a state of equilibrium. We refer to a process of this kind as a *reversible* process. If at any point we make the pressure $P - dP$, the gas will expand infinitely slowly—that is, reversibly. Reversible processes play important roles in thermodynamic arguments. However, all processes that occur naturally are irreversible.

4.4   WORK AND HEAT: THE CONSERVATION OF ENERGY

We come now to the first law of thermodynamics, according to which the total amount of energy in the universe is conserved. In the study of classical mechanics we are accustomed to a particular application of the law of conservation of energy to systems maintained at constant temperature. If, for example, a body is allowed to fall freely, it loses potential energy and gains a corresponding amount of kinetic energy. If we perform work on a body it gains an amount of energy which, if there are no frictional losses producing wasted heat, is exactly equivalent to the amount of work that has been performed.

In thermodynamics we are concerned with a more extended application of the principle of conservation of energy, since we must allow for temperature changes. Such temperature changes are taken into account by introducing a quantity known as the internal energy, and denoted by the symbol U. If we add heat q to a system, and no other change occurs, the internal energy increases by an amount which is exactly equal to the heat supplied:

$$\Delta U = q \tag{4.4}$$

If an amount of work w is performed on the system,[†] and no heat is transferred, the system gains energy by an amount equal to the work done:

$$\Delta U = w \tag{4.5}$$

In general, if heat q is supplied to the system and at the same time an amount of work w is done on the system, the increase in internal energy is given by

$$\Delta U = q + w \tag{4.6}$$

This states the first law of thermodynamics. In applying this equation it is necessary to employ the same units for ΔU, q and w. The two sets of units most commonly employed are *calories* and *joules*. The joule is the SI unit, but since most thermodynamic data are given in calories, we shall make use of calories to a considerable extent. The two sets are connected by the relationship

$$1 \text{ calorie (cal)} = 4.184 \text{ joules (J)} \quad \text{(by definition)}$$
$$1 \text{ kcal} = 4.184 \text{ kJ}$$

We should note that equation (4.6) leaves the energy U indefinite, since we are dealing only with the energy change, ΔU. In practice the energy is defined arbitrarily by assigning it a value of zero for some completely defined state of the system.

The internal energy U is a state function of the system. Consider, for example, the problem of raising the temperature of 1 dm^3 of water, at atmospheric pressure, from 25° C to 26° C; this clearly changes it from one well-defined state to another. This change of state can be brought about in a variety of ways, the simplest being to add heat until the temperature is 26° C. Alternatively, we could stir the water vigorously with a paddle until the desired temperature increase had been achieved; this amounts

---

[†] The SI recommendation is to use the symbol w for the work done *on* the system. The reader is warned that many treatments use the symbol w for the work done *by* the system.

to performing work on the system. We could also bring about a temperature increase by a combination of adding heat q and performing work w. However, *it is found experimentally* that however we bring about the temperature increase, *the sum $q + w$ is always the same*. In other words, for the change in state from 25° C to 26° C the quantity ΔU, equal to $q + w$, is independent of the way in which the change is brought about. This argument proves that the internal energy U is a state function. It also demonstrates that heat, q, and work, w, *are not state functions* since the change can be brought about by various amounts of heat and work; only the sum $q + w$ is fixed.

If U were not a state function we could have violations of the principle of conservation of energy. Consider two states A and B, and suppose that there are two alternative paths, for one of which is ΔU is ten calories, and the other thirty calories:

$$\Delta U_1 = 10 \text{ cal}$$
$$\Delta U_2 = 30 \text{ cal}$$

We could then go from A to B and expend ten calories of heat by going by the first path. If we returned from B to A by the second path we would gain thirty calories. Thus, we would have arrived at the system in its original state with a net gain of twenty calories. Energy would therefore have been created for nothing. Many attempts have been made to do this by the construction of "perpetual-motion machines of the first kind," but all have ended in failure. The inability to make perpetual-motion machines provides convincing evidence that energy cannot be created or destroyed, and that therefore internal energy must be a state function.

In purely thermodynamic studies it is not necessary to consider what internal energy really consists of. However, most of us like to understand internal energy in terms of molecular energies. There are contributions to the internal energy of a substance from

1. The kinetic energy of motion of the individual molecules.

2. The potential energy that arises from interactions between molecules.

3. The kinetic and potential energy of the nuclei and electrons within the individual molecules.

A precise treatment of these factors is very difficult, and it is a great strength of thermodynamics that we can make use of the concept of internal energy without having to deal with it on a detailed molecular basis.

WORK

The question of the *work done* on chemical systems must also be considered. There are various ways in which a system may do work, or by which work may be done on a system. For example, if we pass a current through a solution and electrolyze it we perform one form of work—*electrical* work. Conversely, an electrochemical cell performs work, as will be considered in detail in Chapter 8. Living systems also perform work and this work can be classified into three basic types: chemical work, osmotic work, and mechanical work. *Chemical work* is done by all biological cells, not only during active growth, but also in maintaining themselves. The biosynthesis of large molecules requires an input of energy, and even when growth is not occurring these large molecules are being broken down and must be regenerated. *Osmotic work* is the work required to transport and concentrate chemical substances. It occurs, for example, in the formation of the gastric juice (p. 175), where the acid concentration is much higher than that of the surroundings. *Mechanical work* occurs in higher animals when skeletal muscle contracts—indeed all cells exert intracellular forces by means of contractile filaments.

The simplest way in which work is done is when an external force brings about a compression of a system. Suppose, for example, that we have an arrangement in which a gas or liquid is maintained at *constant* pressure P, which it exerts against a movable piston (Figure 4.2). In order for the system to be at equilibrium we must apply a force

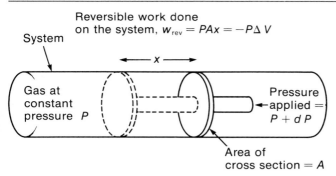

Reversible work done on the system, $w_{rev} = PAx = -P\Delta V$

System

x

Gas at constant pressure P

Pressure applied $= P + dP$

Area of cross section $= A$

Volume decrease, $-\Delta V = Ax$

Figure 4.2

The reversible work done by a constant pressure P moving a piston.

F to the piston, the force being related to the pressure by the relationship

$$F = PA \tag{4.7}$$

where A is the area of the piston. Suppose now that the force is increased by an infinitesimal amount dF, so that the piston moves infinitely slowly, the process being reversible. If the piston moves a distance x the reversible work w_{rev} done on the system is

$$w_{rev} = Fx \tag{4.8}$$

$$= PAx \tag{4.9}$$

However, Ax is the volume swept out by the movement of the piston, i.e., the decrease $-\Delta V$ in volume of the gas. The work done on the system is thus

$$w_{rev} = -P\Delta V \tag{4.10}$$

If, on the other hand, the pressure P varies during the compression, we must calculate the work done by a process of integration. The work done on the system while a pressure P moves the piston so that it sweeps out an infinitesimal volume dV is

$$dw_{rev} = -PdV \tag{4.11}$$

Note that this is a negative quantity when there is an increase in volume, the gas doing a positive amount of work on the surroundings. If, as illustrated in Figure 4.3, the volume changes from V_1 to V_2 the work done is

$$w_{rev} = -\int_{V_1}^{V_2} PdV \tag{4.12}$$

Only if P is constant is it permissible to integrate this directly to give

$$w_{rev} = -P\int_{V_1}^{V_2} dV = -P(V_2 - V_1) = -P\Delta V \tag{4.13}$$

Compare this to equation (4.10). If P is not constant we must express it as a function of V before performing the integration. The case of the isothermal compression of an ideal gas is discussed on p. 173.

We have already noted that work done is not a state function; this is also true of the mechanical work of expansion. The derivation above has shown that the work is related to the process carried out rather than to the initial and final states. We can consider the *reversible expansion* of a gas from volume V_1 to volume V_2, and can also consider an *irreversible* process, in which case *less* work will be done *by* the system.

Reversible work done on the system, $w_{rev} = -\int_{V_1}^{V_2} PdV$

Figure 4.3

The reversible work performed when there is a volume decrease from V_1 to V_2.

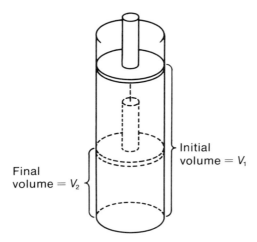

Final volume $= V_2$

Initial volume $= V_1$

This is illustrated in Figure 4.4. The diagram to the left shows the reversible expansion of a gas in which the pressure is falling as the volume increases. The reversible work done *by* the system is given by the integral

$$-w_{rev} = \int_{V_1}^{V_2} P\,dV \tag{4.14a}$$

which is represented by the shaded area in the figure. Suppose instead that we performed the process irreversibly by instantaneously dropping the external pressure to the final pressure P_2. The work done by the system is now against this pressure P_2 throughout the whole expansion, and is given by

$$-w_{irr} = P_2(V_2 - V_1) \tag{4.14b}$$

This work done by the system is represented by the shaded area in Figure 4.4b, and is *less* than the reversible work. Thus, although in both processes the state of the system has changed from A to B, the work done is different.

This argument leads us to another important point. The work done by the system in a *reversible* expansion from A to B represents the *maximum work* that the system can perform in changing from A to B.

If we now return to our expression for the first law of thermodynamics

$$\Delta U = q + w \tag{4.15}$$

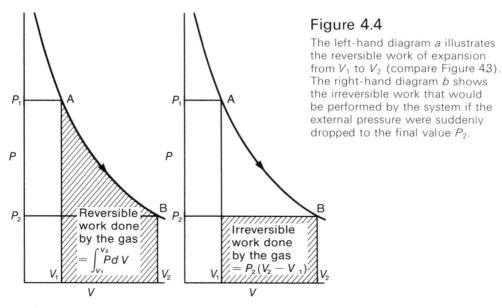

Figure 4.4

The left-hand diagram *a* illustrates the reversible work of expansion from V_1 to V_2 (compare Figure 4.3). The right-hand diagram *b* shows the irreversible work that would be performed by the system if the external pressure were suddenly dropped to the final value P_2.

a. *b.*

we see that whereas ΔU depends only on the initial and final states, w and therefore q depend upon the path by which the process takes place.

EXAMPLE

Suppose that water at its normal boiling point (100° C or 273 K) is maintained in a cylinder that has a frictionless piston. The equilibrium pressure of the vapor will be 1 atm, and an external pressure of 1 atm must therefore be exerted on the piston in order to prevent it from moving. Suppose that we now reduce the external pressure by an infinitesimal amount in order to have a reversible expansion. We allow the piston to sweep out a volume of 2 dm^3, and want to calculate the work done by the system.

SOLUTION

The external pressure remains constant at 1 atm, and therefore the reversible work done by the system is

$$-w_{rev} = P\Delta V = 1 \times 2 \text{ atm dm}^3$$

One atm dm^3† is equal to 101.3 joules or 24.22 calories. The work done by the system is thus 202.6 J = 48.44 cal.

For many purposes it is convenient to express the first law of thermodynamics with respect to an infinitesimal change. In place of equation (4.6) we then have

$$dU = dq + dw \tag{4.16}$$

However, if only PV work is involved and the volume change is reversible, dw may be written as $-PdV$, where dV is the infinitesimal increase in volume. Thus

$$dU = dq - PdV \tag{4.17}$$

PROCESSES AT CONSTANT VOLUME

It follows at once from this equation that if an infinitesimal process occurs at constant volume, and only PV work is involved,

$$dU = dq_V \tag{4.18}$$

---

† The unit atm dm^3, or liter-atmosphere, is not an SI unit but it is very useful for calculations of this kind. 1 atm = 101.3 kPa and 1 dm^3 (liter) = 10^{-3} m^3; thus, 1 atm dm^3 = 101.3 Pa m^3. Since 1 pascal (Pa) = 1 kg m^{-1} s^{-2}, 1 atm dm^3 = 101.3 kg m^2 s^{-2} = 101.3 J (see also Appendix A).

where the subscript V indicates that the heat is supplied at constant volume. This equation integrates to

$$\Delta U = U_2 - U_1 = q_V \tag{4.19}$$

We thus obtain the very important result that in a process occurring at constant volume the increase of internal energy of a system is equal to the heat q_V that is supplied to it.

PROCESSES AT CONSTANT PRESSURE: ENTHALPY

In most chemical systems, including those of interest in biology, we are concerned with processes occurring at constant pressure rather than at constant volume. For example, all of the metabolic and digestive reactions occurring in the body occur at one atmosphere pressure and involve small changes in volume.

The relationships valid for constant-pressure processes may readily be deduced from equation (4.17). For an infinitesimal process at constant pressure the heat absorbed dq_P is given by

$$dq_P = dU + PdV \tag{4.20}$$

provided that no work other than PV work is performed. If the process involves a change from state 1 to state 2 this equation integrates as follows:

$$q_P = \int_{U_1}^{U_2} dU + \int_{V_1}^{V_2} PdV \tag{4.21}$$

Since P is constant,

$$q_P = \int_{U_1}^{U_2} dU + P \int_{V_1}^{V_2} dV \tag{4.22}$$

$$= (U_2 - U_1) + P(V_2 - V_1) \tag{4.23}$$

$$= (U_2 + PV_2) - (U_1 + PV_1) \tag{4.24}$$

This relationship suggests that it would be convenient to give a name to the quantity $E + PV$, which is known as the *enthalpy* (Greek *en*, in; *thalpos*, heat) and is given the symbol H:

$$H = U + PV \tag{4.25}$$

We thus have

$$q_P = H_2 - H_1 = \Delta H \tag{4.26}$$

It is important to recognize that this equation is valid *only if there is no non-PV work*.

This equation, which is analogous to equation (4.19), tells us that the increase in enthalpy ΔH of a system is equal to the heat q_P that is supplied to it at constant pressure. Since U, P, and V are all state functions it follows from equation (4.25) *that enthalpy is also a state function.*

Because q_P and ΔH are equal when there is no non-PV work, enthalpy is sometimes known as the *heat content* of the system. However, this term might cause confusion; enthalpy and heat are equivalent only for processes at constant pressure.

A chemical process occurring at constant pressure for which q_P and ΔH are *positive* is one in which a positive amount of heat is *absorbed* by the system. Such processes are known as *endothermic* processes (Greek *endo*, inside; *therme*, heat). Conversely, processes in which heat is *evolved* (q_P and ΔH are *negative*) are known as *exothermic* processes (Greek *exo*, outside).

In order to find a relationship between ΔH and ΔU, we differentiate equation (4.25), which defines enthalpy:

$$dH = dU + d(PV) \tag{4.27}$$

For a change from state 1 to state 2,

$$\Delta H = \Delta U + \Delta(PV) \tag{4.28}$$
$$= (U_2 - U_1) + P_2 V_2 - P_1 V_1 \tag{4.29}$$

Thus, $\Delta H (= H_2 - H_1)$ and $\Delta U (= U_2 - U_1)$ differ only by the difference in the PV products of the final and initial states. For chemical reactions at constant pressure in which only *solids* and *liquids* are involved there is very little change of volume, and therefore PV changes little during the process; ΔH and ΔU are thus nearly equal. For gas reactions involving a change in the total number of molecules, on the other hand, there is an appreciable change in PV, and ΔH and ΔU therefore differ.

4.5   THERMOCHEMISTRY

We have seen that the heat supplied to a system at constant pressure is equal to the enthalpy increase. For many chemical reactions it is possible to make a direct determination of the heat change at constant pressure; for other reactions indirect methods (to be discussed later) can be used. In these ways it has proved possible to build a considerable body of data on the enthalpy changes of chemical processes of all kinds and these data have been of great value in an understanding of molecular structure and in various other ways. The study of enthalpy changes in chemical systems is known as *thermochemistry*.

STANDARD STATES

The enthalpy change that occurs in a chemical process depends upon the states of the reactants and products. Consider, for example, the complete combustion of ethanol, in which 1 mol is oxidized to carbon dioxide and water:

$$C_2H_5OH + 3O_2 \rightarrow 2CO_2 + 3H_2O$$

The enthalpy change in this reaction depends on whether we start with liquid ethanol or with ethanol in the vapor phase and on whether liquid or gaseous water is produced in the reaction. Another factor is the pressure of the reactants and products. Also, the enthalpy change in a reaction varies with the temperature at which the process occurs. In giving a value for an enthalpy change it is therefore necessary to specify (a) the state of matter of the reactants and products (gaseous, liquid, or solid—if the latter, the allotropic† form), (b) the pressure, and (c) the temperature. If the reaction occurs in solution the concentrations also must be specified.

It has proved convenient in thermodynamic work to define certain *standard states* and to quote data for reactions involving these standard states. By general agreement the standard state of a substance is the state in which it is most stable at 25.00° C (298.15 K) and 1 atmosphere (760 mm or 101.325 kilopascals, kPa) pressure. For example, the standard state of oxygen is the gas, and we specify this by writing O_2 (g). Since mercury, water, and ethanol are liquids at 25° C their standard states are Hg (l), H_2O (l) and C_2H_5OH (l). The standard state of carbon is graphite. These standard states should be specified if there is any ambiguity. For example,

$$C_2H_5OH \text{ (l)} + 3O_2 \text{ (g, 1 atm)} \rightarrow 2CO_2 \text{ (g, 1 atm)} + 3H_2O \text{ (l)}.$$

It is quite legitimate, of course, to consider an enthalpy change for a process not involving standard states. For example,

$$C_2H_5OH \text{ (g, 1 atm)} + 3O_2 \text{ (g, 1 atm)} \rightarrow 2CO_2 \text{ (g, 1 atm)} + 3H_2O \text{ (g, 1 atm)}$$

If a reaction involves species in solution their standard state is 1 molal (1 m), which is one mole per kilogram of solvent (1 mol kg^{-1}). For example,

$$H^+ \text{ (1 } m\text{)} + OH^- \text{ (1 } m\text{)} \rightarrow H_2O \text{ (l)}$$

Enthalpy changes depend somewhat on the temperature at which the process occurs. Standard thermodynamic data are commonly quoted for a temperature of 25.00° C (298.15 K), and this can be given as a subscript or in parentheses. Thus,

$$C_2H_5OH \text{ (l)} + 3O_2 \text{ (g)} \rightarrow 2CO_2 \text{ (g)} + 3H_2O \text{ (l)} \quad \Delta H_c^o \text{ (298)} = -324.5 \text{ kcal}$$

The superscript o on the ΔH specifies that we are dealing with standard states, so that a pressure of 1 atm is assumed and need not be stated. The subscript c is commonly used to indicate complete combustion. In giving thermodynamic values for reactions

---

† Some elements and compounds occur in more than one form. Each form is then known as an *allotrope*; the phenomenon is called *allotropy*. For example, diamond and graphite are allotropes of carbon.

it is better to give them as kcal (or kJ) rather than as kcal mol^{-1} or kJ mol^{-1}. In the above example there is one mole of C_2H_5OH, but more moles of the other reactants and products. In some cases, writing kcal mol^{-1} leads to ambiguity. The value quoted should refer to the reaction *as written*; thus if we wrote a reaction as

$$2A \rightleftharpoons 3B$$

the quoted value would refer to the conversion of two moles of A into three moles of B.

It is quite legitimate to quote standard thermodynamic values for a temperature other than 25° C. For example we could give a value for $\Delta H°$ (100° C), and it would be understood that the pressure was again 1 atm and that reactants and products were in their standard states.

MEASUREMENT OF ENTHALPY CHANGES

The enthalpy changes occurring in chemical processes may be measured by three main methods:

1. *Direct Calorimetry.* Some reactions occur to completion without side reactions, and it is therefore possible to measure their $\Delta H°$ values by causing the reactions to occur in a calorimeter. The neutralization of an aqueous solution of a strong acid by a solution of a strong base is an example of such a process, the reaction which occurs being

$$H^+ (aq) + OH^- (aq) \rightarrow H_2O$$

Combustion processes also frequently occur to completion in a simple manner. When an organic compound is burned in excess of oxygen the carbon is practically all converted into CO_2 and the hydrogen into H_2O, while the nitrogen is present as N_2 in the final products. Usually such combustions of organic compounds occur cleanly, and much thermochemical information has been obtained by burning organic compounds in calorimeters.

2. *Indirect Calorimetry: Use of Hess's Law.* The majority of reactions are accompanied by side reactions, and their enthalpy changes therefore cannot be measured directly. For many of these the enthalpy changes can be calculated from the values for other reactions, by making use of Hess's law. According to this law, it is permissible to write down stoichiometric equations, together with the enthalpy changes, and to treat them as mathematical equations, thereby obtaining a thermochemically valid result. For example, suppose that a substance A reacts with B according to the equation

(a) $A + B \rightarrow X$ $\Delta H_1° = -10$ kcal

Suppose that X reacts with an additional molecule of A to give another product Y:

(b) $A + X \rightarrow Y$ $\Delta H_2° = -20$ kcal

According to Hess's law, it is permissible to add these two equations and obtain

(c) $2A + B \rightarrow Y$ $\Delta H_3° = \Delta H_1° + \Delta H_2° = -30$ kcal

This law was first formulated by the Swiss-Russian chemist Germain Henri Hess (1802–1850), who is generally regarded as the founder of the field of thermochemistry. The law follows from the principle of conservation of energy. Thus, if reactions (a) and (b) occur there is a net *evolution* of 30 kcal when 1 mol of Y is produced. We can then reconvert Y into 2A + B by the reverse of reaction (c). If the heat required to do this differed from 30 kcal, we would have obtained the starting materials with a net gain or loss of heat, which would violate the principle of conservation of energy.

EXAMPLE

The enthalpy changes in the complete combustion of crystalline α-D-glucose and maltose at 298 K, with the formation of gaseous CO_2 and liquid H_2O, are:

$$\Delta H_c^\circ/\text{kcal mol}^{-1}$$

α-D-glucose, $C_6H_{12}O_6$ (c)	−671.4
maltose, $C_{12}H_{22}O_{11}$ (c)	−1349.3

Calculate the enthalpy change accompanying the conversion of 1 mol of glucose (c) into maltose (c).

SOLUTION

The enthalpy changes given relate to the processes

(1) $C_6H_{12}O_6$ (c) $+ 6O_2$ (g) $\rightarrow 6CO_2$ (g) $+ 6H_2O$ (l)

$$\Delta H_c^\circ = -671.4 \text{ kcal mol}^{-1}$$

(2) $C_{12}H_{22}O_{11}$ (c) $+ 12O_2$ (g) $\rightarrow 12CO_2$ (g) $+ 11H_2O$ (l)

$$\Delta H_c^\circ = -1349.5 \text{ kcal mol}^{-1}$$

We are asked to convert 1 mol of glucose into maltose. The reaction is

$$C_6H_{12}O_6 \text{ (c)} \rightarrow \tfrac{1}{2}C_{12}H_{22}O_{11} \text{ (c)} + \tfrac{1}{2}H_2O \text{ (l)}$$

Reaction (2) can be rewritten as

(2′) $6CO_2$ (g) $+ \dfrac{11}{2}H_2O$ (l) $\rightarrow C_{12}H_{22}O_{11}$ (c) $+ 6O_2$ (g)

$$\Delta H_{2'}^\circ = \frac{1349.3}{2} \text{ kcal mol}^{-1}$$

If we add (1) and (2′) we obtain the required equation, with

$$\Delta H^\circ = -671.4 + 1349.3/2 = 3.25 \text{ kcal}$$

3. *Variation of Equilibrium Constant with Temperature.* A third general method of measuring $\Delta H°$ will only be mentioned here very briefly, since it is based on the second law of thermodynamics, to be considered in Chapter 5. This method is based on the equation for the variation of the equilibrium constant K with the temperature:

$$\frac{d\log_{10}K}{d(1/T)} = -\frac{\Delta H°(cal)}{4.57} \tag{4.30}$$

If, therefore, we measure K at a series of temperatures and plot $\log_{10}K$ against $1/T$, the slope of the line at any temperature will be $-\Delta H°(cal)/4.57$. Hence $\Delta H°$ can be calculated. Whenever an equilibrium constant for a reaction can be measured satisfactorily, this method thus provides a useful way of obtaining $\Delta H°$. The method cannot be used for reactions that go essentially to completion (in which case a reliable K cannot be obtained) nor for ones complicated by side reactions.

MICROCALORIMETRY

The direct determination of enthalpy changes in reactions is of great importance as far as biological processes are concerned. Here a special problem frequently arises: biological materials are often present in very low molar concentrations, so the observed heat changes are very small. It is therefore necessary to design calorimeters capable of measuring very small amounts of heat. Such instruments are known as *microcalorimeters*. The prefix *micro* refers to the amount of heat and not to the physical dimensions of the instruments, some of which are very large.

Pioneer work in the field of microcalorimetry has been carried out by the French physical chemists E. Calvet and H. Prat, the American physical chemist Julian M. Sturtevant, and the American biophysicists T. H. Benzinger and V. C. Kitzinger. These workers have designed very sensitive microcalorimeters and have made numerous measurements of the minute amounts of heat associated with enzymic reactions in which the substrate concentrations are low. They have also studied the heats evolved in such processes as the metabolism of microorganisms. Benzinger in particular has been responsible for the technique known as *heat-burst calorimetry*, in which the heat pulse is measured during the course of a reaction. Figure 4.5 shows records obtained by Kitzinger and Benzinger in 1955 for the hydrolysis of adenosine triphosphate (ATP) catalyzed by the enzyme myosin. Myosin is an important constituent of muscle, and this reaction is of great importance in muscle metabolism. The upper curve *a* shows the heat evolved, 86.8×10^{-3} calories, when 5.22 μmol of ATP are hydrolyzed; this amounts to 16.63 kcal mol^{-1}. However, a complication exists in that some of this heat is due to the reaction of protons, released in the reaction, with the buffering substances which are present. The lower curve *b* represents a determination of this amount of heat, which is found to be 11.78 kcal mol^{-1}. The corrected heat *evolved* in the hydrolysis of ATP is thus $16.63 - 11.78 = 4.85$ kcal mol^{-1}. The $\Delta H°$ for the process is thus -4.85 kcal mol^{-1}.

Another type of microcalorimeter is the *continuous flow calorimeter*, developed in 1967 by the American physical chemists P. R. Stoesser and Stanley J. Gill. This instrument permits two reactant solutions to be thermally equilibrated during passage through separate platinum tubes and then brought together in a mixing chamber, a thermopile measuring the heat change in the reaction. J. M. Sturtevant and his coworkers have recently carried out flow-calorimetric work on various biological systems, including the binding of substrates and co-enzymes to enzymes, and processes of enzyme activation.

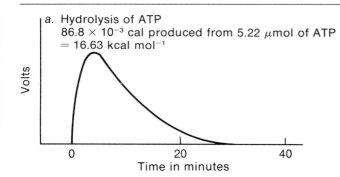

a. Hydrolysis of ATP
86.8×10^{-3} cal produced from 5.22 μmol of ATP
= 16.63 kcal mol^{-1}

Volts

Time in minutes

Figure 4.5

The hydrolysis of ATP studied by heat-burst calorimetry. The upper curve a is the record obtained with 5.22 μmol of ATP in the presence of enzyme (myosin) and buffer (pH 8.0). The lower curve b allows a correction to be made for the heat evolved when the hydrogen ions liberated combine with the buffer solution.

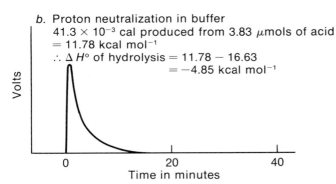

b. Proton neutralization in buffer
41.3×10^{-3} cal produced from 3.83 μmols of acid
= 11.78 kcal mol^{-1}
$\therefore \Delta H°$ of hydrolysis = 11.78 − 16.63
= −4.85 kcal mol^{-1}

Volts

Time in minutes

ENTHALPIES OF FORMATION

The total number of known chemical reactions is enormous, and it would be very inconvenient if one had to tabulate enthalpies of reaction for all of them. We can avoid having to do this by tabulating *enthalpies of formation* of chemical compounds, which are the enthalpy changes associated with the formation of 1 mol of the substance from the elements in their *standard states*. From these enthalpies of formation it is possible to calculate enthalpy changes in chemical reactions. This will be discussed on page 165.

We have seen that the standard state of each element and compound is taken to be the most stable form in which it occurs at 1 atm pressure and 25° C. In addition, there is the convention that an enthalpy of zero is arbitrarily assigned to each *element* in its standard state. The enthalpies of C (graphite), H_2 (g), and O_2 (g) at 1 atm pressure and 25° C are thus taken to be zero.

Suppose now that we form methane, at 1 atm and 25° C, from C (graphite) and

H_2 (g), which are the standard states. The stoichiometric equation is

$$C \text{ (graphite)} + 2H_2 \text{ (g)} \rightarrow CH_4 \text{ (g)}$$

It does not matter that we cannot make this reaction occur cleanly and therefore cannot directly measure its enthalpy change. As seen above, indirect methods can be used. In such ways it is found that $\Delta H°$ for this reaction is -17.9 kcal. Since we have arbitrarily taken the enthalpies of the reactants to be zero, it follows that on this basis the enthalpy of CH_4 (g) is -17.9 kcal mol^{-1}. This is known as the *enthalpy of formation*, $\Delta H_f°$, of methane. It is important to use the term "enthalpy of formation" only for the formation of the compound from elements in their *standard* states.

Enthalpies of formation of organic compounds are commonly obtained from their enthalpies of combustion by application of Hess's law. When, for example, 1 mol of methane is burned in an excess of oxygen at 1 atm pressure and $25°$ C, 191.77 kcal of heat are evolved. Thus we can write

1. $\qquad CH_4 \text{ (g)} + 2O_2 \text{ (g)} \rightarrow CO_2 \text{ (g)} + 2H_2O \text{ (g)} \qquad \Delta H_c° = -191.77$ kcal

In addition we have the following data:

2. $\qquad C \text{ (graphite)} + O_2 \text{ (g)} \rightarrow CO_2 \text{ (g)} \qquad \Delta H° = -94.05$ kcal

3. $\qquad 2H_2 \text{ (g)} + O_2 \text{ (g)} \rightarrow 2H_2O \text{ (g)} \qquad \Delta H° = 2\,(-57.80)$ kcal

If we add (2) and (3) and subtract (1) the result is

$$C \text{ (graphite)} + 2H_2 \text{ (g)} \rightarrow CH_4 \text{ (g)}$$
$$\Delta H_f° \,(CH_4) = 2\,(-57.80) - 94.05 - (-191.71)$$
$$= -17.88 \text{ kcal mol}^{-1}$$

Enthalpies of formation of many other compounds can be deduced in a similar way.

Table 4.1 gives some enthalpies of formation of a small number of compounds, including a few of particular biological interest.[†] The state in which the substance occurs is indicated in the table. The value for liquid ethanol, for example, is a little different from that for ethanol in aqueous solution. For biological work the aqueous values are the most significant, and for the most part these have been given in the table.

Included in Table 4.1 are enthalpies of formation of individual ions. There is an arbitrariness about these values, because thermodynamic quantities can never be determined experimentally for individual ions; it is always necessary to work with a pair of ions. For example, the $\Delta H_f°$ for HCl in aqueous solution is -40.0 kcal mol^{-1}, but there is no way that one can work with the individual H^+ and Cl^- ions. The procedure followed is to assign arbitrarily the value of zero to the proton in its

---

[†] The table also includes, for convenience, values of Gibbs energies of formation; these are considered in the next chapter.

Table 4.1 Standard Enthalpies and Gibbs Energies of Formation

Values, in kcal mol^{-1}, are for 25° C. Standard states: 1 atm pressure; 1 m for substances in aqueous solution. c = crystal; l = liquid; g = gas; aq = aqueous solution.

Compound	Formula	State	ΔH_f°	ΔG_f°
Inorganic Compounds				
Water	H_2O	l	−68.32	−56.69
Water	H_2O	g	−57.80	−54.63
Hydrogen chloride	HCl	aq	−39.95	−31.37
Sodium chloride	NaCl	c	−98.2	−91.8
Carbon dioxide	CO_2	g	−94.05	−94.25
		aq	−98.90	−92.26
Ammonia	NH_3	aq	−19.29	−6.35
Phosphoric acid	H_3PO_4	aq	−307.92	−273.10
Aqueous Ions (conventional values)				
Hydrogen ion	H^+	aq	0	0
Sodium ion	Na^+	aq	−57.39	−62.59
Potassium ion	K^+	aq	−60.32	−67.70
Calcium ion	Ca^{2+}	aq	−129.8	−132.2
Magnesium ion	Mg^{2+}	aq	−110.4	−109.0
Hydroxide ion	OH^-	aq	−54.97	−37.59
Chloride ion	Cl^-	aq	−39.95	−31.37
Phosphate ion	PO_4^{3-}	aq	−305.3	−243.5
Hydrogen phosphate ion	HPO_4^{2-}	aq	−308.83	−260.34
Dihydrogen phosphate ion	$H_2PO_4^-$	aq	−309.82	−260.17
Nitrite ion	NO_2^-	aq	−25.0	−8.9
Nitrite ion	NO_3^-	aq	−49.56	−26.61
Organic Molecules and Ions				
Ethanol	C_2H_5OH	aq	−68.6	−43.44
Glycerol	CH_2OH $\mid$ CHOH $\mid$ CH_2OH	aq	−161.7	−118.9
Acetaldehyde	CH_3CHO	aq	−50.35	−33.24
Acetone	CH_3COCH_3	aq	−52.99	−38.17
Acetic acid	CH_3COOH	—	−116.10	−94.78
Acetate ion	CH_3COO^-	aq	−116.16	−88.29
Oxalic acid	COOH $\mid$ COOH	aq	−197.2	−161.1
Succinic acid	CH_2COOH $\mid$ CH_2COOH	aq	−218.02	−178.45
Hydrogen succinate ion	CH_2COO^- $\mid$ CH_2COOH	aq	−217.23	−172.71
Succinate ion	CH_2COO^- $\mid$ CH_2COO^-	aq	−217.18	−165.02
Fumaric acid	CHCOOH $\parallel$ CHCOOH	aq	−185.2	−154.35
Hydrogen fumarate ion	CHCOO$^-$ $\parallel$ CHCOOH	aq	−185.1	−150.13

Table 4.1 —continued

Values, in kcal mol^{-1}, are for 25°C. Standard states: 1 atm pressure; 1 m for substances in aqueous solution. c = crystal; l = liquid; g = gas; aq = aqueous solution.

Compound	Formula	State	ΔH_f°	ΔG_f°
Fumarate ion	CHCOO$^-$ ‖ CHCOO$^-$	aq	− 185.8	− 143.85
L-lactic acid	CH$_3$CHOHCOOH	aq	− 164.01	− 128.77
L-lactate ion	CH$_3$CHOHCOO$^-$	aq	− 164.11	− 123.50
Pyruvic acid	CH$_3$CO COOH	aq	− 145.2	− 116.3
Pyruvate ion	CH$_3$CO COO$^-$	aq	− 142.5	− 112.9
Citric acid	CH$_2$COOH \ HOCCOOH / CH$_2$COOH	aq	− 364.7	− 297.37
Citrate ion	C$_6$H$_5$O$_7$$^{3-}$	aq	− 362.12	− 277.89
Isocitrate ion	C$_6$H$_5$O$_7$$^{3-}$	aq	no value available	− 276.30
α-Ketoglutarate ion	C$_5$H$_4$O$_5$$^{2-}$	aq	no value available	− 189.63
α-D-glucose	C$_6$H$_{12}$O$_6$	aq	− 301.88	− 218.58
α-lactose	C$_{12}$H$_{22}$O$_{11}$	aq	− 533.55	− 374.02
α-maltose	C$_{12}$H$_{22}$O$_{11}$	aq	− 534.96	− 376.10
Sucrose	C$_{12}$H$_{22}$O$_{11}$	aq	− 529.60	− 370.8
Glycine	H$_2$NCH$_2$COOH	aq	− 122.85	− 88.62
Urea	H$_2$NCONH$_2$	aq	− 75.95	− 48.47

The above data were collected from various sources, in particular: D. D. Wagman et al., "Selected Values of Chemical Thermodynamic Properties," National Bureau of Standards Technical Note 270-3 (1968), and R. C. Wilhoit, "Selected Values of Thermodynamic Properties," being the Appendix (pp. 305–317) of H. D. Brown, ed., *Biochemical Micro-calorimetry*. New York: Academic Press, 1969. These publications contain many additional values. Wilhoit's compilation also includes some values for systems buffered at pH 7.

standard state (1 molal); it then follows that on this basis the value for the Cl$^-$ ion is − 40.0 kcal mol^{-1}. Then, since the ΔH_f° value for NaCl in aqueous solution is − 97.3 kcal mol^{-1}, we have

$$\Delta H_f^\circ (\text{Na}^+) = -57.3 \text{ kcal mol}^{-1}$$

$$\Delta H_f^\circ (\text{Cl}^-) = -40.0 \text{ kcal mol}^{-1}$$

In this way a whole set of values can be built up. Such values are often known as *conventional standard enthalpies of formation*, the word "conventional" referring to the arbitrary value of zero for the proton. In spite of the arbitrariness of the procedure, one always obtains correct values when one uses these conventional values in making calculations for reactions; this follows from the fact that there is always a balancing of charges in a chemical reaction.

One reason why enthalpies of formation are important is that they give us some idea of the stabilities of compounds. Compounds for which ΔH_f° is negative are apt to be stable, since they can only be converted into their elements by the addition of a positive amount of heat. Such compounds are known as *exothermic* compounds.

Conversely, compounds having positive ΔH_f° values, which are known as *endothermic* compounds, will decompose into their elements with evolution of heat and are apt to be unstable with respect to their elements. These relationships are illustrated in Figure 4.6. However, the stability of a compound is related not to enthalpies but to Gibbs energies. This will be dealt with in the next chapter.

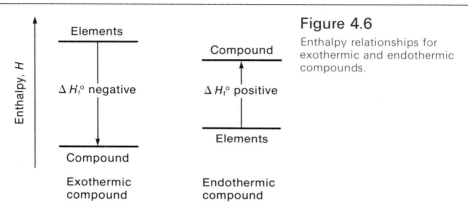

Figure 4.6

Enthalpy relationships for exothermic and endothermic compounds.

In addition, enthalpies of formation allow us to calculate enthalpies of any reaction, provided that we know the ΔH_f° values for all of the reactants and products. The ΔH° of any reaction is the difference between the sum of the ΔH_f° values for all of the products and the sum of the ΔH_f° values for all of the reactants:

$$\Delta H^\circ = \Sigma \Delta H_f^\circ \text{ (products)} - \Sigma \Delta H_f^\circ \text{ (reactants)} \qquad (4.31)$$

This may be illustrated with reference to the hydrolysis of urea to give carbon dioxide and ammonia in aqueous solution:

$$H_2NCONH_2 \text{ (aq)} + H_2O \text{ (l)} \rightarrow CO_2 \text{ (aq)} + 2NH_3 \text{ (aq)}$$

Table 4.1 lists the ΔH_f° values for the reactants and products. These values relate to the following reactions in which the compounds are formed from the elements in their standard states:

1. $\quad C \text{ (graphite)} + 2H_2 \text{ (g)} + \tfrac{1}{2}O_2 \text{ (g)} + N_2 \text{ (g)} \rightarrow H_2NCONH_2 \text{ (aq)}$

    $\Delta H_f^\circ = -75.95 \text{ kcal}$

2. $\quad H_2 \text{ (g)} + \tfrac{1}{2}O_2 \text{ (g)} \rightarrow H_2O \text{ (l)}$ $\qquad \Delta H_f^\circ = -68.32 \text{ kcal}$

3. $\quad C \text{ (graphite)} + O_2 \text{ (g)} \rightarrow CO_2 \text{ (aq)}$ $\qquad \Delta H_f^\circ = -98.90 \text{ kcal}$

4. $\quad \tfrac{1}{2}N_2 \text{ (g)} + \tfrac{3}{2}H_2 \text{ (g)} \rightarrow NH_3 \text{ (aq)}$ $\qquad \Delta H_f^\circ = -19.19 \text{ kcal}$

4'. $\quad N_2 \text{ (g)} + 3H_2 \text{ (g)} \rightarrow 2NH_3 \text{ (aq)}$ $\qquad \Delta H_f^\circ = 2(-19.19) \text{ kcal}$

Subtraction of (1) + (2) from (3) + (4') then leads to the desired equation, so that the enthalpy change in the reaction is

$$\Delta H° = -98.90 + 2(-19.19) - (-75.95 - 68.32)$$
$$= 6.99 \text{ kcal}$$

The fact that it is possible in this way to calculate enthalpies of reactions from the $\Delta H_f°$ values for reactants and products is of very great convenience. Instead of having to tabulate $\Delta H°$ values for a large number of reactions, we can simply tabulate $\Delta H_f°$ values for the much smaller number of chemical compounds. It requires less space to tabulate $\Delta H_f°$ values for, say, a thousand well-known compounds than to list the $\Delta H°$ values for the many millions of reactions that might occur between these compounds.

BOND ENERGIES

One important aspect of thermochemistry relates to the question of the energies of different chemical bonds. As a very simple example we may consider the case of methane, CH_4. The enthalpy of formation of methane is $-17.9 \text{ kcal mol}^{-1}$:

1. C (graphite) + $2H_2$ (g) → CH_4 (g) $\quad\quad\quad\quad$ $\Delta H_f° = -17.9 \text{ kcal}$

We also know the following thermochemical values:

2. C (graphite) → C (gaseous atoms) $\quad\quad\quad\quad$ $\Delta H_f° = 171.7 \text{ kcal}$

3. $\frac{1}{2}H_2$ (g) → H (gaseous atoms) $\quad\quad\quad\quad$ $\Delta H_f° = 52.1 \text{ kcal}$

The former is the enthalpy of sublimation of graphite, and the latter is one-half of the heat of dissociation of hydrogen. We may now apply Hess's law in the following manner:

1. $\quad$ CH_4 (g) → C (graphite) + $2H_2$ (g) $\quad\quad\quad$ $\Delta H° = 17.9 \text{ kcal}$

2. $\quad$ C (graphite) → C (gaseous atoms) $\quad\quad\quad$ $\Delta H° = 171.7 \text{ kcal}$

4 × 3. $\quad$ $2H_2$ (g) → 4H (gaseous atoms) $\quad\quad\quad$ $\Delta H° = 208.4 \text{ kcal}$

Adding:
$\quad\quad$ CH_4 (g) → C + 4H (gaseous atoms) $\quad\quad\quad$ $\Delta H° = 398.0 \text{ kcal}$

This quantity, 398.0 kcal, is known as the *heat of atomization* of methane; it is the heat which has to be supplied to a mole of methane in order to dissociate all of the molecules into gaseous atoms. Since each CH_4 molecule has four C—H bonds, we can divide 398.0 by 4, obtaining 99.5 kcal, and call this average the C—H *bond strength*.

A similar procedure with ethane, C_2H_6, leads to a heat of atomization of 676.2 kcal. This molecule contains one C—C bond, and six C—H bonds. If we subtract 6 × 99.5 = 597.0 as the contribution of the C—H bonds we are left with 79.2 kcal as the C—C bond strength.

However, if we calculate heats of atomization of the higher paraffin hydrocarbons using these values, we find that the agreement with experiment is by no means perfect. In other words, there is not a strict *additivity* of bond strengths. The reason for this is that chemical bonds in a given molecule are not isolated from each other but interact; this is shown by the fact that successive removal of H atoms from CH_4 requires different amounts of energy. On the whole, heats of atomization are more satisfactorily predicted if we use the following bond energies rather than the ones deduced for CH_4 and C_2H_6:

C—H   98.0 kcal

C—C   80.0 kcal

By the use of similar procedures for molecules containing different kinds of bonds it is possible to arrive at a set of bond strengths which will allow us to make approximate estimates of heats of atomization and heats of formation. Such a set is shown in Table 4.2. Values of this kind have proved very useful in deducing thermochemical information when the experimental heats of formation are not available. Various procedures have been suggested for improving these simple additive procedures.

Table 4.2   Bond Energies

Bond	Bond Energy (kcal mol^{-1})
H—H	103
C—C	80
C—H	98
C=C	145
C≡C	198
N—H	92
O—H	109
C—O	79
C=O	173

4.6   MOLAR HEAT CAPACITIES

The amount of heat required to raise the temperature of one gram of a material by one degree Celsius is known as its *specific heat*. The amount of heat required to bring about the same change in *one mole* of a substance is known as the *molar heat capacity*. It is the specific heat multiplied by the molecular weight. The usual units of the molar heat capacity are calories per degree per mole (cal K^{-1} mol^{-1}), or joules per degree per mole (J K^{-1} mol^{-1}), the latter being the SI unit.

Since heat is not a state function, neither is heat capacity. It is therefore always necessary, when stating a heat capacity, to specify the process by which the temperature is raised by one degree. Two heat capacities are of special importance:

1. The molar heat capacity related to a process occurring at constant volume. This is denoted by C_V and is defined by

$$C_V = \frac{\partial q_V}{\partial T} \qquad (4.32)$$

where q_V is the heat supplied per mole at constant volume. Since q_V is equal to ΔU per mole it follows that

$$C_V = \frac{\partial U}{\partial T} \qquad (4.33)$$

2. The molar heat capacity at *constant pressure*, C_P, defined by

$$C_P = \frac{\partial q_P}{\partial T} = \frac{\partial H}{\partial T} \qquad (4.34)$$

On the basis of these definitions the heat required to raise the temperature of one mole of material from T_1 to T_2 at *constant volume* is

$$q_V = \int_{T_1}^{T_2} C_V \, dT \qquad (4.35)$$

If C_V is independent of temperature this integrates to

$$q_V = C_V (T_2 - T_1) \qquad (4.36)$$

Similarly, for a process at constant pressure

$$q_P = \int_{T_1}^{T_2} C_P \, dT \qquad (4.37)$$

This integrates to

$$q_P = C_P (T_2 - T_1) \qquad (4.38)$$

if C_P is independent of temperature. The expressions in equations (4.36) and (4.38) represent ΔU and ΔH respectively.

We have seen that ΔU and ΔH are very close to one another for solids and liquids, since the $\Delta(PV)$ term is very small. Consequently, C_V and C_P are essentially the same for solids and liquids. For gases, however, the $\Delta(PV)$ term is appreciable, and there is a significant difference between C_V and C_P. For an ideal gas, which obeys the relationship

$$PV = nRT \qquad (4.39)$$

where n is the number of moles and R the gas constant, the relationship between C_V and C_P is particularly simple and can be derived as follows: for one mole of gas

$$H = U + PV \tag{4.40}$$

Thus

$$\frac{dH}{dT} = \frac{dU}{dT} + \frac{d(PV)}{dT} \tag{4.41}$$

and

$$C_P = C_V + \frac{d(RT)}{dT} \tag{4.42}$$

$$C_P = C_V + R \tag{4.43}$$

The value of R in cal K^{-1} mol^{-1} is 1.987; the value in SI units is 8.314 J K^{-1} mol^{-1}.

4.7 IDEAL GAS RELATIONSHIPS

The various transformations which can be brought about on ideal gases have played a very important part in the development of thermodynamics. Students of the biological sciences sometimes feel that they need not have to know much of these matters, since biology is primarily concerned with the solid and liquid states. However, there are good reasons why all of us should devote careful study to ideal gases. In the first place, ideal gases represent the simplest systems with which we can deal, and they therefore provide us with valuable and not too difficult exercises for testing our understanding of the subject. In addition, some of the simple conclusions we can draw for ideal gases can readily be adapted to more complicated systems such as solutions; a direct application of thermodynamics to solutions would be difficult if we did not have the ideal-gas equations to guide our way.

REVERSIBLE COMPRESSION AT CONSTANT PRESSURE

As a first example, we will consider the reversible compression of an ideal gas at constant pressure. Suppose that we have 1 mol of an ideal gas confined in a cylinder with a piston, at a pressure of P_1, a volume of V_1, and a temperature (in kelvins) of T_1. The isotherm† for this temperature is shown in the upper curve in Figure 4.7, and the initial state is represented by point A. We now remove heat from the system

---

† This term refers to the P-V relationship at a constant temperature; for an ideal gas, Boyle's law is obeyed and the isotherm is a parabola.

reversibly, at the constant pressure P_1, until the volume has fallen to V_2 (point B). This could be done by lowering the temperature of the surroundings by infinitesimal amounts until the temperature of the system is T_2. The isotherm for T_2 is the lower curve in Figure 4.7.

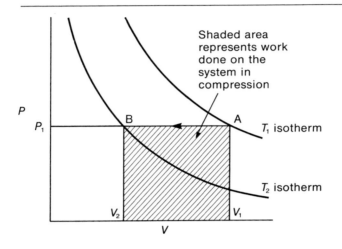

Shaded area represents work done on the system in compression

Figure 4.7

Pressure-volume isotherms for an ideal gas at a lower temperature T_2 and a higher temperature T_1. A reversible compression occurs at constant pressure from state A to state B.

Our problem is to calculate how much work is done on the system during this process, how much heat is lost, and what are the changes in energy and enthalpy. Note that as far as work and heat are concerned there would have been no definite answer to this question unless we had specified the path taken, since work and heat are not state functions. In this case we have specified that the compression is reversible and is occurring at constant pressure. The work done on the system is

$$w_{rev} = -\int_{V_1}^{V_2} P_1 \, dV \tag{4.44}$$

$$= P_1(V_1 - V_2) \tag{4.45}$$

Use of the ideal-gas law $PV = RT$ (for 1 mol) leads to an alternative expression:

$$w_{rev} = P_1\left(\frac{RT_1}{P_1} - \frac{RT_2}{P_1}\right) \tag{4.46}$$

$$= R(T_1 - T_2) \tag{4.47}$$

Since $V_1 > V_2$, and $T_1 > T_2$, a positive amount of work has been done on the system. This work is represented by the shaded area in Figure 4.7. If the system had *expanded* isothermally and at constant pressure from state B to state A, the shaded area would represent the work done *by* the system.

The heat absorbed by the system during the process A → B is given by

$$q_P = \int_{T_1}^{T_2} C_P \, dT \qquad (4.48)$$

since the process occurs at constant pressure. This expression integrates to

$$q_P = C_P(T_2 - T_1) \qquad (4.49)$$

if C_P is independent of temperature (which is the case for an ideal gas). Since $T_1 > T_2$, a negative amount of heat is absorbed; i.e., heat is released by the system. This amount of heat q_P, being absorbed at constant pressure, is the enthalpy change, which is also negative:

$$\Delta H = C_P(T_2 - T_1) \qquad (4.50)$$

The internal change ΔU (also negative for this process) is obtained by use of the first law:

$$\Delta U = q + w \qquad (4.51)$$

$$= C_P(T_2 - T_1) + R(T_1 - T_2) \qquad (4.52)$$

$$= (C_P - R)(T_2 - T_1) \qquad (4.53)$$

$$= C_V(T_2 - T_1) \qquad (4.54)$$

using equation (4.43). It can be verified easily that these expressions for ΔH and ΔU are consistent with the relationship

$$\Delta H = \Delta U + \Delta(PV) \qquad (4.55)$$

REVERSIBLE PRESSURE CHANGE AT CONSTANT VOLUME

Suppose instead that we go from the initial state P_1, V_1, T_1 to the final state P_2, V_1, T_2 as shown in Figure 4.8. The pressure P_1 is taken to be higher than P_2, and to accomplish this at constant volume we must remove heat until the temperature is T_2. Again, we bring about the change reversibly, with 1 mol of the ideal gas.

The work done on the system is the area below the line AC, and is zero. This is confirmed by considering the integral

$$w_{\text{rev}} = -\int_{V_1}^{V_1} P \, dV = 0 \qquad (4.56)$$

Since the process occurs at constant volume, the heat absorbed is given by

$$q_V = \int_{T_1}^{T_2} C_V \, dT = C_V(T_2 - T_1) \tag{4.57}$$

which is negative since $T_1 > T_2$. This expression is also ΔU:

$$\Delta U = C_V(T_2 - T_1) \tag{4.58}$$

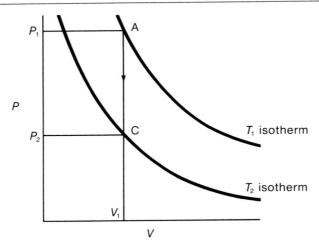

Figure 4.8

Pressure-volume isotherms for an ideal gas at a lower temperature T_2 and a higher temperature T_1. The system at state A is cooled at constant volume V_1, to bring the temperature to T_2 and the pressure to P_2.

It is to be seen that (4.56), (4.57) and (4.58) are consistent with the first law. The value of ΔH is obtained as follows:

$$\Delta H = \Delta U + \Delta(PV) \tag{4.59}$$

$$= \Delta U + \Delta(RT) \tag{4.60}$$

$$= C_V(T_2 - T_1) + R(T_2 - T_1) \tag{4.61}$$

$$= (C_V + R)(T_2 - T_1) \tag{4.62}$$

$$= C_P(T_2 - T_1) \tag{4.63}$$

It is interesting to compare the changes which occur in going from A to B (volume decrease at constant pressure, Figure 4.7) with those in going from A to C (pressure decrease at constant volume, Figure 4.8). The work and heat values are different in the two cases, but the ΔU and ΔH values are the same. This means that the internal energy is the same at point B on the T_2 isotherm as it is at point C; the same is true of the enthalpy. This result can be proved for any two points on an isotherm. We thus reach the very important conclusion:

For an ideal gas the internal energy and the enthalpy depend only on the temperature and remain constant as we move along any isotherm.

REVERSIBLE ISOTHERMAL COMPRESSION

Another process of great importance is that of compression along an isotherm, i.e., at constant temperature. Such a process is illustrated in Figure 4.9, the temperature being written simply as T. The initial conditions are P_1, V_1 and the final P_2, V_2, with

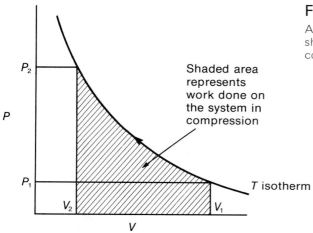

Figure 4.9

A pressure-volume isotherm, showing a reversible isothermal compression from state 1 to state 2.

Shaded area represents work done on the system in compression

T isotherm

$V_1 > V_2$. We have just seen that for an isothermal process with an ideal gas

$$\Delta U = 0 \tag{4.64}$$

and

$$\Delta H = 0 \tag{4.65}$$

The work done on the system in a reversible compression is

$$w_{\text{rev}} = -\int_{V_1}^{V_2} P\,dV \tag{4.66}$$

Since P varies we must express it in terms of V by use of the ideal gas equation, which for 1 mol is $PV = RT$. Thus

$$w_{\text{rev}} = -\int_{V_1}^{V_2} \frac{RT}{V}\,dV \tag{4.67}$$

$$= -RT\left| \ln V \right|_{V_1}^{V_2} \tag{4.68}$$

$$= RT \ln \frac{V_1}{V_2} \tag{4.69}$$

Since $V_1 > V_2$, this is a positive quantity. The heat absorbed per mole is readily found by use of the equation for the first law:

$$\Delta U = q_{rev} + w_{rev} \tag{4.70}$$

thus

$$q_{rev} = \Delta U - w_{rev} = 0 - w_{rev} = RT \ln \frac{V_2}{V_1} \tag{4.71}$$

This is negative, i.e., heat is evolved during the compression.

The concentration of a gas is the number of moles divided by the volume. If we have 1 mol in a volume V_1 the concentration is

$$c_1 = \frac{1}{V_1} \tag{4.72}$$

Similarly

$$c_2 = \frac{1}{V_2} \tag{4.73}$$

The ratio of volumes is therefore the inverse ratio of the concentrations:

$$\frac{V_2}{V_1} = \frac{c_1}{c_2} \tag{4.74}$$

Equation (4.69) for the work done in the isothermal reversible expansion of 1 mol of an ideal gas therefore can be written alternatively as

$$w_{rev} = RT \ln \frac{c_2}{c_1} \tag{4.75}$$

For n mol

$$w_{rev} = nRT \ln \frac{c_2}{c_1} \tag{4.76}$$

This is a useful form of the equation, because certain types of solutions—known as *ideal* solutions—obey exactly the same relationship.

EXAMPLE

Six moles of an ideal gas at 25° C are allowed to expand isothermally and reversibly from an initial volume of 5 dm^3 to a final volume of 15 dm^3. Calculate the work done in kcal and in kJ.

SOLUTION

From equation (4.69), for 6 mol of gas

$$\text{work done by system} = -w_{rev} = -6RT \ln \frac{5}{15} = 6RT \ln \frac{15}{5}$$

To obtain the work in calories, R is taken as 1.99 cal K^{-1} mol^{-1}. Then,

$$-w_{rev} = 6 \times 1.99 \times 298 \times 2.303 \log_{10} 3$$

$$= 3910 \text{ calories}$$

$$= 3.91 \text{ kcal}$$

To obtain kilojoules we multiply by the conversion factor 4.184 (J cal^{-1}):

$$-w_{rev} = 16.4 \text{ kJ}$$

Alternatively, to obtain the result in joules, we use the gas constant R as 8.314 J K^{-1} mol^{-1}.

EXAMPLE

Gastric juice in humans has an acid concentration of about 10^{-1} M (pH ~ 1) and is formed from other body fluids such as blood, which have an acid concentration of about 4.0×10^{-8} M (pH ~ 7.4). On the average, about 3 dm^3 of gastric juice are produced per day. Calculate the minimum work required to produce this quantity at 37° C, assuming the behavior to be ideal.

SOLUTION

Equation (4.76) gives us the reversible work required to produce n mol of acid. Three liters of 10^{-1} M acid contain 0.3 mol, and the reversible work is therefore

$$w_{rev} = 0.3 \times 1.99 \times 310 \ln \frac{10^{-1}}{4.0 \times 10^{-8}}$$

$$= 2720 \text{ cal}$$

The actual work required, of course, will be greater than this because the process, being a natural one, cannot be reversible.

REVERSIBLE ADIABATIC COMPRESSION

The final process we will consider is the compression of an ideal gas which is contained in a vessel whose walls are perfectly insulating, so that no heat can pass through them. Such processes are said to be *adiabatic*. This word comes from the Greek *adiabatos*, impassable, which is derived from the Greek prefix *a-*, not, and the words *dia*, through, and *bainein*, to go.

The pressure-volume diagram for the process is shown in Figure 4.10. Since work is performed on the gas in order to compress it and no heat can leave the system, the final temperature T_2 must be higher than the initial temperature T_1. The figure shows the T_1 and T_2 isotherms, as well as the adiabatic curve AB.

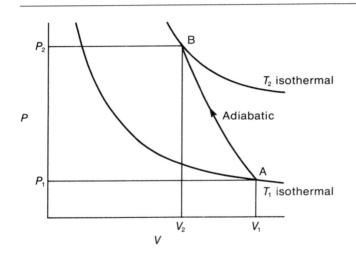

Figure 4.10

A pressure-volume diagram for an ideal gas, showing reversible adiabatic compression, the temperature increasing from T_1 to T_2.

We first need the equation for the adiabatic AB. According to the first law

$$dU = dq - PdV \tag{4.77}$$

Since the process is adiabatic $dq = 0$, and therefore

$$dU + PdV = 0 \tag{4.78}$$

Also, for n mol, $dU = nC_V dT$ so that

$$nC_V dT + PdV = 0 \tag{4.79}$$

For n mol of an ideal gas

$$PV = nRT \tag{4.80}$$

and thus

$$C_V \frac{dT}{T} + R \frac{dV}{V} = 0 \tag{4.81}$$

For the adiabatic AB we integrate this equation between the temperatures T_1 and T_2 and the volumes V_1 and V_2:

$$C_V \int_{T_1}^{T_2} \frac{dT}{T} + R \int_{V_1}^{V_2} \frac{dV}{V} = 0 \tag{4.82}$$

Thus

$$C_V \ln \frac{T_2}{T_1} + R \ln \frac{V_2}{V_1} = 0 \tag{4.83}$$

Since from equation (4.43) $R = C_P - C_V$, this equation may be written as

$$\frac{T_2}{T_1} + \frac{(C_P - C_V)}{C_V} \ln \frac{V_2}{V_1} = 0 \tag{4.84}$$

The ratio of C_P to C_V is often written as γ,

$$\gamma = \frac{C_P}{C_V} \tag{4.85}$$

and equation (4.84) thus becomes

$$\frac{T_2}{T_1} + (\gamma - 1) \ln \frac{V_2}{V_1} = 0 \tag{4.86}$$

or

$$\frac{T_2}{T_1} = \left(\frac{V_1}{V_2}\right)^{\gamma - 1} \tag{4.87}$$

We can eliminate the temperature by making use of the ideal-gas relationship

$$\frac{T_2}{T_1} = \frac{P_2 V_2}{P_1 V_1} \tag{4.88}$$

Equating the right-hand sides of equations (4.87) and (4.88) gives

$$\frac{P_2}{P_1} = \left(\frac{V_1}{V_2}\right)^{\gamma} \tag{4.89}$$

or

$$P_1 V_1^{\gamma} = P_2 V_2^{\gamma} \tag{4.90}$$

This is to be contrasted with the Boyle's law equation $P_1 V_1 = P_2 V_2$ for the isothermal process. Because γ is necessarily greater than unity, the adiabatic is steeper than the isothermal, as is shown in Figure 4.10.

We now consider the various changes in thermodynamic quantities when the process A → B in Figure 4.10 is undergone by 1 mol of an ideal gas. Since the process is adiabatic,

$$q = 0 \tag{4.91}$$

The U and H values are the same at all points on the T_1 isothermal, and the same is true of the T_2 isothermal (see p. 172). The changes are (see equations 4.58 and 4.63):

$$\Delta U = C_V(T_2 - T_1) \tag{4.92}$$

and

$$\Delta H = C_P(T_2 - T_1) \tag{4.93}$$

Since $\Delta U = q + w$, the work done on the system during the adiabatic compression is

$$w = C_V(T_2 - T_1) \tag{4.94}$$

For the compression of n moles, the above expressions would be multiplied by n.

PROBLEMS

4.1 Two moles of oxygen gas, which can be regarded as ideal with $C_V = 5$ cal K^{-1} mol^{-1}, are maintained in a volume of 11.2 dm^3 at 273 K.

   (a) What is the pressure of the gas?

   (b) What is PV in atm dm^3 and in cal mol^{-1}?
        (1 atm dm^3 = 24.2 calories)

   (c) What is C_P?

4.2 Suppose that the temperature of the gas is raised to 373 K at constant *volume*.

   (a) How much work is done on the system?

   (b) What is the increase in internal energy U?

   (c) How much heat was added to the system?

   (d) What is the final pressure?

(e) What is the final value of PV?

(f) What is the increase in enthalpy?

4.3 Suppose that the gas is heated reversibly to 373 K at constant *pressure*.

(a) What is the final volume?

(b) How much work is done by the system?

(c) How much heat has been supplied to the system?

(d) What is the increase in enthalpy?

(e) What is the increase in internal energy of the system?

4.4 Suppose that the gas is compressed reversibly to half its volume at constant T (273 K).

(a) What is the change in U?

(b) What is the final pressure?

(c) How much work has been done on the system?

(d) How much heat has flowed out of the system?

4.5 The bacterium *Acetobacter suboxydans* obtains energy for growth by oxidizing ethanol in two stages, as follows:

(a) C_2H_5OH (l) $+ \frac{1}{2}O_2$ (g) $\rightarrow CH_3CHO$ (l) $+ H_2O$ (l)

(b) CH_3CHO (l) $+ \frac{1}{2}O_2$ (g) $\rightarrow CH_3COOH$ (l)

The enthalpy increases in the complete combustion (to CO_2 gas and liquid H_2O) of the three compounds are:

	ΔH_c° (kcal mol^{-1})
ethanol (l)	-327.6
acetaldehyde (l)	-279.0
acetic acid (l)	-209.4

Calculate the ΔH° values for reactions 1. and 2.

4.6 Calculate ΔH° for the reaction

$$C_2H_5OH \text{ (aq)} + O_2 \text{ (g)} \rightarrow CH_3CO_2H \text{ (aq)} + H_2O \text{ (l)}$$

making use of the enthalpies of formation given in Table 4.1. Is the result consistent with the results obtained for problem 4.5?

4.7 Using the data of Table 4.1, calculate the amount of heat evolved in the complete oxidation of 1 g of aqueous α-D-glucose, assuming the CO_2 to be evolved as gas.

4.8 α-D-glucose can undergo the following fermentation reaction:

$$C_6H_{12}O_6 \text{ (aq)} \rightarrow 2C_2H_5OH \text{ (aq)} + 2CO_2 \text{ (g)}$$

Using the data in Table 4.1, calculate the standard enthalpy change in this reaction.

4.9 The disaccharide α-maltose can be hydrolyzed to glucose according to the equation

$$C_{12}H_{22}O_{11} \text{ (aq)} + H_2O \text{ (l)} \rightarrow 2C_6H_{12}O_6 \text{ (aq)}$$

Using data in Table 4.1, calculate the standard enthalpy change in this reaction.

4.10 The standard enthalpy of formation of the fumarate ion is -185.8 kcal mol^{-1}. If the standard enthalpy change of the reaction

$$\text{fumarate}^{2-} \text{ (aq)} + H_2 \rightarrow \text{succinate}^{2-} \text{ (aq)}$$

is -31.4 kcal, calculate the enthalpy of formation of the succinate ion.

4.11 The ΔH° for the mutarotation of glucose in aqueous solution,

$$\alpha\text{-D-glucose (aq)} \rightarrow \beta\text{-D-glucose (aq)}$$

has been measured in a microcalorimeter and found to be -1.16 kJ at 25° C. The enthalpies of solution of the two forms of glucose have been determined to be

α-D-glucose (s) → α-D-glucose (aq)	$\Delta H^\circ = 10.72$ kJ
β-D-glucose (s) → β-D-glucose (aq)	$\Delta H^\circ = 4.68$ kJ

Calculate ΔH° for the mutarotation of solid α-D-glucose to β-D-glucose, in kilojoules and kilocalories.

4.12 Five moles of sucrose are present in 1 dm^3 of water. If the solution is diluted at 25° C until the concentration is 2 M, what is the maximum work that could be performed, assuming the ideal law in equation (4.76) to be obeyed?

4.13 A bird weighing 1.5 kg leaves the ground and flies to a height of 75 metres, where it attains a velocity of 20 m s^{-1}. What is the minimum amount of metabolic energy that the bird must have expended (acceleration of gravity $= 9.81$ m s^{-2})?

4.14 An average man weighs about 70 kg and produces about 2500 kcal of heat per day. (a) Suppose that a man were an isolated system and that his heat capacity is 1 cal K^{-1} g^{-1}. If his temperature was 37° C at a given time, what would be his temperature 24 hours later? (b) Man is in fact an open system, and the main mechanism for maintaining his temperature constant is evaporation of water. If the enthalpy of vaporization of water at 37° C is 10.35 kcal mol^{-1}, how much water needs to be evaporated per day to keep the temperature constant?

4.15 It is estimated that the adult human brain consumes the equivalent of 10 g of glucose per hour. From the data in Table 4.1, estimate (a) the energy (in calories) utilized per second, and (b) the power output of the brain in watts (1 watt (W) $= 1$ J s^{-1}).

ESSAY QUESTIONS

4.16 Explain clearly what is meant by a thermodynamically reversible process. Why is the reversible work done by a system the maximum work?

4.17 Explain the thermodynamic meaning of a *system*, distinguishing between open, closed, and isolated systems. Which of these is (a) a fish swimming in the sea, or (b) an egg?

SUGGESTED READING

Butler, J. A. V. *The Fundamentals of Chemical Thermodynamics*, 4th ed. London: Macmillan, 1951.

Klotz, I. M. *Energy Changes in Biochemical Reactions.* New York: Academic Press, 1967.

Lehninger, A. L. *Bioenergetics*, 2nd ed. New York: W. A. Benjamin, 1971.

Mahan, B. H. *Elementary Chemical Thermodynamics*, New York: W. A. Benjamin, 1963.

Moore, W. J. *Physical Chemistry*, 4th ed. Englewood Cliffs, N.J.: Prentice-Hall, 1972.

5

THE SECOND LAW AND CHEMICAL EQUILIBRIUM

We have seen that the first law of thermodynamics is concerned with the conservation of energy and with the interrelationship of work and heat. A second important problem with which thermodynamics deals is whether a chemical or physical change can take place spontaneously. This problem is the concern of the second law of thermodynamics.

There are several well-known examples of processes which do not violate the first law, but which do not occur naturally. Suppose, for example, that we have a cylinder separated into two compartments by means of a partition (Figure 5.1), with on one side a gas at high pressure and on the other a gas at low pressure. If the diaphragm is ruptured, an equalization of pressure will occur. However, the reverse of this process

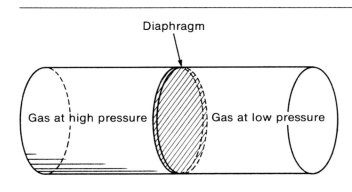

Diaphragm

Gas at high pressure　Gas at low pressure

Figure 5.1

A cylinder separated into two compartments, with gases at different pressures on the two sides of the diaphragm.

does not happen; if we start with gas at uniform pressure, we will never find gas at high pressure on one side of the container and gas at low pressure on the other side. The first process, in which the gas pressures equalize, is known as a *natural* or *spontaneous* process. The reverse process, which does not occur and can only be imagined, is known as an *unnatural* process. Note that the natural, spontaneous process is *irreversible* in the sense in which we used the term in the preceding chapter (p. 148). Instead, we could equalize gas pressures by reversibly expanding the gas at the higher pressure, in which case it would perform work. The reverse process could then occur, but only by virtue of the work we would have to perform on the system. Such a process would not be spontaneous.

Another illustration of a process which occurs naturally and spontaneously in one direction but not in the other is the equalization of temperature. Suppose that we bring together a hot solid and a cold solid. Heat will pass from the hot to the cold solid until the temperature is equalized; this is a spontaneous process. We know from common experience that heat will not flow in the opposite direction, from a cold to a hot body. There would be no violation of the first law if this occurred, but there would be a violation of the second law.

Our third example relates to a chemical reaction. There are many reactions which go spontaneously in one direction but not in the other. An example is the reaction between hydrogen and oxygen to form water:

$$2H_2 + O_2 \rightarrow 2H_2O$$

If we bring together two parts of hydrogen and one part of oxygen we can readily cause their reaction to occur essentially to completion; the remaining uncombined amounts of hydrogen and oxygen would be quite undetectable.[†] The reaction is accompanied by the production of considerable heat ($\Delta H°$ is negative). There would be no violation of the first law if we were to return this heat to water and reconvert it into hydrogen and oxygen. In practice, this cannot be done. The reaction from left to right is spontaneous; the reaction from right to left does not occur naturally.

It is obviously a matter of great importance in chemistry, biology, and other fields to know the factors which determine the direction in which a spontaneous process will occur. This amounts to asking what factors determine the position of equilibrium, since a system will move spontaneously towards the state of equilibrium. These matters are the concern of the second law of thermodynamics. It should be emphasized that thermodynamics is not directly concerned with the *speed* with which systems approach equilibrium, although thermodynamic ideas have made an important contribution to the study of rates of reactions.

The history of chemistry records numerous attempts to deal with this problem of the tendency of a process to occur spontaneously. It was noticed very early that most spontaneous reactions are exothermic, and it was originally thought that the heat evolved was the "driving force" of a reaction. This idea was put forward in particular

---

[†] The reaction is exceedingly slow at room temperature but can be brought about by various catalysts such as platinum, by passing a spark through the mixture, or by a small flame.

by the Danish chemist Peter Jorgen Julius Thomsen (1826–1909) and by the French chemist Pierre E. M. Berthelot (1827–1907), who suggested:

Every chemical change which takes place without the aid of external energy tends to the production of the system which is accompanied by the development of the maximum amount of heat.

However, more recent work has shown this idea to be untenable. There are many spontaneous endothermic reactions—for example, the solution in water of salts such as lithium chloride, where there is a cooling of the system. Furthermore there are many examples in which a reaction can be made to proceed in either direction to a definite state of equilibrium. Therefore heat evolved is not the sole driving force, although it does make an important contribution to the driving force.

The answer which finally emerged is that the true measure of the tendency of a reaction to occur is the *maximum work* that can be performed by the system. In the first example—the equalization of pressure of two gases—the process could have been conceived of as occurring in a reversible fashion, with work being performed *by* the system; in the reverse reaction, which is unnatural and does not occur spontaneously, work would have to be done *on* the system. Similarly, in the hydrogen-oxygen reaction, the reverse reaction can be made to occur by electrolyzing water, in which case work is done on the system. With the equalization of temperature, the situation is not quite as clear, but again it would be possible to cause this process to occur in reverse by the performance of work on the system.

In order to arrive at the conclusion that the ability of a system to do work is the true criterion of spontaneity, it is necessary to go through a number of fairly detailed arguments. It is most convenient to develop these arguments by studying the behavior of ideal gases, because these obey the simplest laws. We shall find that certain logical and simple relationships emerge for an ideal gas and that these can be applied without difficulty to the more complicated systems with which chemists and biologists are concerned.

5.1   THE CARNOT CYCLE

The arguments relating to the second law of thermodynamics are based upon a cycle of operations first suggested in 1824 by the French physicist Sadi Carnot (1796–1832). Suppose that we have one mole of an ideal gas contained in a cylinder with a piston and having an initial pressure of P_1, an initial volume of V_1, and an initial temperature of T_h (h standing for "hotter"). We refer to this gas as being in state A. Figure 5.2 is a pressure-volume diagram indicating the initial state A.

Now we bring about four reversible changes in the system, which will eventually bring it back to the initial state A. First we bring about an *isothermal* expansion A → B, changing the pressure and volume to P_2 and V_2 and keeping the temperature constant at T_h. To do this, we keep the cylinder immersed in a bath of liquid at temperature T_h. Second, we bring about an *adiabatic* expansion, i.e., one in which no heat is allowed to leave or enter the system. We could accomplish this by surrounding the cylinder with insulating material. Since the gas does work during expansion and no heat is supplied, the temperature must fall. We call the final temperature T_c (c for "cooler") and the pressure and volume P_3 and V_3. Third, we compress the gas isothermally (at temperature T_c) until the pressure and volume are P_3 and V_3. Finally, the gas is compressed adiabatically until it returns to its original state A (P_1, V_1, T_h). The performance of work on the system, with no heat transfer permitted, raises the temperature from T_c to T_h.

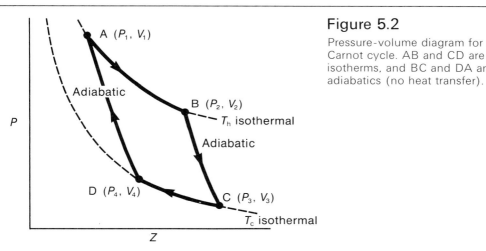

Figure 5.2

Pressure-volume diagram for the Carnot cycle. AB and CD are isotherms, and BC and DA are adiabatics (no heat transfer).

Let us consider these four steps in further detail. In particular we want to know the ΔU values for each step, the amounts of heat absorbed (q) and the work done (w). The expressions for these quantities are summarized in Table 5.1.

Table 5.1 Values of ΔU, q and w for the four reversible steps in the Carnot cycle

Step	ΔU	q_{rev}	w_{rev}
A → B	0	$RT_h \ln \dfrac{V_2}{V_1}$	$RT_h \ln \dfrac{V_1}{V_2}$
B → C	$C_V(T_c - T_h)$	0	$C_V(T_c - T_h)$
C → D	0	$RT_c \ln \dfrac{V_4}{V_3}$	$RT_c \ln \dfrac{V_3}{V_4}$
D → A	$C_V(T_h - T_c)$	0	$C_V(T_h - T_c)$
Net	0	$R(T_h - T_c) \ln \dfrac{V_2}{V_1}$	$R(T_h - T_c) \ln \dfrac{V_1}{V_2}$

$$\left(\text{since } \frac{V_1}{V_2} = \frac{V_4}{V_3} \right)$$

STEP 1

Step A $\rightarrow$ B is the reversible isothermal expansion at T_h. On p. 172 we showed that for the isothermal expansion of an ideal gas there is no change of internal energy:

$$\Delta U_{A \rightarrow B} = 0 \tag{5.1}$$

In equation (4.69) we showed that the work done on the system in an isothermal reversible process is $RT \ln (V_{initial}/V_{final})$:

$$w_{A \rightarrow B} = RT_h \ln \frac{V_1}{V_2} \tag{5.2}$$

Since by the first law

$$\Delta U_{A \rightarrow B} = q_{A \rightarrow B} + w_{A \rightarrow B} \tag{5.3}$$

it follows that

$$q_{A \rightarrow B} = RT_h \ln \frac{V_2}{V_1} \tag{5.4}$$

STEP 2

Step B $\rightarrow$ C involves surrounding the cylinder with an insulating jacket and allowing the system to expand reversibly and adiabatically to a volume of V_3. Since the process is adiabatic

$$q_{B \rightarrow C} = 0 \tag{5.5}$$

We saw in equation (4.92) that ΔU for an adiabatic process involving 1 mol of gas is

$$\Delta U = C_V(T_{final} - T_{initial}) \tag{5.6}$$

so that for the process B $\rightarrow$ C

$$\Delta U_{B \rightarrow C} = C_V(T_c - T_h) \tag{5.7}$$

Application of the first law then gives

$$w_{B \rightarrow C} = C_V(T_c - T_h) \tag{5.8}$$

STEP 3

Step C $\rightarrow$ D involves placing the cylinder in a heat bath at temperature T_c, and compressing the gas reversibly until the volume and pressure are V_4 and P_4 respectively.

The state D must lie on the adiabatic which passes through A (see Figure 5.2). Since process C → D is an isothermal one,

$$\Delta U_{C\to D} = 0 \tag{5.9}$$

The work done on the system is

$$w_{C\to D} = RT_c \ln \frac{V_3}{V_4} \tag{5.10}$$

and this is a positive quantity since $V_3 > V_4$; i.e., we must do work to compress the gas. By the first law

$$q_{C\to D} = RT_c \ln \frac{V_4}{V_3} \tag{5.11}$$

which means that a negative amount of heat is absorbed; i.e., heat is actually rejected.

STEP 4

The gas is finally compressed reversibly and adiabatically from D to A. The heat absorbed is zero:

$$q_{D\to A} = 0 \tag{5.12}$$

The $\Delta U_{D\to A}$ value is

$$\Delta U_{D\to A} = C_V(T_h - T_c) \tag{5.13}$$

By the first law,

$$w_{D\to A} = C_V(T_h - T_c) \tag{5.14}$$

Table 5.1 gives the net contributions as well as the individual contributions for the entire cycle. We see that ΔU for the cycle is zero. The contributions for the isothermal are zero, while those for the adiabatics are equal and opposite to each other. This result—that ΔU is zero for the entire cycle—is necessary in view of the fact that the internal energy is a state function: in completing the cycle the system is returned to its original state, and the internal energy is therefore unchanged.

On p. 177 we derived equation 4.87 for an adiabatic process. If we apply this equation to the two processes B → C and D → A in Figure 5.2 we have

$$\frac{T_h}{T_c} = \left(\frac{V_4}{V_1}\right)^{\gamma} \qquad \text{and} \qquad \frac{T_h}{T_c} = \left(\frac{V_3}{V_2}\right)^{\gamma} \tag{5.15}$$

Thus

$$\frac{V_4}{V_1} = \frac{V_3}{V_2} \tag{5.16}$$

The net q_{rev} value (see Table 5.1) is

$$q_{rev} = RT_h \ln \frac{V_2}{V_1} + RT_c \ln \frac{V_4}{V_3} \tag{5.17}$$

and since $V_4/V_3 = V_1/V_2$ this becomes

$$q_{rev} = R(T_h - T_c) \ln \frac{V_2}{V_1} \tag{5.18}$$

which is a positive quantity. Since by the first law $q = -w$ it follows that

$$w_{rev} = R(T_h - T_c) \ln \frac{V_1}{V_2} \tag{5.19}$$

which is a negative quantity, i.e., a positive amount of work has been done *by* the system.

Note that the net work done *by* the system (hence the net heat absorbed) is represented by the area within the Carnot diagram. This is illustrated in Figure 5.3. Diagram *a* shows the processes A → B and B → C, both of which are expansions. The work done *by* the system is represented by the area below the lines, which is shaded. Diagram *b* shows the processes C → D and D → A, in which work is done on the system in the amount shown by the shaded area. The net work done by the system

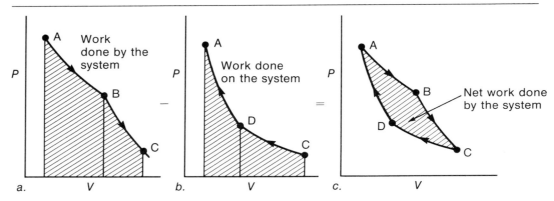

Figure 5.3

Diagram *a* shows the work done by the system in going from A → B and then from B → C; *b* shows the work done on the system in the return process, via D. The net work, obtained by subtracting the shaded area in *b* from that in *a*, is thus the area enclosed by the cycle *c*.

is represented by the area in *a* minus the area *b*, and is thus the area within the cycle, shown in *c*.

The important thing to note about the Carnot cycle is that we have returned the system, originally at state A, to its original state by processes in the course of which a net amount of work has been done by the system. This work has been performed at the expense of heat absorbed, as is required by the first law. Since work and heat are not state functions, the net work can be done even though the system returns to its original state.

EFFICIENCY OF THE REVERSIBLE CARNOT ENGINE

The efficiency of the reversible Carnot engine is the work done by the system during the cycle, divided by the work that would have been done if all of the heat absorbed at the higher temperature had been converted into work. Thus

$$\text{Efficiency} = \frac{R(T_h - T_c) \ln \dfrac{V_2}{V_1}}{R T_h \ln \dfrac{V_2}{V_1}} \tag{5.20}$$

$$= \frac{T_h - T_c}{T_h} \tag{5.21}$$

This efficiency is unity (i.e., 100%) only if the lower temperature T_c is zero—i.e., if the heat is rejected at absolute zero. This gives us a definition of absolute zero. The efficiency is also the net heat absorbed, $q_h + q_c$, divided by the heat absorbed at the higher temperature, q_h.

$$\text{Efficiency} = \frac{q_h + q_c}{q_h} \tag{5.22}$$

From equations (5.21) and (5.22) it follows that, for the reversible engine,

$$\frac{T_h - T_c}{T_h} = \frac{q_h + q_c}{q_h} \tag{5.23}$$

CARNOT'S THEOREM

Carnot's cycle has been thus far discussed for an ideal gas. Similar reversible cycles can be performed on other materials, including solids and liquids, and the efficiency of these cycles determined. The importance of the reversible cycle for the ideal gas is that, as has just been seen, it gives us an extremely simple expression for the efficiency, namely $(T_h - T_c)/T_h$. Similarly, a theorem of Carnot shows that the efficiency of *all* reversible cycles operating between the temperatures T_h and T_c is the same, namely $(T_h - T_c)/T_h$. This theorem leads to important quantitative formulations of the second law of thermodynamics.

Carnot's theorem employs the method of *reductio ad absurdum*. It supposes that there are two reversible engines which operate between T_h and T_c and which have different efficiencies. The theorem then shows that this would lead to the possibility of heat flowing from a lower to a higher temperature, which is contrary to experience.

Suppose that there are two engines, A and B, operating reversibly between T_h and T_c, and that A has a higher efficiency than B. By allowing B to use a larger amount of material it is possible to arrange for it to do exactly the same amount of work in a cycle. Then suppose the engines to be coupled together, so that A just works B backwards, B rejecting heat at T_h and absorbing it at T_c. During each complete cycle the more efficient engine A draws a quantity of heat q_h from the heat reservoir at T_h, and rejects $-q_c$ at T_h. The less efficient engine B, which is being driven backwards, rejects q_h' at T_h and absorbs $-q_c'$ at T_c. (If it were operating in its normal manner it would absorb q_h' at T_h and reject $-q_c'$ at T_c.)

Since the engines were adjusted to perform equal amounts of work in a cycle, engine A only just operates B in reverse with no extra energy available. The work performed by A is $q_h + q_c$, while that performed *on* B is $q_h' + q_c'$, and these quantities are equal:

$$q_h + q_c = q_h' + q_c' \tag{5.24}$$

Since engine A is more efficient than B,

$$\frac{q_h + q_c}{q_h} > \frac{q_h' + q_c'}{q_h'} \tag{5.25}$$

From (5.24) and (5.25) it follows that

$$q_h' > q_h \tag{5.26}$$

and

$$q_c > q_c' \tag{5.27}$$

During the operation of the cycle, in which A has forced B to work backwards, A has absorbed q_h at T_h and B has rejected q_h' at T_h. The combined system A + B has therefore absorbed $q_h - q_h'$ at T_h. We see from (5.26) that this is a negative quantity; i.e., the system has rejected a positive amount of heat at the higher temperature. At the lower temperature T_c, A has absorbed q_c while B has absorbed $-q_c'$. The combined system has therefore absorbed $q_c - q_c'$ at this temperature, and this according to (5.27) is a positive quantity (equal to $q_h' - q_h$). In performing the cycle, the A + B system has thus absorbed heat at a lower temperature and rejected it at a higher temperature. It is contrary to experience that heat can flow uphill in this way, in a complete cycle of operations in which the system ends up in the same state. Therefore it must be concluded that the initial postulate is invalid and there cannot be two engines, A and B, operating reversibly between two fixed temperatures and having different efficiencies. Thus, the efficiencies of all reversible engines must be the same as that for the ideal-gas reversible engine, namely

$$\frac{T_h - T_c}{T_h}$$

THE GENERALIZED CYCLE: THE CONCEPT OF ENTROPY

From equation (5.22) for the efficiency of a reversible Carnot cycle, it follows that

$$\frac{q_h}{T_h} + \frac{q_c}{T_c} = 0 \qquad (5.28)$$

This equation applies to any reversible cycle which has distinct isothermal and adiabatic parts and can be put into a more general form to apply to any reversible cycle. Consider the cycle represented by ABA in Figure 5.4. During the operation of the

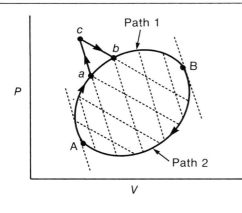

Figure 5.4

A generalized cycle ABA. The cycle can be traversed via infinitesimal adiabatics and isothermals; thus we can go from *a* to *b* by the adiabatic *ac* followed by the isothermal *cb*.

engine from A to B and back to A, there is heat exchange between the system and its environment at various temperatures. The full cycle may be split up into elements, such as *ab* shown in the figure. The distance between *a* and *b* should be infinitesimally small, but is enlarged in the diagram for clarity. During the change from *a* to *b*, the pressure, volume, and temperature have increased and a quantity of heat has been absorbed. Let the temperature corresponding to *a* be T, and that corresponding to *b* be $T + dT$. The isothermal corresponding to T is shown as *bc*, and the adiabatic at *a* as *ac*, the two intersecting at *c*. The change from *a* to *b* may therefore be carried out by means of an adiabatic change *ac*, followed by an isothermal change *cb*. During the isothermal process an amount of heat dq has been absorbed by the system. If the whole cycle is completed in this manner, quantities of heat dq_1, dq_2, etc., will have been absorbed by the system during isothermal changes carried out at the temperatures T_1, T_2, etc. Equation (5.28) is therefore replaced by

$$\frac{dq_1}{T_1} + \frac{dq_2}{T_2} + \frac{dq_3}{T_3} + \cdots = 0 \qquad (5.29)$$

the summation being made around the entire cycle. Since the *ab* elements are all infinitesimal, the cycle consisting of reversible isothermal and adiabatic steps is equivalent to the original cycle A → B → A. It follows that for the original cycle

$$\oint \frac{dq_{rev}}{T} = 0 \qquad\qquad (5.30)$$

where the symbol $\oint$ denotes integration over a complete cycle.

The result that the integral of the quantity dq_{rev}/T over the entire cycle is equal to zero is a very important one. We have seen earlier that certain functions are *state functions*, which means that their value is a true property of the system. The change in a state function in passing from state A to state B is independent of the path, and therefore a state function will not change when we traverse a complete cycle. For example, pressure, P, volume, V, temperature, T, and internal energy, U, are state functions, and thus we can write

$$\oint dP = 0 \qquad \oint dV = 0 \qquad \oint dT = 0 \qquad \oint dU = 0 \qquad\qquad (5.31)$$

The relationship expressed in equation (5.30) is therefore a significant one, and it is convenient to write dq_{rev}/T as dS, so that

$$\oint \frac{dq_{rev}}{T} = \oint dS = 0 \qquad\qquad (5.32)$$

The property S is known as the *entropy* of the system, and it is a state function.

Since entropy is a state function its value is independent of the path by which the state is reached. Thus if we consider a reversible change from state A to state B and back again (Figure 5.4) it follows from equation (5.32) that

$$\int_A^B dS + \int_B^A dS = 0 \qquad\qquad (5.33)$$

or

$$\Delta S^{(1)}_{A \to B} + \Delta S^{(2)}_{B \to A} = 0 \qquad\qquad (5.34)$$

where $\Delta S^{(1)}_{A \to B}$ denotes the change of entropy in going from A to B by path (1), and $\Delta S^{(2)}_{B \to A}$ is the change in going from B to A by path (2). The change of entropy in going from B to A by path (2) is the negative of the change in going from A to B by this path,

$$\Delta S^{(2)}_{B \to A} = -\Delta S^{(2)}_{A \to B} \qquad\qquad (5.35)$$

It therefore follows from (5.34) and (5.35) that

$$\Delta S^{(1)}_{A \to B} = \Delta S^{(2)}_{A \to B} \qquad\qquad (5.36)$$

This means that the change of entropy is the same whatever path is followed.

5.2 IRREVERSIBLE PROCESSES

The treatment of thermodynamically reversible processes is of great importance in connection with the second law. However, in practice we are concerned with thermodynamically irreversible processes, since these are the processes that occur in nature. Therefore it is important to consider the relationships which apply to irreversible processes.

A simple example of an irreversible process is the transfer of heat from a warmer to a colder body. Suppose that we have two reservoirs, a warm one at temperature T_h, and a cooler one at temperature T_c. We might imagine connecting these together by a metal rod, as shown in Figure 5.5a, and waiting until an amount of heat q has flowed from the warmer to the cooler reservoir. To simplify the argument, let us

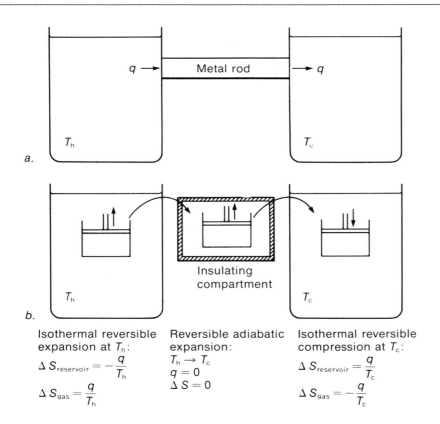

Isothermal reversible expansion at T_h:
$$\Delta S_{reservoir} = -\frac{q}{T_h}$$
$$\Delta S_{gas} = \frac{q}{T_h}$$

Reversible adiabatic expansion:
$$T_h \rightarrow T_c$$
$$q = 0$$
$$\Delta S = 0$$

Isothermal reversible compression at T_c:
$$\Delta S_{reservoir} = \frac{q}{T_c}$$
$$\Delta S_{gas} = -\frac{q}{T_c}$$

Figure 5.5

The transfer of heat q from a hot reservoir at temperature T_h to a cooler one at temperature T_c:
a. Irreversible transfer through a metal rod.
b. Reversible transfer by use of an ideal gas.

suppose that the reservoirs are so large that the transfer of heat does not change their temperatures appreciably.

In order to calculate the entropy changes in the two reservoirs after this irreversible process has occurred, we must devise a way of transferring the heat reversibly, since an entropy change can be calculated directly only for a reversible process. We can make use of an ideal gas to carry out the heat transfer process, as shown in Figure 5.5b. The gas is contained in a cylinder with a piston. We first place it in the warm reservoir, at temperature T_h, and expand it reversibly and isothermally until it has taken up heat equal to q. The gas is then removed from the hot reservoir, placed in an insulated container, and allowed to expand reversibly and adiabatically until its temperature has fallen to T_c. Finally, the gas is placed in contact with the colder reservoir at T_c and compressed isothermally until it has given up heat equal to q.

The entropy changes that occur in the two reservoirs and in the gas are shown in the figure. We see that the two reservoirs have experienced a net entropy change of

$$\Delta S_{reservoirs} = -\frac{q}{T_h} + \frac{q}{T_c} \tag{5.37}$$

This is a positive quantity, since $T_h > T_c$. The gas has experienced an exactly equal and opposite entropy change:

$$\Delta S_{gas} = \frac{q}{T_h} - \frac{q}{T_c} \tag{5.38}$$

Thus there is no overall entropy change, as is necessarily the case for reversible changes in an isolated system.

On the other hand, for an irreversible change in which the reservoirs are in thermal contact (Figure 5.5a), there is no compensating entropy decrease in the ideal gas. The entropy increase in the two reservoirs is the same as for the reversible process (5.37), and this is the overall entropy increase.

The result that a spontaneous (and therefore irreversible) process occurs with an overall increase of entropy in the system and its surroundings is universally true. The proof of this is based on the fact that the efficiency of a Carnot cycle in which some of the steps are irreversible must be less than that of a purely reversible cycle, since the maximum work is performed by systems which are undergoing reversible processes. Thus in place of equation (5.23) we have, for an irreversible cycle,

$$\frac{q_h^{irr} + q_c^{irr}}{q_h^{irr}} < \frac{T_h - T_c}{T_h} \tag{5.39}$$

This relationship reduces to

$$\frac{q_h^{irr}}{T_h} + \frac{q_c^{irr}}{T_c} < 0 \tag{5.40}$$

so that in general, for any cycle that is not completely reversible,

$$\oint \frac{dq_{irr}}{T} < 0 \tag{5.41}$$

This is known as the *inequality of Clausius*, after the German physicist Rudolf Julius Emmanuel Clausius (1822–1888), who suggested the relationship in 1850. The word "entropy" was coined by Clausius in 1865, and comes from the Greek *en*, in, and *trope*, transformation.

Consider an irreversible change from state A to state B in an *isolated* system, as represented by the dashed line in Figure 5.6. Suppose that the conditions are now

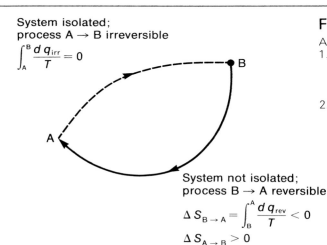

System isolated;
process A → B irreversible

$$\int_A^B \frac{d\,q_{irr}}{T} = 0$$

B

System not isolated;
process B → A reversible

$$\Delta S_{B \to A} = \int_B^A \frac{d\,q_{rev}}{T} < 0$$

$$\Delta S_{A \to B} > 0$$

Figure 5.6

A cyclic process in two stages:
1. The system is isolated and changes its state from A to B by an irreversible process (dashed line).
2. The system is not isolated and changes its state by a reversible process (solid line).

changed so that the system is no longer isolated. Then it returns to its initial state, A, by a reversible path represented by the solid line in Figure 5.6. During this reversible process the system is not isolated and can exchange heat and work with the environment. Since the entire cycle A → B → A is in part reversible, equation (5.41) applies, which means that

$$\int_A^B \frac{dq_{irr}}{T} + \int_B^A \frac{dq_{rev}}{T} < 0 \tag{5.42}$$

The first integral is equal to zero since the system was isolated during the irreversible process, so that any heat change in one part of the system is exactly compensated by an equal and opposite change in another part. The second integral is the entropy change when the process B → A occurs, so that

$$\Delta S_{B \to A} < 0 \tag{5.43}$$

It thus follows that

$$\Delta S_{A \to B} > 0 \qquad\qquad (5.44)$$

The entropy of the final state B is thus always greater than that of the initial state A if the process A → B occurs irreversibly in an isolated system.

Any change which occurs in nature is spontaneous and is therefore accompanied by a net increase in entropy. This conclusion led Clausius to his famous concise statement of the laws of thermodynamics:

The energy of the universe is a constant; the entropy of the universe tends always towards a maximum.

EXAMPLE

(a) One mole of water is frozen at $0°$ C and 1 atm pressure. Calculate the entropy change in the water and in the surroundings if the latent heat of fusion of water is 1.44 kcal mol^{-1}.

(b) One mole of supercooled water at $-10°$ C and 1 atm pressure turns into ice. Calculate the entropy change in the system and in the surroundings, and the net entropy change. Take the heat capacities of water and ice at 1 atm pressure as constant at 18.0 and 9.0 cal K^{-1} mol^{-1}.

SOLUTION

(a) Water and ice are at equilibrium at $0°$ C, and the freezing is therefore a reversible process. The entropy change when the water freezes is therefore

$$\Delta S_{\text{fusion}} = -\frac{q_{\text{fusion}}}{T} = -\frac{1440 \text{ cal mol}^{-1}}{273 \text{ K}}$$

$$= -5.27 \text{ cal K}^{-1} \text{mol}^{-1}$$

The surroundings experience an entropy *increase* of 5.27 cal K^{-1} mol^{-1}. There is no net entropy change during this reversible process.

(b) Supercooled water at $-10°$ C and its surroundings at $-10°$ C are not at equilibrium, and the freezing is therefore not reversible. To calculate the entropy change we must devise a series of reversible processes by which supercooled water at $-10°$ C is converted into ice at $-10°$ C. We can (1) heat the water reversibly to $0°$ C, (2) bring about the reversible freezing, and then (3) cool the ice reversibly to $-10°$ C.

(1) The heat required to raise the temperature of the supercooled

water by dT is

$$dq_{rev} = C_p dT$$

The entropy increase during this process is

$$dS = \frac{C_p dT}{T}$$

The entropy change in raising the temperature from 263 K to 273 K is therefore

$$\Delta S_1 = \int_{263}^{273} \frac{C_p dT}{T} = C_p \ln T \Big|_{263}^{273}$$

$$= 18.0 \ln \frac{273}{263}$$

$$= 0.672 \text{ cal K}^{-1} \text{ mol}^{-1}$$

(2) The entropy change in the reversible freezing of the water at $0°$ C is

$$\Delta S_2 = -\frac{q_{fusion}}{T} = \frac{-1440}{273} = -5.27 \text{ cal K}^{-1} \text{ mol}^{-1}$$

(3) The entropy change in cooling the ice from 273 K to 263 K is

$$\Delta S_3 = 9.0 \int_{273}^{263} \frac{dT}{T}$$

$$= 9.0 \ln \frac{263}{273}$$

$$= -0.336 \text{ cal K}^{-1} \text{ mol}^{-1}$$

The entropy change when the water freezes at $-10°$ C is therefore

$$\Delta S_{syst} = \Delta S_1 + \Delta S_2 + \Delta S_3$$

$$= 0.672 - 5.27 - 0.336$$

$$= -4.93 \text{ cal K}^{-1} \text{ mol}^{-1}$$

The entropy increase in the surroundings is obtained by first calculating the net amount of heat that has been transferred to the surroundings. The sum of three terms is:

(1) Heat lost to the surroundings in heating the water from $-10°$ C is $10 \times 18 = 180 \text{ cal mol}^{-1}$.

(2) Heat gained by the surroundings when the water freezes at $0°C$ is 1440 cal mol^{-1}.

(3) Heat gained by the surroundings when the ice is cooled to $-10°C$ is $10 \times 9 = 90$ cal mol^{-1}.

The net heat transferred to the surroundings when the water freezes at $-10°C$ is therefore

$$-180 + 1440 + 90 = 1350 \text{ cal mol}^{-1}$$

This heat is taken up by the surroundings at the constant temperature of $-10°C$, and the entropy increase is

$$\Delta S_{surr} = \frac{1350}{263} = 5.13 \text{ cal K}^{-1} \text{ mol}^{-1}$$

The overall entropy change in the system and the surroundings is therefore

$$\Delta S_{overall} = \Delta S_{syst} + \Delta S_{surr}$$
$$= -4.93 + 5.13$$
$$= 0.20 \text{ cal K}^{-1} \text{ mol}^{-1}$$

A net entropy increase in the system and surroundings is, of course, what we expect for an irreversible process.

5.3 MOLECULAR INTERPRETATION OF ENTROPY

We emphasized earlier that thermodynamics is a branch of science which can be developed without any regard to the molecular nature of matter. The logical arguments employed do not require any knowledge of molecules, yet in the understanding of thermodynamics, many of us find it helpful to interpret thermodynamic principles in the light of molecular structure.

In specifying a thermodynamic state we ignore the positions and velocities of individual atoms and molecules. However, any macroscopic property is in fact the result of the position and motion of these particles. At any instant we could define the *microscopic* state of a system *in principle*, which means that we would specify the position and momentum of each atom. An instant later, even though the system might remain in the same *macroscopic* state, the *microscopic* state would be completely different, since at ordinary temperatures molecules change their positions at speeds of the order of 10^5 cm per second.

Thus, a system at equilibrium remains in the same macroscopic state, even though its microscopic state is changing rapidly. There are an enormous number of microscopic states consistent with any given macroscopic state. This concept leads us at once to a molecular interpretation of entropy: *entropy is a measure of how many different microscopic states are consistent with a given macroscopic state.*

A deck of cards provides us with a useful analogy here. The cards may be arranged in a particular specified order (separate suits, cards arranged from ace to king, etc.), or completely shuffled. There are many sequences (analogous to microscopic states) which correspond to the shuffled (macroscopic) state, but there is only one microscopic state for a specified order. Thus, the shuffled state has a higher entropy than the unshuffled state. If we start with an ordered deck and shuffle it, the deck moves towards a state of greater randomness or lower order—the entropy increases. The reason why the random state is approached when the ordered deck is shuffled is simply that there are many microscopic states consistent with the shuffled condition, but only one microscopic state consistent with the ordered condition. The chance of producing an ordered deck by shuffling a disordered one is obviously very small.

Thus far we have considered a mere fifty-two cards; if we consider the vast number of molecules (6.02×10^{23} in a mole) the likelihood of a net decrease in entropy is obviously much more remote.

In light of this, it is easy to predict what kinds of entropy changes will occur when various processes take place. Suppose, for example, that we increase the temperature of a gas. The range of molecular speeds becomes more extended, since a larger proportion of the molecules have speeds which differ from the most probable value. Thus there is more disorder at a higher temperature, and the entropy is greater.

Entropy also increases if a solid melts. We know this from formal thermodynamics, since an amount of heat ΔH_f—the enthalpy of fusion—can be supplied reversibly to the solid at the melting temperature T_f. Thus

$$\Delta S_f = \frac{\Delta H_f}{T_f} \tag{5.45}$$

ΔH_f is always positive, so that there is an entropy increase. This is the case because in the solid state the molecules occupy fixed sites in a crystal lattice, while in a liquid there is much less restriction as to position. A similar argument applies to the evaporation of a liquid, for which

$$\Delta S_{vap} = \frac{\Delta H_{vap}}{T_b} \tag{5.46}$$

where ΔH_{vap} is the enthalpy of vaporization and T_b the boiling point.[†] This entropy increase is due to the large increase in disorder in going from the liquid to the gaseous state, since there are many more microscopic states for a gas compared to a liquid.

[†] ΔS_{vap} for many liquids is around 23 cal K^{-1} mol^{-1}. This is the basis of Trouton's rule, according to which ΔH_{vap} in calories is roughly 23 times the boiling point in kelvins.

The sublimation of a solid into a gas is also accompanied by an entropy increase for the same reason.

Entropy changes in chemical reactions also can be understood on a molecular basis. Consider, for example, the process

$$H_2 \rightarrow 2H$$

If we convert a mole of hydrogen molecules into two moles (gram-atoms) of hydrogen atoms, a considerable increase in entropy will be found. The reason for this is that there are more microscopic states associated with the separated hydrogen atoms than with the molecules, in which the atoms are paired together. Again, an analogy is provided by a deck of cards. The hydrogen atoms are like a completely shuffled deck, while the molecular system is like a deck in which all the aces, twos, etc., must be paired. The latter restriction means that there are fewer permissible states, and therefore a lower entropy.

In general, we find that for a gaseous chemical reaction there will be an increase of entropy in the system if there is an increase in the number of molecules. The dissociation of ammonia, for example,

$$2NH_3 \rightarrow N_2 + 3H_2$$

is accompanied by an entropy increase, because we are imposing a smaller restriction on the system by pairing the atoms as N_2 and H_2, as compared with organizing them as NH_3 molecules.

The situation with reactions in solution, however, is more complicated. It might be thought that an ionization process of the type

$$MX \rightarrow M^+ + X^-$$

occurring in aqueous solution would be accompanied by an entropy increase, analogous to the dissociation of H_2 into 2H. However, there is now an additional factor, arising from the fact that ions interact with surrounding water molecules, which tend to orient themselves in such a manner that there is an electrostatic attraction between the ion and the dipolar water molecules. This effect is known as *electrostriction*, or more simply as the *binding* of water molecules.† This electrostriction leads to a considerable reduction in entropy, since the bound water molecules have a restricted freedom of motion. As a result, ionization processes in solution always involve an entropy *decrease*.

A considerable entropy *increase* occurs when adenosine triphosphate (ATP) becomes bound to myosin, a protein which is an important constituent of muscle. Myosin is an extended protein molecule which bears a number of positive charges, while ATP under normal physiological conditions bears four negative charges. The binding process

---

† This is discussed in further detail in Chapter 7 (see especially Figures 7.4 and 7.5, and pp. 298–301).

can be represented schematically as

In the free myosin and ATP there is considerable binding of water molecules, but when the complex is formed there is a certain amount of charge neutralization and therefore a reduction in the electric field and a decrease in the binding of water molecules. Thus there is an entropy increase, and this plays a significant role in the mechanism of muscular contraction.

The entropy change which occurs on the contraction of a muscle is also of interest. A stretched strip of muscle or piece of rubber contracts spontaneously. When stretched, muscle or rubber is in a state of lower entropy than in the contracted state. Both muscle and rubber consist of very long molecules. If a long molecule is stretched as far as possible without breaking its bonds, there are few conformations available to it. However, if the ends of the molecule are brought closer together, the molecules can assume a large number of conformations, and they therefore have more freedom of motion. The entropy then will be higher. In 1913 the British physiologist Archibald Vivian Hill made accurate measurements of the heat produced when a muscle contracts, and found it to be extremely small. The same is true of a piece of rubber. When muscle or rubber contracts there is very little entropy change in the surroundings, so that the overall entropy is essentially that in the material itself, which is positive. We can thus understand why muscular contraction, or the contraction of a stretched piece of rubber, occurs spontaneously. Processes of this kind which are largely controlled by the entropy change in the system are referred to as *entropic processes*.

5.4   ENTROPY OF EXPANSION AND MIXING

We have seen that if one mole of an ideal gas performs a reversible isothermal expansion at temperature T, from volume V_1 to volume V_2, the heat absorbed is

$$q_{rev} = RT \ln \frac{V_2}{V_1} \tag{5.47}$$

The entropy increase, ΔS, is the heat absorbed divided by the constant temperature T:

$$\Delta S = R \ln \frac{V_2}{V_1} \tag{5.48}$$

As an example, consider the expansion of 1 mol of an ideal gas from 1 dm³ to

10 dm^3 (see Figure 5.7a). The entropy increase will be

$$\Delta S = R \ln 10 = 1.99 \ln 10 \text{ cal K}^{-1} \text{ mol}^{-1}$$
$$= 4.57 \text{ cal K}^{-1} \text{ mol}^{-1}$$
$$= 19.1 \text{ J K}^{-1} \text{ mol}^{-1}$$

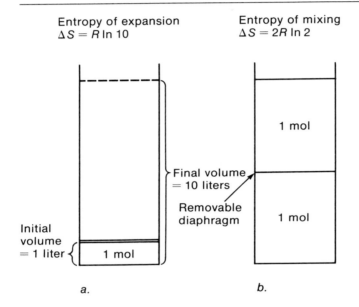

Entropy of expansion
$\Delta S = R \ln 10$

Entropy of mixing
$\Delta S = 2R \ln 2$

Initial volume = 1 liter

1 mol

Final volume = 10 liters

Removable diaphragm

1 mol

1 mol

a.

b.

Figure 5.7

a. Entropy increase on the expansion of an ideal gas from a volume of 1 liter to 10 liters.

b. Entropy increase on the mixing of two gases at equal volumes and pressures.

Suppose, instead, that we have 1 mol each of two gases, at equal pressures and volumes, separated by a diaphragm (see Figure 5.7b). If the diaphragm is removed, *each* gas will undergo an entropy increase of

$$\Delta S = R \ln 2 = 1.99 \ln 2$$
$$= 1.38 \text{ cal K}^{-1} \text{ mol}^{-1}$$
$$= 5.78 \text{ J K}^{-1} \text{ mol}^{-1}$$

The total entropy increase is thus

$$2 \times 1.38 = 2.76 \text{ cal K}^{-1} \text{ mol}^{-1}$$
$$= 11.56 \text{ J K}^{-1} \text{ mol}^{-1}$$

This entropy increase is known as the *entropy of mixing*. There is an increase in entropy because the mixed state is more probable than the unmixed state.

5.5   THE THIRD LAW OF THERMODYNAMICS

The definition of entropy

$$\Delta S = \int_A^B \frac{dq_{rev}}{T} \tag{5.49}$$

defines only changes in entropy, not entropy itself. The question of absolute entropy values is dealt with by the third law of thermodynamics, which was first formulated in 1906 by the German physical chemist Walther Hermann Nernst (1864–1941). This law is related to the experimental study of the behavior of matter at very low temperatures, a subject known as *cryogenics* (Greek *kryos*, frost; *genes*, become).

Various techniques are used to produce very low temperatures. The most familiar one, used in commercial refrigerators, is based on the fact that under certain circumstances gases become cooler when they expand, as a result of the work done by the gas in overcoming the mutual attraction of the molecules. Liquid nitrogen, which boils at 77 K, is manufactured in large quantities by the application of this principle. By performing successive expansions with nitrogen, then with hydrogen, and then with helium, it is possible to reach temperatures somewhat lower than 1 K.

The attainment of lower temperatures than 1 K requires the use of another principle. For example, one can make use of the temperature changes that occur during magnetization and demagnetization. Certain salts, such as those of the rare earths, have high paramagnetic susceptibilities. The cations act as little magnets, which line up when a magnetic field is applied; the substance is then in a state of lower entropy. When the magnetic field is decreased the magnets adopt a more random arrangement, and the entropy increases.

Figure 5.8 illustrates a procedure that can be employed to achieve a low temperature. A paramagnetic salt, such as gadolinium sulfate octahydrate, is placed between the poles of an electromagnet and is cooled to about 1 K, which can be done by the expansion techniques mentioned above. In step 1, the magnetic field is then applied, and the heat produced is allowed to flow into the surrounding liquid helium. In step 2, the system is isolated and the magnetic field removed; the adiabatic process leads to a cooling. Temperatures of about 0.005 K are produced in this way.

The attainment of still lower temperatures, down to about 10^{-6} K, is achieved by making use of nuclear magnetic properties. The nuclear magnets are about 2000 times smaller than the magnet in a paramagnetic substance such as gadolinium sulfate, yet there is still a significant difference between the nuclear entropies even at temperatures as low as 10^{-6} K.

The fact that the absolute zero of temperature cannot be attained and that it becomes more and more difficult to approach that temperature suggests that the entropies of all materials at absolute zero are the same. However, this statement must be qualified slightly by adding that the substance must be in its thermodynamically most stable state. For example, many substances are frozen into a metastable glassy state as the temperature is reduced. This state will persist indefinitely because of the slowness with which processes occur at low temperatures. Therefore we must exclude noncrystalline systems, and conclude that the entropy of all perfect crystalline substances must be the same at absolute zero. This is the *heat theorem* proposed by Nernst in 1906.

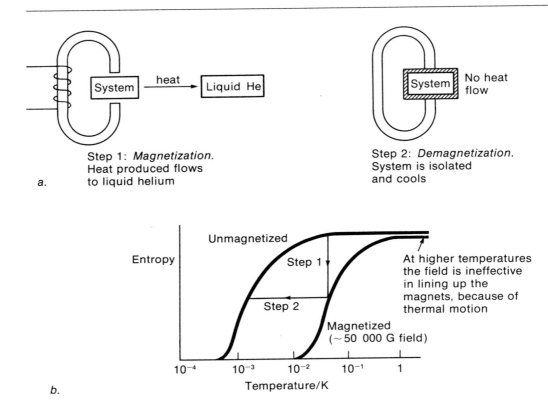

Figure 5.8
The production of very low temperatures by adiabatic demagnetization:
a. The magnetization and demagnetization steps.
b. The variation of entropy with temperature for the magnetized and demagnetized material.

The convention that has been adopted is to assign a value of zero to every crystalline substance at absolute zero. Entropies then can be determined for other temperatures by considering a series of reversible processes by which the temperature is raised from absolute zero to the temperature in question. Table 5.2 lists some absolute entropies obtained in this way.

If the absolute entropies of all the substances in a chemical reaction are known, it is then a simple matter to calculate the entropy change in the reaction, the relationship being

$$\Delta S^{\circ} = \Sigma S^{\circ} \text{ (products)} - \Sigma S^{\circ} \text{ (reactants)} \tag{5.50}$$

Table 5.2 Absolute Entropies at 25° C†

Substance	State	Absolute entropy, $S°/$ cal K^{-1} mol^{-1}
Carbon	graphite‡	1.36
Hydrogen	gas‡	31.2
Oxygen	gas‡	49.0
Nitrogen	gas‡	45.8
Carbon dioxide	gas	47.3
Water	liquid	16.7
Ammonia	gas	46.0
Ethane	gas	54.9
Ethylene	gas	52.5
Methanol	liquid	30.3
Ethanol	liquid	38.4
Acetic acid	liquid	38.2
Acetaldehyde	gas	63.5
Urea	solid	25.0

† These data are from F. D. Rossini et al., *Selected Values of Chemical Thermodynamic Properties*, National Bureau of Standards, Circular 500 (1952); and F. D. Rossini et al., *Selected Values of Physical and Thermodynamic Properties of Hydrocarbons and Related Compounds*, American Petroleum Institute Research Project 44 (1953).
‡ These are the standard states of the elements.

EXAMPLE

From the data given in Table 5.2 calculate the standard entropy of formation, $\Delta S_f°$, of liquid ethanol at 25° C.

SOLUTION

The equation for the formation of ethanol from its elements is

$$2C \text{ (graphite)} + 3H_2 \text{ (g)} + \tfrac{1}{2}O_2 \text{ (g)} \rightarrow C_2H_5OH \text{ (l)}$$

The entropy of formation is thus

$$\Delta S_f° = 38.4 - (2 \times 1.36 + 3 \times 31.2 + \tfrac{1}{2} \times 49.0)$$

$$= -82.4 \text{ cal } K^{-1} \text{ mol}^{-1}$$

5.6 CONDITIONS FOR CHEMICAL EQUILIBRIUM

We have seen that, according to the second law, any spontaneous or natural process must be accompanied by an increase in the total entropy of the system and its surroundings. This immediately gives us a condition for equilibrium, since when a system is at equilibrium it cannot undergo any spontaneous change. If a system is at equilibrium in state A, we can imagine an infinitesimal change to state B, where there is still equilibrium (see Figure 5.9). The change from A to B cannot involve a total entropy increase, since otherwise the change would be spontaneous. Similarly, a change from B to A cannot involve an entropy increase, since if it did that process would be spontaneous. It follows that the states A and B, separated by an infinitesimal amount, must have the same total entropies. Therefore, the condition for equilibrium is

$$dS = 0 \tag{5.51}$$

It must be emphasized that the entropy S which appears in this equation is the *total entropy of the system and its surroundings*. The position of equilibrium must correspond to a state of *maximum* total entropy, since the total entropy increases in any spontaneous process.

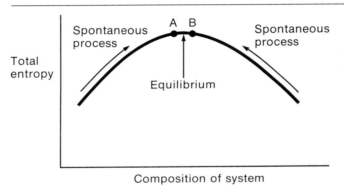

Figure 5.9

A condition for equilibrium. The total entropy of the system and its surroundings is plotted against the composition.

In practice, it is not very convenient to define equilibrium with reference to total entropy. In most problems in chemistry and biology we are interested in the properties of the system itself and it is inconvenient to have to consider the properties of the surroundings. For example, if a process is made to occur in a beaker it is usually a relatively simple matter to determine the thermodynamic properties of the system itself, but rather troublesome to determine the properties of the surroundings. For this reason, it is important to derive conditions for equilibrium which relate to the system itself and which ignore the properties of the surroundings. This is conveniently done by combining the first and second laws of thermodynamics. If a system is at equilibrium and we provide an infinitesimal amount of heat dq, the process of heat transfer must be reversible. From the second law, for a reversible process

$$dS = \frac{dq_{\text{rev}}}{T} \tag{5.52}$$

where S is now the entropy of the system itself, without the surroundings. Since for a system at equilibrium any small change is a reversible change, equation (5.52) provides us with a description of a system which must be at equilibrium. For an infinitesimal change, the first law tells us that, for PV work,

$$dU = dq - PdV \tag{5.53}$$

where dU is the increase in internal energy, dq the heat supplied, and PdV the work done by the system. From equations (5.52) and (5.53)

$$dU = TdS - PdV \tag{5.54}$$

This expression, since it involves equation (5.52), which is only true for a reversible change, is a condition for equilibrium. It tells us that if we maintain the entropy and volume constant, i.e., if

$$dS = 0 \qquad dV = 0 \tag{5.55}$$

it is necessary that

$$dU = 0 \tag{5.56}$$

In other words, as shown schematically in Figure 5.10a, a system maintained at constant entropy and volume will come to equilibrium at a state of minimum energy U.

Another condition for equilibrium is obtained if we hold the entropy and pressure constant. Since

$$d(PV) = PdV + VdP \tag{5.57}$$

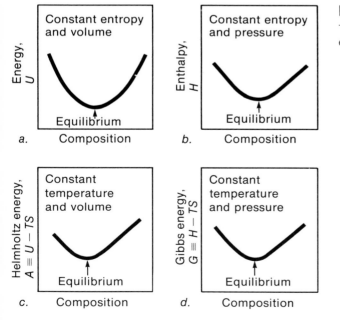

a. Constant entropy and volume — Energy, U — Equilibrium — Composition

b. Constant entropy and pressure — Enthalpy, H — Equilibrium — Composition

c. Constant temperature and volume — Helmholtz energy, $A \equiv U - TS$ — Equilibrium — Composition

d. Constant temperature and pressure — Gibbs energy, $G \equiv H - TS$ — Equilibrium — Composition

Figure 5.10

The conditions for chemical equilibrium.

we can transform equation (5.54) as follows:

$$dU = TdS - d(PV) + VdP \qquad (5.58)$$

or

$$d(U + PV) = TdS + VdP \qquad (5.59)$$

However, $U + PV$ is the enthalpy H, so that

$$dH = TdS + VdP \qquad (5.60)$$

It follows that at constant entropy and pressure, when

$$dS = 0 \qquad dP = 0 \qquad (5.61)$$

the enthalpy must have a minimum value for a system at equilibrium:

$$dH = 0 \qquad (5.62)$$

This is illustrated in Figure 5.10b.

5.7   THE HELMHOLTZ AND GIBBS ENERGIES

Thus far we have been concerned with holding the entropy of the system constant. In practice this cannot be done in any simple manner. On the other hand, we can keep the temperature of a system constant by holding it in a temperature-controlled bath. We are therefore more interested in knowing the conditions for equilibrium when a system is maintained either (a) at constant temperature and volume, or (b) at constant temperature and pressure.

The first situation is dealt with if we combine equation (5.54) with the relationship

$$d(TS) = TdS + SdT \qquad (5.63)$$

We obtain

$$dU = d(TS) - SdT - PdV \qquad (5.64)$$

or

$$d(U - TS) = -SdT - PdV \qquad (5.65)$$

Therefore, if we maintain a system at constant temperature and volume, i.e.,

$$dT = 0 \qquad dV = 0 \qquad (5.66)$$

it will come to equilibrium at a state of minimum $U - TS$:

$$d(U - TS) = 0 \qquad (5.67)$$

(see Figure 5.10c). In view of the significance of this conclusion, the quantity $U - TS$ is given a special symbol, A, and is called the *Helmholtz energy*,† after the German physiologist and physicist Hermann Ludwig Ferdinand von Helmholtz (1821–1894):

$$A = U - TS \qquad (5.68)$$

We can therefore write

$$dA = -SdT - PdV \qquad (5.69)$$

Of even greater interest is the condition for chemical equilibrium when we maintain a system at constant temperature and pressure. For example, most biological systems are well thermostatted, and the processes occur at a constant one-atmosphere pressure. This condition is readily obtained from equation (5.69) by combining it with (5.57):

$$dA = -SdT - d(PV) + VdP \qquad (5.70)$$

or

$$d(A + PV) = -SdT + VdP \qquad (5.71)$$

Since A is $U - TS$ we can also write equation (5.71) as

$$d(U + PV - TS) = -SdT + VdP \qquad (5.72)$$

or as

$$d(H - TS) = -SdT + VdP \qquad (5.73)$$

since H is $U + PV$. The condition for chemical equilibrium at constant temperature and pressure (see Figure 5.10d) is therefore that the function $H - TS$ has a minimum value. This function is known as the *Gibbs energy*, after the American physicist Josiah Willard Gibbs (1839–1903), and is usually given the symbol G:

$$G = H - TS \qquad (5.74)$$

Thus, from (5.73)

$$dG = -SdT + VdP \qquad (5.75)$$

---

† It has long been known as the Helmholtz *free* energy. However, the International Union of Pure and Applied Chemistry has recommended that the "free" be dropped and that we call it the Helmholtz energy. The same recommendation applies to the Gibbs energy.

The Helmholtz and Gibbs energies play an enormously important role in thermodynamics, in view of their significance in leading to equilibrium conditions. It is evident that both the Helmholtz and the Gibbs energies are *state functions*, since they are composed of quantities U, H, T, and S, which are themselves state functions. In the remainder of this book we will often make use of the Gibbs energy, since we will be concerned with processes occurring at constant temperature and pressure and with equilibria established at constant temperature and pressure.

5.8   MOLECULAR INTERPRETATION OF GIBBS ENERGY

As previously stated, although thermodynamic arguments can be developed without any reference to the existence and behavior of atoms and molecules, it is nevertheless informative to interpret the arguments in terms of molecular structure.

Consider first the dissociation of hydrogen molecules into hydrogen atoms

$$H_2 \rightleftharpoons 2H$$

We know that this process occurs only to a very slight extent from left to right at ordinary temperatures, but that if we start with hydrogen atoms the combination will occur spontaneously. We have seen that at constant temperature and pressure a natural or spontaneous process is one in which there is a decrease in Gibbs energy, with the system approaching an equilibrium state in which the Gibbs energy is at a minimum. Clearly, therefore, for the process

$$2H \rightarrow H_2$$

at ordinary temperatures, ΔG is negative. Let us consider how this can be interpreted in terms of enthalpy and entropy changes, in light of the molecular structures.

We know that when hydrogen atoms are brought together and combine there is evolution of heat, which means that the enthalpy goes to a lower level; that is

$$\Delta H(2H \rightarrow H_2) \text{ is negative.}$$

The entropy change is also negative when hydrogen atoms combine, because the atoms have a less ordered arrangement than the molecules:

$$\Delta S(2H \rightarrow H_2) \text{ is negative.}$$

The Gibbs energy change for the combination process at constant temperature is made up as follows:

$$\Delta G = \Delta H - T\Delta S \qquad (5.76)$$
$$(<0) \quad (<0)$$

If T is small enough, ΔG will be negative. This is in fact the situation at room temperature; indeed up to quite high temperatures the negative ΔH term dominates the situation and ΔG is negative, which means that the process occurs spontaneously.

If, however, we go to very high temperatures,[†] the $T\Delta S$ term will become dominant; since ΔS is negative and $T\Delta S$ is *subtracted* from ΔH, the net value of ΔG becomes positive when T is large enough. Therefore we predict that at very high temperatures hydrogen atoms will not spontaneously combine; instead hydrogen molecules will spontaneously dissociate into atoms. This is indeed the case.

In this example the ΔH and $T\Delta S$ terms (except at very high temperatures) work in opposite directions, both being negative. In almost all reactions ΔH and $T\Delta S$ work against each other. In the reaction

$$2H_2 + O_2 \rightarrow 2H_2O$$

ΔH is negative, the reaction being exothermic. $T\Delta S$ is also negative, since there is a decrease in the number of molecules and an increase of order. Thus, at a fixed temperature

$$\Delta G = \Delta H - T\Delta S \qquad (5.77)$$
$$(<0) \quad (<0)$$

At ordinary temperatures $T\Delta S$ is negligible compared with ΔH; ΔG is therefore negative and reaction occurs spontaneously from left to right. As the temperature is raised $T\Delta S$ becomes more negative, so that at sufficiently high temperatures $\Delta H - T\Delta S$ becomes positive. The spontaneous reaction will then be from right to left.

It follows from the relationship

$$\Delta G = \Delta H - T\Delta S \qquad (5.78)$$

that temperature is a weighting factor which determines the relative importance of enthalpy and entropy. At absolute zero $\Delta G = \Delta H$, and the direction of spontaneous change is determined solely by the enthalpy change. At very high temperatures, on the other hand, entropy is the driving force which determines the direction of spontaneous change.

EXAMPLE

As an example of the relationship between the sign of a Gibbs-energy change and the question of whether a process is spontaneous or not, we may consider the vaporization of water at 100° C, the normal boiling point. Suppose first that we have liquid water at 100° C in equilibrium with water vapor, which will be at 1 atm

---

[†] It must be emphasized that ΔH and ΔS do vary somewhat with temperature, so that this discussion is oversimplified.

pressure (Figure 5.11a). Since equilibrium is established

$$\Delta G = 0$$

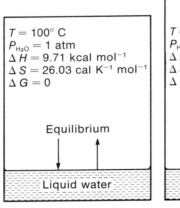
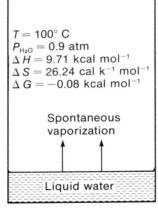

T = 100° C
P_{H_2O} = 1 atm
ΔH = 9.71 kcal mol^{-1}
ΔS = 26.03 cal K^{-1} mol^{-1}
ΔG = 0

Equilibrium

Liquid water

a.

T = 100° C
P_{H_2O} = 0.9 atm
ΔH = 9.71 kcal mol^{-1}
ΔS = 26.24 cal k^{-1} mol^{-1}
ΔG = −0.08 kcal mol^{-1}

Spontaneous
vaporization

Liquid water

b.

Figure 5.11

The vaporization of water at 100° C. In a, liquid water at 100° C is in equilibrium with water vapor at 1 atm pressure. In b, liquid water at 100° C is in contact with water vapor at 0.9 atm pressure, and there is spontaneous vaporization.

The enthalpy increase associated with the vaporization of 1 mol of liquid water is 9.71 kcal:

$$\Delta H = 9.71 \text{ kcal mol}^{-1}$$

Since

$$\Delta G = \Delta H - T\Delta S$$

it follows that

$$\Delta S = \frac{9710}{373} = 26.03 \text{ cal K}^{-1} \text{ mol}^{-1}$$

Suppose now that the volume of the gas phase is increased so that the pressure becomes 0.9 atm (Figure 5.11b). We have seen in equation (5.48) that the entropy increase for the expansion of 1 mol of gas from 1 atm pressure to 0.9 atm is

$$\Delta S = R \ln \frac{V_2}{V_1} = R \ln \frac{P_1}{P_2}$$

$$= R \ln \frac{1.0}{0.9} = 1.99 \ln \frac{1.0}{0.9}$$

$$= 0.21 \text{ cal K}^{-1} \text{ mol}^{-1}$$

The entropy increase when 1 mol of liquid water evaporates to give vapor at 0.9 atm pressure is thus

$$\Delta S = 26.03 + 0.21 = 26.24 \text{ cal K}^{-1} \text{ mol}^{-1}$$

The value of $T\Delta S$ is

$$26.24 \times 373 = 9.79 \text{ kcal mol}^{-1}$$

The value of the Gibbs energy increase is thus

$$\Delta G = 9.71 - 9.79 = -0.08 \text{ kcal mol}^{-1}$$

Since this is a negative quantity, the vaporization process is spontaneous.

5.9   GIBBS ENERGIES OF FORMATION

In Chapter 4 we dealt with enthalpy changes in chemical reactions and found that it was very useful to determine enthalpies of formation of compounds; the standard enthalpy of formation, ΔH_f°, is the enthalpy increase for the process in which a compound is formed, at one atmosphere pressure, from the elements in their standard states. If we know the standard enthalpies of formation of all compounds in a chemical reaction (it is a simple matter, see equation 4.31) to calculate the enthalpy change in the reaction.

Exactly the same procedure is followed with Gibbs energies. We assign every element in its *standard state* (e.g., gaseous H_2, gaseous O_2) a Gibbs energy of zero. The standard Gibbs energy of formation of any *compound* is then simply the Gibbs energy change, ΔG_f°, which accompanies the formation of the element in its standard state from its elements in their standard states. Then we can calculate the standard Gibbs energy change for any reaction, ΔG°, by adding the Gibbs energies of formation of all of the products and subtracting the sum of the Gibbs energies of formation of all of the reactants:

$$\Delta G^\circ = \Sigma \Delta G_f^\circ \text{ (products)} - \Sigma \Delta G_f^\circ \text{ (reactants)} \tag{5.79}$$

Table 4.1 lists Gibbs energies of formation of a number of compounds, as well as of ions in aqueous solution. In the case of ions, it is impossible to measure Gibbs energies of formation of individual ions, since experiments always are done with systems involving ions of opposite signs. To overcome this difficulty the same procedure is adopted as with enthalpies, the arbitrary assumption being made that the Gibbs energy of formation of the proton in water is zero, and the Gibbs energies of formation of all of the other ions are calculated on that basis. The ionic values obtained in this way are known as *conventional* Gibbs energies of formation.

EXAMPLE

Calculate $\Delta G°$ at 25° C for the following fermentation reaction,

$$C_6H_{12}O_6 \text{ (aq)} \rightarrow 2C_2H_5OH \text{ (aq)} + 2CO_2 \text{ (g)}$$
$$\text{(glucose)} \qquad \text{(ethanol)}$$

The standard Gibbs energies of formation of glucose, ethanol, and carbon dioxide are given in Table 4.1.

SOLUTION

From equation (5.77)

$$\Delta G° = (2 \times -43.44) + (2 \times -94.25) - (-218.58)$$
$$= -56.8 \text{ kcal}†$$

Gibbs energies of formation are useful in providing information about the *thermodynamic stability* of a compound. For example, ethylene has a positive Gibbs energy of formation of 16.3 kcal mol^{-1}, which means that the decomposition of ethylene into its elements will occur with a negative standard Gibbs energy change:

$$C_2H_4 \text{ (g)} \rightarrow 2C \text{ (graphite)} + 2H_2$$
$$\Delta G° = -16.3 \text{ kcal}$$

A negative $\Delta G°$ for a reaction means that the process is spontaneous. Ethylene is therefore thermodynamically unstable with respect to decomposition into its elements. A compound whose standard Gibbs energy of formation is positive is known as an *endergonic*‡ compound (compare *endothermic*, for a compound formed with a positive $\Delta H_f°$). Conversely, a compound having a negative $\Delta G_f°$ value is known as an *exergonic* compound (compare *exothermic*). Most compounds are exergonic.

The terms *exergonic* and *endergonic* also are employed with respect to other chemical processes. Thus any reaction having a negative $\Delta G°$ value (i.e., accompanied by a liberation of Gibbs energy) is said to be *exergonic*. A reaction having a positive $\Delta G°$ is said to be *endergonic*.

---

† Again, it is best to express Gibbs energies for reactions as J or kcal for the reaction as written in the equation, since kcal mol^{-1} or J mol^{-1} can be ambiguous.
‡ From the Greek *ergon*, work.

5.10 GIBBS ENERGY AND REVERSIBLE WORK

When an ideal gas is reversibly compressed at constant temperature, the work done is equal to the increase in Gibbs energy. We have seen in equation (4.69) that the reversible work done in compressing one mol of an ideal gas at temperature T from a volume V_1 to a volume V_2 is

$$w_{\text{rev}} = RT \ln \frac{V_1}{V_2} \tag{5.80}$$

During this isothermal process there is no change in internal energy, since the internal energy of an ideal gas is a function of temperature only, not a function of pressure or volume:

$$\Delta U = 0 \tag{5.81}$$

It follows from the first law that the heat absorbed by the system is the negative of the work done on the system:

$$q_{\text{rev}} = RT \ln \frac{V_2}{V_1} \tag{5.82}$$

The entropy change (numerically a decrease since $V_1 > V_2$) is therefore

$$\Delta S = \frac{q_{\text{rev}}}{T} = R \ln \frac{V_2}{V_1} \tag{5.83}$$

There is no change in enthalpy, since enthalpy (like the internal energy) is for an ideal gas a function only of temperature:

$$\Delta H = 0 \tag{5.84}$$

The increase in Gibbs energy is thus

$$\Delta G = \Delta H - T\Delta S \tag{5.85}$$

$$= RT \ln \frac{V_1}{V_2} \tag{5.86}$$

which is the same expression as in (5.80). The reversible work done on the system is thus the increase in Gibbs energy.

Of special importance are the Gibbs energy changes in *processes at constant temperature and pressure*. The Gibbs energy is defined as

$$G = U + PV - TS \tag{5.87}$$

so that in general

$$dG = dU + PdV + VdP - TdS - SdT \qquad (5.88)$$

At constant temperature and pressure

$$dG = dU + PdV - TdS \qquad (5.89)$$

and according to the first law

$$dU = dq + dw \qquad (5.90)$$

Then

$$dG = dq + dw + PdV - TdS \qquad (5.91)$$

We consider first the case in which the *only work performed is that done as a result of a volume change.* In this case

$$dw = -PdV \qquad (5.92)$$

so that equation (5.91) reduces to

$$dG = dq - TdS \qquad (5.93)$$

The equations (5.87) to (5.93) apply whether the process is reversible or not. If the process is reversible,

$$dq_{rev} = TdS \qquad (5.94)$$

Thus, for a reversible process occurring at constant temperature and pressure, with no work other than PV work,

$$dG_{rev} = 0 \qquad (5.95)$$

This relationship does not apply if the process is *irreversible.* In this case

$$TdS > q_{irr}$$

and it then follows from (5.93) that

$$dG_{irr} < 0 \qquad \text{(only } PV \text{ work)} \qquad (5.96)$$

We have seen in equation (5.95) that there is no change in Gibbs energy for a reversible process occurring at constant temperature and pressure, provided that the work done results only from a volume change. However, there are other kinds of work, e.g., electrical work and osmotic work, and we can refer to this as non-PV work.

It is of importance to see how the Gibbs energy change is related to non-PV work. We can show, by an argument which is a slight extension of that just given in equations (5.87) to (5.95), that for a reversible process the Gibbs energy increases by an amount equal to the non-PV work done on the system.

In place of equation (5.92) we now have

$$dw = -PdV + dw_{\text{non-}PV} \qquad (5.97)$$

where $dw_{\text{non-}PV}$ is the non-PV work done on the system. Then from (5.91),

$$dG = dq + dw_{\text{non-}PV} - TdS \qquad (5.98)$$

Contrast this with (5.93). If the process is reversible, (5.94) applies, and now

$$dG_{\text{rev}} = dw_{\text{non-}PV} \qquad (5.99)$$

For a finite change

$$\Delta G_{\text{rev}} = w_{\text{non-}PV} \qquad (5.100)$$

Thus, *for a reversible process at constant temperature and pressure, the Gibbs energy changes by an amount equal to the non-PV work done on the system.* This important result is applied to electrochemical cells on p. 342.

5.11   THE EQUILIBRIUM CONSTANT

If we consider 1 mol of a gas, the concentration c is simply the reciprocal of the volume:

$$c = \frac{1}{V} \qquad (5.101)$$

Thus the Gibbs energy change ΔG when 1 mol of an ideal gas changes its volume V_1 to V_2 at constant temperature is given not only by equation (5.86) but also by

$$\Delta G = RT \ln \frac{c_2}{c_1} \qquad (5.102)$$

where c_1 and c_2 are the concentrations in states 1 and 2 respectively. This equation can also be written as

$$\Delta G = RT \ln c_2 - RT \ln c_1 \qquad (5.103)$$

If the initial state is taken to be 1 mol dm^{-3}, i.e., $c_1 = 1$ M, we have

$$\Delta G = RT \ln c_2 \qquad (5.104)$$

The Gibbs energy at concentration c (dropping the subscript 2) is thus greater than that at 1 M by $RT \ln c$. If the Gibbs energy at 1 M is denoted as G°, the Gibbs energy at c mol dm^{-3} is given by

$$G = G^{\circ} + RT \ln c \qquad (5.105)$$

Now consider a gaseous chemical reaction having the stoichiometric equation

$$aA + bB \rightleftharpoons xX + yY$$

the concentrations of A, B, X, and Y being [A], [B], [X], and [Y]. The Gibbs energy of a moles of A is

$$aG_A = aG_A^{\circ} + aRT \ln [A] \qquad (5.106)$$

where G_A° is the Gibbs energy of 1 mol of A at a concentration of 1 mol dm^{-3}. Similarly

$$bG_B = bG_B^{\circ} + bRT \ln [B] \qquad (5.107)$$

$$xG_X = xG_X^{\circ} + xRT \ln [X] \qquad (5.108)$$

$$yG_Y = yG_Y^{\circ} + yRT \ln [Y] \qquad (5.109)$$

The increase in Gibbs energy when a mol of A (at concentration [A]) react with b mol of B (at concentration [B]) to give x mol of X (at concentration [X]) and y mol of Y (at concentration [Y]) is thus

$$\Delta G = xG_X + yG_Y - aG_A - bG_B \qquad (5.110)$$

$$= xG_X^{\circ} + yG_Y^{\circ} - aG_A^{\circ} - bG_B^{\circ} + RT \ln \frac{[X]^x[Y]^y}{[A]^a[B]^b} \qquad (5.111)$$

$$= \Delta G^{\circ} + RT \ln \frac{[X]^x[Y]^y}{[A]^a[B]^b} \qquad (5.112)$$

where ΔG°, equal to $xG_X^{\circ} + yG_Y^{\circ} - aG_A^{\circ} - bG_B^{\circ}$, is the *standard Gibbs-energy change*.

If the concentrations [A], [B], [X], and [Y] correspond to equilibrium, the Gibbs-energy change ΔG is equal to zero and thus

$$\Delta G^{\circ} = -RT \ln \left(\frac{[X]^x[Y]^y}{[A]^a[B]^b} \right)_{\text{eq}} \qquad (5.113)$$

ΔG° is a constant, and therefore the ratio of concentrations is also a constant:

$$\left(\frac{[X]^x[Y]^y}{[A]^a[B]^b} \right)_{\text{eq}} = K \qquad (5.114)$$

The constant K is known as the *equilibrium constant* for the reaction. We thus have

$$\Delta G^\circ = -RT \ln K \qquad\qquad (5.115)$$

If, on the other hand, the initial and final concentrations do not correspond to equilibrium,

$$\Delta G = -RT \ln K + RT \ln \frac{[X]^x[Y]^y}{[A]^a[B]^b} \qquad\qquad (5.116)$$

$$= -RT\left[\ln K - \ln \frac{[X]^x[Y]^y}{[A]^a[B]^b}\right] \qquad\qquad (5.117)$$

This is the Gibbs-energy increase that occurs when a mol of A, at concentration $[A]$, react with b mol of B, at concentration $[B]$, to produce x mol of X, at concentration $[X]$, and y mol of Y, at concentration $[Y]$.

The relationship between ΔG° and K, as expressed by equation (5.115), is illustrated in Figure 5.12. We see from (5.115) that if ΔG° is positive, $\ln K$ is negative, which means that K is less than unity; this case is illustrated in Figure 5.12a. If, on the other hand, ΔG° is negative, $\ln K$ is positive, which means that K is greater than unity (Figure 5.12b). If ΔG° is zero, $\ln K$ is zero, which means that $K = 1$ (Figure 5.12c).

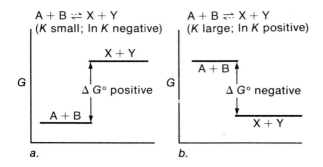

a.

b.

c.

Figure 5.12

A standard Gibbs energy diagram, illustrating the relationship between ΔG° for a reaction and the equilibrium constant K. In a, ΔG° is positive and K is small; in b, ΔG° is negative and K is large; in c, ΔG° is zero and K is unity.

REACTIONS IN SOLUTION

The above relationships have been derived for equilibria involving ideal gases. In practice they are often obeyed very satisfactorily by reactions occurring in solution, especially if the concentrations are fairly low. At higher concentrations of solutes there may be significant deviations from the ideal equations, which are dealt with by the use of activity coefficients.

We have seen in equation (5.105) that for an ideal gas, the Gibbs energy of a component is related to its concentration c by

$$G = G^\circ + RT \ln c \tag{5.118}$$

where G° is the Gibbs energy at unit concentration. For a gas the concentration unit may be 1 mol dm^{-3}, and we say that the *standard state* is 1 M. For a substance in solution it is customary to take the standard state as 1 mol kg^{-1}, i.e., 1 molal (1 m), and the relationship for an ideal system is

$$G = G^\circ + RT \ln m \tag{5.119}$$

where m is the molality. Deviations from ideality are taken care of by multiplying the molality m by an activity coefficient γ:

$$G = G^\circ + RT \ln \gamma m \tag{5.120}$$

The product γm is referred to as the *activity* and we give it the symbol a. We see that G° is the Gibbs energy of the solute at unit activity.

If we follow through the arguments on p. 218 using equation (5.120) instead of (5.105), the resulting expression for the equilibrium constant for a reaction

$$A + B \rightleftharpoons X + Y$$

is

$$K = \frac{a_X a_Y}{a_A a_B} = \frac{m_X m_Y}{m_A m_B} \times \frac{\gamma_X \gamma_Y}{\gamma_A \gamma_B} \tag{5.121}$$

This is the true equilibrium constant, and the ratio

$$\frac{m_X m_Y}{m_A m_B}$$

may show some variation as the concentrations are changed. For uncharged species in aqueous solution it is usually a satisfactory approximation to express equilibrium constants in terms of molalities or molarities. However, when ions are involved there may be serious deviations, and *activities* must be used in place of *concentrations*. This matter is described in further detail in Chapters 7 and 8.

Thus far we have been concerned with solute species. If we wish to deal with Gibbs energies of solvents, the usual procedure is a little different. We now work with the mole fraction x_1 of the solvent,† and for an ideal system express the Gibbs energy as

$$G = G^o + RT \ln x_1 \tag{5.122}$$

where G^o is the Gibbs energy at unit mole fraction, i.e., for the pure solvent. Nonideal states are taken care of by an activity coefficient, which for mole fractions is given the symbol f:

$$G = G^o + RT \ln x_1 f_1 \tag{5.123}$$

We shall later see an application of equation (5.122) to the elevation of the boiling point and the depression of the freezing point.

Now let us consider some examples of the relationship between the standard Gibbs energy change, ΔG^o, and the equilibrium constant K, making the assumption that the behavior is ideal.

EXAMPLE

The enzyme *aldolase* catalyzes the breakdown of fructose-1, 6-diphosphate and the reverse reaction :‡

$$\text{Fructose-1,6-diphosphate} \rightleftharpoons \text{glyceraldehyde-3-phosphate}$$

$$+ \text{ dihydroxyacetone phosphate}$$

At 25° C the equilibrium constant for this reaction is given by

$$\frac{[\text{glyceraldehyde-3-phosphate}][\text{dihydroxyacetone phosphate}]}{[\text{fructose-1,6-diphosphate}]}$$

$$= 8.9 \times 10^{-5} \ M$$

Calculate ΔG^o for this reaction.

SOLUTION

We know that

$$\Delta G^o = -RT \ln K = -2.303 \ R \log_{10} K$$

---

† The mole fraction, x_B, of a substance B is the ratio of the number of moles of B, n_B, to the total number of moles present in the system:

$$x_B = \frac{n_B}{\sum_i n_i}$$

‡ We shall see in Chapter 9 that if catalysts such as enzymes catalyze a reaction in one direction, they must also do so in the reverse direction.

where $R = 1.99$ cal K^{-1} mol^{-1}; $T = 298$ K; $K = 8.9 \times 10^{-5}$ M. Thus

$$\Delta G° = -(1.99 \times 298 \times 2.303)(-4.05)$$

$$= 5520 \text{ cal}$$

$$= 5.52 \text{ kcal}$$

EXAMPLE

Suppose that, for the above reaction, we mix fructose-1,6-diphosphate (FDP) at 0.01 M with glyceraldehyde-3-phosphate (G-3-P) and dihydroxyacetone phosphate (DHAP), both at 10^{-5} M. In which direction will the reaction occur?

SOLUTION

To answer this, we must calculate ΔG for the reaction under these conditions. Equation (5.112) for this reaction is

$$\Delta G = \Delta G° + RT \ln \frac{[\text{G-3-P}][\text{DHAP}]}{[\text{FDP}]}$$

For this example $\Delta G° = 5520$ cal, and the concentrations are as given above. Then

$$\Delta G = 5520 + (2.303 \times 1.99 \times 298) \log_{10} \frac{10^{-5} 10^{-5}}{10^{-2}}$$

$$= 5520 + (1364)(-8)$$

$$= -5390 \text{ cal}$$

Since this is negative, the reaction will proceed spontaneously from left to right under these conditions. It is important to note that this situation exists in spite of the positive value of $\Delta G°$. The positive value of $\Delta G°$ tells us only that reaction is spontaneous from right to left if the reactant and product concentrations are initially 1 M; for other conditions ΔG must be calculated from equation (5.112) or (5.117).

TESTS FOR CHEMICAL EQUILIBRIUM

It follows from the arguments presented above that any reaction

$$a\text{A} + b\text{B} + \cdots \rightleftharpoons x\text{X} + y\text{Y} + \cdots$$

will eventually reach a state of equilibrium at which

$$\frac{[X]^x[Y]^y}{[A]^a[B]^b}\cdots = K \qquad\qquad (5.124)$$

where K is a constant. Obviously, it is of great practical importance to determine whether a chemical system is in a state of equilibrium.

It is not enough to simply establish that the composition of the system does not change as time goes on. Reaction may be proceeding so slowly that no detectable change will occur over a very long period of time. A good example is the reaction

$$2H_2 + O_2 \rightarrow 2H_2O$$

The equilibrium for this reaction lies almost completely to the right. However, the reaction is so slow that if we bring hydrogen and oxygen together at room temperature, there will be no detectable change even over many hundreds of years. In fact, even after 5×10^9 years (the estimated age of the solar system) only an insignificant amount of reaction will have taken place!

Obviously, more practical tests for equilibrium are required. One of these consists of adding a condition that speeds up the reaction. For example, if to the hydrogen-oxygen mixture we introduce a lighted match, or if we add certain substances known as *catalysts* (an example is powdered platinum), the reaction will occur with explosive violence. This shows that the system was not at equilibrium. It simply appeared to be at equilibrium because of the slowness of the reaction. The function of catalysts in speeding up reactions will be considered later; here we may simply note:

Catalysts do not affect the position of equilibrium; they merely decrease the time required for equilibrium to be attained.

A second and more fundamental test for equilibrium is as follows. If a system

$$A + B \rightleftharpoons X + Y$$

is truly at equilibrium, the addition of a small amount of A or B will cause the equilibrium to shift to the right, with the formation of more X or Y (this is further discussed later, p. 226). Similarly, the addition of more X or Y will cause a shift to the left. If the system is not in a state of true equilibrium, these shifts either will not occur at all, or they will not occur in the manner predicted by the equilibrium equations.

UNITS OF THE EQUILIBRIUM CONSTANT

The practical equilibrium constant has units, which depend upon the type of reaction. Consider, for example, a reaction of the type

$$A + B \rightleftharpoons X$$

in which there has been a decrease, by one, in the number of molecules. The equilibrium constant is then

$$K = \frac{[X]}{[A][B]} \qquad (5.125)$$

If the concentrations are expressed in mol dm^{-3}, we have

$$K = \frac{[X](\text{mol dm}^{-3})}{[A][B](\text{mol dm}^{-3})^2} \qquad (5.126)$$

so that the units of K are dm^3 mol^{-1}.

If there is no change in the number of molecules, as in a reaction of the type

$$A + B \rightleftharpoons X + Y$$

the units cancel out, and K is dimensionless. If the stoichiometric equation involves an increase Δn in the number of moles, the units of K are

$$(\text{mol dm}^{-3})^{\Delta n}$$

If moles cm^{-3} are used to express concentrations, the units will be correspondingly different.

Students are sometimes puzzled by the fact that although practical equilibrium constants have units, we have to make use of relationships such as equation (5.115) in which we are taking the logarithm of an equilibrium constant. If we trace back the argument leading to (5.115) we see that the apparent difficulty begins with (5.105), where we have taken the logarithm of a concentration, which is again a quantity having units. In fact, (5.105) comes from (5.102), where there is no difficulty since c_2/c_1 is dimensionless. In arriving at (5.105) we have taken c_1 as unity, so that the c in (5.105) is really the ratio of a concentration to a unit concentration. Similarly, in (5.115), K is in reality the ratio of the *practical* equilibrium constant to the corresponding function of concentrations, all of which have been set equal to unity. In other words, the *thermodynamic* equilibrium constant is dimensionless. There is thus no difficulty about the logarithms in equations (5.113) and (5.115), and in related equations. However, it is important to realize that the value of the thermodynamic equilibrium constant does depend upon the standard state employed.

Another important point about equilibrium constants is that their value depends upon how the stoichiometric equation is written. Consider, for example, the dissociation

(a) $A \rightleftharpoons 2X$

the equilibrium constant K for which is

$$K = \frac{[X]^2}{[A]} \text{ mol dm}^{-3} \qquad (5.127)$$

If the concentrations are expressed as mol dm^{-3}, the units will be as shown above. Alternatively, we could write the reaction as

(b) $\frac{1}{2}A \rightleftharpoons X$

and express the equilibrium constant as

$$K' = \frac{[X]}{[A]^{\frac{1}{2}}} \, \text{mol}^{\frac{1}{2}} \, \text{dm}^{-\frac{3}{2}} \tag{5.128}$$

This latter constant K' is obviously the square root of K:

$$K' = K^{\frac{1}{2}} \tag{5.129}$$

The standard Gibbs-energy change for the reaction written as (a) will be

$$\Delta G_a^\circ = -RT \ln K \tag{5.130}$$

while that for the reaction written as (b) will be

$$\Delta G_b^\circ = -RT \ln K^{\frac{1}{2}} \tag{5.131}$$

Since

$$\ln K^{\frac{1}{2}} = \frac{1}{2} \ln K \tag{5.132}$$

we see that

$$\Delta G_b^\circ = \frac{1}{2} \Delta G_a^\circ \tag{5.133}$$

This is as it should be, since if we write the reaction as (b), the reaction to which ΔG_b° refers is for the conversion of *one-half mole* of A into X; for (a) we refer to the conversion of *one mole* of A into 2X.

SHIFTS OF EQUILIBRIUM AT CONSTANT TEMPERATURE

One of the important consequences of the theory of equilibrium relates to the way in which equilibria shift as the volume of the system changes. It is important to realize that the equilibrium constant is a function of temperature only, and does not change as we change the volume. However, because of this very constancy of the equilibrium constant, the position of equilibrium does shift as we vary the volume. Why is this so?

Consider the equilibrium

$$AB \rightleftharpoons A^+ + B^-$$

in which a species is dissociating into ions. Assuming ideal behavior, the equilibrium constant K is

$$K = \frac{[A^+][B^-]}{[A]} \tag{5.134}$$

Obviously, if we add A^+ to the system the ratio will be temporarily increased; the equilibrium must shift from right to left to keep K constant. Similarly, addition of B^- will cause a shift to the left. Addition of AB will cause a shift to the right.

The effect of changing the *volume* can be seen if we express each concentration in (5.134) as the ratio of the number of moles to the volume; i.e.,

$$[AB] = \frac{n_{AB}}{V} \tag{5.135}$$

$$[A^+] = \frac{n_{A^+}}{V} \tag{5.136}$$

$$[B^-] = \frac{n_{B^-}}{V} \tag{5.137}$$

so that (5.134) becomes

$$\frac{(n_{A^+}/V)(n_{B^-}/V)}{n_{AB}/V} = K \tag{5.138}$$

$$\frac{n_{A^+} n_{B^-}}{n_{AB}} \times \frac{1}{V} = K \tag{5.139}$$

Suppose that we dilute the system by adding solvent and making V larger. Since K must remain constant, the ratio

$$\frac{n_{A^+} n_{B^-}}{n_{AB}}$$

must become larger in proportion to V; that is, there will be more dissociation into ions at the larger volume. Increasing the volume causes an increase in the total number of solute species present.

It is easy to show by this type of argument that for any reaction, increasing the volume will cause a shift in the direction of producing more molecules. If the reaction is such that the number of species is the same on both sides of the equation, the volume cancels out in K, and changing the volume brings about no shift in equilibrium. These effects were predicted in 1884 by the *principle of Le Chatelier*.

COUPLING OF REACTIONS

We have seen in the last section that for a chemical reaction

$$(1) \qquad A + B \rightleftharpoons X + Y$$

removal of either of the products X or Y will lead to a shift of equilibrium to the right. Frequently in biological systems a product is removed by the occurrence of another reaction; for example, X might isomerize into Z:

(2) $X \rightleftharpoons Z$

We can write an equilibrium constant for reaction (1) as

$$K_1 = \left(\frac{[X][Y]}{[A][B]} \right)_{eq} \tag{5.140}$$

and a corresponding standard Gibbs-energy change

$$\Delta G_1^{\circ} = -RT \ln K_1 \tag{5.141}$$

Similarly for the second reaction

$$K_2 = \left(\frac{[Z]}{[X]} \right)_{eq} \tag{5.142}$$

and

$$\Delta G_2^{\circ} = -RT \ln K_2 \tag{5.143}$$

It should be emphasized that the concentrations appearing in equations (5.140) and (5.142) are those corresponding to equilibrium having been reached.
    We can add together reactions (1) and (2) to obtain

(3) $A + B \rightleftharpoons Z + Y$

Suppose that we multiply together K_1 and K_2:

$$K_1 K_2 = \left(\frac{[X][Y]}{[A][B]} \right)_{eq} \left(\frac{[Z]}{[X]} \right)_{eq} \tag{5.144}$$

$$= \left(\frac{[Z][Y]}{[A][B]} \right)_{eq} \tag{5.145}$$

Obviously $K_1 K_2$ is the equilibrium constant K_3 for the combined reaction (3):

$$K_3 = K_1 K_2 \tag{5.146}$$

We can take natural logarithms of both sides of this equation

$$\ln K_3 = \ln K_1 + \ln K_2 \tag{5.147}$$

and then multiply by $-RT$:

$$-RT \ln K_3 = -RT \ln K_1 - RT \ln K_2 \qquad (5.148)$$

so that

$$\Delta G_3^\circ = \Delta G_1^\circ + \Delta G_2^\circ \qquad (5.149)$$

This result is easily shown to be quite general: if we add together two reactions to obtain a third, the ΔG° of the third reaction is simply the sum of the ΔG° values of the component reactions. This is an extension of Hess's law to standard Gibbs-energy changes.

As a result of this coupling of chemical reactions, it is quite possible for a reaction to occur extensively, even though it has a positive value of ΔG°. Thus, for the scheme given above, suppose that reaction (1) has a positive value:

$$\Delta G_1^\circ = \text{positive} \qquad (5.150)$$

Reaction (1) by itself therefore will not occur to a considerable extent. Suppose, however, that reaction (2) has a negative value. It is then possible for ΔG_3° to be negative, so that A + B will react to a considerable extent to give Z + Y. In terms of the Le Chatelier principle this simply means that X is constantly removed by the isomerization reaction (2), so that the equilibrium of (1) is shifted over to the right. This principle is illustrated by the following examples.

EXAMPLE
A reaction

(1) $A + B \rightleftharpoons X + Y$

has a ΔG_1° value at 25° C of 4.09 kcal. Calculate its equilibrium constant. If the reaction is coupled with

(2) $X \rightleftharpoons Z$

which has a ΔG_2° value of -5.46 kcal, is there now a more extensive reaction of A with B?

SOLUTION
The equilibrium constant for (1) is

$$K_1 = e^{-\Delta G_1^\circ/RT}$$
$$= 10^{-4090/(2.303 \times 1.99 \times 298)}$$
$$= 10^{-3}$$
$$= 0.001$$

When reactions (1) and (2) are coupled the net reaction is

(3) $A + B \rightleftharpoons Z + Y$

and the standard Gibbs-energy change is

$$\Delta G_3^\circ = \Delta G_1^\circ + \Delta G_2^\circ$$
$$= 4.09 - 5.46$$
$$= -1.37 \text{ kcal}$$

The equilibrium constant for (3) is thus

$$K_3 = 10^{1370/(2.303 \times 1.99 \times 298)}$$
$$= 10^{1.00}$$

The equilibrium (3) therefore lies much more to the right.

EXAMPLE

The aminotransferases (also known as transaminases) are enzymes which perform the important function of transferring amino groups from one amino acid with the formation of another. For example,

$$\alpha\text{-ketoglutarate} + \text{alanine} \rightleftharpoons \text{glutamine} + \text{pyruvate}$$

At 25° C, ΔG° for this reaction is 0.06 kcal. Suppose that pyruvate is further oxidized in a process for which ΔG° is -61.8 kcal. Does this enhance the formation of glutamine?

SOLUTION

Qualitatively we can see that this will be the case, since the oxidation reaction with $\Delta G^\circ = -61.8$ kcal will occur essentially to completion. Pyruvate therefore is completely removed as it is formed, and the reaction will go completely in favor of glutamine.

Quantitatively, we have for reaction (1) the glutamine synthesis,

$$K_1 = 10^{-60/(2.303 \times 1.99 \times 298)}$$
$$= 10^{-0.043} = 10^{0.957} \times 10^{-1}$$
$$= 9.057 \times 10^{-1} = 0.91$$

However ΔG_3° for the combined reaction is $0.06 - 61.8 = -61.74$ kcal so that

$$K_3 = 10^{61\ 740/(2.303 \times 1.99 \times 298)} = 10^{45.20} = 1.6 \times 10^{45}$$

There is another way in which chemical reactions can be coupled. In the cases we have considered, the product of one reaction is removed in a second reaction. Alternatively, two reactions might be coupled by a catalyst. Suppose, for example, that there are two independent reactions having no common reactants or products:

(1) $A + B \rightleftharpoons X + Y$

(2) $C \rightleftharpoons Z$

There might exist some catalyst which brings about the reaction

(3) $A + B + C \rightleftharpoons X + Y + Z$

This catalyst would couple reactions (1) and (2) together.

EXAMPLE

A good example is provided by the synthesis of glutamine in living systems. The reaction is

(1)   glutamate $+ NH_4^+ \rightleftharpoons$ glutamine

The ΔG_1^0 for this reaction is 3.75 kcal at $37°$ C, so that the equilibrium lies to the left. In fact

$$K_1 = 10^{-3750/(4.57 \times 310)}$$
$$= 10^{-2.650} = 10^{0.350} \times 10^{-3} = 2.24 \times 10^{-3}$$

Another important biological reaction is the hydrolysis of adenosine triphosphate (ATP) into adenosine diphosphate (ADP) and phosphate (P)

(2) $ATP \rightleftharpoons ADP + P$

The ΔG_2^0 value for this at $37°$ C under certain ionic conditions is $- 7.4$ kcal, so that

$$K_2 = 10^{7400/(4.57 \times 310)}$$
$$= 10^{5.226} = 1.68 \times 10^5$$

These two reactions are coupled together by the enzyme glutamine synthetase, which catalyzes the reaction

(3)   glutamate $+ NH_4^+ + ATP \rightleftharpoons$ glutamine $+ ADP + P$

Since this reaction is the sum of (1) and (2), its Gibbs-energy change,

ΔG_3°, is

$$\Delta G_3^\circ = \Delta G_1^\circ + \Delta G_2^\circ = 3.75 - 7.40 = -3.65 \, \text{kcal}$$

Its equilibrium therefore lies to the right, the equilibrium constant being

$$K_3 = 10^{3650/(4.57 \times 310)}$$

$$= 10^{2.580} = 3.80 \times 10^2$$

Note that $K_3 = K_1 K_2$.

The net result is that although the glutamine synthesis, reaction (1), is endergonic, so that it does not occur to any great extent by itself, the addition of ATP and the enzyme leads to an exergonic reaction, the equilibrium of which lies to the right. This type of coupling of reactions is quite common in biological systems.

5.12   TEMPERATURE DEPENDENCE OF EQUILIBRIUM CONSTANTS

The way in which equilibrium constants vary with temperature is a matter of considerable importance in thermodynamics. It leads us to a very convenient experimental procedure for measuring enthalpy changes in chemical reactions.

An equation for the effect of temperature on the Gibbs energy change in a reaction can be derived if we start with equation (5.75):

$$dG = -SdT + VdP \tag{5.151}$$

We also have, from differential calculus, the general relationship

$$dG = \left(\frac{\partial G}{\partial T}\right)_P dT + \left(\frac{\partial G}{\partial P}\right)_T dP \tag{5.152}$$

This equation merely says that if G is a function of T and P (see Figure 5.13) the effect of an infinitesimal temperature increase dT is to increase G by $(\partial G/\partial T)_P \, dT$, the partial derivative being merely the slope of the G versus T curve along the T axis. Similarly, the effect of an infinitesimal pressure increase dP is to increase G by $(\partial G/\partial P)_T \, dP$. From equations (5.151) and (5.152) it follows that

$$-S = \left(\frac{\partial G}{\partial T}\right)_P \qquad \text{and} \qquad V = \left(\frac{\partial G}{\partial P}\right)_T \tag{5.153}$$

It is the first of these relations with which we are concerned. For a change in entropy

ΔS we have

$$-\Delta S = \left(\frac{\partial \Delta G}{\partial T}\right)_P \tag{5.154}$$

However, since

$$\Delta G = \Delta H - T\Delta S \tag{5.155}$$

we can write ΔS as

$$\Delta S = \frac{\Delta H - \Delta G}{T} \tag{5.156}$$

and introduction of this into (5.154) gives

$$\left(\frac{\partial \Delta G}{\partial T}\right)_P = \frac{\Delta G - \Delta H}{T} \tag{5.157}$$

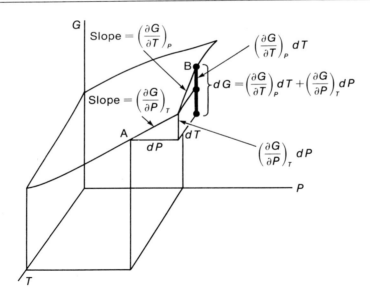

Figure 5.13

The Gibbs energy G as a function of T and P; an interpretation of equation 5.152. The slope of the surface in the T direction, at constant P, is $(\partial G/\partial T)_P$, so that this multiplied by dT gives the effect of increasing the temperature alone by dT. Similarly, at constant T the slope along the P axis is $(\partial G/\partial P)_T$, and this multiplied by dP is the increase in G due to an increase in pressure of dP. The net dG is the sum of the two terms.

or

$$\left(\frac{\partial \Delta G}{\partial T}\right)_P - \frac{\Delta G}{T} = -\frac{\Delta H}{T} \tag{5.158}$$

From differential calculus we have

$$\frac{\partial}{\partial T}\left(\frac{\Delta G}{T}\right) = \frac{1}{T}\frac{\partial \Delta G}{\partial T} - \frac{\Delta G}{T^2} \tag{5.159}$$

and the left-hand side of (5.158) can therefore be written as

$$T\frac{\partial}{\partial T}\left(\frac{\Delta G}{T}\right)$$

Thus

$$T\frac{\partial}{\partial T}\left(\frac{\Delta G}{T}\right) = -\frac{\Delta H}{T} \tag{5.160}$$

or

$$\frac{\partial}{\partial T}\left(\frac{\Delta G}{T}\right) = -\frac{\Delta H}{T^2} \tag{5.161}$$

This is an important thermodynamic relationship, which is known as the *Gibbs-Helmholtz equation*. For processes involving substances in their standard states the equation takes the form

$$\frac{\partial}{\partial T}\left(\frac{\Delta G^\circ}{T}\right) = -\frac{\Delta H^\circ}{T^2} \tag{5.162}$$

This equation easily leads to an equation for the variation of equilibrium constants with temperature, in view of the fact that

$$\Delta G^\circ = -RT \ln K \tag{5.163}$$

Introduction of this into (5.162) gives

$$\frac{\partial}{\partial T}(-R \ln K) = -\frac{\Delta H^\circ}{T^2} \tag{5.164}$$

or to

$$\frac{\partial \ln K}{\partial T} = \frac{\Delta H^\circ}{RT^2} \tag{5.165}$$

In the last few equations, for convenience we have omitted the brackets around the partial derivatives, but it is to be understood that we are concerned with processes at constant pressure.

Equation (5.165) can be written as

$$\frac{\partial \ln K}{\partial(1/T)} = -\frac{\Delta H^\circ}{R} \tag{5.166}$$

since we know from differential calculus that $\partial(1/T) = -\partial T/T^2$. Equation (5.166) tells us that a plot of $\ln K$ against $1/T$ will have a slope equal to $\Delta H^\circ/R$. This is shown in Figure 5.14a, which applies to the general situation in which ΔH° varies with temperature, so that the slope varies with temperature (i.e., we do not have a straight-line relationship). However, it is frequently found that plots of $\ln K$ against $1/T$ are linear (i.e., are straight lines); this is shown in Figure 5.14b and tells us that ΔH° is independent of temperature over the range investigated.

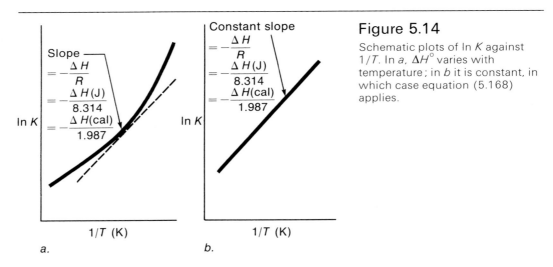

Figure 5.14

Schematic plots of $\ln K$ against $1/T$. In a, ΔH° varies with temperature; in b it is constant, in which case equation (5.168) applies.

If instead of plotting $\ln K$ against $1/T$ we plot $\log_{10} K$, the slope is now

$$-\frac{\Delta H^\circ}{2.303\,R}$$

Since $R = 1.987$ cal K^{-1} mol^{-1} = 8.314 J K^{-1} mol^{-1}, the slopes are equal to

$$-\frac{\Delta H^\circ(\text{cal})}{4.57} \qquad \text{or to} \qquad -\frac{\Delta H^\circ(\text{J})}{19.147}$$

From the plots the ΔH° values can therefore be calculated. This is a very important way of determining enthalpy changes for chemical reactions.

When ΔH° is independent of temperature (as indicated by linear plots such as that in Fig. 5.10b) it is possible to integrate (5.166) as follows:

$$d \ln K = -\frac{\Delta H^\circ}{R} d(1/T) \qquad (5.167)$$

Thus

$$\ln K = -\frac{\Delta H^\circ}{RT} + I \qquad (5.168)$$

where I is the constant of integration. However, we know that

$$\ln K = -\frac{\Delta G^\circ}{RT} \qquad (5.169)$$

$$= -\frac{\Delta H^\circ}{RT} + \frac{\Delta S^\circ}{R} \qquad (5.170)$$

since $\Delta G^\circ = \Delta H^\circ - T\Delta S^\circ$. The constant of integration I is therefore identified as $\Delta S^\circ/R$.

EXAMPLE
An enzyme inhibitor Q is bound to an enzyme E, the process being

$$E + Q \rightleftharpoons EQ$$

The equilibrium constant for this association reaction was found to be $1.80 \times 10^3 \text{ M}^{-1}$ at 25° C, and $3.45 \times 10^3 \text{ M}^{-1}$ at 40° C. Assuming ΔH° to be constant over the temperature range, calculate its value and that of ΔS°.

SOLUTION
If the lower temperature, 298 K, is written as T_1 and the corresponding equilibrium constant as K_1, we have from (5.170)

$$\ln K_1 = -\frac{\Delta H^\circ}{RT_1} + \frac{\Delta S^\circ}{R} \qquad (5.171)$$

Similarly for the higher temperature, 313 K,

$$\ln K_2 = -\frac{\Delta H^\circ}{RT_2} + \frac{\Delta S^\circ}{R} \qquad (5.172)$$

Subtraction of (5.171) from (5.172) gives

$$\ln K_2 - \ln K_1 = -\frac{\Delta H^\circ}{R} \left(\frac{1}{T_2} - \frac{1}{T_1} \right) \qquad (5.173)$$

or

$$\ln \frac{K_2}{K_1} = \frac{\Delta H^\circ}{R} \left(\frac{T_2 - T_1}{T_1 T_2} \right)$$

(5.174)

In our particular example we have

$$2.303 \log_{10} \frac{3.45}{1.80} = \frac{\Delta H^\circ}{1.99} \times \frac{15}{298 \times 313}$$

or

$$2.303 \times 0.282 = \frac{\Delta H^\circ}{1.99} \times \frac{15}{298 \times 313}$$

Thus

$$\Delta H^\circ = \frac{2.303 \times 0.282 \times 1.99 \times 298 \times 313}{15}$$

$$= 8036 \text{ cal}$$

$$= 8.036 \text{ kcal}$$

$$= 33.62 \text{ kJ}$$

The entropy change ΔS° can be calculated by inserting the values of K, ΔH° and T into either equation (5.171) or (5.172). For example

$$2.303 \log_{10} (3.45 \times 10^3) = - \frac{8036}{1.99 \times 298} + \frac{\Delta S^\circ}{1.99}$$

thus

$$\Delta S^\circ = 4.57 \times 3.538 + \frac{8036}{298}$$

$$= 43.14 \text{ cal K}^{-1}$$

$$= 180.5 \text{ J K}^{-1}$$

5.13   THE COLLIGATIVE PROPERTIES

We have seen in Chapter 3 that measurements of the colligative properties are of great value in obtaining molecular weights. The colligative properties of solutions depend largely on the relative amounts of solvent and solute present, and only to a small extent on the nature of the solute species. We can derive the relationship between the colligative properties from thermodynamics.

RAOULT'S LAW

We shall start with Raoult's law as an empirical relationship. For a system of two components, solvent and solute, the partial vapor pressure p_1 of the solvent in a solution of solvent mole fraction x_1 is related to the vapor pressure of the pure solution, p_1^0, by

$$\frac{p_1}{p_1^0} = x_1 \qquad\qquad (5.175)$$

In 1886 the French chemist Francois Marie Raoult (1830–1907) reported extensive vapor-pressure results on solutions, and found them to agree very satisfactorily with this equation. Solutions that obey this law are said to be *ideal*. For solutions that do not, the mole fraction is corrected by multiplying it by an activity coefficient.

The law can be understood easily if we consider that the tendency of solvent molecules to escape from the surface of the solution is proportional to the fraction of solvent molecules present at the surface, i.e., to the mole fraction.

BOILING-POINT ELEVATION

The boiling point is the temperature at which the vapor pressure becomes equal to the external pressure. If the external pressure is 1 atm (760 mm), the temperature of boiling is known as the *normal boiling point*. The addition of an involatile solute lowers the vapor pressure of the solvent and thus raises its boiling point, since a higher temperature must be reached before the vapor pressure becomes equal to the external pressure. This is shown schematically in Figure 5.15.

The equilibrium between a solution of mole fraction x and its vapor at 1 atm pressure can be expressed as

Solution (mole fraction of solvent $= x_1$)$\rightleftharpoons$ vapor (1 atm)

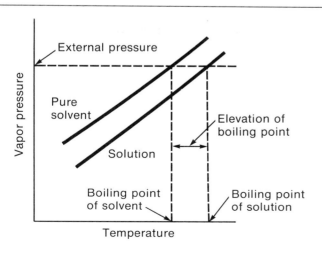

Figure 5.15

Vapor-pressure curves for a pure liquid, and for a solution of an involatile solid in the liquid, showing the elevation of the boiling point.

and we can write the equilibrium constant as

$$K = \frac{1}{x_1} \tag{5.176}$$

The general equation for the temperature variation of an equilibrium constant is (5.165) and we can apply this to (5.176):

$$\frac{d \ln (1/x_1)}{dT} = \frac{\Delta H^{\circ}_{vap}}{RT^2} \tag{5.177}$$

where ΔH°_{vap} is the enthalpy of vaporization, assumed to be independent of temperature. Since $d \ln (1/x_1) = -d \ln x_1$ we have

$$d \ln x_1 = -\frac{\Delta H^{\circ}_{vap}}{RT^2} dT \tag{5.178}$$

This gives us the increase dT in the boiling point of a solution when the mole fraction increases by an amount represented by $d \ln x_1$. To find the boiling point T that corresponds to a solvent mole fraction of x_1 we integrate (5.178), using as lower limits the condition corresponding to the pure solvent, $\ln x_1 = 0$ (i.e., $x_1 = 1$) and the boiling temperature T_b of the pure solvent:

$$\int_0^{\ln x_1} d \ln x_1 = -\frac{\Delta H^{\circ}_{vap}}{R} \int_{T_b}^{T} \frac{dT}{T^2} \tag{5.179}$$

Thus

$$\ln x_1 = \frac{\Delta H^{\circ}_{vap}}{R} \left(\frac{1}{T} - \frac{1}{T_b} \right) \tag{5.180}$$

$$= \frac{\Delta H^{\circ}_{vap}}{R} \frac{(T_b - T)}{T T_b} \tag{5.181}$$

Since for dilute solutions the elevation of the boiling point $\Delta T_b \, (= T - T_b)$ is small, we can replace $T T_b$ by T_b^2:

$$\ln x_1 = -\frac{\Delta H^{\circ}_{vap}}{R T_b^2} \Delta T_b \tag{5.182}$$

The mole fraction x_1 is equal to $1 - x_2$, and if the solution is dilute

$$\ln x_1 = \ln (1 - x_2) \approx -x_2 \tag{5.183}$$

Equation (5.182) thus simplifies to

$$\Delta T_b = \frac{R T_b^2}{\Delta H^{\circ}_{vap}} x_2 \tag{5.184}$$

Experimentally, a somewhat better agreement is obtained if one uses molalities, m, rather than mole fractions. The solute mole fraction x_2 is

$$x_2 = \frac{n_2}{n_1 + n_2} \tag{5.185}$$

where n_1 is the number of moles of solvent and n_2 the number of moles of solute. The former is the weight w_1 of solvent in grams divided by the molecular weight M_1:

$$n_1 = \frac{w_1}{M_1} \tag{5.186}$$

The molality m is the number of moles of solute per kilogram of solvent, and thus, since w_1 is in grams,

$$m = \frac{1000\, n_2}{w_1} \tag{5.187}$$

For a dilute solution $n_1 \gg n_2$, so that from (5.185)

$$x_2 \approx \frac{n_2}{n_1} = \frac{w_1\, m/1000}{w_1/M_1} = \frac{m M_1}{1000} \tag{5.188}$$

Equation (5.184) thus becomes

$$\Delta T_b = \frac{R T_b{}^2 M_1}{1000\, \Delta H^\circ_{vap}} \times m \tag{5.189}$$

which can be written as

$$\Delta T_b = K_b m \tag{5.190}$$

The molal boiling point elevation constant, K_b, thus can be calculated from the properties of the solvent, namely T_b, M_1, and ΔH°_{vap}. Table 5.3 gives K_b values calculated in this way, as well as experimental K_b values obtained by direct measurement of ΔT_b for solutions of known molality. The agreement is quite satisfactory.

Table 5.3 Molal Boiling-Point Elevation Constants

Solvent	Boiling Point, T_b/K	Enthalpy of vaporization, ΔH°_{vap}/kcal mol^{-1}	K_b (calc)	K_b (exp)
Water	373.1	9.72	0.51	0.51
Acetone	329.1	7.34	1.70	1.71
Benzene	353.3	7.35	2.62	2.53
Chloroform	334.3	7.02	3.67	3.63

FREEZING-POINT DEPRESSION

The way in which an involatile solute depresses the freezing point by lowering the vapor pressure is illustrated in Figure 5.16. The lines AB and BC represent vapor-pressure curves for the pure solvent; AB is for the solid and BC for the liquid. The point B at which the curves intersect corresponds to the melting point of the pure solvent at the pressure in question, P_1. The curve BF shows how the melting point varies with the pressure, the point F corresponding to the melting point at 1 atm pressure. When solute is added, the vapor pressure of the liquid is lowered and the new curve is DE. When freezing occurs, usually only the solvent becomes solid and curve AB therefore is unchanged. The curves AB and DE intersect at point D, which is

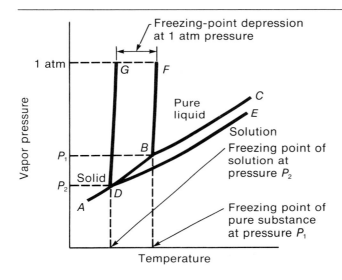

Figure 5.16

Vapor-pressure curves for a solvent in its solid and liquid states, and for a solution of an involatile solute. The diagram shows how the addition of the solute depresses the freezing point.

the freezing point at this particular pressure, P_2. The curve DG shows how the freezing point of the solution varies with the pressure, so that GF represents the freezing-point depression at 1 atm pressure.

The equilibrium which exists at the freezing point is

$$\text{Pure solid} \rightleftharpoons \text{solution (mole fraction of solvent} = x_1)$$

Since the concentration of a pure solid is a constant, the equilibrium constant can be written simply as

$$K = x_1 \tag{5.191}$$

From equation (5.165) we then have

$$\frac{d \ln x_1}{dT} = \frac{\Delta H_f^\circ}{RT^2} \tag{5.192}$$

where ΔH_f° is the molar enthalpy of fusion. We integrate from T_f, the melting point of the pure liquid ($x_1 = 1$ or $\ln x_1 = 0$), to T, the melting point of the solution where the mole fraction is x_1:

$$\int_0^{\ln x_1} d \ln x_1 = \frac{\Delta H_f^\circ}{R} \int_{T_f}^T \frac{dT}{T^2} \tag{5.193}$$

Thus

$$\ln x_1 = - \frac{\Delta H_f^\circ}{R} \left(\frac{1}{T} - \frac{1}{T_f} \right) \tag{5.194}$$

$$= - \frac{\Delta H_f^\circ}{R} \frac{(T_f - T)}{T T_f} \tag{5.195}$$

If we make the approximation that $T T_f = T_f{}^2$, and call $T_f - T = \Delta T_f$ the freezing point depression, we have

$$\ln x_1 = - \frac{\Delta H_f^\circ}{R T_f{}^2} \Delta T_f \tag{5.196}$$

Expansion of the logarithm (as was done with (5.183)) leads to

$$\Delta T_f = \frac{R T_f{}^2}{\Delta H_f^\circ} x_2 \tag{5.197}$$

and conversion to molality as in (5.188) gives

$$\Delta T_f = \frac{R T_f{}^2 M_1}{1000 \, \Delta H_f^\circ} \times m \tag{5.198}$$

$$= K_f m \tag{5.199}$$

Again, K_f values can be calculated from T_f, ΔH_f°, and M_1 for the solvent. Some results are shown in Table 5.4. The agreement with the direct determination of K_f from freezing-point depressions is very satisfactory. The very large value of K_f for camphor means that this substance is frequently a very suitable solvent for molecular weight determinations.

Table 5.4 Molal Freezing-Point Depression Constants

Solvent	Freezing Point, T_f/K	Enthalpy of Fusion, ΔH_f°/kcal mol^{-1}	K_f (calc)	K_f (exp)
Water	273.1	1.44	1.85	1.86
Benzene	278.5	2.35	5.09	5.12
Naphthalene	353.1	4.55	6.9	6.8
Camphor	451.1	1.38	47.5	48.5

OSMOTIC PRESSURE

We have seen in Chapter 3 that osmotic pressure is the most useful of the colligative properties for determining the molecular weights of high-molecular-weight material. Osmotic effects also play a very important role in physiological systems; for example, the flow of water into and out of cells is controlled to some extent by osmotic effects, although other effects such as active transport also are involved.

Consider an experiment (see Figure 5.17) in which a solution and the pure solvent are separated from each other by a *semipermeable* membrane, which will permit the

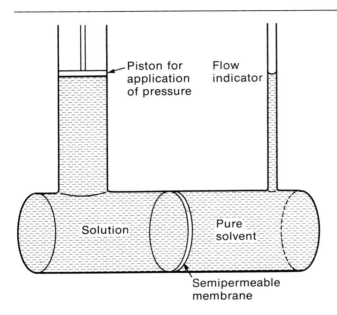

Figure 5.17

Schematic representation of apparatus for measuring osmotic pressure. The flow of solvent through the semipermeable membrane is followed by observing the movement of the meniscus in the flow indicator. The osmotic pressure π is the pressure that must be applied to the solution to stop the flow.

solvent but not the solute molecules to pass through it. The concentration of the solvent molecules is greater on the pure solvent side of the membrane than on the solution side, since on the solution side some of the volume is occupied by solute molecules. The tendency of molecules to pass through a membrane, like their tendency to evaporate from a surface, increases with increasing concentration. Therefore, although solvent molecules will flow in both directions through the membrane, there will be a more rapid flow from the pure solvent into the solution than from the solution into the pure solvent. Thus there is a net flow from the pure solvent into the solution.

This net flow of solvent molecules can be stopped by applying a hydrostatic pressure to the solution side (see Figure 5.17). The effect of this pressure is to increase the tendency of the solvent molecules to flow from solution into pure solvent. The particular pressure which causes the rate of flow from the solution side to be equal to that from the solvent side and thus reduce the net flow to zero is known as the *osmotic pressure* of the solution.

Experiments on osmotic pressure were carried out in 1877 by the German botanist Wilhelm Pfeffer (1845–1920). The osmotic pressure law was derived on the basis of purely thermodynamic reasoning by the Dutch physical chemist J. H. van 't Hoff (1852–1911). The role of the semipermeable membrane is to allow the solvent in the solution to come to equilibrium with the pure solvent; thus equilibrium is reached when the molar Gibbs energy of the solvent in the solution, G_1, is equal to the molar Gibbs energy of the pure solvent, G_1°:

$$G_1 = G_1^{\circ} \tag{5.200}$$

If no pressure is imposed on the solution the Gibbs energy of the solvent in the solution is given by equation (5.122):

$$G_1 = G_1^{\circ} + RT \ln x_1 \tag{5.201}$$

The effect of hydrostatic pressure P on the Gibbs energy, at constant temperature, is obtained from (5.153):

$$dG = V_1 \, dP \tag{5.202}$$

where V_1 is the molar volume of the solvent. If V_1 is assumed to be constant, the effect of a pressure π on the Gibbs energy is thus

$$\Delta G = V_1 \int_0^{\pi} dP = \pi V_1 \tag{5.203}$$

The Gibbs energy of the solution of mole fraction x_1, subjected to a pressure π, is thus

$$G_1 = G_1^{\circ} + RT \ln x_1 + \pi V_1 \tag{5.204}$$

At this pressure, however, $G_1 = G_1^{\circ}$, so that

$$\pi V_1 = -RT \ln x_1 \tag{5.205}$$

Substituting $x_2 = 1 - x_1$ and expanding the logarithm as before gives

$$\pi V_1 = RT x_2 \tag{5.206}$$

If the solution is dilute

$$x_2 \approx \frac{n_2}{n_1} \qquad \text{and} \qquad V_1 = \frac{V}{n_1} \tag{5.207}$$

where V is the total volume. Thus

$$\pi = \frac{RT}{V} n_2 \tag{5.208}$$

$$= cRT \tag{5.209}$$

where c, equal to n_2/V, is the molar concentration of the solution.

It is interesting to note the equivalence between equations (5.208) and (5.209) and the ideal gas law $PV = nRT$. However, no direct significance can be attached to this similarity. It is sometimes thought that osmotic pressure can be visualized as a bombardment of solute molecules against the semipermeable membrane just as gas pressure is due to bombardment of gas molecules, but this view is incorrect. The phenomenon must be explained in terms of the flow of solvent molecules or interpreted in terms of thermodynamic arguments, as has been done above.

5.14   THERMODYNAMICS AND BIOLOGY

Throughout this and the preceding chapter we have considered the thermodynamic aspects of a number of processes which occur in biological systems. Let us now summarize the main contributions that the study of thermodynamics makes to our understanding of the functioning of a living organism.

Again it should be emphasized that there is no evidence to suggest that biological systems violate the laws of thermodynamics in any way. Statements such as "Life cheats in the game of entropy" are misleading. A correct statement of the situation was made by Erwin Schrödinger, who said that

Living systems live on negative entropy absorbed from their environment.

In other words, they decrease their own entropy by increasing that of their surroundings.

Consider, for example, the development and reproduction of bacterial cells. These are *open* systems in the thermodynamic sense; the bacterial cell continuously acquires food from its environment and excretes waste products into its environment. If we were to consider the bacterial cells alone, without reference to the environment, we would find that the process was endergonic, largely as a result of the substantial decrease in entropy associated with the increase in order. Such an endergonic process would not be thermodynamically possible *in itself*. However, the metabolic reactions which are coupled with the development of cells are sufficiently exergonic to enable the development to occur, in full accord with the second law.

The development and hatching of a hen's egg after it has been laid presents an interesting situation. Here there is a *closed* system; it is not *isolated*, since there can be transfer of heat between the egg and its surroundings but no appreciable transfer of material. There is undoubtedly a decrease in entropy as the chicken is formed, since there is an increase in order. Since the process occurs spontaneously it must be exergonic (ΔG is negative). In light of the equation

$$\Delta G = \Delta H - T\Delta S \qquad (5.210)$$
$$(<0) \qquad (<0)$$

which applies at a fixed temperature, it follows that ΔH must be sufficiently negative to overcome the negative value of $T\Delta S$. In other words, exothermic reactions must be occurring inside the egg during the hatching process. It is true that the egg needs to be kept warm (in an incubator or under the brooding hen), but this is in order to keep the temperature high enough for the exothermic reactions to occur sufficiently rapidly; the heat supplied by the incubator or the hen to the egg is less than that evolved in the metabolic reactions.

COUPLING OF BIOLOGICAL REACTIONS

The coupling of reactions in biological systems as a device to drive endergonic processes is a question worth considering in more detail. Suppose that there occurs in a living system a substance Y which is important for driving endergonic processes essential to the system. Suppose also that the synthesis of this substance Y from a precursor substance X is an endergonic process:

$$X \rightarrow Y \qquad \Delta G^{\circ} > 0$$

This means, of course, that when Y is converted into X the process is exergonic:

$$Y \rightarrow X \qquad \Delta G^{\circ} < 0$$

The occurrence of this latter reaction can bring about the essential endergonic processes. Therefore, if there are certain exergonic processes, such as oxidative processes, which will convert X into Y, these exergonic processes will be coupled by the $X \rightleftharpoons Y$ system with the essential endergonic reactions. The situation is represented schematically in Figure 5.18.

Exergonic (e.g., oxidation) processes

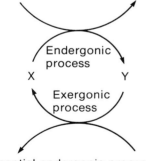

Endergonic process

X   Y

Exergonic process

Essential endergonic processes

Figure 5.18

A diagram showing how the $X \rightleftharpoons Y$ system allows a coupling of exergonic (e.g., oxidation) processes with endergonic processes that are essential to the functioning of the biological systems.

Studies of a wide variety of living systems have shown that the $X \rightleftharpoons Y$ system in most cases is the synthesis of adenosine triphosphate (ATP) from adenosine diphosphate (ADP) and inorganic phosphate (P):

$$ADP + P \rightleftharpoons ATP \qquad \Delta G^{\circ} > 0$$

The energy required to drive this reaction is supplied by oxidative processes occurring in metabolism. These may produce an oxidized species (Ox) which can be converted into its reduced form (Red) by an electron transfer (e):

$$Ox + e^- \rightarrow Red \qquad \Delta G^o < 0$$

This can be coupled by an enzyme to the formation of ATP, and the ATP hydrolysis in turn can be coupled to an essential endergonic task, which we can represent as

$$Y \rightarrow Z \qquad \Delta G^o > 0$$

The coupling processes are represented in Figure 5.19.

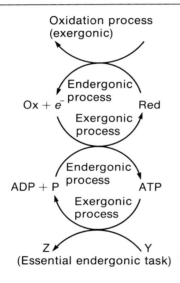

Figure 5.19

Schematic representation of the coupling of a dehydrogenation reaction with an essential endergonic task, by the ADP-ATP system.

HIGH-ENERGY METABOLITES

A very important matter in connection with the ATP system is that the ΔG^o associated with the hydrolysis of ATP to ADP and phosphate has an abnormally large negative value, -7.0 kcal mol^{-1} at 25° C. By contrast, substances like glycerol-1-phosphate and fructose-6-phosphate are hydrolyzed with the release of much less Gibbs energy. Some ΔG^o values associated with the hydrolysis of organic phosphates are included in Table 5.5, in which it can be seen that there is a group of substances having ΔG^o values ranging from -2 to -4 kcal mol^{-1}, and another group with significantly higher values.

When it was discovered that there is a group of phosphates having abnormally large $-\Delta G^o$ values, it was suggested that these compounds contained a "high-energy phosphate bond." This bond was represented by the symbol $\sim$. Thus acetyl phosphate would be represented as acetyl $\sim$ P and ATP as ADP $\sim$ P. However, this terminology is misleading for several reasons and should be avoided. In the first place, there is the obvious confusion with the chemist's concept of the bond energy, which we have discussed on p. 166. If a bond were in reality a "high-energy" bond, it would be a strong bond and therefore more difficult to hydrolyze. Such compounds therefore would have abnormally *low* $-\Delta G^{\circ}$ values. In addition, one cannot relate the ΔG° values for hydrolysis to the strength of one particular bond; much more is involved. For example, the hydrolysis of a phosphate might result in a product in which there is much resonance stabilization, and this would lead to a large $-\Delta G^{\circ}$ value. In view of these objections it is much more satisfactory to refer to substances having large $-\Delta G^{\circ}$ values as *high-energy metabolites*. This term correctly implies that, on the hydrolysis of these compounds, Gibbs energy is released and can be utilized for the carrying out of endergonic tasks.

Table 5.5 Gibbs Energies of Hydrolysis of Phosphates at 25° C and pH 7.0†

Compound	$\Delta G^{\circ}/\text{kcal mol}^{-1}$
Acetyl adenylate	− 13.3
Phosphoenol pyruvate	− 12.8
Creatine phosphate	− 10.3
Acetyl phosphate	− 10.1
Adenosine triphosphate (ATP)	− 7.0
Glucose-1-phosphate	− 5.0
Fructose-6-phosphate	− 3.8
Fructose-1-phosphate	− 3.1
3-phosphoglycerate	− 3.0
Glycerol-1-phosphate	− 2.3

† Data taken from M. R. Atkinson and R. K. Morton, Chapter I of *Comparative Biochemistry*, ed. H. S. Mason and M. Florkin, Vol. II. New York: Academic Press, 1960.

One reaction used by many organisms as a source of ATP is the coupling of ATP production from ADP + P with the hydrolysis of phosphoenol pyruvate (PEP). The two reactions are

(1) $ADP + P \rightleftharpoons ATP + H_2O$ $\Delta G_1^{\circ} = 7.0$ kcal
(2) $PEP + H_2O \rightleftharpoons$ pyruvate + P $\Delta G_2^{\circ} = -12.8$ kcal

The net reaction is

(3) $PEP + ADP \rightleftharpoons$ pyruvate + ATP $\Delta G_3^{\circ} = -5.8$ kcal

The two reactions are coupled by the enzyme pyruvate kinase, which catalyzes reaction (3). There is direct transfer of phosphate from PEP to ADP, no free phosphate being liberated. In view of the substantial Gibbs-energy decrease in reaction (3), this process will occur to a considerable extent. Note that reaction (3) is not highly efficient: 12.8 kcal have been utilized to drive a reaction requiring 7.0 kcal. The *efficiency* is thus $7.0/12.8 = 0.55$ or 55%.

An example of a reaction in which the Gibbs energy stored in ATP is used to drive an endergonic reaction is

Glycerol + ATP $\rightleftharpoons$ Glycerol-1-phosphate + ADP

This reaction is catalyzed by glycerol kinase. The reaction

$$\text{Glycerol} + \text{P} \rightleftharpoons \text{Glycerol-1-phosphate} \qquad \Delta G^\circ = 2.3 \text{ kcal}$$

is endergonic and will occur by itself only to a small extent. However, the hydrolysis of ATP is exergonic, the ΔG° value being -7.0 kcal. The coupled reaction thus is sufficiently exergonic ($\Delta G^\circ = -4.7$ kcal) that it will occur to a considerable extent. The high $-\Delta G^\circ$ value for the ATP hydrolysis makes ATP an excellent phosphorylating agent.

For the sake of simplicity we have thus far considered only the coupling of single pairs of reactions. In metabolic processes a large series of reactions may be coupled together in this way by a number of different enzymes. Many cells, for example, oxidize glucose,

$$\text{C}_6\text{H}_{12}\text{O}_6 + 6\text{O}_2 \rightarrow 6\text{CO}_2 + 6\text{H}_2\text{O} \qquad \Delta G^\circ = -686.0 \text{ kcal}$$

This reaction is coupled, by dozens of steps, to the phosphorylation of 38 molecules of ADP:

$$38 \text{ ADP} + 38 \text{ P} \rightarrow 38 \text{ ATP} \qquad \Delta G^\circ = 38 \times 7.0 \text{ kcal}$$

The complete reaction is thus

$$\text{C}_6\text{H}_{12}\text{O}_6 + 6\text{O}_2 + 38 \text{ ADP} + 38 \text{ P} \rightarrow 6\text{CO}_2 + 6\text{H}_2\text{O} + 38 \text{ ATP}$$
$$\Delta G^\circ = -420.0 \text{ kcal}$$

The overall efficiency is now $38 \times 7.0/686.0 = 38.8\%$.

CONFORMATIONAL CHANGES IN PROTEINS

A final matter interesting to consider in connection with the thermodynamics of living systems is that of changes in shape of proteins and other macromolecules. The three-dimensional shape of a macromolecule has an important influence on its biological activity, and chemical processes frequently bring about shape changes. There is considerable evidence for shape changes in a variety of systems, and some of this has come from thermodynamic studies. For example, certain compounds known as *allosteric effectors* become attached to protein molecules at a site different from that occupied by the normal substrate. In so doing they alter the conformation of the protein and hence alter the protein's catalytic activity.

For example, adenosine-5'-phosphate (AMP) becomes attached to a respiratory enzyme, NADH_2 dehydrogenase (E):

$$\text{AMP} + \text{E} \rightleftharpoons \text{AMP—E}$$

Measurements of the equilibrium constant for this reaction at various temperatures have led to the following values at 30° C:

$$\Delta G^\circ = -4.79 \text{ kcal}$$

$$\Delta H^\circ = 12.5 \text{ kcal}$$

The entropy change is thus

$$\Delta S^\circ = \frac{\Delta H^\circ - \Delta G^\circ}{T}$$

$$= \frac{12000 + 4790}{303}$$

$$= 55.4 \text{ cal K}^{-1}$$

This large increase in entropy on binding suggests a very substantial conformational change in the protein molecule.

Similar large entropy changes, suggesting important conformational changes, have been found for the denaturations of protein molecules and for the attachment of ATP to the muscle enzyme myosin.

THE HYDROPHOBIC EFFECT

We have seen that hydrophobic bonds play a very important role in biology—for example, in connection with the conformations of protein molecules in aqueous solution (p. 131). There is very little bonding between nonpolar groups in the absence of water, but in an aqueous environment there is an indirect bonding which arises from the peculiar structure of liquid water. The thermodynamic aspects are of some importance, since they throw considerable light on the nature of this bond.

Table 5.6 gives some enthalpy, entropy, and Gibbs-energy values for the transfer of nonpolar molecules from a nonpolar solvent into water. Note that the ΔG° values are all positive, which means that there is a tendency for polar molecules or groups to remove themselves from an aqueous environment, which they can do by becoming attached to each other. However, the positive ΔG° values in Table 5.6 are not associated with positive ΔH° values, but rather with negative ΔH° values. Instead, the positive ΔG° values are due to the large negative ΔS° values for the process.

Table 5.6 Changes in Thermodynamic Values in the Transfer of Hydrocarbons from a Nonpolar Environment to Water at 25° C.

Process	$\Delta H^\circ /$ kcal mol^{-1}	$\Delta S^\circ /$ cal K^{-1} mol^{-1}	$\Delta G^\circ /$ kcal mol^{-1}
CH_4 in benzene → CH_4 in water	−2.8	−18	2.6
CH_4 in CCl_4 → CH_4 in water	−2.5	−18	2.9
C_2H_6 in benzene → C_2H_6 in water	−2.2	−20	3.6
C_2H_6 in CCl_4 → C_2H_6 in water	−1.8	−19	3.8
C_2H_6 (pure liquid) → C_2H_6 in water	−2.5	−21	3.9

These thermodynamic values give us considerable insight into what occurs when a nonpolar group enters an aqueous environment. It might have been thought that there would be *less* hydrogen bonding in the surrounding water molecules, because of the interference of the nonpolar groups. The negative $\Delta H°$ values, however, strongly suggest that there is *more* hydrogen bonding. An explanation for this was first given in 1945 by the American physical chemists Henry S. Frank and M. W. Evans. They suggested that when a nonpolar molecule or group is present in water, the neighboring water molecules arrange themselves into a structure which has more of a crystalline character than ordinary liquid water. They used the word "icebergs" to describe these structures, but emphasized that the structures do not correspond exactly to those of ice. One important distinction is that ice has a lower density than water, whereas the "icebergs" have a higher density.

The reason for the formation of these "icebergs" is that the water molecules next to the polar groups, being deprived of the opportunity of forming hydrogen bonds with the molecules on the other side of the polar groups, become oriented in a special way with respect to their neighboring water molecules and form hydrogen bonds with them. The result is even more hydrogen bonding than in ordinary liquid water. Because the water molecules surrounding the polar groups are held more rigidly, there is a lowering of entropy.

The formation of hydrophobic bonds is similar to the reverse of the processes shown in Table 5.6. It has a negative ΔG (i.e., is spontaneous), a positive ΔH, and a positive ΔS. Thus, the tendency for hydrophobic bonds to be formed results *not* because water can form more hydrogen bonds. On the contrary, there is a decrease in hydrogen bonding and a corresponding increase in entropy, which allows the process to be spontaneous. Hydrophobic bonding thus provides another example of an *entropic* process.

5.15   THE BOLTZMANN PRINCIPLE

We have seen that, strictly speaking, thermodynamics is not concerned with the atomic and molecular nature of matter, i.e., with microscopic properties. Instead, it deals with the interrelationships of macroscopic properties such as temperature, pressure, and work.

A completely separate branch of science is *statistical mechanics*, which is concerned with microscopic properties. This subject makes use of what quantum mechanics tells us about the energy levels of molecules, and allows us to calculate macroscopic properties on the basis of this information. The area of overlap between statistical mechanics and thermodynamics is known as *statistical thermodynamics*, which allows us, for example, to calculate equilibrium constants for chemical reactions using the molecular properties obtained from quantum mechanics.

A treatment of these matters is outside the scope of the present book, but there is one aspect which must be referred to, namely the way in which systems distribute themselves between different energy states. This is dealt with by the Boltzmann principle, which was developed about 1876 by the Austrian physicist Ludwig Boltzmann (1844–1906).

Suppose that certain energy states $\varepsilon_1, \varepsilon_2, \varepsilon_3, \ldots \varepsilon_j, \ldots$ are possible for a molecule. The Boltzmann principle tells us that the number of molecules N_j in the jth state of energy ε_j is related to the total number of molecules N by the expression

$$\frac{N_j}{N} = \frac{e^{-\varepsilon_j/kT}}{\sum\limits_{i=0}^{\infty} e^{-\varepsilon_j/kT}} \tag{5.211}$$

where the summation is taken over all the energy states. The constant $\mathbf{k}$ is known as the Boltzmann constant, and has the value 1.381×10^{-23} J K^{-1}. The ratio N_j/N is the *fraction* of all the molecules that exist in the jth state, and is the *probability* that a given molecule is in the jth state.

A modification to equation (5.211) is required if there is more than one state having a particular energy value ε_i. If this is the case, the level is said to be *degenerate* and is assigned a *statistical weight*, g_i, equal to the number of superimposed levels. Equation (5.211) is then modified to

$$N_j = \frac{g_j e^{-\varepsilon_j/kT}}{\sum\limits_{i=0}^{\infty} g_i e^{-\varepsilon_i/kT}} \tag{5.212}$$

It follows from (5.212) that if we consider two levels of energies, ε_i and ε_j, the ratio of their populations will be

$$\frac{n_i}{n_j} = \frac{g_i e^{-\varepsilon_i/kT}}{g_j e^{-\varepsilon_j/kT}} \tag{5.213}$$

$$= \frac{g_i}{g_j} e^{-(\varepsilon_i - \varepsilon_j)/kT} \tag{5.214}$$

Instead, we may express the energy per mole rather than the energy per molecule; if the two energies per mole are E_i and E_j, (5.214) becomes

$$\frac{n_i}{n_j} = \frac{g_i}{g_j} e^{-(E_i - E_j)/RT} \tag{5.215}$$

where R is the gas constant $N_A k$ (8.314 J K^{-1} mol^{-1}, or 1.987 cal K^{-1} mol^{-1}).

Independently of, and earlier (1860) than Boltzmann, the Scottish mathematician and physicist James Clerk Maxwell (1831–1879) developed a similar theory, which he applied particularly to the distribution of velocities of gas molecules. Because of this, equations (5.211) to (5.215) are sometimes referred to as Maxwell-Boltzmann equations, although this expression is perhaps best reserved for the velocity-distribution equations.

The Boltzmann principle will later be employed in two places in this book. On p. 269 it is used to give us the average number of positive and negative charges in a volume element close to an ion. Suppose that the average concentration of positive ions in a solution is n_+ ions per dm^3; then if there were no electrostatic forces, dn_+ in a volume dV would be $n_+ dV$. However, if the volume element were close to a positive ion, the positive ions in the element would be at a positive energy level, because of their electrostatic repulsions. If this electrostatic energy level is denoted by E_ϕ, the Boltzmann principle tells us that the number of ions in the element will be reduced to the fraction

$$e^{-E_\phi/\mathbf{k}T}$$

that is

$$dn_+ = n_+ \, e^{-E_\phi/\mathbf{k}T} \, dV \tag{5.216}$$

The electrostatic energy E_ϕ is equal to the charge on the positive ions, $z_+ e$, multiplied by the electric potential produced by the neighboring central ion, ϕ. Thus

$$E_\phi = z_+ e\phi \tag{5.217}$$

and we at once have equation (6.14).

The second application of the Boltzmann principle in this book is in relation to chemical kinetics (p. 388). It can be shown from (5.212) that in any system the *fraction f* of molecules having energy *in excess* of a specified value E (per mole) is equal to

$$f = e^{-E/RT} \tag{5.218}$$

(The gas constant R is now used since E is energy per mole; if it were energy per molecule we would use the Boltzmann constant $\mathbf{k}$.) Equation (5.218) at once gives an interpretation of the Arrhenius law, as discussed on p. 388.

PROBLEMS

5.1 The diagram below represents a reversible Carnot cycle for an ideal gas:

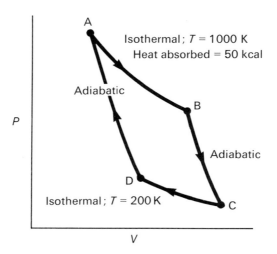

(a) What is the thermodynamic efficiency of the engine?

(b) How much heat is rejected at the lower temperature, 200 K, during the isothermal compression?

(c) What is the entropy increase during the isothermal expansion at 1000 K?

(d) What is the entropy decrease during the isothermal expansion at 200 K?

(e) What is the entropy change during the adiabatic expansion $B \rightarrow C$?

(f) What is the overall entropy change for the entire cycle?

(g) What is the increase in Gibbs energy during the process $A \rightarrow B$?

5.2 Calculate the entropies of vaporization in cal K^{-1} mol^{-1} of the following substances, from their boiling points and enthalpies of vaporization:

	Boiling Point/K	ΔH_{vap}/kcal mol^{-1}
C_6H_6	353	7.35
$CHCl_3$	334	7.02
H_2O	373	9.70
C_2H_5OH	351	9.20

In terms of the structures of the liquids, suggest reasons for the higher values observed for H_2O and C_2H_5OH.

5.3 Obtain a general expression, in terms of the molar heat capacity C_p and the temperatures T_1 and T_2, for the entropy increase of 1 mol of a gas which is heated at constant pressure so that its temperature changes from T_1 to T_2.

5.4 Calculate the standard entropies of formation of (1) liquid methanol, and (2) solid urea, making use of the absolute entropies listed in Table 5.2.

5.5 Predict the signs of the entropy changes in the following reactions:

(a) Hydrolysis of urea: $H_2NCONH_2 + H_2O \rightarrow CO_2 + 2NH_3$

(b) Conversion of glucose into its open-chain form:

(c) Condensation of amino-acid molecules to form a protein.

(d) Binding of a positively-charged substrate at a negatively-charged group on an enzyme molecule, in aqueous solution.

(e) Binding of a substrate to an enzyme, with the approach of like charges, in aqueous solution.

5.6. The hydrolysis of adenosine triphosphate to give adenosine diphosphate and phosphate can be represented as

$$ATP \rightleftharpoons ADP + P$$

The following values have been obtained for the reaction at $37°C$:

$$\Delta G° = -7.4 \text{ kcal}$$

$$\Delta H° = -4.8 \text{ kcal}$$

(a) Calculate the standard entropy change $\Delta S°$.
(b) Calculate the equilibrium constant for the reaction at $37°C$.

5.7 In the enzymic conversion of L-aspartate to fumarate $+ NH_4^+$ the equilibrium constant was found to be 1.60×10^{-2} at $39°C$ and 0.74×10^{-2} at $29°C$.

(a) What is $\Delta H°$, assuming it to be independent of T?
(b) What is $\Delta G°$ at $29°C$?
(c) What is $\Delta S°$ at $29°C$?

5.8 The $\Delta G°$ value for the reaction

$$C_6H_{12}O_6 \rightleftharpoons 2CH_3CHOHCOO^- + 2H^+$$

$$\text{(glucose)} \qquad \text{(lactate)}$$

is -27.9 kcal at $37°C$. Calculate the equilibrium constant for the reaction at that temperature.

5.9 The bacterium *acetobacter* catalyzes the reaction

$$C_2H_5OH \text{ (aq)} + O_2 \text{ (g)} \rightleftharpoons CH_3COOH \text{ (aq)} + H_2O \text{ (l)}$$

The standard Gibbs energies of formation of C_2H_5OH, CH_3COOH, and H_2O are given in Table 4.1.

(a) Calculate $\Delta G°$.
(b) Calculate the equilibrium constant for this process at $25°C$.

5.10 In the Krebs cycle, citrate is converted to isocitrate,

citrate $\rightleftharpoons$ isocitrate

(a) Using the $\Delta G_f°$ values given in Table 4.1, calculate $\Delta G°$ for the process at $25°C$.
(b) Which is the favored direction of the isomerization?
(c) Calculate the equilibrium constant at $25°C$.

The following reaction also occurs during the Krebs cycle:

Isocitrate $+ \frac{1}{2}O_2 + H^+ \rightleftharpoons \alpha$-ketoglutarate $+ H_2O + CO_2$

(d) Using $\Delta G_f°$ values in Table 4.1, calculate $\Delta G°$ for this oxidation.
(e) Will the occurrence of this process have an effect on the conversion of citrate to isocitrate?

5.11 The oxidation of α-ketoglutarate proceeds according to the equation

α-ketoglutarate $+ \frac{1}{2}O_2 \rightleftharpoons$ succinate $+ CO_2$

The value of $\Delta G°$ at $25°C$ is -68.5 kcal.

(a) What is the equilibrium constant at $25°C$?
(b) Are we justified in referring to the reaction as irreversible?

5.12 The bacterium *nitrobacter* plays an important role in the "nitrogen cycle," by oxidizing nitrite to nitrate. It obtains the energy it requires for growth from the reaction

$$NO_2^- \text{ (aq)} + \tfrac{1}{2}O_2 \text{ (g)} \rightleftharpoons NO_3^- \text{ (aq)}$$

Calculate $\Delta G°$ for this reaction at $25°\,C$, making use of the standard Gibbs energies of formation given in Table 4.1.

5.13 The enzyme triose phosphate isomerase catalyzes the reaction glyceraldehyde-3-phosphate $\rightleftharpoons$ dihydroxyacetone phosphate. If the equilibrium constant is 22.0 at $25°\,C$, calculate $\Delta G°$ for the reaction.

5.14 An enzyme-catalyzed reaction has a thermodynamic equilibrium constant (K) of 1 at $25°\,C$.

(a) What is the value of $\Delta G°$ of the reaction?
(b) What will be the value of K at $25°\,C$ in the absence of the enzyme?
(c) If $\Delta H°$ has a positive value, will K at $37°\,C$ be (a) < 1, (b) 1, or (c) > 1?

5.15 The reaction,

$$A + B \rightleftharpoons X + Y \qquad \Delta H° > 0$$

was allowed to attain equilibrium at $25°\,C$, and K was found to equal 10.

(a) Is the formation of A and B from X plus Y an exergonic reaction under standard conditions?
(b) Will an increase in the temperature of reaction enhance the yield of X and Y?
(c) Will the addition of A to the equilibrium mixture increase the yield of X and Y?

5.16 The enzyme phosphoglucomutase catalyzes the interconversion of glucose-1-phosphate and glucose-6-phosphate,

$$\text{Glucose-1-phosphate} \rightleftharpoons \text{Glucose-6-phosphate}$$

If at chemical equilibrium at $25°\,C$, 95% glucose-6-phosphate is present, calculate:

(a) K and $\Delta G°$ for the reaction
(b) ΔG for the reaction in the presence of $10^{-2}\,M$ glucose-1-phosphate and $10^{-4}\,M$ glucose-6-phosphate.

5.17 The enzyme L-glutamate-pyruvate aminotransferase catalyzes a transamination reaction between glutamate and pyruvate to yield α-ketoglutarate and L-alanine,

$$\text{L-glutamate} + \text{pyruvate} \rightleftharpoons \alpha\text{-ketoglutarate} + \text{L-alanine}$$

(a) If the equilibrium constant K for the synthesis of L-alanine is 1.107 at $25°\,C$, calculate the value of $\Delta G°$.
(b) Does the result obtained in (1) mean that the synthesis of alanine from pyruvate plus glutamate is always spontaneous?
(c) If $10^{-4}\,M$ each of L-glutamate and pyruvate were mixed with $10^{-2}\,M$ each of α-ketoglutarate and L-alanine at $25°\,C$ in the presence of the amino-transferase, (1) what is the value of ΔG for L-alanine formation?, and (2) will alanine be formed spontaneously under these conditions?

5.18 The standard Gibbs energies for the hydrolyses of glucose-6-phosphate (G-6-P) and creatine phosphate (creatine-P) are

$$G\text{-}6\text{-}P + H_2O \rightleftharpoons G + P \qquad \Delta G^\circ = -3.0 \text{ kcal}$$

$$\text{Creatine-P} + H_2O \rightleftharpoons \text{creatine} + P \qquad \Delta G^\circ = -7.0 \text{ kcal}$$

where P = phosphate. Calculate ΔG° for the reaction

$$G\text{-}6\text{-}P + \text{creatine} \rightleftharpoons G + \text{creatine-P}$$

5.19 The conversion of malate into fumarate,

$$\text{malate} \rightleftharpoons \text{fumarate} + H_2O$$

is endergonic, the ΔG° being 0.7 kcal. In metabolism this reaction is coupled with

$$\text{fumarate} \rightleftharpoons \text{aspartate}$$

for which ΔG° is -3.7 kcal.

(a) Does this coupling aid the removal of malate?
(b) Calculate ΔG° for the overall reaction

$$\text{malate} \rightleftharpoons \text{aspartate}$$

(c) Calculate K for the malate-aspartate equilibrium at 37° C.

5.20 The equilibrium constant for the reversible denaturation of trypsin

$$\text{trypsin} \rightleftharpoons \text{denatured trypsin}$$

has been found to be 1.00 at 44.0° C and 7.20 at 50.0° C. Calculate ΔG°, ΔH° and ΔS° at 44.0° C.

5.21 The latent heat of vaporization of water at 100° C is 9.70 kcal mol^{-1}. When 1 mol of water is vaporized at 100° C and 1 atm pressure the volume increase is 30.19 dm^3. Calculate the work done by the system, the change in internal energy ΔU, the change in Gibbs energy ΔG, and the entropy change ΔS.

5.22 200 g of mercury at 100° C are added to 80 g of water at 20° C in a vessel which has a water equivalent of 20 g. The specific heats of water and mercury may be taken as constant at 1.00 and 0.0334 respectively. Calculate the entropy change of

(a) The mercury.
(b) The water and vessel.
(c) The mercury, water and vessel together.

5.23 A typical sample of seawater contains 31.6 g dm^{-3} of sodium chloride. Other substances are present in much smaller amounts and can be neglected. If seawater is contained in a cylinder with a semipermeable membrane at one end and a piston at the other, application of a pressure greater than the osmotic pressure will cause pure water to pass through the membrane. Estimate the minimum work that is required to produce 1 mol (18 cm^3) of water by the desalination of seawater at 25° C. (This process is known as *reverse osmosis*, and when suitable membranes are developed this type of desalination may be economic. Note that the work required is very much less than that required for purification by distillation, which is about 9.7 kcal mol^{-1} or 40.6 kJ mol^{-1}.)

5.24 The kidneys remove about 180 dm^3 of fluid from the blood each day by reverse osmosis; 99% of this is returned to the blood, the remainder being secreted as urine. Calculate the minimum work that the kidneys must perform each day in the filtration process, taking the osmotic pressure of blood to be 28 mm Hg ($T = 37°$ C).

5.25 Bacterial cells placed in a series of NaCl solutions at different concentrations at 25° C were found to shrink in solutions more concentrated than 0.7%, and to swell in solutions less concentrated than 0.7%. An NaCl solution of concentration 0.7% freezes at $-0.406°$ C (K_f for water $= 1.86°$). Estimate the osmotic pressure of the cell cytoplasm at 25° C.

5.26 In the human body the concentration of K$^+$ ions is about 155 mM inside the cells and 4 mM outside. An adult contains about 150 g of K$^+$. Calculate the total Gibbs energy ΔG associated with this concentration difference ($T = 37°$ C).

ESSAY QUESTIONS

5.27 Give an account of the most important methods that are employed to determine the Gibbs-energy changes of chemical reactions.

5.28 Explain what is meant by the Gibbs energy. Show, by means of a few examples, the relevance of the Gibbs energy to the understanding of biological reactions.

5.29 Discuss the effect of temperature on the equilibrium constant for a chemical reaction.

5.30 The frying of a hen's egg is a spontaneous reaction and has a negative Gibbs energy change. The process can apparently be reversed by feeding the fried egg to a hen and waiting for it to lay another egg. Does this constitute a violation of the Second Law? Discuss.†

5.31 Consider the following statements:

(a) In a reversible process there is no change in the entropy;
(b) In a reversible process the entropy change is

$$\int \frac{dq_{rev}}{T}$$

How must these statements be qualified so that they are correct, and not contradictory?

---

† In answering this question, a student commented that a hen would never eat a fried egg. I suspect she would if she were hungry and had no alternative. In any case, let us postulate a hen sufficiently eccentric to eat a fried egg.

5.32 Consider the following statements:

(a) The solution of certain salts in water involves a decrease in entropy.

(b) For any process to occur spontaneously there must be an increase in entropy.

Qualify these statements so that they are correct and not contradictory, and suggest a molecular explanation for the behavior.

SUGGESTED READING

See also the list of books for Chapter 4 (p. 181).

Kauzmann, W. "Some Factors in the Interpretation of Protein Denaturation," in *Advances in Protein Chemistry*, Vol. 14. New York: Academic Press, 1959. (Pages 37–47 give an excellent account of the thermodynamics of the formation of hydrophobic bonds.)

Lauffer, M. A. *Entropy-Driven Processes in Biology*. New York: Springer-Verlag, 1975.

Morowitz, H. J. *Entropy for Biologists*. New York: Academic Press, 1970.

Morowitz, H. J. *Energy Flow in Biology*. New York: Academic Press, 1968.

Tanford, C. *The Hydrophobic Effect: Formation of Micelles and Biological Membranes*. New York: Wiley-Interscience, 1973.

6

ELECTROLYTIC CONDUCTIVITY

The most abundant substance in any biological system is water. The human body, for example, is approximately 80% water. This is of great significance for biology, because water has very unusual properties. In particular, it is a solvent which tends to encourage the existence of electrically charged particles, or ions. For example, when acetic acid (a covalent substance in the liquid and solid states) is dissolved in water, there is a certain amount of dissociation into acetate and hydrogen ions. Protein and other important biological molecules also tend to exist in ionized states when in an aqueous environment.

The biologist's study of physical chemistry must therefore include a good deal of the subject known as *electrochemistry*, which is particularly concerned with the interaction between electricity and chemical systems. Much can be learned about the behavior of ions in solvents such as water by investigations of electrical effects. Thus, measurements of the conductivities of aqueous solutions at various concentrations have led to an understanding of the extent to which substances are ionized in water, the association of ions with surrounding water molecules, and the way in which ions move in water. All of these topics have a great significance in biology. For example, the functioning of a nerve depends to a great extent on the movement of sodium, potassium, and other ions. This chapter is concerned with the conductivities of aqueous solutions and with the information derived from a study of these conductivities. Further aspects of the behavior of ions in water will be considered in Chapter 7.

6.1   FARADAY'S LAWS OF ELECTROLYSIS

During the years 1833 and 1834 the great English scientist Michael Faraday (1791–1867) published the results of an extended series of investigations on the relationships between the quantity of electricity passing through a solution and the amount of material liberated at the electrodes. For example, when a current is passed through an acidified solution of water, hydrogen collects at the negative electrode (known as the *cathode*) and oxygen at the positive electrode (the *anode*). Similarly, the passage of a current through molten sodium chloride yields sodium and chlorine at the two electrodes.

Faraday made quantitative studies of the amounts of materials produced in electrolysis and formulated two laws of electrolysis, which may be summarized as follows:

1. The weight of an element produced at an electrode is proportional to the quantity of electricity passed through the liquid. The quantity of electricity (Q) in coulombs (C) is defined as equal to the current (I) in amperes (A) multiplied by the time (t) in seconds (s).

$$Q = It \tag{6.1}$$

2. The weight of an element liberated at an electrode is proportional to the *equivalent weight* of the element.

The SI recommendation, supported by the International Union of Pure and Applied Chemistry and other international scientific bodies, is to abandon the use of the word "equivalent," and to refer to moles instead. However, the concept of the equivalent is still employed, even though the name has been dropped. Suppose, for example, that we pass one ampere of electricity for one hour through a dilute sulfuric acid solution and through solutions of silver nitrate, $AgNO_3$, and cupric sulfate, $CuSO_4$. The weights liberated at the respective cathodes are:

0.038 g of H_2 (at. wt. 1.008, mol. wt. 2.016)

4.025 g of Ag (at. wt. 107.9)

1.186 g of Cu (at. wt. 63.6)

These weights 0.038, 4.025, and 1.186 are approximately in the ratio $1.008:107.9:31.8$. Therefore, the amount of electricity which liberates 1 mol of Ag liberates 0.5 mol of H_2 and 0.5 mol of Cu. These quantities were formerly referred to as one equivalent, but the modern practice is to speak instead of 1 mol of $\frac{1}{2}H_2$ and 1 mol of $\frac{1}{2}Cu$. These are the quantities liberated by one electron:

$$e^- + H^+ \rightarrow \tfrac{1}{2}H_2$$
$$e^- + \tfrac{1}{2}Cu^{2+} \rightarrow \tfrac{1}{2}Cu$$

In what follows, when we refer to 1 mol we may mean 1 mol of a fraction of an entity, e.g., to $\frac{1}{2}Cu$ or $\frac{1}{2}SO_4^{2-}$.

In honor of Faraday the proportionality factor which relates the number of moles deposited to the quantity of electricity passed through the solution is known as the *Faraday constant* and given the symbol F. In modern terms, the charge carried by 1 mol of ions bearing z unit charges is zF. According to the latest measurements the Faraday constant F is equal to 96 485 coulombs per mole ($C\ mol^{-1}$); in this discussion the figure will be rounded to 96 500 $C\ mol^{-1}$. In other words, 96 500 C will liberate 1 mol of Ag, 1 mol of $\frac{1}{2}H_2$ (i.e., 0.5 mol of H_2), etc. If a constant current of 1 amperes is passed for t seconds, the number of moles deposited is $It/96\ 500$.

It follows that the (negative) charge on 1 mol of electrons is 96 500 C. The charge on one electron is this quantity divided by the Avogadro number N_A, and is thus

$$\frac{96\ 500\ (C\ mol^{-1})}{6.023 \times 10^{23}\ (mol^{-1})} = 1.602 \times 10^{-19}\ C$$

EXAMPLE

An aqueous solution of gold (III) nitrate, $Au(NO_3)_3$, was electrolyzed with a current of 0.250 A until 1.200 g of Au (at. wt. 197.0 $g\ mol^{-1}$) had been deposited at the cathode. Calculate (1) the number of coulombs passed, (2) the duration of the experiment, and (3) the volume of O_2 (at. S.T.P.) liberated at the anode.

SOLUTION

(1) The reaction at the cathode is

$$3e^- + Au^{3+} \rightarrow Au \qquad or \qquad e^- + \tfrac{1}{3}Au^{3+} \rightarrow \tfrac{1}{3}Au$$

Therefore, 96 500 C of electricity liberates 1 mol of $\frac{1}{3}Au$, i.e., 197.0/3 g. The quantity of electricity required to produce 1.200 g of Au is thus

$$\frac{96\ 500 \times 1.200}{197.0/3} = 1.76 \times 10^3\ C.$$

(2) Since the current was 0.250 A, the time was

$$\frac{1.76 \times 10^3\ (A\ s)}{0.025\ (A)} = 7.04 \times 10^3\ s$$

(3) The equation for the liberation of O_2 is $2H_2O \rightarrow O_2 + 4H^+ + 4e^-$ or $\frac{1}{2}H_2O \rightarrow \frac{1}{4}O_2 + H^+ + e^-$. Therefore, 96 500 C liberates $\frac{1}{4}$ mol O_2, so that 1.76×10^3 C produces

$$\frac{1.76 \times 10^3}{96\ 500 \times 4} = 4.56 \times 10^{-3}\ mol\ O_2$$

The volume of this at S.T.P. is

$$4.56 \times 10^{-3}\ (mol) \times 22.4\ (dm^3\ mol^{-1}) = 0.102\ dm^3$$

Faraday's work on electrolysis was of great importance in that it was the first to suggest a relationship between matter and electricity. Dalton had shown earlier in the 19th century that matter consists of atoms, and Faraday's work indicated that atoms might contain electrically-charged particles. Faraday's laws further suggested that discrete particles of electricity may be components of the atoms. Later work led to the conclusion that an electric current is a stream of electrons and that electrons are universal components of atoms.

6.2   MOLAR CONDUCTIVITY

Important information about the nature of solutions has been provided by measurements of their conductivities. A solution of sucrose in water has the same electrical conductivity as water itself, while a solution of sodium chloride or acetic acid has a much higher conductivity. The latter substances are known as *electrolytes*, since their solutions in water and in other solvents contain ions. By contrast, sucrose is a nonelectrolyte, its molecules being present intact in solution.

Aqueous solutions of sodium chloride and cupric sulfate of, say, 1 M concentration conduct electricity much better than a 1 M solution of acetic acid. This suggests that the sodium chloride and cupric sulfate exist in solution to a greater extent as ions than does acetic acid. More detailed investigations have indicated that certain substances, including sodium chloride and cupric sulfate, occur almost entirely as ions when in aqueous solution. Such substances are known as *strong* electrolytes. Other substances are present only partially as ions. Acetic acid, for example, exists in solution partly as CH_3COOH and partly as $CH_3COO^- + H^+$. These are known as *weak* electrolytes.

Further information is provided by studies of the way in which the electrical conductivities of solutions vary with the concentration of solute. The pioneering work in this field was carried out during the last century by the German physicist Friedrich Wilhelm Georg Kohlrausch (1840–1910), who also clarified the theory by introducing the concept of *equivalent conductivity*. This is now referred to as *molar conductivity*. Again, note that the mole referred to may relate to a fraction of a molecule, such as $\frac{1}{2}H_2SO_4$.

According to Ohm's law, the resistance R of a slab of material is equal to the potential V in volts (V), divided by the current I in amperes (A):

$$R(\Omega) = \frac{V(V)}{I(A)} \tag{6.2}$$

The unit of electrical resistance is the ohm, given the symbol Ω (omega), and the reciprocal of the resistance is the *conductance*, the unit of which is the siemen (S or Ω^{-1}). The electrical conductance of a material of length l and cross-sectional area A is proportional to A and inversely proportional to l (see Figure 6.1a). It is usual to express l in cm and A in cm², so that we have

$$\text{conductance } (\Omega^{-1}) = \kappa \frac{A(\text{cm}^2)}{l(\text{cm})} \tag{6.3}$$

The proportionality constant κ is now known as the *electrolytic conductivity* (formerly *specific conductivity*), and its units are Ω^{-1} cm^{-1}. It is the conductance of a 1 cm cube.

Kohlrausch realized that the electrolytic conductivity is not a suitable quantity for comparing the conductivities of different solutions. If a solution of one electrolyte is much more concentrated than another, it may well have a higher conductivity simply because it contains more ions; the value of the electrolytic conductivity thus does not immediately tell us anything of significance about the solution. What is needed instead is a property in which there has been some compensation for the differences in concentrations. Such a property, first defined and used by Kohlrausch, is now called the *molar conductivity* and given the symbol Λ (lambda). It can be

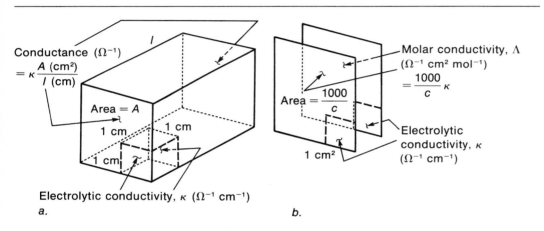

Figure 6.1
a. Relationship between (specific) electrolytic conductivity and conductivity.
b. Relationship between molar conductivity and (specific) electrolytic conductivity.

visualized as follows. Suppose that we construct a cell having parallel plates 1 cm apart, the plates being of such an area that for a particular solution 1 mol of the electrolyte (e.g., HCl, $\frac{1}{2}$CuSO$_4$) is present (see Figure 6.1b). Then, by definition, the molar conductivity Λ is the conductance (the reciprocal of the resistance) across the plates.

We need not actually construct cells for each solution for which we need to know the molar conductivity, since we can easily calculate it from the specific conductivity. If the concentration of the solution is c mol dm^{-3}, which is $c/1000$ mol cm^{-3}, the area of the plates in our hypothetical molar conductivity cell would be $1000/c$ cm^2 in order for 1 mol to be between the plates (see Figure 6.1b). The molar conductivity Λ is thus the conductivity across plates $1000/c$ cm^2 in area, whereas the specific conductivity κ is the conductivity across plates 1 cm^2 in area. The relationship between Λ and κ is thus

$$\Lambda = \frac{1000}{c}\kappa \qquad\qquad (6.4)$$

Since κ has the units $\Omega^{-1}\,cm^{-1}$, and $c/1000$ has the units $mol\,cm^{-3}$, the units of Λ are $\Omega^{-1}\,cm^2\,mol^{-1}$

The importance of molar conductivity is that it gives us information about the conductivity of ions produced in solution by 1 mol of a substance. Studies of the variation of molar conductivity have revealed important results, such as those shown schematically in Figure 6.2. In all cases, molar conductivity diminishes as the concentration is raised, but two patterns of behavior can be distinguished. The magnitudes of conductivities suggest that there are two extreme classes of electrolytes: the *strong*, which produce many ions, and the *weak*, which produce few.

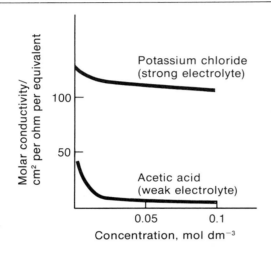

Figure 6.2

Variations of molar conductivity Λ with concentration for strong and weak electrolytes.

The strong electrolytes, which comprise almost all salts and certain acids such as hydrochloric acid, are found to behave as shown in Figure 6.2; their molar conductivities fall only slightly as the concentration is raised. On the other hand, the weak electrolytes, which include acids such as acetic acid and bases such as ammonia, exhibit a much more pronounced decline of Λ with increasing concentration, as shown by the lower curve in Figure 6.2.

We can extrapolate the curves back to zero concentration and obtain a quantity known as Λ_0, the molar conductivity at infinite dilution, or zero concentration. With weak electrolytes this extrapolation may be unreliable, and an indirect method, explained on p. 285, is usually employed. It is convenient to denote the ratio of Λ at any concentration to Λ_0 by the symbol α:

$$\alpha = \frac{\Lambda}{\Lambda_0} \tag{6.5}$$

The pioneering work in 1884 of the Swedish scientist Svante August Arrhenius (1859–1927) suggested that the same extrapolation applied to the variation of Λ with c for both strong and weak electrolytes. However, it was realized later that Arrhenius's explanation only applies to weak electrolytes and that a completely different explanation is required for strong electrolytes. Arrhenius's treatment is considered in the following section. The theory of strong electrolytes is dealt with in Section 6.4.

6.3   WEAK ELECTROLYTES

According to the theory of Arrhenius, the variations of Λ with concentration are due to shifts in equilibrium between undissociated and dissociated species. This idea was expressed quantitatively by the Russian-German physical chemist Friedrich Wilhelm Ostwald (1853–1932) in terms of a *dilution law*. Consider an electrolyte AB which exists in solution partly as the undissociated species AB and partly as the ions A^+ and B^-

$$AB \rightleftharpoons A^+ + B^-$$

The equilibrium constant, on the assumption of ideal behavior, is

$$K = \frac{[A^+][B^-]}{[AB]} \tag{6.6}$$

Suppose that n mol of the electrolyte are present in V dm^3, and that the fraction dissociated is α. The fraction not dissociated is $1 - \alpha$. The amounts of the three species present at equilibrium, and the corresponding concentrations, are therefore

	AB $\rightleftharpoons$	A$^+$ +	B$^-$
Number of moles at equilibrium:	$n(1 - \alpha)$	$n\alpha$	$n\alpha$
Concentration at equilibrium/mol dm^{-3}:	$\dfrac{n(1 - \alpha)}{V}$	$\dfrac{n\alpha}{V}$	$\dfrac{n\alpha}{V}$

The equilibrium constant is therefore

$$K = \frac{\left(\dfrac{n\alpha}{V}\right)^2}{\dfrac{n(1 - \alpha)}{V}} \tag{6.7}$$

and thus

$$K = \frac{n\alpha^2}{V(1 - \alpha)} \tag{6.8}$$

Therefore, for a given number of moles the degree of dissociation α must vary with the volume V as

$$\frac{\alpha^2}{1 - \alpha} = \text{constant} \times V \tag{6.9}$$

Alternatively, we can write

$$\frac{\alpha^2}{1 - \alpha} = \frac{\text{constant}}{\text{concentration}} \tag{6.10}$$

The larger the V, the lower the concentration, and the larger the degree of dissociation. Thus, if we start with a solution containing 1 mol of electrolyte and dilute it, the degree of dissociation increases and the number of moles of the ionized species increases. As V becomes very large (the concentration approaching zero), the degree of dissociation α approaches unity; that is, dissociation approaches 100% as infinite dilution is approached. The experimental value of Λ_0 corresponds to complete dissociation; at finite concentrations the molar concentration Λ is therefore lower by the factor

$$\alpha = \frac{\Lambda}{\Lambda_0} \tag{6.11}$$

The dilution law can thus be expressed as

$$KV = \frac{n(\Lambda/\Lambda_0)^2}{1 - \Lambda/\Lambda_0} \tag{6.12}$$

or

$$K = \frac{c(\Lambda/\Lambda_0)^2}{1 - \Lambda/\Lambda_0} \tag{6.13}$$

where $c\ (=n/V)$ is the concentration. Equation 6.13 has been found to give a satisfactory interpretation of the variation of Λ with c for a number of weak electrolytes.

6.4   STRONG ELECTROLYTES

Arrhenius originally believed that his theory and the formulation of Ostwald would explain the conductivity behavior of both strong and weak electrolytes. However, it soon became apparent that strong electrolytes required another explanation. Several lines of evidence indicated this. One was that the plots of Λ against c for strong electrolytes, unlike those for weak ones, could not be fitted to Ostwald's equations. Another was that the heats of neutralization of solutions of strong acids and bases (e.g., the neutralization of HCl and NaOH) could be explained only if it was assumed that these electrolytes are completely dissociated over a considerable concentration range.

If strong electrolytes are indeed fully dissociated, we need to explain why their molar conductivities decrease as concentration increases. Admittedly the decrease is much less than with the weak electrolytes (see Figure 6.2), but it is significant. Several workers in the early years of the present century suggested explanations, but a satisfactory detailed treatment was not given until 1923, when P. J. W. Debye and E. Hückel published a very important mathematical paper on the subject. According to them, the decrease in molar conductivity of a strong electrolyte is due to the mutual interference of the ions, which becomes more pronounced as the concentration increases. Because of the strong attractive forces between ions of opposite signs, the arrangement of ions in solution is not completely random. In the immediate neighborhood of any positive ion there tend to be more negative ions than positive ions, whereas for a negative ion there are more positive ions than negative ions. This is shown schematically in Figure 6.3 for a sodium chloride solution. In *solid* sodium chloride, there is a regular array of sodium and chloride ions. As shown in Figure 6.3*a*, each sodium ion has six chloride ions as its nearest neighbors, and each

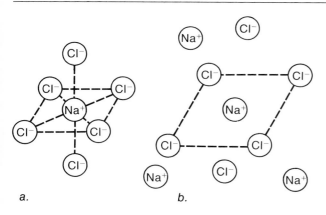

a.   b.

Figure 6.3

The distribution of chloride ions round a sodium ion (*a*) in the crystal lattice, (*b*) in a solution of sodium chloride. In the solution the interionic distances are greater, and the distribution is not as regular, but near to the sodium ion there are more chloride ions than sodium ions.

chloride ion has six sodium ions. When the sodium chloride is dissolved in water this ordering is still preserved to a very slight extent (Figure 6.3*b*). The ions are much further apart than in the solid; the electrical attractions are therefore much smaller and the thermal motions cause irregularity. The small amount of ordering which does exist, however, is sufficient to exert an important effect on the conductivity of the solution.

The way in which this ionic distribution affects the conductivity is as follows. If an electric potential is applied, a positive ion will move towards the negative electrode and must drag along with it an entourage of negative ions. The more concentrated the solution the closer these negative ions are to the positive ion under consideration, and the greater is the drag. The ionic "atmosphere" around a moving ion is therefore not symmetrical (the charge density behind being greater than that in front) and this will result in a retardation in the motion of the ion. This influence on the speed of an ion is called the *relaxation*, or *asymmetry*, effect.

A second factor which retards the motion of an ion in solution is the tendency of the applied potential to move the ionic atmosphere itself. This in turn will tend to drag solvent molecules, because of the attractive forces between ions and solvent molecules. As a result, the ion at the center of the ionic atmosphere is required to move upstream, and this is an additional retarding influence. This effect is known as the *electrophoretic* effect, since it is analogous to the resistance acting against the movement of a colloidal particle in an electric field (see Section 11.9).

6.5   THE IONIC ATMOSPHERE

The Debye-Hückel theory is very complicated mathematically and a detailed treatment is outside the scope of this book. Here, an account will be given of the theoretical treatment of the ionic atmosphere, and this account will be followed by a brief qualitative discussion of the way in which the theory of the ionic atmosphere explains conductivity behavior.

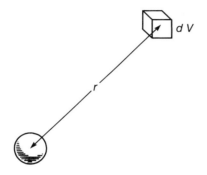

Figure 6.4

An ion in solution of charge $z_c e$, with an element of volume dV situated at a distance r.

Figure 6.4 shows an ion of charge $z_c e$ situated at a point A; e is the unit positive charge (the negative of the electronic charge) and z_c the valence of the ion. Consider a small element of volume dV situated at a distance r from the positive ion. As a result of thermal motion there will sometimes be an excess of positive ions in the volume element, and sometimes an excess of negative ions. On the average the charge density will be negative because of the electrostatic forces. In other words, the probability that there is a negative ion in the volume element is greater than the probability that there is a positive ion. The negative charge density will obviously be greater if r is small than if it is large, but we will see later that it is possible to define an effective thickness of the ionic atmosphere.

Suppose that the average electric potential in the volume element dV is ϕ. The work required to bring a positive ion of charge $z_+ e$ from infinity up to this point is then $z_+ e\phi$. (Note that in this derivation z_+ and z_- are taken to be the numerical values of the valences of the ions; e.g., for Na^+, $z_+ = 1$; for Cl^-, $z_- = 1$.) According to the Boltzmann principle, the time-average numbers of positive and negative ions present in the volume element are (see Section 5.15, especially equation (5.216)).

$$dn_+ = n_+ e^{-z_+ e\phi/\mathbf{k}T}\, dV \tag{6.14}$$

and

$$dn_- = n_- e^{-(-z - e\phi/\mathbf{k}T)}\, dV \tag{6.15}$$

where n_+ and n_- are the respective total numbers of positive and negative ions in unit volume of solution, $\mathbf{k}$ is the Boltzmann constant (the gas constant R divided by the Avogadro number), and T is the absolute temperature. The charge density in the volume element, i.e., the net charge per unit volume, is given by

$$\rho = \frac{e\,(z_+ dn_+ - z_- dn_-)}{dV} \tag{6.16}$$

$$= e\,(n_+ z_+ e^{-n_+ e\phi/\mathbf{k}T} - n_- z_- e^{n_- e\phi/\mathbf{k}T}) \tag{6.17}$$

For a univalent electrolyte (one in which both ions are univalent; an example is NaCl) z_+ and z_- are unity, and n_+ and n_- must be equal because the entire solution is neutral; equation (6.17) then becomes

$$\rho = n\,e(e^{-e\phi/\mathbf{k}T} - e^{e\phi/\mathbf{k}T}) \tag{6.18}$$

where $n = n_+ = n_-$. If $e\phi/\mathbf{k}T$ is sufficiently small (which requires that the volume element is not too close to the central ion, so that ϕ is small) the exponential terms are given by[†]

$$e^{-e\phi/\mathbf{k}T} \approx 1 - \frac{e\phi}{\mathbf{k}T} \tag{6.19}$$

and

$$e^{e\phi/\mathbf{k}T} \approx 1 + \frac{e\phi}{\mathbf{k}T} \tag{6.20}$$

Equation (6.18) thus reduces to

$$\rho = -\frac{2ne^2\phi}{\mathbf{k}T} \tag{6.21}$$

In the more general case, in which there are a number of different types of ions,

---

[†] The expansion of e^x is $e^x = 1 + x + \dfrac{x^2}{2!} + \dfrac{x^3}{3!} + \dfrac{x^4}{4!} + \cdots$ and when x is sufficiently small we can neglect terms beyond x; thus $e^x \approx 1 + x$ and $e^{-x} \approx 1 - x$.

the expression is

$$\rho = -\frac{e^2\phi}{\mathbf{k}T}\sum_i n_i z_i^2 \tag{6.22}$$

where n_i and z_i represent the number per unit volume and the positive value of the valence of the ions of the ith type. The summation is taken over all the types of ions present.

Equation (6.22) relates the charge density to the average potential ϕ. In order to obtain these quantities separately it is necessary to have another relationship between ρ and ϕ. This is provided by the equation due to Poisson, which can be derived from electrostatic theory on the assumption that Coulomb's law of force applies. In the present situation there must be spherical symmetry about the central ion, since no one direction can be favored; in these circumstances Poisson's equation takes the form

$$\frac{1}{r^2}\frac{\partial}{\partial r}\left(r^2\frac{\partial\phi}{\partial r}\right) = -\frac{\rho}{\varepsilon_o\varepsilon} \tag{6.23}$$

where ε is the dielectric constant (or relative permittivity) of the medium and ε_o is the permittivity of a vacuum (see Appendix A). Insertion of the expression for ρ in equation (6.22) then gives

$$\frac{1}{r^2}\frac{\partial}{\partial r}\left(r^2\frac{\partial\phi}{\partial r}\right) = \frac{e^2\phi}{\varepsilon_o\varepsilon\mathbf{k}T}\sum_i n_i z_i^2 \tag{6.24}$$

This equation can be written as

$$\frac{1}{r^2}\frac{\partial}{\partial r}\left(r^2\frac{\partial\phi}{\partial r}\right) = \kappa^2\phi \tag{6.25}$$

where the quantity κ (not to be confused with electrolytic conductivity, p. 263) is given by

$$\kappa^2 = \frac{e^2}{\varepsilon_o\varepsilon\mathbf{k}T}\sum_i n_i z_i^2 \tag{6.26}$$

The general solution of (6.25) is

$$\phi = \frac{Ae^{-\kappa r}}{r} + \frac{Be^{\kappa r}}{r} \tag{6.27}$$

where A and B are constants. Since ϕ must approach zero as r becomes very large, and $e^{\kappa r}/r$ becomes very large as r becomes large, the multiplying factor B must be zero, and equation (6.27) thus becomes

$$\phi = \frac{Ae^{-\kappa r}}{r} \tag{6.28}$$

The constant A can be evaluated by considering the situation when the solution is infinitely dilute, when the central ion can be considered to be isolated. In such a solution $\sum_i n_i z_i^2$ approaches zero, so that κ also approaches zero and the potential is therefore

$$\phi = \frac{A}{r} \tag{6.29}$$

However, in these circumstances the potential at a distance r will simply be the potential due to the ion itself, since there is no interference by the atmosphere. The potential at a distance r due to an ion of charge z_c is

$$\phi = \frac{z_c e}{4\pi\varepsilon_o \varepsilon r} \tag{6.30}$$

Equating these two expressions for ϕ leads to

$$A = \frac{z_c e}{4\pi\varepsilon_o \varepsilon} \tag{6.31}$$

and insertion of this in equation (6.28) gives

$$\phi = \frac{z_c e \, e^{-\kappa r}}{4\pi\varepsilon_o \varepsilon r} \tag{6.32}$$

If κ is sufficiently small (i.e., if the solution is sufficiently dilute) the exponential $e^{-\kappa r}$ is approximately $1 - \kappa r$, and (6.32) therefore becomes

$$\phi = \frac{z_c e}{4\pi\varepsilon_o \varepsilon r} - \frac{z_c e \kappa}{4\pi\varepsilon_o \varepsilon} \tag{6.33}$$

The first term on the right-hand side of this equation is the potential at a distance r, due to the central ion itself. The second term is therefore the potential produced by the ionic atmosphere:

$$\phi_a = -\frac{z_c e \kappa}{4\pi\varepsilon_o \varepsilon} \tag{6.34}$$

There is now no dependence on r; this potential is therefore uniform and exists at the central ion. If the ionic atmosphere were replaced by a charge $-z_c e$ situated at a distance $1/\kappa$ from the central ion, the effect due to it at the central ion would be exactly the same as that produced by the ionic atmosphere. The distance $1/\kappa$ is therefore referred to as the *thickness of the ionic atmosphere*.

The quantity κ is given by (6.26); the thickness of the ionic atmosphere is thus

$$\frac{1}{\kappa} = \left(\frac{\varepsilon_o \varepsilon \mathbf{k} T}{e^2 \sum_i n_i z_i^2} \right)^{1/2} \tag{6.35}$$

Instead of the numbers of ions n_i per unit volume (m^3 in SI) it is more convenient to use the concentration c_i in mol dm^{-3}; $n_i = 10^3 c_i N_A$, where N_A is the Avogadro number. Thus

$$\frac{1}{\kappa} = \left(\frac{\varepsilon_o \varepsilon \mathbf{k} T}{10^3 e^2 \sum_i c_i z_i^2 N_A} \right)^{1/2} \tag{6.36}$$

This equation allows values of the ionic atmosphere to be estimated.

EXAMPLE

Estimate the thickness of the ionic atmosphere for a solution of
(1) 0.01 M NaCl and (2) 0.001 M $ZnCl_2$, both in water at 25° C,
with $\varepsilon = 78$.

SOLUTION

(1) The summation in equation (6.36) is $\sum_i c_i z_i^2 = 0.01 \times 1^2 +$

$0.01 \times 1^2 = 0.02$ mol dm^{-3}. Insertion of the appropriate values in
(6.36) then gives

$$\frac{1}{\kappa} = \left(\frac{8.85 \times 10^{-12} (C^2 N^{-1} m^{-2}) \times 78 \times 1.381 \times 10^{-23} (J\ K^{-1}) 300(K)}{10^3 (dm^3\ m^{-3}) \times (1.602 \times 10^{-19})^2\ (C^2) \times 0.02\ (\text{mol dm}^{-3})\ \times 6.022 \times 10^{23}\ (\text{mol}^{-1})} \right)^{1/2}$$

$$= 3.04 \times 10^{-9}\ (J\ N^{-1}\ m)^{1/2}$$

Since $J = kg\ m^2\ s^{-2}$ and $N = kg\ m\ s^{-2}$ the units are metres;
the thickness is thus 3.04 nanometres (nm).

(2) The summation is now

$$\sum_i c_i z_i^2 = 0.001 \times 2^2 + 0.002 \times 1^2$$
$$= 0.006\ \text{mol dm}^{-3}$$

and the result is

$$\frac{1}{\kappa} = 5.56 \times 10^{-9}\ m = 5.56\ nm$$

Table 6.1 shows values of $1/\kappa$ for various types of electrolytes at different molar concentrations in aqueous solution. We see from equation (6.36) that the thickness of the ionic atmosphere is inversely proportional to the square root of the concentration, the atmosphere moving further from the central ion as the solution is diluted.

Table 6.1 Thickness in Nanometres of Ionic Atmospheres in Water at 25° C

Type of Electrolyte	Molar Concentration		
	0.10 M	0.01 M	0.001 M
Uni-univalent	0.962	3.04	9.62
Uni-bivalent and bi-univalent	0.556	1.76	5.56
Bi-divalent	0.481	1.52	4.81
Uni-tervalent and ter-univalent	0.392	1.24	3.92

6.6   MECHANISM OF CONDUCTIVITY

The treatment of conductivity is very difficult mathematically and only a very general account of the main ideas can be given here.

We have seen that the effect of the ionic atmosphere is to exert a drag on the movement of a given ion. If the ion is stationary, the atmosphere is arranged symmetrically about it and does not tend to move it in either direction (see Figure 6.5a). However, if a potential which tends to move the ion to the right is applied, the atmosphere will decay to some extent on the left of the ion and build up more on the right (Figure 6.5b). Since it takes time for these *relaxation* processes to occur there will be an excess of ionic atmosphere to the left of the ion (i.e., behind it), and a deficit to the right (in front of it). This asymmetry of the atmosphere will have the effect of dragging the central ion back. This is the relaxation or asymmetry effect.

There is a second reason why the existence of the ionic atmosphere impedes the

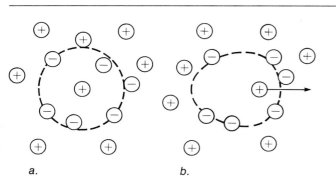

a.   b.

Figure 6.5

a. A stationary central ion with a spherically-symmetrical ion atmosphere.

b. A positive ion moving to the right. The ion atmosphere behind it is relaxing, while that in front is building up. The distribution of negative ions around the positive ion is now asymmetric.

motion of an ion. Ions are attracted to solvent molecules mainly by ion-dipole forces; therefore when they move they drag solvent along with them. The ionic atmosphere, having a charge opposite to that of the central ion, moves in the opposite direction to it, and therefore drags solvent in the opposite direction. This means that the central ion has to travel upstream, and therefore it travels more slowly than if there were no effect of this kind. This is the *electrophoretic* effect.

Debye and Hückel carried through a theoretical treatment which led them to expressions for the forces exerted upon the central ion by the relaxation effect. They supposed the ions to travel through the solution in straight lines, neglecting the zig-zag Brownian motion brought about by the collisions of surrounding solvent molecules. This theory was improved in 1926 by the Norwegian-American physical chemist Lars Onsager (1903–1976), who took Brownian motion into account and whose expression for the relaxation force f_r was

$$f_r = \frac{e^2 z_i \kappa}{24\pi\varepsilon_0 \varepsilon \mathbf{k} T} wV'$$
(6.37)

where V' is the applied potential gradient and w is a number whose magnitude depends upon the type of electrolyte; for a uni-univalent electrolyte w is $2 - \sqrt{2} = 0.586$.

The electrophoretic force, f_e, was given by Debye and Hückel as

$$f_e = \frac{e z_i \kappa}{6\pi\eta} K_c V'$$
(6.38)

where K_c is the coefficient of frictional resistance of the central ion with reference to the solvent, and η is the viscosity of the medium. The viscosity enters into this expression because we are concerned with the motion of the ion past the solvent molecules, which depends on the viscosity of the solvent.

The final expression obtained on this basis for the molar conductivity Λ of an electrolyte is

$$\Lambda = \Lambda_0 - \left[\frac{29.15(z_+ + z_-)}{(\varepsilon T)^{1/2}\eta} + \frac{9.90 \times 10^5}{(\varepsilon T)^{3/2}}\Lambda_0 w\right]\sqrt{c(z_+ + z_-)}$$
(6.39)

where c is the concentration of the electrolyte in mol dm^{-3}. In the simple case of a uni-univalent electrolyte, w is $2 - \sqrt{2}$ and z_+ and z_- are unity, and the result is

$$\Lambda = \Lambda_0 - \left[\frac{82.4}{(\varepsilon T)^{1/2}\eta} + \frac{8.20 \times 10^5}{(\varepsilon T)^{3/2}}\Lambda_0\right]\sqrt{c}$$
(6.40)

These equations are based on an assumption of complete dissociation. If an electrolyte is neither a strong nor a weak electrolyte, these equations must be modified by a factor involving the degree of dissociation. For weak electrolytes the ions are generally so far apart that the ionic atmospheres are very diffuse; the relaxation and electrophoretic effects therefore are negligible and the above equations need not be applied. The Arrhenius-Ostwald treatment (Section 6.3) is then satisfactory.

A number of experimental tests have been made of equations (6.39) and (6.40), which are known as Debye-Hückel-Onsager equations. Equation (6.40), for example, which applies to a uni-univalent electrolyte, can be written in the form

$$\Lambda = \Lambda_0 - (P + Q\Lambda_0)\sqrt{c} \qquad (6.41)$$

where P and Q are constants the values of which can be calculated from the properties (ε and η) of the particular solvent. Equation (6.41) can be tested by seeing whether a plot of Λ against $\sqrt{c}$ is linear, and whether the slope of the line is as predicted from the values of P and Q. For aqueous solutions of uni-univalent electrolytes, equation (6.41) is obeyed very satisfactorily up to a concentration of about 2×10^{-3} M; at higher concentrations deviations are found. The corresponding equations for other types of electrolytes in water are also obeyed satisfactorily at very low concentrations, but deviations are found at lower concentrations than with uni-univalent electrolytes. With $CaCl_2$ and $LaCl_3$, for example, the equation is obeyed up to about 4×10^{-5} mol dm^{-3}. Work with nonaqueous solvents has also shown agreement at very low concentrations and deviations at higher ones. If the dielectric constant is fairly high, as with methanol ($\varepsilon = 32$), the equation is obeyed over a wider range than if the solvent is of low dielectric constant and is nonhydroxylic.

Various explanations have been given for deviations from the Debye-Hückel-Onsager equations. A common type of behavior is for the negative slopes of the Λ versus $\sqrt{c}$ plots to be greater than predicted by the equation; that is, the experimental conductivities are *lower* than predicted by the theory. This has been explained in terms of *ion pairing*, a concept which was developed by the Danish physical chemist Niels Bjerrum (1879–1958) in 1926. Although most salts, such as sodium chloride, are present in the solid state and in solution as ions and not as covalent species, there is a tendency for them to come together from time to time to form ion pairs. This is quite different from covalent bond formation, in which the ion pairs have only a temporary existence, since there is a constant interchange between the various ions in the solution. At any instant of time a certain number of ions are paired together in this way, and these ions therefore do not contribute to the transport of electric current. Solvents of low dielectric constant give rise to higher electrostatic attractions between the ions than do high-dielectric-constant solvents such as water, and thus favor the formation of ion pairs.

6.7   CONDUCTIVITY AT HIGH FREQUENCIES AND POTENTIALS

An important consequence of the existence of the ionic atmosphere is that the conductivity should depend upon the frequency if an alternating potential is applied to the solution. Suppose that the alternating potential is of sufficiently high frequency that the time of oscillation is small compared with the time it takes for the ionic atmosphere to relax. There will not be time for the atmosphere to relax behind the ion and to form in front of it; the ion will be virtually stationary and its ionic atmosphere will remain symmetrical. Therefore, as the frequency of the potential increases the relaxation and electrophoretic effects will become less and less important and there will be an increase in the molar conductivity. This effect was first predicted theoretically by P. J. W. Debye and H. Falkenhagen in 1928, and has been confirmed experimentally. The frequency region over which the conductivity is observed to change is consistent with what is predicted by the theory.

A related effect is observed if conductivities are measured at very high potential gradients. For example, if the applied potential is 20 000 V cm^{-1}, an ion will move at a speed of about 1 m s^{-1} and will travel several times the thickness of the effective ionic atmosphere in the time of relaxation of the atmosphere. Consequently, the moving ion is essentially free from the effect of the ionic atmosphere, which does not have time to build up around it to any extent. Therefore, at sufficiently high voltages the relaxation and electrophoretic effects will diminish and eventually disappear and the molar conductivity will increase. This effect was first observed experimentally by the German physicist Max Carl Wien (1866–1938) in 1928, and is known as the *Wien effect*. It was originally thought that the Wien effect would provide a method of determining degrees of dissociation of incompletely dissociated electrolytes, since the results at high potentials should provide Λ_o, with no complications due to the ionic atmosphere. However, molar conductivities of weak electrolytes at high potentials are anomalously large. It appears that very high potentials bring about a dissociation of the molecules into ions. This phenomenon, known as the *dissociation field effect*, unfortunately invalidates this method of determining degrees of dissociation.

6.8   INDEPENDENT MIGRATION OF IONS

In principle the plots of Λ against concentration (Figure 6.2, p. 264) can be extrapolated back to zero concentration to give the Λ_0 value. In practice this extrapolation can only satisfactorily be made with strong electrolytes. With weak electrolytes there is rather a strong variation of Λ with c at low concentrations and therefore the extrapolations do not lead to reliable Λ_0 values. An indirect method for obtaining Λ_0 for weak electrolytes is considered later.

Between 1879 and 1885 Kohlrausch made a number of determinations of the Λ_0 values, and observed that they exhibited certain regularities. Some values are given in Table 6.2 for corresponding sodium and potassium salts. We see that the difference between the molar conductivities of a potassium and a sodium salt of the same anion is independent of the nature of the anion. Similar results were obtained for a variety of pairs of salts with common cations or anions, both in aqueous and nonaqueous solvents.

Kohlrausch explained this behavior in terms of his *law of independent migration of ions*. Each ion is assumed to make its own contribution to the molar conductivity, irrespective of the nature of the other ion with which it is associated. In other words,

$$\Lambda_0 = \lambda_+^o + \lambda_-^o \tag{6.42}$$

where λ_+^o and λ_-^o are the *ion conductivities* of cation and anion respectively at infinite dilution. Thus for potassium chloride (see Table 6.2)

$$\Lambda_0(KCl) = 149.9 = \lambda_{K^+}^o + \lambda_{Cl^-}^o \tag{6.43}$$

and for sodium chloride

$$\Lambda_0(NaCl) = 126.5 = \lambda_{Na^+}^o + \lambda_{Cl^-}^o \tag{6.44}$$

The difference, 23.4, is thus the difference between the λ_+^o values for K^+ and Na^+:

$$23.4 = \lambda_{K^+}^o - \lambda_{Na^+}^o \tag{6.45}$$

and this difference will be the same whatever the nature of the anion.

Table 6.2   Molar Conductivities at Infinite Dilution for Various Sodium and Potassium Salts in Aqueous Solution at 25° C

Electrolyte	$\Lambda_0/\Omega^{-1}\,cm^2\,mol^{-1}$	Electrolyte	$\Lambda_0/\Omega^{-1}\,cm^2\,mol^{-1}$	Difference
KCl	149.9	NaCl	126.5	23.4
KI	150.3	NaI	126.9	23.4
$\frac{1}{2}K_2SO_4$	153.5	$\frac{1}{2}Na_2SO_4$	130.1	23.4

IONIC MOBILITIES

It is not possible from the Λ_0 values alone to determine the individual λ_+^o and λ_-^o values; however, once one λ_+^o or λ_-^o value (such as $\lambda_{Na^+}^o$) has been determined, the rest can be calculated. The λ_+^o and λ_-^o values are proportional to the speeds with which the ions move under standard conditions, and this is the basis of the methods by which the individual conductivities are determined. The *mobility* of an ion, u, is defined as the velocity with which the ion moves under a unit potential gradient, and it is common to express it as the velocity in centimetres per second in a potential gradient of one volt per centimetre; the units of the mobility are then $cm\,s^{-1}/V\,cm^{-1} = cm^2\,V^{-1}\,s^{-1}$. Suppose that the potential drop across the opposite faces of a unit cube (1 cm^3) is V volts, and that the concentration of univalent positive ions in the cube is c_+ M, or $c_+/1000$ mol cm^{-3} (see Figure 6.6). If the mobility of the positive ions is u cm^2 V^{-1} s^{-1}, the velocity of the ions is uV cm s^{-1}. All positive ions within a distance of uV cm of the negative plate will therefore reach that plate each second, and the number of such ions is $uVc_+/1000$. The charge they carry is $FuVc_+/1000$ coulombs, and the current is thus $FuVc_+/1000$ amperes. The electrolytic conductivity due to the positive ions, κ_+, is thus

$$\kappa_+ = \frac{current}{voltage} = \frac{Fuc_+}{1000} \tag{6.46}$$

The molar conductivity due to these positive ions, Λ^o_+, is therefore

$$\Lambda^o_+ = \frac{\kappa_+}{c_+/1000} = Fu \tag{6.47}$$

Concentration of positive ions $= \dfrac{c_+}{1000}$ mol cm^{-3}

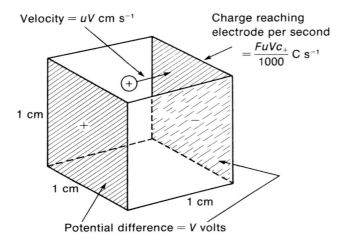

Figure 6.6

A unit cube containing a solution in which the concentration of unit positive charges is c_+ M, and where there is a drop of V volts between the opposite faces.

Velocity $= uV$ cm s^{-1}

Charge reaching electrode per second

$= \dfrac{FuVc_+}{1000}$ C s^{-1}

1 cm

1 cm

1 cm

Potential difference $= V$ volts

EXAMPLE

The mobility of a sodium ion in water at 25° C is 51.9×10^{-4} cm^2 V^{-1} s^{-1}. Calculate the molar conductivity of the sodium ion.

SOLUTION

From equation (6.47),

$$\Lambda^o_{Na^+} = 96\,500 \text{ (A s mol}^{-1}) \times 51.9 \times 10^{-4} \text{ (cm}^2 \text{ V}^{-1} \text{ s}^{-1})$$
$$= 50.1 \text{ A V}^{-1} \text{ cm}^2 \text{ mol}^{-1}$$

A V^{-1} is Ω^{-1}, and therefore

$$\Lambda^o_{Na^+} = 50.1 \; \Omega^{-1} \text{ cm}^2 \text{ mol}^{-1}$$

6.9   TRANSPORT NUMBERS

In order to split the Λ^o values into the λ^o_+ and λ^o_- values for the individual ions, use is made of a property known variously as the *transport number*, the *transference number*, and the *migration number*. It is the fraction of the current carried by each of the ions present in solution.

The quantity of electricity q_i carried through unit volume of solution by ions of the ith kind is proportional to:

(1) the concentration of the ions, c_i, in mol dm^{-3},
(2) the charge number z_i of each ion, and
(3) the mobility u_i (cm^2 V^{-1} s^{-1}).

Thus, for a particular type of ion,

$$q_i = kc_i z_i u_i \tag{6.48}$$

where the proportionality constant k includes the time. The total quantity of electricity q carried by all of the ions in the solution is thus the sum of the q_i values for each type of ion:

$$q = kc_1 z_1 u_1 + kc_2 z_2 u_2 + \cdots \tag{6.49}$$

$$= k \sum_i c_i z_i u_i \tag{6.50}$$

The fraction of current carried by an ion of the ith kind is thus

$$t_i = \frac{q_i}{q} = \frac{c_i z_i u_i}{\sum_i c_i z_i u_i} \tag{6.51}$$

This fraction t_i is the *transport, transference,* or *migration number* of the given ion in the particular solution. For the simplest case of a solution containing a single electrolyte yielding two ions the transport numbers of the two ions, t_+ and t_-, are

$$t_+ = \frac{c_+ z_+ u_+}{c_+ z_+ u_+ + c_- z_- u_-} \quad \text{and} \quad t_- = \frac{c_- z_- u_-}{c_+ z_+ u_+ + c_- z_- u_-} \tag{6.52}$$

However, $c_+ z_+$ is equal to $c_- z_-$, since the solution must be electrically neutral; therefore it follows that

$$t_+ = \frac{u_+}{u_+ + u_-} \quad \text{and} \quad t_- = \frac{u_-}{u_+ + u_-} \tag{6.53}$$

Note that

$$t_+ + t_- = 1 \tag{6.54}$$

and in general the sum of the transport numbers for all of the ions in a solution must be unity.

The speed of an ion at any concentration is proportional to the individual ion conductivity at that concentration, and transport numbers may alternatively be expressed as

$$t_+ = \frac{\lambda_+}{\lambda_+ + \lambda_-} \qquad \text{and} \qquad t_- = \frac{\lambda_-}{\lambda_+ + \lambda_-} \qquad (6.55)$$

or as

$$t_+ = \frac{\lambda_+}{\Lambda} \qquad \text{and} \qquad t_- = \frac{\lambda_-}{\Lambda} \qquad (6.56)$$

If t_+ and t_- can be measured over a range of concentrations the values at zero concentration, t_+^o and t_-^o, can be obtained by extrapolation, and these are

$$t_+^o = \frac{\lambda_+^o}{\Lambda_0} \qquad \text{and} \qquad t_-^o = \frac{\lambda_-^o}{\Lambda_0} \qquad (6.57)$$

This therefore allows us to split the Λ_0 value into λ_+^o and λ_-^o.

At first sight it may appear surprising that Faraday's laws, according to which equivalent quantities of different ions are liberated at the two electrodes, can be reconciled with the fact that the ions are moving at different speeds towards the electrodes. Figure 6.7, however, shows how these two facts can be reconciled. The

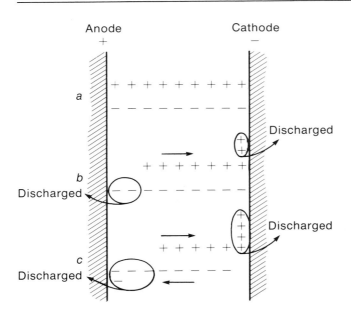

Figure 6.7

Schematic representation of the movement of ions during an electrolysis experiment, showing that equivalent amounts are neutralized at the electrodes in spite of differences in ionic velocities.

diagram shows in a very schematic way an electrolysis cell in which there are equal numbers of positive and negative ions of unit charge. The situation before electrolysis occurs is shown in *a*. Suppose that the cations only were able to move, the anions having zero mobility; after some motion has occurred the situation will be as represented at *b*. At each electrode two ions remain unpaired and are discharged, two electrons at the same time traveling in the outer circuit from the anode to the cathode. Thus, although only the cations have moved through the bulk of the solution, equivalent amounts have been discharged at the two electrodes. It is easy to extend this argument to the case in which both ions are moving but at different speeds. Thus *c* shows the situation when the speed of the cation is three times that of the anion, four ions being discharged at each electrode.

THE HITTORF METHOD

Three experimental methods have been employed for the determination of transport numbers. One of them, developed by the German physicist Johann Wilhelm Hittorf (1824–1914) in 1853, involves measuring the changes of concentration in the vicinity of the electrodes. In the second, the *moving boundary method*, a study is made of the rate of movement, under the influence of a current, of the boundary between two solutions. This method is described on p. 283. A third method, which we will not consider in this book, involves the measurement of the electromotive force of certain electrochemical cells.

A very simple type of apparatus used in the Hittorf method is shown in Figure 6.8. The solution to be electrolyzed is placed in the cell, and a small current is passed between the electrodes for a short period of time. The solution then is run out through the stopcocks and the samples analyzed for concentration changes.

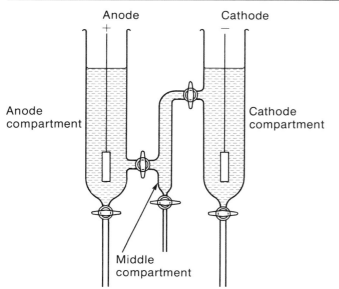

Anode **Cathode**

Anode compartment

Cathode compartment

Middle compartment

Figure 6.8

Simple type of apparatus used for measuring transport numbers by the Hittorf method.

The theory of the method is as follows: Suppose that the solution contains the ions M^+ and A^- which are not necessarily univalent but are denoted as such for simplicity. The fraction of the total current carried by the cations is t_+, and that carried by the anions is t_-.

Thus, when 96 500 C of electricity passes through the solution, 96 500 t_+ C are carried in one direction by t_+ moles of M^+ ions, and 96 500 t_- C are carried in the other direction by t_- moles of A^- ions. At the same time 1 mol of each ion is discharged at an electrode.

Suppose that, as represented in Figure 6.9, the cell containing the electrolyte is

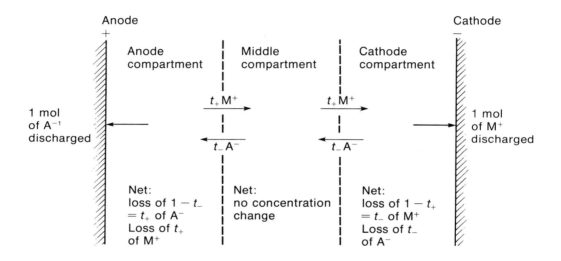

Figure 6.9

The concentration changes in three compartments of a conductivity cell (Hittorf method).

divided into three compartments by hypothetical partitions. One is a compartment near the cathode, one is near the anode, and the middle compartment is one in which no concentration change occurs. It is supposed that the two ions are discharged simply at the electrodes which are not attacked. The changes that occur when 96 500 C is passed through the solution are then as shown in Figure 6.9. The net results are:

At the cathode—A loss of $1 - t_+ = t_-$ mol of M^+; a loss of t_- mol of A^-; i.e., a loss of t_- mol of the electrolyte MA.

In the middle compartment—No concentration change.

At the anode—Loss of $1 - t_- = t_+$ mol of A^-; loss of t_+ mol of M^+. The net loss is therefore t_+ mol of MA.

It follows that

$$\frac{\text{Moles lost from anode compartment}}{\text{Moles lost from cathode compartment}} = \frac{t_+}{t_-} \tag{6.58}$$

Then, since $t_+ + t_- = 1$, the values of t_+ and t_- can be calculated from the experimental results. Alternatively,

$$\frac{\text{Moles lost from anode compartment}}{\text{Moles deposited}} = t_+ \tag{6.59}$$

and

$$\frac{\text{Moles lost from cathode compartment}}{\text{Moles deposited}} = t_- \tag{6.60}$$

Any of these three equations allows t_+ and t_- to be determined.

In the above discussion we have assumed that the electrodes are inert—i.e., are not attacked as electrolysis proceeds—and that the ions M^+ and A^- are simply deposited. If instead, for example, the anode were to pass into solution during electrolysis, the concentration changes would be correspondingly different; the treatment can readily be modified for this and other situations.

THE MOVING BOUNDARY METHOD

The moving boundary method was developed in 1886 by the British physicist Sir Oliver Joseph Lodge (1851–1940) and in 1893 by the British physicist Sir William Cecil Dampier (formerly Whetham) (1867–1952). The method is illustrated in Figure 6.10. Suppose that it is necessary to measure the transport numbers of the ions in the electrolyte MA. Two other electrolytes M′A and MA′ are selected as "indicators"; each has an ion in common with MA, and the electrolytes are such that M'^+ moves more slowly than M^+, and A'^- moves more slowly than A^-. The solution of MA is placed in the electrolysis tube with the solution of M′A on one side of it and that of MA′ on the other; the electrode in M′A is the anode, that in MA′ is the cathode. As the current flows the boundaries remain distinct, since the M'^+ ions cannot overtake the M^+ at the boundary a, and the A'^- ions cannot overtake A^- at boundary b. The slower ions M'^+ and A'^- are known as the *following* ions. The boundaries a and b move, as shown in Figure 6.10a, to positions a' and b'. The distances aa' and bb' are proportional to the ionic velocities, and therefore

$$\frac{aa'}{aa' + bb'} = t_+ \qquad \text{and} \qquad \frac{bb'}{aa' + bb'} = t_- \tag{6.61}$$

Alternatively, the movement of only one boundary may be followed, as shown by the example in Figure 6.10b. If Q coulombs of electricity are passed through the system, $t_+ Q$ is the quantity of electricity carried by the positive ions, and this corresponds to t_+Q/F mol of positive ions. If the concentration of positive ions is c mol dm^{-3}, t_+Q/F mol occupy a volume of t_+Q/Fc dm^3, or $1000t_+Q/Fc$ cm^3. This volume is equal to the area of cross section of the tube, A cm^2, multiplied by the distance aa' through which the boundary moves. Thus it follows that

$$aa' = \frac{1000t_+Q}{FcA} \text{ cm} \tag{6.62}$$

Since aa', Q, F, c, and A are known, the transport number t_+ can be calculated.

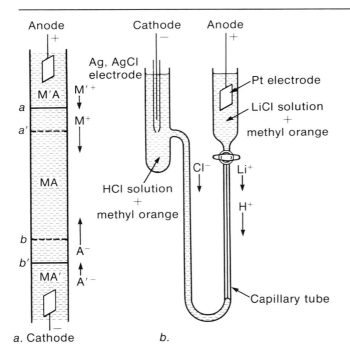

Figure 6.10

The determination of transport numbers by the moving-boundary method:

a. Schematic diagram showing the movement of the ions and the boundaries.

b. Simple apparatus for measuring the transport number of H$^+$ using Li$^+$ as the following cation. A clear boundary is formed between the receding acid solution and the LiCl solution, and is easily observed if methyl orange is present.

6.10   ION CONDUCTIVITIES

With the use of the transport numbers the Λ_0 values can be split into λ_+^o and λ_-^o, the contributions for the individual ions. The transport numbers extrapolated to infinite dilution are

$$t_+ = \frac{\lambda_+^o}{\Lambda_0} \qquad \text{and} \qquad t_- = \frac{\lambda_-^o}{\Lambda_0} \tag{6.63}$$

Once a single λ_+^o or λ_-^o value has been determined a complete set can be calculated from the available Λ_0 values. For example, suppose that a transport number study on NaCl led to $\lambda^o_{Na^+}$ and $\lambda^o_{Cl^-}$. If Λ_0 for KCl is known, the value of $\lambda^o_{K^+}$ $(= \Lambda_0(KCl) - \lambda^o_{Cl^-})$ can be obtained; from $\lambda^o_{K^+}$ and Λ_0 (KBr) the value of $\lambda^o_{Br^-}$ can be calculated, and so on. A set of values obtained in this way is given in Table 6.3. Symbols such as $\frac{1}{2}Ca^{2+}$, $\frac{1}{2}SO_4^{2-}$ are used in the case of the polyvalent ions.

Table 6.3 Individual Molar Ionic Conductivities (Ω^{-1} cm^2 mol^{-1}) in Water at 25°C

Cation	λ^o_+/Ω^{-1} cm^2 mol^{-1}	Anion	λ^o_-/Ω^{-1} cm^2 mol^{-1}
H^+	349.8	OH^-	198.6
Li^+	38.6	F^-	55.4
Na^+	50.1	Cl^-	76.4
K^+	73.5	Br^-	78.1
Rb^+	77.8	I^-	76.8
Cs^+	77.2	CH_3COO^-	40.9
Ag^+	61.9	$\frac{1}{2}SO_4^{2-}$	80.0
Tl^+	74.7	$\frac{1}{2}CO_3^{2-}$	69.3
$\frac{1}{2}Mg^{2+}$	53.1		
$\frac{1}{2}Ca^{2+}$	59.5		
$\frac{1}{2}Sr^{2+}$	59.5		
$\frac{1}{2}Ba^{2+}$	63.6		
$\frac{1}{2}Cu^{2+}$	56.6		
$\frac{1}{2}Zn^{2+}$	52.8		
$\frac{1}{3}La^{3+}$	69.7		

An important use for individual ion conductivities is in determining the Λ_0 values for weak electrolytes. We have seen (Figure 6.2) that for a weak electrolyte the extrapolation to zero concentration is rather a long one, and a reliable figure cannot be obtained. For acetic acid, however, Λ_0 is simply $\lambda^o_{H^+} + \lambda^o_{CH_3COO^-}$; from the values in Table 6.3 the value of Λ_0 is thus $349.8 + 40.9 = 390.7$ Ω^{-1} cm^2 mol^{-1}.

An equivalent but alternative procedure avoids the necessity of splitting the Λ_0 values into individual ionic contributions. The molar conductivity Λ_0 of any electrolyte MA can be expressed, for example, as

$$\Lambda_0(MA) = \Lambda_0(MCl) + \Lambda_0(NaA) - \Lambda_0(NaCl) \tag{6.64}$$

If MCl, NaA, and NaCl are all strong electrolytes, their Λ_0 values are readily obtained by extrapolation, and therefore permit the calculation of Λ_0 for MA, which may be a weak electrolyte. The following example illustrates the use of this method for acetic acid:

$$\Lambda_0(CH_3COOH) = \Lambda_0(HCl) + \Lambda_0(CH_3COONa) - \Lambda_0(NaCl) \tag{6.65}$$

$$= 426.2 + 91.0 - 126.5$$

$$= 390.7 \ \Omega^{-1} \ cm^2 \ mol^{-1} \ at \ 25°C.$$

This procedure is also useful for a highly insoluble salt, for which direct determinations of Λ_0 values might be impractical.

IONIC SOLVATION

The magnitudes of the individual ion conductivities (see Table 6.3) are of considerable interest. There is no simple dependence on the conductivity and the size of the ion. It might be thought that the smallest ions, such as Li^+, would be the fastest moving, but the values for Li^+, Na^+, K^+, Rb^+, and Cs^+ show that this is by no means true; K^+ moves faster than either Li^+ or Na^+. The explanation for this is that Li^+, because of its small size, becomes strongly attached to about four surrounding water molecules by ion-dipole and other forces, so that when the current passes it is actually $Li(H_2O)_4^+$, and not Li^+, which moves. With Na^+ the binding of water molecules is less strong, and with K^+ still less. The solvation of ions is discussed in greater detail in Chapter 7.

MOBILITIES OF HYDROGEN AND HYDROXIDE IONS

Of particular interest is the very high conductivity of the hydrogen ion. This ion has an unexpectedly high conductivity in a number of hydroxylic solvents such as water, methanol and ethanol, but behaves more normally in nonhydroxylic solvents such as nitrobenzene and liquid ammonia. At first sight the high values might seem to be due to the small size of the proton. However, there is a powerful electrostatic attraction between a water molecule and the proton, which because of its small size can come very close to the water molecule. As a consequence the equilibrium

$$H^+ + H_2O \rightleftharpoons H_3O^+$$

lies very much to the right. In other words, there are very few free protons in water; the ions exist as H_3O^+, which are hydrated by other water molecules (see Figure 7.8 on p. 301). Thus there remains the difficulty in explaining the high conductivity and mobility of the hydrogen ion. By virtue of its size it would be expected to move about as fast as the Na^+ ion, which in fact is the case in nonhydroxylic solvents.

In order to explain the high mobilities of hydrogen ions in hydroxylic solvents such as water, a special mechanism must be invoked. As well as moving through the solution in the way that other ions do, the H_3O^+ ion can also transfer its proton to a neighboring water molecule:

$$H_3O^+ + H_2O \rightarrow H_2O + H_3O^+$$

The resulting H_3O^+ ion can now transfer a proton to another H_2O molecule. In other words, protons, although not free in solution, can be passed from one water molecule to another. Calculations from the known structure of water show that the proton must jump a distance of 0.086 nm (0.86 Å) from an H_3O^+ ion to a water molecule, but that as a result the proton moves effectively through a distance of 0.31 nm (3.1 Å). The conductivity by this mechanism therefore will be much greater than by the normal mechanism. This proton transfer must be accompanied by some rotation of H_3O^+ and H_2O molecules in order for them to be correctly positioned for the proton transfers. Similar types of mechanisms have been proposed for other hydroxylic solvents. In methyl alcohol, for example, the process is:

$$
\begin{array}{cccc}
CH_3 & CH_3 & CH_3 & CH_3 \\
| & | & | & | \\
O^+ + O & \rightarrow & O + O^+ \\
\diagup\;\diagdown \quad \diagup\;\diagdown & & \diagup\;\diagdown \quad \diagup\;\diagdown \\
H \quad\; H \quad\; H \; H & & H \quad\; H \quad\; H \quad\; H
\end{array}
$$

Mechanisms of this type bear some resemblance to a mechanism which was suggested in 1806 by the German physicist Christian Johan Dietrich von Grotthuss (1785–1822) as an explanation of electrolytic conductance in general. At that time it was believed that electrolytes were present in solution not as ions but as neutral molecules. Grotthuss proposed a mechanism in which species were progressively passed from one neutral molecule to another. Because of its resemblance to the original Grotthuss proposal, the mechanism of conductance of hydrogen ions is frequently referred to as a *Grotthuss mechanism*.

Hydroxide ions also have abnormally high conductivities, and the explanation in their case is similar to that for hydrogen-ion conductivity. Protons are transferred from water molecules to hydroxide ions:

$$
\begin{array}{ccccccc}
H & & H & H & & & H \\
| & & | & | & & & | \\
O & + & O^- & \rightarrow O^- & + & & O \\
\diagdown & & & & & \diagup \\
H & & & & & H
\end{array}
$$

The net result is effectively the transfer of OH^- from one position to the next.

6.11   WALDEN'S RULE

An important relationship between molar conductivity and viscosity was discovered in 1906 by the Russian-German chemist Paul Walden (1863–1957). In the course of a study of the conductivity of tetramethylammonium iodide in various solvents, Walden noticed that the product of the molar conductivity at infinite dilution and the viscosity η of the solvent was approximately constant:

$$\Lambda_0\eta = \text{constant} \tag{6.66}$$

Similarly, it has been shown that the product of individual ion conductivities and viscosity is also a constant. Such relationships are most satisfactory for ions which are approximately spherical.

These relationships are not particularly surprising, since the speed with which a species diffuses through a liquid is inversely proportional to the viscosity of the liquid. Viscosity is a property which is a measure of the reciprocal of the speed with which molecules move past one another in the liquid state (see Section 11.8).

Behavior similar to that described by Walden's rule is observed in a solvent as the temperature is varied. As an example, Table 6.4 gives values for the product of λ_-^0 and η for the acetate ion in water at different temperatures.

Table 6.4 Molar Conductivity—Viscosity Products for the Acetate Ion in Water at Various Temperatures

Temperature/°C	Molar ionic conductivity, $\lambda_-^o/\Omega^{-1}\,cm^2\,mol^{-1}$	Viscosity, $\eta/kg\,m^{-1}\,s^{-1}$	$\lambda_-^o\,\eta/kg\,m^{-1}\,cm^2\,\Omega^{-1}\,s^{-1}\,mol^{-1}$
0	18.8	0.00179	0.0366
18	34.7	0.00106	0.0368
25	41.1	0.00089	0.0366
59	76.7	0.00048	0.0368
75	97.1	0.00038	0.0369
100	131.4	0.00028	0.0368
128	167.7	0.00022	0.0369
156	205.0	0.00018	0.0369

PROBLEMS

6.1 A constant current was passed through a solution of cupric sulfate, $CuSO_4$, for 1 hour, and 0.040 g of copper was deposited. Calculate the current in amperes. (At. wt. of Cu = 63.5; Faraday constant = 96 500 C mol^{-1}).

6.2 After passage of a constant current for 45 minutes, 7.19 mg of silver (at. wt. 107.9) were deposited from a solution of silver nitrate. Calculate the current.

6.3 The following are the molar conductivities of chloroacetic acid in aqueous solution at 25° C:†

Concentration/10^{-4} M	625	312.5	156.3	78.1	39.1	19.6	9.8
$\Lambda/\Omega^{-1}\,cm^2\,mol^{-1}$	53.1	72.4	96.8	127.1	164.0	205.8	249.2

Plot Λ against c. If $\Lambda_0 = 362$, are these values in accord with the Ostwald dilution law? What is the value of the dissociation constant?

6.4 The (specific) electrolytic conductivity of a saturated solution of silver chloride, AgCl, in pure water at 25° C is $1.26 \times 10^{-6}\,\Omega^{-1}\,cm^{-1}$ higher than that of the water used. Calculate the solubility of AgCl in water if the molar ionic conductivities are: Ag$^+$, 61.8; Cl$^-$, 76.4 $\Omega^{-1}\,cm^2\,mol^{-1}$.

6.5 The thickness of the ionic atmosphere $(1/\kappa)$ for a univalent electrolyte is 0.962 nanometres at a concentration of 0.10 M in water at 25° C $(\varepsilon = 78)$. Estimate the thickness

---

† We are using the modern convention in setting up this and other tables. The numbers are the physical quantity divided by the unit. For example, the number 625 is concentration/10^{-4} M, and therefore:

$$\text{concentration} = 6.25 \times 10^{-4}\,M$$

$$= 0.0625\,M$$

(a) in water at a concentration of 0.0001 M

(b) in a solvent of $\varepsilon = 38$ at a concentration of 0.10 M.

6.6 The molar conductivities of 0.001 M solutions of potassium chloride, sodium chloride, and potassium sulfate ($\frac{1}{2}K_2SO_4$) are 149.9, 126.5, and 153.3 cm^2 Ω^{-1} mol^{-1} respectively. Calculate an approximate value for the molar conductivity of a solution of sodium sulfate of the same concentration.

6.7 The molar conductivity at 18° C of a 0.010 M aqueous solution of ammonia is 9.6 cm^2 Ω^{-1} mol^{-1}. For NH$_4$Cl, $\Lambda_0 = 129.8$, and the molar ionic conductivities of OH$^-$ and Cl$^-$ are 174 and 65.6 respectively. Calculate Λ_0 for NH$_3$ and the degree of ionization in 0.01 M solution.

6.8 A solution of LiCl was electrolyzed in a Hittorf cell. After a current of 0.79 A had been passed for 2 hours, the mass of LiCl in the anode compartment had decreased by 0.793 g. (1) Calculate the transport numbers of the Cl$^-$ and Li$^+$ ions. (2) If Λ^o (LiCl) is 115.0 Ω^{-1} cm^2 mol^{-1}, what are the individual molar ionic conductivities?

ESSAY QUESTIONS

6.9 State Faraday's two laws of electrolysis, and discuss their significance in connection with the electrical nature of matter.

6.10 Discuss the main ideas that lie behind the Debye-Hückel theory, as applied to the conductivities of solutions of strong electrolytes.

6.11 Outline two important methods for determining transport numbers of ions.

6.12 Explain why Li$^+$ has a lower ionic conductivity than Na$^+$, and why the value for H$^+$ is so much higher than the values for both of these ions.

SUGGESTED READING

Fuoss, R. M., and Accascina, F. *Electrolytic Conductance*. New York: Interscience, 1959.

Glasstone, S. *An Introduction to Electrochemistry*. New York: Van Nostrand, 1942.

Harned, H. S., and Owen, B. B. *The Physical Chemistry of Electrolytic Solutions*. New York: Reinhold, 1950.

Lyons, E. H. *Introduction to Electrochemistry*. Boston: D. C. Heath, 1967.

Moore, W. J. *Physical Chemistry*, 4th ed. Englewood Cliffs, N.J.: Prentice-Hall, 1972.

Onsager, Lars. "The Motion of Ions: Principles and Concepts," *Science 166* (1969): 1359.

Robbins, J. *Ions in Solution*. Oxford: Clarendon Press, 1972.

7

IONS IN AQUEOUS SOLUTION

The behavior of individual ions in aqueous solution is a matter of great importance to the biologist, since ions have important functions in living systems. The biologist is also concerned with the various ionic equilibria which exist in solution. Changes in acidity, for example, will bring about important changes in the nature of protein molecules, and these may have a profound biological effect. This chapter is concerned with the thermodynamic and other properties of individual ions, and with ionic equilibria.

7.1 THERMODYNAMIC PROPERTIES OF IONS

We have seen in Chapters 4 and 5 that in thermodynamic work it is customary to obtain values of standard enthalpies and Gibbs energies of species. It will be remembered that these quantities relate to the formation of 1 mol of substance from elements in their standard states, usually at 25.0° C. In the case of entropies it is common to obtain absolute values, based on the third law of thermodynamics.

It is a comparatively straightforward matter to determine, for pairs of ions in solution, standard enthalpies of formation, Gibbs energies of formation, and absolute entropies. However, one cannot carry out experiments on ions of only one kind. The conventional procedure, as explained in Chapters 4 and 5, is to set the value for H^+ as zero. This allows complete sets of values to be built up, and one then speaks of *conventional standard enthalpies of formation*, of *conventional Gibbs energies of formation*, and of *conventional absolute entropies*.

A few such values for ions were included in Table 4.1 (p. 163). More complete compilations are given here, in Tables 7.1, 7.2, and 7.3. These tables include values of enthalpies, Gibbs energies, and entropies of *hydration*, the subscript h being used to identify these quantities. These values relate the thermodynamic properties of the ions in water to their properties in the gas phase, and are therefore important in leading to an understanding of the effect of the surrounding water molecules. Thus, the Gibbs energy of hydration is the increase in Gibbs energy when an ion is transferred from the gas phase into aqueous solution. Again, the convention is that the values for the hydrogen ion H^+ are all zero.

Table 7.1 Conventional Standard Enthalpies of Formation, ΔH_f^o, and of Hydration, ΔH_h^o, for Individual Ions†

Ion	$\Delta H_f^o/\text{kcal mol}^{-1}$	$\Delta H_h^o/\text{kcal mol}^{-1}$
H^+	0	0
Li^+	−66.6	137.7
Na^+	−57.3	163.8
K^+	−60.0	184.0
Ag^+	25.3	147.1
Ca^{2+}	−129.8	140.8
Ba^{2+}	−128.7	210.1
Cu^{2+}	15.4	19.9
Zn^{2+}	−36.4	32.8
Cl^-	−40.0	−351.1
Br^-	−28.9	−347.7
I^-	−13.4	−333.9
OH^-	−55.0	−371.0

† Heats of formation are from a National Bureau of Standards compilation, "Selected values of chemical thermodynamic properties," N.B.S. Circular 500 (1952). Heats of hydration are from D. R. Rosseinsky, *Chem. Rev.* 65 (1965): 467.

Whereas absolute thermodynamic values for individual ions cannot be measured directly, they can be estimated on the basis of theory. Unfortunately, owing to the large number of interactions involved, the theoretical treatment of an ion in aqueous solution is very difficult. The following absolute values are generally agreed to be not far from the actual values, for the proton:

$$\Delta H_h^o \text{ (absolute)} = -260.7 \text{ kcal mol}^{-1}$$

$$\Delta S_h^o \text{ (absolute)} = -31.5 \text{ cal K}^{-1} \text{ mol}^{-1}$$

$$\Delta G_h^o \text{ (absolute)} = -251.3 \text{ kcal mol}^{-1}$$

Once these values are accepted, absolute values for the other ions can be calculated from their conventional values.

Table 7.2   Conventional Standard Gibbs Energies of Formation of Ions, ΔG_f^o, and of Hydration, ΔG_h^o, at 25° C[†]

Ion	ΔG_f^o/kcal mol^{-1}	ΔG_h^o/kcal mol^{-1}
H^+	0	0
Li^+	−70.2	138.4
Na^+	−62.6	162.3
K^+	−67.5	179.6
Rb^+	−67.5	185.0
Cs^+	−67.4	192.7
Ag^+	18.43	146.0
Mg^{2+}	−109.0	65.5
Ca^{2+}	−132.2	140.2
Sr^{2+}	−133.2	175.1
Ba^{2+}	−134.0	205.9
Al^{3+}	−115.0	−321.8
F^-	−66.1	−364.3
Cl^-	−31.4	−336.3
Br^-	−24.6	−333.0
I^-	−12.4	−321.9

† Standard Gibbs energies of formation are from a National Bureau of Standards compilation, "Selected values of chemical thermodynamical properties," N.B.S. Circular 500 (1952). Gibbs energies of hydration are from D. R. Rosseinsky, *Chem. Rev. 65* (1965): 467.

Table 7.3   Conventional Standard Entropies of Ions, and Conventional Entropies of Hydration at 25° C[†]

Ion	Standard entropy, S^o/cal K^{-1} mol^{-1}	ΔS_h^o/cal K^{-1} mol^{-1}
H^+	0	0
Li^+	3.4	−2.4
Na^+	14.4	5.1
K^+	24.5	13.6
Rb^+	29.7	16.5
Cs^+	31.8	17.2
Ag^+	18.4	3.7
Mg^{2+}	−28.2	−11.7
Ca^{2+}	−13.2	1.8
Sr^{2+}	−9.4	3.4
Al^{3+}	−74.9	−32.7
F^-	−2.3	−63.1
Cl^-	13.2	−49.5
Br^-	19.3	−45.8
I^-	26.1	−40.3

† The above values are taken from D. R. Rosseinsky, *Chem. Rev. 65* (1965): 467.

EXAMPLE
Calculate the absolute enthalpies of hydration of Li^+, I^-, and Cu^{2+} on the basis of a value of -260.7 kcal mol^{-1} for the proton, using the values listed in Table 7.1.

SOLUTION
The enthalpy of hydration of a uni-univalent electrolyte M^+X^- is independent of whether conventional or absolute ionic values are used. Thus, for H^+I^-, if we lower the value for H^+ by 260.7, we must raise that for I^- by the same amount. Thus the absolute value for I^- will be

$$I^-: -333.9 + 260.7 = -73.2 \text{ kcal mol}^{-1}$$

Similarly, the value for Li^+ must be lowered by 260.7; the absolute value is thus

$$137.7 - 260.7 = -123.0 \text{ kcal mol}^{-1}$$

To obtain the value for Cu^{2+}, we must consider a salt like CuI_2 ($Cu^{2+} + 2I^-$). Since we have raised the value for *each* I^- ion by 260.7, we must lower the Cu^{2+} value by 2×260.7. Thus, Cu^{2+}: $19.9 - (2 \times 260.7) = -501.5$ kcal mol^{-1}.

7.2   THEORIES OF IONS IN SOLUTION
A SIMPLE MODEL

A simple interpretation of the thermodynamic quantities for ions in solution was suggested by the German-British physicist Max Born (1882–1970). In Born's model, the solvent is assumed to be a continuous dielectric and the ion a conducting sphere. Born obtained an expression for the work of charging such a sphere, which is the Gibbs energy increase during the charging process. Suppose that the spherical ion is of radius r and has a final charge equal to ze. The charging process may be carried out by transporting from an infinite distance small increments of charge equal to dQ. Suppose that at a particular instant the charge on the sphere is Q; then the work done in transporting the increment through a distance dx is

$$dw = -\frac{QdQdx}{4\pi\varepsilon_o\varepsilon x^2} \tag{7.1}$$

where x is the distance between the increment and the center of the sphere, ε_o is the permittivity of a vacuum (see Appendix A) and ε is the dielectric constant (or relative permittivity). The work done in transporting the increment from infinity to the surface of the sphere is thus

$$dw = -\frac{1}{4\pi\varepsilon_o\varepsilon} \int_\infty^r \frac{QdQ}{x^2} dx \tag{7.2}$$

$$= \frac{QdQ}{4\pi\varepsilon_o\varepsilon x}\bigg|_\infty^r \tag{7.3}$$

$$= \frac{QdQ}{4\pi\varepsilon_o\varepsilon r} \tag{7.4}$$

The total reversible work in transporting increments until the sphere has a charge of ze is thus

$$w_{rev} = \frac{1}{4\pi\varepsilon_o\varepsilon} \int_0^{ze} \frac{QdQ}{r} = \frac{1}{4\pi\varepsilon_o\varepsilon}\left[\frac{Q^2}{2r}\right]_0^{ze} = \frac{z^2e^2}{8\pi\varepsilon_o\varepsilon r} \tag{7.5}$$

This work, being non-PV work, is the electrostatic contribution to the Gibbs energy of the ion:

$$G^o_{es} = \frac{z^2e^2}{8\pi\varepsilon_o\varepsilon r} \tag{7.6}$$

If the same charging process is carried out *in vacuo* ($\varepsilon = 1$) the electrostatic Gibbs energy is

$$G^o_{es}\,(vac) = \frac{z^2e^2}{8\pi\varepsilon_o r} \tag{7.7}$$

The electrostatic Gibbs energy of hydration is therefore

$$\Delta G^o_{hyd} = G^o_{es} - G^o_{es}\,(vac) = \frac{z^2e^2}{8\pi\varepsilon_o r}\left(\frac{1}{\varepsilon} - 1\right) \tag{7.8}$$

For water at 25° C, $\varepsilon = 78$, and the above expression leads to

$$\Delta G^o_{hyd} = -16.4\frac{z^2}{r}\ \text{kcal mol}^{-1} \tag{7.9}$$

where r is the radius in nanometres.

Figure 7.1 depicts a plot of absolute ΔG^o_{hyd} values against z^2/r. In view of the simplistic nature of the above model, the agreement with the predictions of the treatments is not unsatisfactory; the theory accounts quite well for the main effects. Note, however, that agreement with the theory is less the higher the charge and the smaller the radius. On the whole, the percentage difference is greatest with trivalent ions and least with univalent ions. Of the univalent ions Li^+, the smallest, shows the greatest deviation. Of the divalent ions Mg^{2+} is furthest from the line. Of the trivalent ions Al^{3+} is predicted least satisfactorily. Figure 7.1 also shows the prediction for $\varepsilon = 2$, which is approximately the effective dielectric constant of water when it is subjected to an intense field. The line for $\varepsilon = 2$ is closer to the points for ions of high charge and small radius. The significance of this will be discussed later.

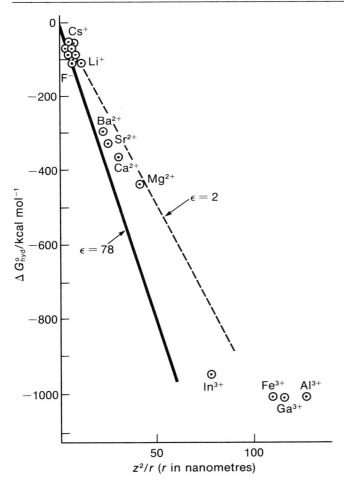

Figure 7.1

A plot of the absolute free energy of hydration of ions against z^2/r. The firm line is the theoretical line for a dielectric constant of 78, the dashed line for one of 2.

The Born equation also leads to a simple interpretation of entropies of hydration and of absolute entropies of ions. The thermodynamic relationship between entropy and Gibbs energy (see equation (5.153)) is

$$S = -\left(\frac{\partial G}{\partial T}\right)_P \tag{7.10}$$

The only quantity in (7.8) which is temperature dependent is the dielectric constant ε, and therefore

$$S_{es}^o = \Delta S_{hyd}^o = \frac{z^2 e^2}{8\pi\varepsilon_o \varepsilon r} \frac{\partial \ln \varepsilon}{\partial T} \tag{7.11}$$

The reason that the theory leads to the same expression for the absolute entropy of the ion as for its entropy of hydration is that it gives zero entropy for the ion in the gas phase. The Gibbs-energy expression in equation (7.6) contains no temperature-dependent terms. For water at 25° C, ε is approximately 78, and $\partial \ln \varepsilon / \partial T$ is about -0.0046 over a considerable temperature range. Insertion of these values into (7.11) leads to

$$S_{es}^o = \Delta S_{hyd}^o = -0.98 \frac{z^2}{r} \text{ cal K}^{-1} \text{ mol}^{-1} \tag{7.12}$$

r being expressed in nanometres.

Figure 7.2 shows a plot of the experimental ΔS_{hyd}^o values, together with a line of slope -0.98. The line has been drawn through a value of -10 cal K^{-1} mol^{-1} at $z^2/r = 0$, since it is estimated that there will be a ΔS_{hyd}^o value of about -10 cal

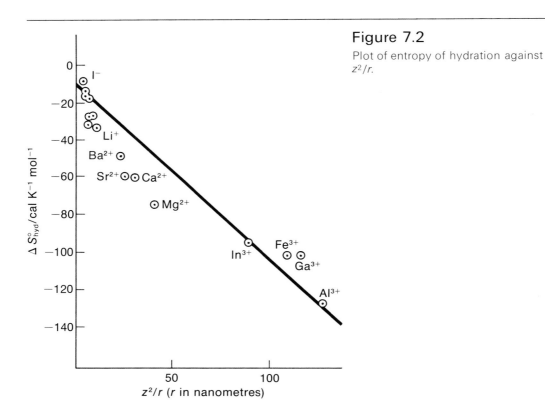

Figure 7.2

Plot of entropy of hydration against z^2/r.

K^{-1} mol^{-1} due to nonelectrostatic effects. In particular, there is an entropy loss resulting from the fact that the ion is confined in a small solvent "cage"—an effect that is ignored in the electrostatic treatment.

Again, the simple model of Born accounts satisfactorily for the main effects. The deviations are primarily for small ions.

MORE ADVANCED THEORIES

Various attempts have been made to improve Born's simple treatment while still regarding the solvent as continuous. One way of doing this is to consider how the effective dielectric constant of a solvent varies in the neighborhood of an ion. The dielectric behavior of a liquid is related to the tendency of ions to orient themselves in an electric field. At very high field strengths the molecules become fully aligned in the field and there can be no further orientation. The dielectric constant then falls to a very low value of about 2. This effect is known as *dielectric saturation*.

In this way it is possible to estimate that the effective dielectric constant in the neighborhood of ions of various types will be as shown in Figure 7.3. For example, for a ferric ion (Fe^{3+}) the dielectric constant has a value of about 1.78 up to a distance of about 0.3 nm (3 Å) from the center of the ion, owing to the high field produced by the ion. At greater distances, where the field is less, the effective dielectric constant rises towards its limiting value of 78. However, even at a separation of 1 nm (10 Å) the dielectric constant is still significantly below the limiting value.

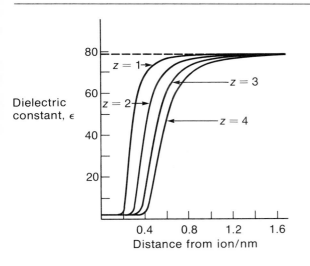

Figure 7.3

The variation of dielectric constant of water in the neighborhood of ions.

When the Born treatments of Gibbs energies and entropies of hydration of ions are modified by taking into consideration these dielectric saturation effects, the agreement with experiment is considerably better. The experimental values in Figure 7.1 are consistent with a dielectric constant of somewhat less than the true value of 78;

this can be understood in terms of the reduction in dielectric constant arising from dielectric saturation.

Attempts also have been made to develop *discontinuous* theories for ions in solution. In these, the solvent is not treated as a continuum; instead, its detailed molecular structure is considered, and estimates are made of the various attractive and repulsive forces that act between ions and solvent molecules. These are quite numerous, and it is therefore difficult to develop theoretical treatments of this kind. However, reliable estimates of the thermodynamic values have been obtained in this way.

QUALITATIVE TREATMENTS

On the basis of the above quantitative theories, we can arrive at useful qualitative pictures of ions in solution. The main attractive force between an ion and a surrounding water molecule is the ion-dipole attraction. The way in which a water molecule is expected to orient itself in the neighborhood of a positive or negative ion is shown in Figure 7.4. There are two possible orientations for a negative ion; it appears that both may play a role. Quite strong binding can result from ion-dipole forces. The binding energy for an Na^+ ion and a neighboring water molecule, for example, is about 20 kcal mol^{-1} ($\approx$ 80 kJ mol^{-1}).

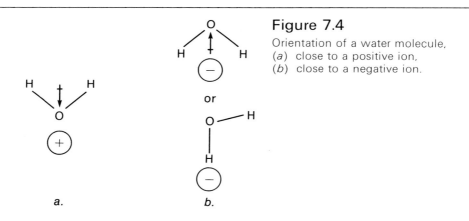

Figure 7.4

Orientation of a water molecule,
(*a*) close to a positive ion,
(*b*) close to a negative ion.

a.   *b.*

The smaller the ion, the greater the binding energy between the ion and a water molecule, because the water molecule can approach the center of the ion more closely. Thus, the binding energy is particularly strong for the Li^+ ion, and calculations suggest that a Li^+ ion in water is surrounded tetrahedrally by four water molecules oriented as in Figure 7.4; this is depicted in Figure 7.5. Thus we can say that Li^+ has a hydration number of 4. The four water molecules are sufficiently strongly attached to the Li^+ ion that during electrolysis they are dragged along with it. For this reason, Li^+ has a lower mobility than Na^+ and K^+ (Table 6.3); the latter ions, being larger, have a smaller tendency to drag water molecules. However, the situation is not as clear-cut as implied above; a moving Li^+ ion not only drags its four neighboring water molecules, but also some of the water molecules which are further away from it.

A very useful description of ions in solution has been developed by H. S. Frank and his coworkers. They have concluded that for ions in aqueous solution it is possible to distinguish three different zones in the neighborhood of the solute molecule, as

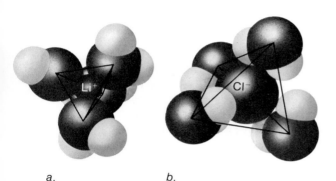

a.   b.

Figure 7.5

The tetrahedral arrangement of four water molecules: (*a*) around a Li$^+$ ion, (*b*) around a Cl$^-$ ion (which can probably also have 5 or 6 water molecules around it).

shown in Figure 7.6. In the immediate vicinity of the ion there is a shell of water molecules which are more or less immobilized by the very high field due to the ion. These water molecules can be described as constituting the *inner hydration shell*, and are sometimes described as an *iceberg*, since their structure has some ice-like characteristics. However, this analogy should not be pressed too far, since ice is less dense than liquid water, whereas the "iceberg" around the ion is more compressed than normal liquid water.

Surrounding this iceberg there is a second region, which Frank and his coworkers refer to as a region of *structure-breaking*. Here the water molecules are oriented more randomly than in ordinary water, where there is considerable ordering due to hydrogen

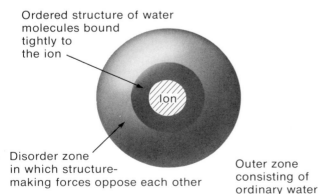

Ordered structure of water molecules bound tightly to the ion

Disorder zone in which structure-making forces oppose each other

Outer zone consisting of ordinary water

Figure 7.6

H. S. Frank's concept of the structure of water in the neighborhood of an ion.

bonding. The occurrence of this structure-breaking region results from two competing orienting influences on each water molecule, which approximately balance. One is the normal structural effect of the neighboring water molecules; the other is the orienting influence upon the dipole of the spherically-symmetrical ionic field.

The third region around the ion comprises all the water sufficiently far from the ion that its effect is not felt.

The behavior of the proton, H^+, in water deserves some special discussion. Because of its small size, the proton attaches itself very strongly to a water molecule:

$$H^+ + H_2O \rightleftharpoons H_3O^+$$

The equilibrium for this process, which is strongly exothermic, lies very far to the right. Thus, the hydrogen ion is frequently regarded as existing as the H_3O^+ ion, which is known as the *hydronium, hydroxonium,* or *oxonium* ion. The use of *oxonium* has been recommended by the International Union of Pure and Applied Chemistry. Experimental studies and theoretical calculations have indicated that the angle between the O—H bonds in H_3O^+ is about 115°. As a result the ion is almost, but not quite, flat (Figure 7.7).

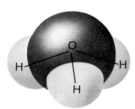

Figure 7.7

The structure of the oxonium (H_3O^+) ion.

However, the state of the hydrogen ion in water is more complicated than implied by this description. Because of ion-dipole attractions, other molecules are held quite closely to the H_3O^+ species. The most recent calculations tend to support the view that three water molecules are held particularly strongly, in the manner shown in Figure 7.8. These three molecules are held by hydrogen bonds involving three hydrogen atoms of the H_3O^+ species. The hydrated proton may thus be written as $H_3O^+ \cdot 3H_2O$, or as $H_9O_4^+$. Alternatively, some think that the species $H_9O_4^+$ has the same kind of structure as the hydrated Li^+ ion (Figure 7.5a), the proton being surrounded tetrahedrally by four water molecules. Other water molecules will also be held to $H_9O_4^+$, but not as strongly. Thus, in Figure 7.8 an additional H_2O molecule is held by ion-dipole forces but not by hydrogen bonding, and therefore is not held as strongly as the three other water molecules. The International Union of Pure and Applied Chemistry has recommended that the term *hydronium ion* be employed for these hydrated H_3O^+ ions, in contrast to *oxonium ion* for H_3O^+ itself.

The situation is hardly clear-cut: we can describe the hydrated proton as H^+, H_3O^+, $H_9O_4^+$ or $H_{11}O_5^+$. It is satisfactory to write it simply as H^+, as long as we remember that the proton is strongly hydrated.

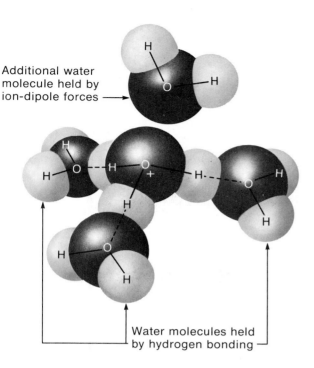

Figure 7.8

A possible structure for the hydrated proton. The hydrated H_3O^+ ion, e.g., $H_9O_4^+$ is called the hydronium ion.

Additional water molecule held by ion-dipole forces →

Water molecules held by hydrogen bonding

7.3 ACTIVITY COEFFICIENTS

Thus far we have been concerned with isolated ions and with their interaction with surrounding solvent molecules. In Chapter 6 we saw that for strong electrolytes the interactions between ions have an important effect on the conductances of solutions at all except exceedingly low concentrations. These interactions also have a significant effect on the thermodynamic properties of ions. A very convenient way of dealing with this matter is in terms of *activity coefficients*.

If there were no electrostatic interactions the behavior of ions would be ideal and the Gibbs energy of an ion of type i would be given by the relationship

$$G_i = G_i^\circ + \mathbf{k}T \ln c_i \tag{7.13}$$

where c_i is the concentration of the ion. In reality the Gibbs energy is

$$G_i = G_i^\circ + \mathbf{k}T \ln c_i y_i \tag{7.14}$$

$$= G_i^\circ + \mathbf{k}T \ln c_i + \mathbf{k}T \ln y_i \tag{7.15}$$

The additional term, $\mathbf{k}T \ln y_i$, is due to the presence of the ionic atmosphere; y is the activity coefficient used with concentrations.

We have seen that the work of charging an isolated ion, on the basis of the Born model, is

$$\frac{z_i^2 e^2}{4\pi\varepsilon_o \varepsilon r}$$

We also need the work of charging the ionic atmosphere, because this is the required correction to the Gibbs energy. This work is equal to $\mathbf{k}T \ln y_i$. If ϕ is the potential due to the ionic atmosphere, the work of transporting a charge dQ to the ion is ϕdQ. The net work of charging the ionic atmosphere is thus

$$w = \int_o^{z_i e} \phi dQ \tag{7.16}$$

The potential ϕ corresponding to the charge Q is given by equation (6.34) as $-Q\kappa/4\pi\varepsilon_o\varepsilon$, where $1/\kappa$ is the radius of the ionic atmosphere. The work is thus†

$$w = -\int_o^{z_i e} \frac{Q\kappa}{4\pi\varepsilon_o\varepsilon} dQ \tag{7.17}$$

$$= -\left[\frac{Q^2\kappa}{8\pi\varepsilon_o\varepsilon}\right]_o^{z_i e} \tag{7.18}$$

$$= -\frac{z_i^2 e^2 \kappa}{8\pi\varepsilon_o\varepsilon} \tag{7.19}$$

The expression for the activity coefficient y_i is therefore

$$\mathbf{k}T \ln y_i = -\frac{z_i^2 e^2 \kappa}{8\pi\varepsilon_o\varepsilon} \tag{7.20}$$

whence

$$\ln y_i = -\frac{z_i^2 e^2 \kappa}{8\pi\varepsilon_o\varepsilon \mathbf{k}T} \tag{7.21}$$

or

$$\log_{10} y_i = -\frac{z_i^2 e^2 \kappa}{8 \times 2.303\pi\varepsilon_o\varepsilon \mathbf{k}T} \tag{7.22}$$

---

† Equation (7.19) can be obtained from (7.5) by replacing r by $1/\kappa$ and changing the sign. The reciprocal $1/\kappa$ plays the same role for the atmosphere as does r for the ion. The change of sign is required because the net charge on the atmosphere is opposite to that on the ion.

Equation (6.26) shows that κ is proportional to the square root of the quantity $\sum_i n_i z_i^2$, where n_i is the concentration of the ions of the ith type and z_i is its valency.

It is convenient to define a quantity known as the *ionic strength, I*, of a solution. This is defined as

$$I = \tfrac{1}{2} \sum_i c_i z_i^2 \qquad\qquad (7.23)$$

where c_i is the *molar* concentration of the ion of type i (in contrast to n_i which is the molecular concentration).† The ionic strength is obviously proportional to $\sum_i n_i z_i^2$, and the reciprocal of the radius of the ionic atmosphere, κ, is proportional to $\sqrt{I}$. Equation (7.22) may thus be written as

$$\log_{10} y_i = - z_i^2 B \sqrt{I} \qquad\qquad (7.24)$$

where B is a quantity which depends upon properties such as ε and T. When water is the solvent at 25° C the value of B is 0.51.

Experimentally one cannot measure the activity coefficient—or indeed any thermodynamic property—of a simple ion, since at least two types of ions must be present in any solution. To circumvent this difficulty we can define a *mean activity coefficient* $y_\pm$ in terms of the individual values for y_+ and y_- by the relationship

$$y_\pm^{v_+ + v_-} = y_+^{v_+} y_-^{v_-} \qquad\qquad (7.25)$$

where v_+ and v_- are the numbers of ions of the two kinds produced by the electrolyte. For example, for $ZnCl_2$, $v_+ = 1$ and $v_- = 2$. For a uni-univalent electrolyte $(v_+ = v_- = 1)$ the mean activity coefficient is the geometric mean, $(y_+ y_-)^{\frac{1}{2}}$, of the individual values.

In order to express $y_\pm$ in terms of the ionic strength we proceed as follows. From equation (7.25)

$$(v_+ + v_-) \log_{10} y_\pm = v_+ \log_{10} y_+ + v_- \log_{10} y_- \qquad\qquad (7.26)$$

Insertion of expression (7.24) for y_+ and y_- gives

$$(v_+ + v_-) \log_{10} y_\pm = -(v_+ z_+^2 + v_- z_-^2) B \sqrt{I} \qquad\qquad (7.27)$$

---

† For a uni-univalent electrolyte such as NaCl the ionic strength is equal to the molar concentration. Thus for a 1 M solution $c_+ = 1$, $c_- = 1$, $z_+ = 1$, and $z_- = -1$; hence $\tfrac{1}{2} \sum_i c_i z_i^2 = \tfrac{1}{2}(1 + 1) = 1$ M. For a 1 M solution of a uni-bivalent electrolyte such as K_2SO_4, $c_+ = 2$, $c_- = 1$, $z_+ = 1$, and $z_- = -2$; hence the ionic strength is $\tfrac{1}{2}(2 \times 1 + 1 \times 4) = 3$ M. Similarly, for a 1 M solution of a uni-trivalent electrolyte such as Na_3PO_4, the ionic strength is 6 M.

For electrical neutrality†

$$v_+ z_+ = v_- |z_-| \qquad \text{or} \qquad v_+{}^2 z_+{}^2 = v_-{}^2 z_-{}^2 \qquad (7.28)$$

and therefore

$$(v_+ + v_-) \log_{10} y_\pm = - v_+{}^2 z_+{}^2 (1/v_+ + 1/v_-) B \sqrt{I} \qquad (7.29)$$

Thus

$$\log_{10} y_\pm = - \frac{v_+ z_+{}^2}{v_-} B \sqrt{I} \qquad (7.30)$$

$$= - z_+ |z_-| B \sqrt{I} \qquad (7.31)$$

Thus for aqueous solutions at 25° C

$$\log_{10} y_\pm = - 0.51 \, z_+ |z_-| \sqrt{I} \qquad (7.32)$$

Equation (7.31) is known as the *Debye-Hückel limiting law* (DHLL).

DEVIATIONS FROM THE DEBYE-HÜCKEL LIMITING LAW

Experimentally, the DHLL is found to apply satisfactorily only at extremely low concentrations‡; at higher concentrations there are very significant deviations, as shown schematically in Figure 7.9. Equation (7.31) predicts that a plot of $\log_{10} y_\pm$ against $\sqrt{I}$ will have a negative slope, the magnitude of the slope being $- z_+ |z_-| B$. The results with a number of electrolytes have shown that at very low I values $\log_{10} y_\pm$ does indeed fall linearly with $\sqrt{I}$, within the correct slope; for this reason equation (7.31) is satisfactory as a *limiting* law. However, at higher $\sqrt{I}$ values the value of $\log_{10} y_\pm$ becomes significantly less negative than predicted by the law, and at sufficiently high ionic strengths it may actually attain positive values. The significance of this in connection with the solubilities of salts is considered later in this chapter in Section 7.5.

Various theories have been put forward to explain these deviations from the DHLL, but here can be considered only briefly. One thing that has not been taken into account in the development of the equations is the fact that the ions occupy space; thus far they have been treated as point charges with no restrictions on how closely

---

† Note that in this treatment one uses the positive sign for the valency of negative ions; e.g., $|z_{Cl^-}| = +1$.

‡ For example, for a uni-univalent electrolyte such as KCl, the law is obeyed satisfactorily up to a concentration of about 0.01 M. For other types of electrolytes, deviations begin to appear at even lower concentrations.

they can come together. If the theory is modified in such a way that the centers of the ions cannot approach one another more closely than distance a, the expression for the activity coefficient of an individual ion (compare (7.21)) becomes

$$\ln y_i = - \frac{z_c^2 e^2 \kappa}{8\pi\varepsilon_o \varepsilon \mathbf{k}t} \cdot \frac{1}{1 + \kappa a} \tag{7.33}$$

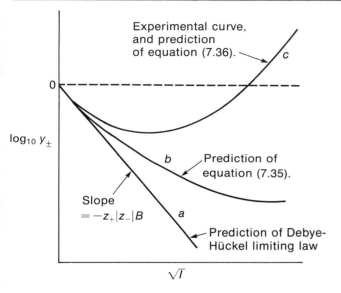

Experimental curve, and prediction of equation (7.36).

c

0

$\log_{10} y_\pm$

b

Prediction of equation (7.35).

Slope $= -z_+ |z_-| B$

a

Prediction of Debye-Hückel limiting law

$\sqrt{I}$

Figure 7.9

Variation of $\log_{10} y_\pm$ with the square root of the ionic strength.

Since κ is proportional to $\sqrt{I}$, this equation can be written as

$$\log_{10} y_i = - \frac{z_i^2 B \sqrt{I}}{1 + aB'\sqrt{I}} \tag{7.34}$$

Compare this with equation (7.24), where B is the same constant as used previously and B' is a new constant. The corresponding equation for the mean activity coefficient $y_\pm$ is

$$\log_{10} y_\pm = - \frac{z_+ |z_-| B \sqrt{I}}{1 + aB'\sqrt{I}} \tag{7.35}$$

Compare with equation (7.31). Equation (7.35) leads to the curve shown as b in Figure 7.9; however, it cannot explain the positive values of $\log_{10} y_\pm$ that are obtained at high ionic strengths.

In order to explain these positive values it is necessary to take account of the orientation of solvent molecules by the ionic atmosphere. This was considered by

Hückel, who showed that it gives a term linear in I in the expression for $\log_{10} y_\pm$:

$$\log_{10} y_\pm = -\frac{Bz_+ z_- \sqrt{I}}{1 + aB'\sqrt{I}} + CI \qquad (7.36)$$

where C is a constant. At sufficiently high ionic strengths the CI term predominates, and $\log_{10} y_\pm$ is approximately linear in I, as is found experimentally. The CI term is often known as the "salting-out" term since, as will be seen in Section 7.5, it accounts for the lowered solubilities of salts at high ionic strengths.

Various other modifications to the DHLL have been made. For example, consideration has been given to the effects of ion pairing (Section 6.6).

7.4   IONIC EQUILIBRIA

Equilibrium is usually established very rapidly between ionic species in solution. In this section an account will be given of the more important types of equilibria that are established in solution, and of how the relevant equilibrium constants are measured experimentally.

ACIDS AND BASES

As an example, let us consider the dissociation of acetic acid in water:

$$CH_3COOH \rightleftharpoons CH_3COO^- + H^+$$

The equilibrium constant is

$$\frac{[CH_3COO^-][H^+]}{[CH_3COOH]} = K_a \qquad (7.37)$$

At $25°C$ the value of K_a is about 1.80×10^{-5} mol dm^{-3} (M).

EXAMPLE
Calculate the concentrations of the species CH_3COOH, CH_3COO^-, and H^+ in an 0.100 M solution of acetic acid in water at $25°C$.

SOLUTION
Let x be the concentration of CH_3COO^- and H^+. The concentration of CH_3COOH must therefore be $0.100 - x$. The

equilibrium constant equation is thus

$$\frac{x^2}{0.100 - x} = 1.80 \times 10^{-5}$$

An approximate solution, sufficiently accurate for most purposes, is obtained by noting that x must be much less than 0.100, so that $0.100 - x$ may be replaced by 0.100. Thus

$$\frac{x^2}{0.100} = 1.80 \times 10^{-5}$$

whence

$$x^2 = 1.80 \times 10^{-6}$$

and

$$x = 1.34 \times 10^{-3}$$

The required concentrations are thus

$$[H^+] = 1.34 \times 10^{-3} \text{ M}$$
$$[CH_3COO^-] = 1.34 \times 10^{-3} \text{ M}$$
$$[CH_3COOH] = 0.100 - 1.34 \times 10^{-3}$$
$$= 0.0986 \text{ M}$$

Thus the acid is 1.4% dissociated into ions and 98.6% undissociated.

The dissociation constant K_a for an acid is known as the *acid strength*. It is a measure of the extent to which the acid produces hydrogen ions, and therefore of its strength as an acid. Bases dissociate in water as follows,

$$B + H_2O \rightleftharpoons BH^+ + OH^-$$

and the corresponding dissociation constant

$$K_b = \frac{[BH^+][OH^-]}{[B]} \tag{7.38}$$

is known as the *base strength*.

IONIZATION OF WATER

The behavior of acids and bases in water is influenced considerably by the fact that water ionizes into hydrogen and hydroxide ions:

$$H_2O \rightleftharpoons H^+ + OH^-$$

The equilibrium constant for this dissociation is

$$\frac{[H^+][OH^-]}{[H_2O]} = K \tag{7.39}$$

However, since the concentration of water hardly varies from solution to solution, the convention is to incorporate this concentration into the equilibrium constant. Thus we write

$$[H^+][OH^-] = K_w (\text{mol}^2\,\text{dm}^{-6}) \tag{7.40}$$

The constant K_w is known as the *ionic product of water*. Its value at $25°C$ is almost exactly $10^{-14}\,\text{mol}^2\,\text{dm}^{-6}$. Although this ionization occurs only to a very small extent, it has a very important effect on the properties of aqueous solutions.

In chemically pure water the concentrations of H^+ and OH^- ions produced by this ionization are equal, so that

$$[H^+] = [OH^-] = 10^{-7}\,\text{mol dm}^{-3} \tag{7.41}$$

at $25°C$. It is convenient to refer to any aqueous solution in which $[H^+]$ and $[OH^-]$ are equal as a *neutral* solution. If $[H^+] > 10^{-7}$ M (in which case $[OH^-] < 10^{-7}$ M) the solution is said to be acidic, while if $[H^+] < 10^{-7}$ M (and $[OH^-] > 10^{-7}$ M) the solution is alkaline or basic.

The reaction between H^+ and OH^- is very rapid. Therefore, if we try to prepare a solution in which the product $[H^+][OH^-]$ differs from 10^{-14}, the reaction $H_2O \rightleftharpoons H^+ + OH^-$ will shift rapidly and equilibrium will be established within a small fraction of a second. Thus, if an acid solution is mixed with a basic solution the process $H^+ + OH^- \rightarrow H_2O$ will occur until again $[H^+][OH^-] = 10^{-14}$.

Various methods have been employed for the experimental determination of the ionic product of water. We will mention them here briefly:

Conductivity Method. The original method, employed by Kohlrausch in 1894, was to measure the conductivity of pure water. Since the molar conductivities of H^+ and OH^- are known (see Table 6.3), the conductivity allows the concentrations to be calculated. The difficulty with this method is ensuring the purity of the water, which is essential because even minute traces of impurities have a relatively large effect on conductivity. In spite of this, Kohlrausch obtained a very reliable value, having purified water by forty-eight distillations!

Kinetic Method. A kinetic method was employed in 1893 by J. J. A. Wijs, which also gave a reliable value. The method depended upon studying catalytic effects in solutions of various hydrogen-ion and hydroxide-ion concentrations, and is explained briefly in Chapter 9 (p. 419).

EMF Method. The ionic product can also be obtained from emf measurements on voltaic cells. The principle of the method is to prepare an alkaline solution of known hydroxide-ion concentration, and then to determine the hydrogen-ion concentration by dipping into the solution a hydrogen electrode and a reference electrode, then measuring the emf (Chapter 8). Instead of the hydrogen electrode one may, of course, use a glass electrode or any other electrode which responds to the hydrogen-ion concentration.

THE pH SCALE

A 1 M HCl solution has a hydrogen ion concentration of 1 M, while a 1 M NaOH solution has a hydrogen ion concentration of 10^{-14} M. Therefore there is a very wide variation in $[H^+]$ and $[OH^-]$ as acidic and basic solutions are mixed. We sometimes need to plot a particular property, such as electrode potential, against a function of the hydrogen-ion concentration, and to cover the range from acidic to basic solutions. A plot against $[H^+]$ itself is impractical in these circumstances. A logarithmic scale, the *pH scale*, is generally used to avoid this difficulty.

The pH of a solution is defined by the relationship

$$pH = -\log_{10} a_{H^+} \tag{7.42}$$

and for sufficiently dilute solutions the activity a_{H^+} can be replaced by $[H^+]$. Thus, if we express the hydrogen-ion concentration as 10^{-x}, the pH is x. Since the negative sign has been dropped, the pH is large when the acidity is low and small when the acidity is high. In pure water the hydrogen-ion concentration is 10^{-7} M; the pH is thus 7. In an acidic solution the hydrogen-ion concentration is greater than 10^{-7} M, and the pH is less than 7; in an alkaline solution the pH is greater than 7.

CONJUGATE ACIDS AND BASES

The Danish physical chemist Johannes Nicolaus Brønsted (1874–1947) defined an *acid as a substance which can donate a proton to another molecule.* Similarly, he defined a *base* as *a substance which can accept a proton.* These definitions have proved to be very useful, particularly for the aqueous solutions with which we are mainly concerned. For nonaqueous systems it is preferable to use a more general definition, first proposed by G. N. Lewis, according to which an *acid is a substance which can accept electrons,* and *a base is a substance which can donate electrons.* Here we shall employ the Brønsted definition, which is somewhat more convenient for aqueous solutions.

When an acid HA ionizes in aqueous solution we can write the process as

$$HA \rightleftharpoons H^+ + A^-$$

The ions are solvated—the H^+ ion particularly strongly—so that the ionization process really involves the transfer of a proton from HA to a water molecule, giving the species H_3O^+ which is still further hydrated (p. 300). When we wish to emphasize this transfer process it is convenient to write the ionization equation as

$$HA + H_2O \rightleftharpoons H_3O^+ + A^-$$

In this process the water molecule is accepting a proton and therefore is acting as a base. When the reverse reaction takes place the species H_3O^+ acts as an acid, since it is transferring a proton to A^-, which acts as a base. Thus the process is of the type

$$\text{acid 1} + \text{base 2} \rightleftharpoons \text{acid 2} + \text{base 1}$$
$$example: \quad (HA) \quad (H_2O) \quad (H_3O^+) \quad (A^-)$$

The species A^- is a base produced from HA by the removal of a proton and is referred to as the *conjugate base* of the acid HA. Similarly, HA is the *conjugate acid* of the base A^-, having been produced from it by the addition of a proton. Every acid has a conjugate base, which it produces by giving up a proton, and every base has a conjugate acid, formed by the addition of a proton.

The ionization of a base B in water involves a proton transfer from the water to the base:

$$B + H_2O \rightleftharpoons BH^+ + OH^-$$

For example,

$$NH_3 + H_2O \rightleftharpoons NH_4^+ + OH^-$$
$$\text{base} \quad \text{acid} \quad \text{acid} \quad \text{base}$$

Here the ammonium ion is the conjugate acid of the base NH_3, and H_2O is the conjugate acid of the base OH^-. The fact that H_2O is now acting as an acid, whereas in the HA ionization it acts as a base, is of particular importance. On account of its ability to act both as an acid and a base, H_2O is said to be *amphoteric* (Greek *amphoteros*, in both ways). In the self-ionization of water one molecule is acting as an acid and another as a base, as is clear if we write the reaction as

$$H_2O + H_2O \rightleftharpoons H_3O^+ + OH^-$$

One reason why Brønsted's definition is so useful is that the ability of species to donate and accept protons is an important factor in connection with their ability to catalyze chemical reactions. For example, the ammonium ion acts as an acid catalyst for certain reactions which are catalyzed by acids, and it does so by donating protons. Similarly the species CH_3COO^-, which is the conjugate base of acetic acid, is a base catalyst for certain reactions.

Because every base has a conjugate acid we can conveniently quote the acid dissociation constant of the conjugate acid instead of the basic dissociation constant of the base. Consider, for example, the ionization of the base NH_3 in water:

$$NH_3 + H_2O \rightleftharpoons NH_4{}^+ + OH^-$$

The base dissociation constant is defined by

$$K_b = \frac{[NH_4{}^+][OH^-]}{[NH_3]} \qquad (7.43)$$

The ionization of the conjugate acid $NH_4{}^+$ follows the equation

$$NH_4{}^+ \rightleftharpoons NH_3 + H^+ \qquad (7.44)$$

and the acid dissociation constant is

$$K_a = \frac{[NH_3][H^+]}{[NH_4{}^+]} \qquad (7.45)$$

From (7.43) and (7.45)

$$K_a K_b = [H^+][OH^-] = K_w \qquad (7.46)$$

This equation applies to all acid-base systems. Thus, if K_a values are listed, the K_b values can readily be calculated if needed. Table 7.4 shows some K_a and K_b values for both acids and bases. It also includes the pK values which are defined, by analogy with pH, as equal to $-\log_{10} K$.

If we take logarithms of (7.46) the result is

$$\log_{10} K_a + \log_{10} K_b = \log_{10} K_w \qquad (7.47)$$

Table 7.4   Dissociation Constants of Acids and Bases, and pK Values, at 25° C

Acid	K_a/mol dm^{-3}	pK$_a$
Acetic acid, CH$_3$COOH	1.8×10^{-5}	4.75
Formic acid, HCOOH	1.8×10^{-4}	3.75
Hydrogen cyanide, HCN	4.8×10^{-10}	9.32
Hydrogen fluoride, HF	6.8×10^{-4}	3.17
Benzoic acid, C$_6$H$_5$COOH	6.3×10^{-5}	4.20

Base	K_b/mol dm^{-3}	K_a of conjugate acid/mol dm^{-3}	pK$_a$
Ammonium hydroxide, NH$_4$OH	1.8×10^{-5}	5.6×10^{-10}	9.25
Aniline, C$_6$H$_5$NH$_2$	4.6×10^{-10}	2.2×10^{-5}	4.66
Ethylamine, C$_2$H$_5$NH$_2$	5.6×10^{-4}	1.8×10^{-11}	10.75
Methylamine, CH$_3$NH$_2$	5.0×10^{-4}	2.0×10^{-11}	10.70
Pyridine, C$_5$H$_5$N	2.3×10^{-9}	4.3×10^{-6}	5.37

With the definitions

$$pK_a = -\log_{10} K_a \tag{7.48}$$

$$pK_b = -\log_{10} K_b \tag{7.49}$$

$$pK_w = -\log_{10} K_w \tag{7.50}$$

we see that

$$pK_a + pK_b = pK_w \tag{7.51}$$

HYDROLYSIS CONSTANTS

Salts are usually fully dissociated in water, but the ions produced may interact with water, forming acidic and basic species. As a result, an aqueous solution of a salt may be on the acidic or basic side of neutrality.

Consider, for example, a solution of sodium acetate in water. The ions produced are Na^+ and CH_3COO^-, and the following equilibria are established by the interaction of these ions with water:

$$Na^+ + H_2O \rightleftharpoons NaOH + H^+$$

$$CH_3COO^- + H_2O \rightleftharpoons CH_3COOH + OH^-$$

Since NaOH is a very strong base, the first equilibrium lies almost completely over to the left and the Na^+ ions therefore remain in the solution. However, acetic acid is not a strong acid and the second reaction occurs to a significant extent. The second equilibrium must therefore be taken into account. Its equilibrium constant is given by

$$K_h = \frac{[CH_3COOH][OH^-]}{[CH_3COO^-]} \tag{7.52}$$

and is referred to as the *hydrolysis constant* for the salt. Note that K_h is K_b for the conjugate base CH_3COO^-.

There is a simple relationship between the hydrolysis constant of a salt of a weak acid and the acid strength, which in this case is defined by

$$K_a = \frac{[CH_3COO^-][H^+]}{[CH_3COOH]} \tag{7.53}$$

The product $K_a K_h$ is

$$K_a K_h = [H^+][OH^-] = K_w \tag{7.54}$$

In a similar way it can be shown that for the salt of a strong and a weak base, the product of the hydrolysis constant and the base strength K_b is equal to K_w.

EXAMPLE

Calculate the pH of an 0.05 M solution of sodium acetate at $25°\,C$; the dissociation constant of acetic acid is 1.80×10^{-5}.

SOLUTION

Suppose that, after the hydrolysis equilibrium is established, the concentrations of CH_3COOH and OH^- are x. The concentration of unhydrolyzed CH_3COO^- is thus $0.05 - x$:

$$CH_3COO^- + H_2O \rightleftharpoons CH_3COOH + OH^-$$
$$0.05 - x \qquad\qquad\qquad x \qquad\quad x$$

The equilibrium constant for this hydrolysis reaction, by equation (7.54), is K_w/K_a, i.e., $10^{-14}/1.80 \times 10^{-5}$. Thus

$$\frac{x^2}{0.05 - x} = \frac{10^{-14}}{1.80 \times 10^{-5}} = 5.56 \times 10^{-10}$$

Since x is obviously small, an approximate solution is obtained by neglecting x in comparison with 0.05. Thus

$$x^2 = 0.05 \times 5.56 \times 10^{-10} = 2.77 \times 10^{-11}$$

from which we get

$$x = 5.27 \times 10^{-6}$$

which is the concentration of OH^- ions. The concentration of H^+ ions is

$$[H^+] = \frac{10^{-14}}{5.27 \times 10^{-6}} = 1.90 \times 10^{-9}$$
$$= 10^{0.28} \times 10^{-9}$$

The pH is thus 8.72.

POLYBASIC ACIDS

A number of acids have more than one ionizable proton. Oxalic acid, for example, ionizes as follows:

$$\begin{array}{ccc} COOH & COO^- & \\ | & \rightleftharpoons & | & + H^+ \\ COOH & COOH & \end{array} \qquad K_1 = 5.0 \times 10^{-2}; \ pK_1 = 1.3$$

$$\begin{matrix} COO^- \\ | \\ COOH \end{matrix} \rightleftharpoons \begin{matrix} COO^- \\ | \\ COO^- \end{matrix} + H^+ \qquad K_2 = 5.2 \times 10^{-5}; pK_2 = 4.3$$

The second dissociation constant K_2 is very much smaller than K_1. The reason for this is easily understood if we consider the different electrostatic forces involved. When a proton is removed from the ion $\begin{matrix} COO^- \\ | \\ COOH \end{matrix}$ the resulting ion $\begin{matrix} COO^- \\ | \\ COO^- \end{matrix}$ attracts it with a more powerful force than in the removal of a proton from the neutral molecule.

It is important to be able to recognize the predominant species which will be present in solutions of polybasic acids at various pH values. The two ionizations for oxalic acid may be written as

$$\begin{matrix} COOH \\ | \\ COOH \end{matrix} \xrightarrow{pK_1=1.3} \begin{matrix} COO^- \\ | \\ COOH \end{matrix} \xrightarrow{pK_2=4.3} \begin{matrix} COO^- \\ | \\ COO^- \end{matrix}$$

or as $\qquad ox \rightleftharpoons ox^- \rightleftharpoons ox^{2-}$

The first dissociation constant is

$$\frac{[ox^-][H^+]}{[ox]} = K_1 = 5.0 \times 10^{-2} \tag{7.55}$$

and we see that when $[H^+] = 5.0 \times 10^{-2}$ M, the concentrations of ox^- and ox are equal to one another. That is, when the pH is equal to the pK these two concentrations are equal. However, when the hydrogen-ion concentration is greater than 5.0×10^{-2} M, it follows from equation (7.55) that

$$[ox] > [ox^-]$$

That is, when the pH is less than 1.3 the predominant species is the undissociated oxalic acid.

From the same type of argument it follows that when the hydrogen-ion concentration lies between the values of K_1 and K_2, i.e., when the pH lies between 1.3 and 4.3, the predominant species will be ox^-. At pH values higher than 4.3, the predominant species will be ox^{2-}.

EXAMPLE
The pK values for the successive dissociations of phosphoric acid, H_3PO_4, are:

$$H_3PO_4 \xrightarrow{pK_1=2.1} H_2PO_4^- \xrightarrow{pK_2=7.2} HPO_4^{2-} \xrightarrow{pK_3=12.3} PO_4^{3-}$$

What will be the predominant species at the following pH values: 1.5, 3.0, 7.5, 11.5, and 14.0?

SOLUTION

The predominant species below pH 2.1 is H_3PO_4; that between
2.1 and 7.2 is $H_2PO_4^-$; that between 7.2 and 12.3 is HPO_4^{2-},
and that above 12.3 is PO_4^{3-}. Therefore, the predominant species
at the pH values specified are

$$
\begin{array}{ll}
pH = 1.5 & H_3PO_4 \\
pH = 3.0 & H_2PO_4^- \\
pH = 7.5 & HPO_4^{2-} \\
pH = 11.5 & HPO_4^{2-} \\
pH = 14.0 & PO_4^{3-}
\end{array}
$$

AMPHOLYTES

The term *amphoteric* is applied to substances like water which are capable of acting
both as acids and as bases. Certain electrolytes contain both an acidic and a basic
group; a simple example is glycine, with the structure

$$H_2NCH_2COOH$$

The —COOH group is acidic since it can give up a proton to form —COO$^-$;
the —NH$_2$ group is basic since it can add a proton to form —NH$_3^+$.

Molecules of this type, having both an acidic and a basic group, are known as
amphoteric electrolytes, or *ampholytes*. It was originally thought that when a molecule
such as glycine is dissolved in water it exists mainly as H_2NCH_2COOH, and that
the following ionizations occur:

$$H_3N^+CH_2COOH \underset{+H^+}{\overset{-H^+}{\rightleftharpoons}} H_2NCH_2COOH \underset{+H^+}{\overset{-H^+}{\rightleftharpoons}} H_2NCH_2COO^-$$

However, it is now realized that at intermediate pH values the predominant species
is $H_3N^+CH_2COO^-$, not H_2NCH_2COOH, the ionizations being

$$H_3N^+CH_2COOH \overset{pK_1=2.3}{\rightleftharpoons} H_3N^+CH_2COO^- \overset{pK_2=9.7}{\rightleftharpoons} H_2NCH_2COO^-$$

The pK values are noted above. The species $H_3N^+CH_2COO^-$ is known as a
zwitterion (double ion).

Several lines of evidence support the latter point of view. One argument relates to
the magnitudes of dissociation constants. The splitting-off of a proton from a —NH$_3^+$
group typically has a dissociation constant of about 1.6×10^{-10}:

$$-NH_3^+ \rightleftharpoons -NH_2 + H^+ \qquad K_a = 1.6 \times 10^{-10} \qquad (pK = 9.8)$$

while the dissociation of a —COOH group has a constant of about 4.0×10^{-3}:

$$-COOH \rightleftharpoons -COO^- + H^+ \qquad K_a = 4.0 \times 10^{-3} \qquad (pK = 2.4)$$

Thus if we consider the ionization of fully protonated glycine, $H_3N^+CH_2COOH$, we expect that

(1) $$\frac{[H_3N^+CH_2COO^-][H^+]}{[H_3N^+CH_2COOH]} \approx 4.0 \times 10^{-3}$$

and

(2) $$\frac{[H_2NCH_2COOH][H^+]}{[H_3N^+CH_2COOH]} \approx 1.6 \times 10^{-10}$$

(These estimates will be approximate because of interactions between the different ionizing groups; as seen above, the values for glycine are close to these values). Division of (2) by (1) leads to

$$\frac{[H_2NCH_2COOH]}{[H_3N^+CH_2COO^-]} \approx \frac{1.6 \times 10^{-10}}{4.0 \times 10^{-3}} = 4.0 \times 10^{-8}$$

so that H_2NCH_2COOH is present in insignificant amounts compared with $H_3N^+CH_2COO^-$. Similar conclusions have been reached for other ampholytes.

Secondly, studies in solvents of different dielectric constants have strongly supported the new interpretation. If an ionization is of the type

$$-NH_3^+ \rightleftharpoons -NH_2 + H^+$$

a change in dielectric constant will have little effect on the pK value, since there is no large change in the electric field, because a positively charged group has produced another positively-charged ion (the proton). On the other hand, if the ionization occurring at low pH is

$$-COOH \rightleftharpoons -COO^- + H^+$$

as in the modern mechanism, there is a large change in electric field, because the neutral species —COOH has produced a positive and a negative ion. An increase in dielectric constant should, therefore, greatly favor the ionization by aiding the separation of charges and thus lead to an increase in dissociation constant and a lowering in pK. Experimentally this is found to be the case, and the modern view is therefore supported.

EXAMPLE

What will be the predominant species in glycine at the following pH values: 1.5; 2.3; 5.0; 7.0; 11.5?

SOLUTION

At pH 1.5 the predominant species is $H_3N^+CH_2COOH$. At 2.3, $H_3N^+CH_2COOH$ and $H_3N^+CH_2COO^-$ are present in equal amounts, with a small amount of $H_2NCH_2COO^-$. At 5.0 and 7.0, the zwitterion $H_3N^+CH_2COO^-$ is predominant. At 11.5, $H_2NCH_2COO^-$ is predominant.

POLYELECTROLYTES

A number of molecules contain a large number of ionizing groups; these are known as *polyelectrolytes*. Examples are the protein molecules which play so important a role in biological systems. At one end of each protein chain there is a free $-NH_3^+$ group; at the other a free $-COO^-$ group. In addition, a number of the amino-acid side chains, R, contain ionizing groups. For example, one amino acid is arginine,

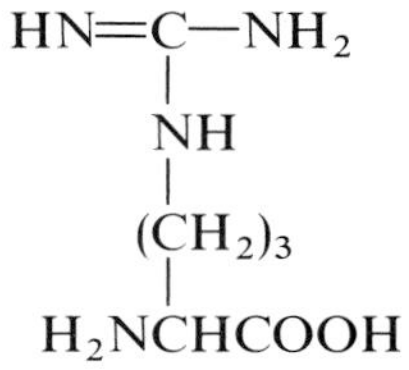

$$HN=C-NH_2$$
$$|$$
$$NH$$
$$|$$
$$(CH_2)_3$$
$$|$$
$$H_2NCHCOOH$$

which contains the $=NH$ group in the side chain. In a sufficiently acid solution this group will exist as $=NH_2^+$. Some of the groups which exist on the side chains of proteins are listed in Table 7.5, with their pK values given. These values are approximate, and vary appreciably from molecule to molecule.

Table 7.5   Approximate pK Values for Groups on Protein Side Chains

Group	Approximate pK
$-COOH$	4
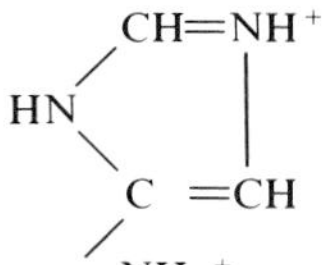 (imidazole ring: HN with CH=NH$^+$ and C=CH)	7
$-NH_3^+$	8–10
$-OH$	10
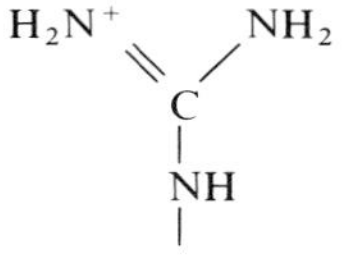 (guanidinium: H_2N^+, NH_2 on C, NH)	12

Such polyelectrolytes will tend to bear a net positive charge when the pH is low, since neutral groups such as —NH_2 will have been protonated to become —NH_3^+, and negatively charged groups such as —O^- and —COO^- will have become the neutral —OH and —COOH species. Consequently, at sufficiently low pH values, polyelectrolytes in an applied electric field will tend to move towards the negatively-charged electrode (the cathode). On the other hand, in basic solutions (i.e., at high pH values), polyelectrolytes will bear a net negative charge, since protons will have left positive groups such as —NH_3^+ and will have left neutral groups to form negative groups (e.g., —COOH → —COO^-). An applied electric field will therefore produce movement towards the anode, or positive electrode. This type of motion is referred to as *electrophoresis*, and is considered further in Chapter 11.

There is a certain pH at which the number of positive charges on the polyelectrolyte is equal to the number of negative charges. This pH is known as the *isoionic point* (Greek *iso*, equal). If we were concerned only with proton ionizations, the isoionic point would be the pH corresponding to a zero charge on the molecule, i.e., to no movement in an electric field. However, other ions may interact with the ionizing groups; thus the pH where there is no net charge, known as the *isoelectric point*, may not exactly correspond to the isoionic point.

For a simple zwitterion such as glycine, in which there are two ionizing groups, the isoionic point is simply related to the pK values for the two groups. The ionization of glycine is shown on p. 315, and the corresponding dissociation constants are

$$K_1 = \frac{[H_3N^+CH_2COO^-][H^+]}{[H_3N^+CH_2COOH]} \tag{7.56}$$

$$K_2 = \frac{[H_2NCH_2COO^-][H^+]}{[H_3N^+CH_2COO^-]} \tag{7.57}$$

At the isoionic point

$$[H_3N^+CH_2COOH] = [H_2NCH_3COO^-] \tag{7.58}$$

Therefore, from (7.56) and (7.57)

$$\frac{[H_3N^+CH_2COO^-][H^+]}{K_1} = \frac{K_2[H_3N^+CH_2COO^-]}{[H^+]} \tag{7.59}$$

Thus

$$[H^+] = \sqrt{K_1K_2} \tag{7.60}$$

Taking logarithms of both sides

$$\log_{10}[H^+] = \tfrac{1}{2}(\log_{10}K_1 + \log_{10}K_2) \tag{7.61}$$

so that, if pI is the pH at the isoionic point,

$$pI = \frac{pK_1 + pK_2}{2} \tag{7.62}$$

Thus, for the particular case of glycine, for which $pK_1 = 2.3$ and $pK_2 = 9.7$, the isoionic point is $pI = \frac{1}{2}(2.3 + 9.7) = 6.0$.

Equation (7.62) applies to any species having two ionizing groups. If there are more, as with certain other amino acids or the proteins, the situation is usually not much more complicated, since two ionizations commonly dominate the situation. Consider, for example, the case of arginine, for which the ionizations are

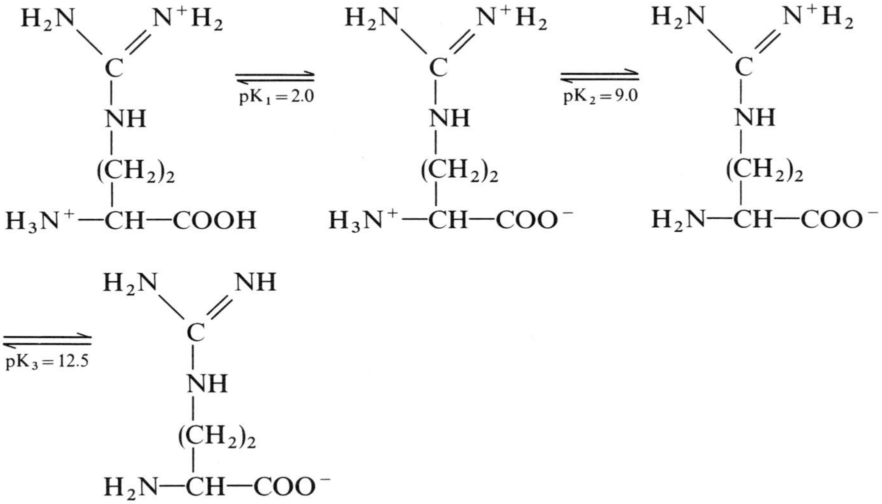

We see, on inspection of this system, that at a pH of $\frac{1}{2}(9.0 + 12.5) = 10.75$ the concentrations of the second and fourth species, with net charges of $+1$ and -1 respectively, will be the same. Moreover, at this pH the concentration of the first species, with a net charge of $+2$, will be quite negligible. The isoionic point of arginine, therefore, corresponds to a pH of 10.75.

BUFFER SOLUTIONS

Let us consider the interesting case of solutions which contain weak electrolytes (acids or bases), and in addition contain a salt having a common ion. For example, a solution might be made up to the following specifications:

0.3 M acetic acid

0.2 M sodium acetate

In calculating the pH of such a solution we must remember that acetic acid is a weak electrolyte ($K_a = 1.80 \times 10^{-5}$), but that sodium acetate is a strong electrolyte, being completely dissociated into CH_3COO^- and H^+ ions. Therefore, the concentration of acetate ions is very close to 0.2 M, since hardly any acetate ions are generated by the acetic acid; the concentration of CH_3COOH is 0.3 M, since hardly any of this species is dissociated. That is,

$$[CH_3COO^-] \approx 0.2 \text{ M}$$

$$[CH_3COOH] \approx 0.3 \text{ M}$$

The dissociation constant for the acetic acid ionization is

$$\frac{[CH_3COO^-][H^+]}{[CH_3COOH]} = 1.80 \times 10^{-5} \tag{7.63}$$

and therefore

$$[H^+] = 1.80 \times 10^{-5} \frac{[CH_3COOH]}{[CH_3COO^-]} \tag{7.64}$$

Insertion of the concentrations leads to

$$[H^+] = \frac{0.3}{0.2} \times 1.80 \times 10^{-5} = 2.7 \times 10^{-5} \text{ M}$$

The pH is 4.57.

A more accurate solution of the problem is obtained as follows. If the acetic acid produces x mol dm^{-3} of CH_3COO^- and H^+, there are $0.3 - x$ mol dm^{-3} of acetate ions left. The total concentration of CH_3COO^- is $0.2 + x$, so that the equilibrium constant equation is

$$x = \frac{0.3 - x}{0.2 + x} \times 1.80 \times 10^{-5} \tag{7.65}$$

The solution of this quadratic equation is

$$[H^+] = x = 2.699 \times 10^{-5} \text{ M} \qquad (\text{pH} = 4.57)$$

which is very close to the approximate solution. The smaller K_a is, the more accurate will be the solution obtained by the approximate method.

EXAMPLE

An aqueous solution is 0.6 M with respect to acetic acid and 0.4 M with respect to sodium hydroxide. Calculate the pH.

SOLUTION

In each liter 0.4 mol of acetic acid will have been neutralized by the 0.4 mol of sodium hydroxide, to yield 0.4 mol of sodium acetate. Consequently, 0.2 mol of acetic acid will remain free. The solution is equivalent to the following:

0.2 M acetic acid

0.4 M sodium acetate

To a good approximation

$$[CH_3COOH] = 0.2 \text{ M}$$
$$[CH_3COO^-] = 0.4 \text{ M}$$

The hydrogen-ion concentration is thus

$$[H^+] = \frac{[CH_3COOH]}{[CH_3COO^-]} \times 1.80 \times 10^{-5}$$

$$= \frac{0.2}{0.4} \times 1.80 \times 10^{-5} = 9.0 \times 10^{-6} \text{ M}$$

The pH is 5.05.

Solutions containing a mixture of a weak acid and one of its salts, or a weak base and one of its salts, are known as *buffer solutions*. The important property of a buffer solution is that its hydrogen-ion concentration does not change very much if small amounts of acids or bases are added to the solution. In the acetate-acetic acid buffer, for example, the concentrations of CH_3COOH and CH_3COO^- are a good deal larger than the concentration of H^+ ions. Thus, for the buffer solution just considered,

$$[CH_3COOH] = 0.20 \text{ M}$$
$$[CH_3COO^-] = 0.40 \text{ M}$$
$$[H^+] = 9.0 \times 10^{-6} \text{ M}$$

There is therefore a very large "reservoir" of the species CH_3COOH and CH_3COO^-, and any added acid or base will largely be removed by a shift in the equilibrium

$$CH_3COOH \rightleftharpoons CH_3COO^- + H^+$$

Addition of acid will cause a shift to the left, such that most of the added acid will be removed; added base will be largely removed by a shift to the right.

The following calculation will demonstrate this point. A solution which is 9.0×10^{-6} M in H^+ ions could be prepared in two ways:

1. We could make a 9.0×10^{-6} M solution of hydrochloric acid.

2. We could make a buffer solution that is 0.20 M in acetic acid and 0.40 M in sodium acetate.

Suppose that 1.00×10^{-5} mol of acid were added to 1 dm^3 of each solution. In the HCl solution the hydrogen ion concentration would become

$$9.05 \times 10^{-6} + 1.00 \times 10^{-5} = 1.9 \times 10^{-5} \text{ M}$$

and would be *more than doubled*. In the case of the buffer solution, however, the buffer equilibrium would shift, the concentrations of the various species being as follows:

	CH_3COOH	$\rightleftharpoons$	CH_3COO^-	$+$	H^+
Initially	0.20		0.40		9.0×10^{-6}
Immediately after adding 1.00×10^{-5} mol of H^+ to 1 dm^3	0.20		0.40		19.0×10^{-6}
After equilibrium is reestablished	$0.20 + x$		$0.40 - x$		$19.0 \times 10^{-6} - x$

The value of x is found by solving the equation

$$\frac{(0.40 - x)(1.90 \times 10^{-5} - x)}{0.20 + x} = 1.8 \times 10^{-5}$$

The solution is

$$x = 0.999935 \times 10^{-5} \text{ M}$$

which differs from the amount of added acid by only 0.0065%. In other words, all except this tiny fraction of the added H^+ ions have been removed by the shift in equilibrium, and the pH remains almost exactly the same.

Buffer systems are particularly important in biological systems. Blood and other body fluids are buffer solutions which prevent small additions of acids and bases from causing drastic changes in the pH. The functioning of the enzymes—the biological catalysts—is very sensitive to pH changes, and if biological systems were not buffered there would be serious variations in the rates of metabolic processes.

7.5   SOLUBILITY PRODUCTS

An important type of equilibrium in solution involves a solid salt and its ions when the salt is in contact with its saturated solution. For example, consider the case of solid silver chloride, AgCl (s), in contact with its saturated aqueous solution, which contains the hydrated ions Ag^+ and Cl^-. The equilibrium is

$$AgCl\ (s) \rightleftharpoons Ag^+ + Cl^-$$

Here we shall neglect activity coefficient effects, which will be introduced in equation (7.72). The equilibrium constant can then be written as

$$[Ag^+][Cl^-] = K_s = 1.7 \times 10^{-10}\ mol^2\ dm^{-6} \tag{7.66}$$

The concentration of AgCl (s) is constant and is incorporated into the constant K_s. The constant K_s, known as the *solubility product*, has a characteristic value for each salt. If the solution is less than saturated, the product $[Ag^+][Cl^-]$ will be less than K_s, because equilibrium is not established. However, the product cannot be greater than K_s for any period of time. If solutions containing Ag^+ and Cl^- ions are mixed, solid AgCl will rapidly precipitate out until the product $[Ag^+][Cl^-]$ has reached the value of K_s.

In a saturated aqueous solution of AgCl, with no other substance present, the concentrations of Ag^+ and Cl^- ions will be equal. Since their product is K_s, the individual concentrations are $\sqrt{K_s}$, which is therefore the solubility:

$$Solubility = [Ag^+] = [Cl^-] = \sqrt{K_s} \tag{7.67}$$

Thus, the solubility of AgCl in pure water is approximately $\sqrt{1.7 \times 10^{-10}} = 1.3 \times 10^{-5}$ M. This result, that the solubility is the square root of the solubility product, is true for any electrolyte of the type AB which ionizes into one ion of A and one of B. Thus it is true for uni-univalent electrolytes such as AgCl and for bi-bivalent ones such as CdS. The relationship is different for other types of electrolytes, such as uni-bivalent ones like Ag_2CrO_4, which will be considered later.

The solubility of a salt is less if a common ion is added. For example, suppose that we make a saturated solution of AgCl in 0.1 M $AgNO_3$ solution. At equilibrium

$$[Ag^+][Cl^-] = 1.7 \times 10^{-10}\ mol^2\ dm^{-6}$$

However, the Ag^+ concentration is 0.1 M†; it thus follows that

$$[Cl^-] = \frac{1.7 \times 10^{-10}\ mol^2\ dm^{-6}}{0.1\ mol\ dm^{-3}}$$

$$= 1.7 \times 10^{-9}\ M$$

The concentration of AgCl in the solution is thus 1.7×10^{-9} M; this is the solubility under these conditions. In view of equation (7.67), the solubility will decrease as the common ion concentration increases.

---

† Actually it is slightly greater because of the solution of AgCl, but this is a negligible correction.

The relationship between solubility s and solubility product K_s is different for a salt which is not of the AB type. Consider, for example, silver chromate, Ag_2CrO_4, which is of the type A_2B. This salt ionizes as follows:

$$Ag_2CrO_4 \text{ (s)} \rightleftharpoons 2Ag^+ + CrO_4{}^{2-}$$

and the solubility product is

$$[Ag^+]^2[CrO_4{}^{2-}] = K_s = 9.0 \times 10^{-12} \text{ mol}^3 \text{ dm}^{-9} \text{ at } 25°C \qquad (7.68)$$

When the salt is dissolved in pure water the concentration of Ag^+ is twice that of $CrO_4{}^{2-}$. The solubility s is the concentration of $CrO_4{}^{2-}$ and one-half that of Ag^+:

$$[CrO_4{}^{2-}] = s \qquad [Ag^+] = 2s \qquad (7.69)$$

Insertion of these expressions into (7.68) gives

$$(2s)^2 s = K_s \qquad (7.70)$$

or

$$4s^3 = K_s \qquad (7.71)$$

The solubility of silver chromate is thus

$$s = \sqrt[3]{\frac{9.0 \times 10^{-12}}{4}} = 1.31 \times 10^{-4} \text{ M}$$

Similar relationships can be worked out for other types of electrolytes, such as A_3B, A_2B_2 (e.g., Hg_2Cl_2), and A_3B_2.

ACTIVITY COEFFICIENTS

So far we have ignored activity coefficients, the solubility products being expressed as products of concentrations instead of activities. The solubility product for silver chloride should be written more accurately as

$$K_s = [Ag^+][Cl^-]y_+ y_- \qquad (7.72)$$

where y_+ and y_- are the activity coefficients of Ag^+ and Cl^- respectively. The product $y_+ y_-$ is equal to $y_\pm^2$, where $y_\pm$ is the mean activity coefficient, so that

$$K_s = [Ag^+][Cl^-]y_\pm^2 \qquad (7.73)$$

One matter of interest which can be understood in terms of this equation is the effect of inert electrolytes on solubilities. An inert electrolyte is one which does not contain a common ion, i.e., Ag^+ or Cl^- in this instance, and also does not contain any ion which will complicate the situation by forming a precipitate with either the Ag^+ or the Cl^- ions. In other words, the added inert electrolyte does not bring about a chemical effect; its influence arises only because of its ionic strength.

The influence of ionic strength I on the activity coefficient of an ion is given according to the DHLL by the equation

$$\log_{10} y_{\pm} = -z_{+}|z_{-}|B\sqrt{I} \tag{7.74}$$

(see equation (7.24)). Figure 7.9 on p. 305 shows that this equation satisfactorily accounts for the drop in $y_{\pm}$ which occurs at very low ionic strengths, but that considerable deviations occur at higher ones, the value of $\log y_{\pm}$ becoming positive (i.e., $y_{\pm}$ is greater than unity) at sufficiently high values of I.

It follows that as a result of this behavior there are two qualitatively different ionic-strength effects on solubilities, one arising at low I values when the $y_{\pm}$ falls with increasing I, and the other found when $y_{\pm}$ increases with increasing I. Thus, at low ionic strengths the product $[Ag^+][Cl^-]$ will increase with increasing I, because the product $[Ag^+][Cl^-]\,y_{\pm}^2$ remains constant and $y_{\pm}$ decreases. Under these conditions added salt increases solubility, and we speak of *salting in*.

At higher ionic strengths, however, $y_{\pm}$ rises as I increases, and $[Ag^+][Cl^-]$ therefore diminishes. Thus there is a decrease in solubility, and we speak of *salting out*. Of particular interest in biology are the salting-in and salting-out effects found with protein molecules, a matter of considerable practical importance since proteins are conveniently classified in terms of their solubility behavior. As we have seen, proteins are polyelectrolytes (p. 317), and the influence of ionic strength on their solubility is therefore expected to be qualitatively the same as with simple salts such as silver chloride. Quantitatively, of course, the behavior will be very much more complicated because of the numerous ionizations involved. Figure 7.10 shows the type of effect that various salts have on the solubility of a protein such as hemoglobin. At low ionic

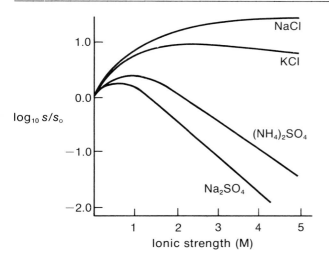

Figure 7.10

Influence of the ionic strengths of various salt solutions on the solubility s of a protein such as hemoglobin; s_0 is the solubility at zero ionic strength.

strengths there is a salting-in effect, while at higher ones there is a salting-out effect.

We can see from this diagram that plotting against ionic strength by no means eliminates differences between various salts. For example, at a given ionic strength, sulfates are significantly more effective than chlorides. Many years ago, the German chemist Fritz Hofmeister arranged inorganic ions in a series which predicts their salting-out effects on a great many protein molecules. This series, known as the *Hofmeister series*, corresponds to some extent with the order of hydration of the ions. This suggests that the salting-out effect, like hydration, depends upon the tendency of the ions to become coordinated. A precise theory of these effects is very difficult to work out, but a certain amount of progress has been made.

7.6   THE DONNAN EQUILIBRIUM

A matter of considerable importance in biological systems is the ionic equilibrium which exists between two solutions separated by a membrane, such as a cell wall. Complications arise with ions that are too large to diffuse through the membrane, and the diffusible ions then reach a special type of equilibrium known as the Donnan equilibrium. Its theory was first worked out in 1911 by the British physical chemist Frederick George Donnan (1870–1956).

Consider first the equilibrium established when the ions all can diffuse through the membrane. A very simple case is when solutions of sodium chloride are separated by the membrane, as shown in Figure 7.11a. Suppose that at equilibrium the concentrations are $[Na^+]_1$ and $[Cl^-]_1$ on the left-hand side, and $[Na^+]_2$ and $[Cl^-]_2$ on the right-hand side. Intuitively, we know that in this simple case these concentrations must be all the same at equilibrium. In thermodynamic terms we can arrive at this conclusion by saying that at equilibrium

$$\Delta G = \Delta G_{Na^+} + \Delta G_{Cl^-} = 0 \qquad (7.75)$$

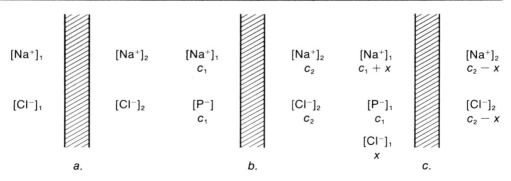

Figure 7.11

a. Sodium and chloride ions separated by a membrane.
b. Na^+P^- and Na^+Cl^- separated by a membrane: initial conditions.
c. The final Donnan equilibrium conditions arising from (*b*).

where the terms are the Gibbs energy differences across the membrane (e.g., ΔG is the change in Gibbs energy in going from left to right). The expressions for the individual molar ionic Gibbs-energy terms are (see equation (5.102))

$$\Delta G_{Na^+} = RT \ln \frac{[Na^+]_2}{[Na^+]_1} \tag{7.76}$$

$$\Delta G_{Cl^-} = RT \ln \frac{[Cl^-]_2}{[Cl^-]_1} \tag{7.77}$$

Thus

$$RT \ln \frac{[Na^+]_2}{[Na^+]_1} + RT \ln \frac{[Cl^-]_2}{[Cl^-]_1} = 0 \tag{7.78}$$

from which it follows that

$$\frac{[Na^+]_2[Cl^-]_2}{[Na^+]_1[Cl^-]_1} = 1 \tag{7.79}$$

Since, for electrical neutrality, $[Na^+]_1 = [Cl^-]_1$ and $[Na^+]_2 = [Cl^-]_2$, the result is that

$$[Na^+]_1 = [Na^+]_2 = [Cl^-]_1 = [Cl^-]_2 \tag{7.80}$$

In other words, at equilibrium we have equal concentrations of the electrolyte on each side of the membrane.

This rather trivial case is useful as an introduction to the Donnan equilibrium, since equation (7.79) will still be obeyed even if the system contains a nondiffusible ion in addition to sodium and chloride ions. For example, suppose that initially we have the situation represented in Figure 7.11b. On the left-hand side, there are sodium ions and nondiffusible anions, P^-. On the right-hand side there are sodium and chloride ions. Since there are no chloride ions on the left-hand side, spontaneous diffusion of chloride ions from right to left will occur. Since there must always be electrical neutrality on each side of the membrane, an equal number of sodium ions must also diffuse from right to left. Figure 7.11c shows the situation at equilibrium: x mol dm^{-3} of $[Na^+]$ and $[Cl^-]$ have diffused from right to left, the initial concentrations on the two sides being c_1 and c_2.

Application of equation (7.79) then leads to

$$(c_2 - x)^2 = (c_1 + x)x \tag{7.81}$$

Thus

$$x = \frac{c_2{}^2}{c_1 + 2c_2} \tag{7.82}$$

As an example, suppose that $c_1 = 0.01$ M and $c_2 = 0.05$ M. Use of equation (7.82) leads to the result that $x = 0.023$ M. The final concentrations of the Na^+ and Cl^- ions on the right-hand side are thus 0.027 M. The Na^+ concentration on the left-hand side is 0.033 M, while the Cl^- concentration on the left is 0.023 M.

Equations for more complicated Donnan equilibria (e.g., those involving divalent ions) can easily be worked out using the same principles. Equilibria of the Donnan type are relevant to many types of biological systems; the theory is particularly important with reference to the passage of ions across the membranes of nerve fibres. However, under physiological conditions the significance of the Donnan effect is not easily assessed because of the complication of *active transport* (p. 487), which is a phenomenon in which ions are transported against concentration gradients by processes requiring the expenditure of energy. One straightforward example of the Donnan effect in which there is little or no active transport is found with the erythrocytes (red blood cells). Here the concentration of chloride ions within the cells is significantly smaller than the concentration in the plasma surrounding the cells. This effect can be attributed to the much higher concentration of protein anions retained within the erythrocyte. In fact, hemoglobin itself accounts for a third of the dry weight of the cell.

Under certain circumstances, the establishment of the Donnan equilibrium can lead to other effects, such as changes in pH. Suppose, for example, that an electrolyte NaP (where P is a large anion) is on one side of a membrane, with pure water on the other. The Na^+ ions will tend to cross the membrane and, to restore the electrostatic balance, H^+ ions will cross in the other direction, leaving an excess of OH^- ions. Dissociation of water molecules will occur as required. There will thus be a lowering of pH on the NaP side of the membrane, and a raising on the other side.

PROBLEMS

7.1 The following are some conventional standard enthalpies of ions in aqueous solution, at 25° C:

Ion	ΔH_f°/kcal mol^{-1}
H^+	0
Na^+	-57.3
Ca^{2+}	-129.8
Zn^{2+}	-36.4
Cl^-	-40.0
Br^-	-28.9

Calculate the enthalpy of formation in aqueous solution of 1 mol of NaCl, $CaCl_2$, and $ZnBr_2$, assuming complete dissociation.

7.2 One estimate for the absolute Gibbs energy of hydration of the H^+ ion in aqueous solution is -251.3 kcal mol^{-1}. On this basis, calculate the absolute Gibbs energies of hydration of the following ions, whose conventional standard Gibbs energies of hydration are as follows:

Ion	ΔG_h° (conv)/kcal mol^{-1}
H$^+$	0
Na$^+$	162.3
Mg^{2+}	65.5
Al^{3+}	-321.8
Cl$^-$	-336.3
Br$^-$	-333.0

7.3 Calculate the concentration of NH_4^+ ions in an 0.01 M solution of ammonia (K_a for $NH_4^+ = 5.6 \times 10^{-10}$; $K_w = 10^{-14}$).

7.4 Calculate the ionic strengths of 0.1 M solutions of KNO_3, K_2SO_4, $ZnSO_4$, $ZnCl_2$, $K_4Fe(CN)_6$; assume complete dissociation.

7.5 Calculate the mean activity coefficient, $y_\pm$, for the Ba^{2+} and SO_4^{2-} ions in a saturated solution of $BaSO_4$ ($K_{sp} = 9.2 \times 10^{-11}$) in 0.2 M K_2SO_4, assuming the Debye-Hückel limiting law to apply.

7.6 The concentration of carbonic acid, H_2CO_3, in blood plasma is about 0.00125 M, and the pK_a value is about 6.1. Calculate the concentration of HCO_3^- ions if the pH is 7.4.

7.7 What is the pH at 25° C (p$K_w = 10^{-14}$) of a solution which contains twice as many hydroxide ions as pure water?

7.8 Tyrosine, $H_2NCHCOOH$, has three dissociation constants

corresponding to the following pK values:

	pK
—COOH	2.4
—NH$_3^+$	9.6
—C$_6$H$_4$OH	10.1

(a) Write down the equilibria corresponding to these three pK values.

(b) What is the isoionic point?

(c) What will be the predominant species at the following pH values: 2.0, 3.0, 9.0, 9.8, 10.5?

7.9 The dissociation constant of hydrocyanic acid, HCN, is 7.2×10^{-10} at 25° C. Calculate the molarity of a solution of HCN that is 0.1% ionized.

7.10 If the acid dissociation constants of the amino acid serine ($CH_2OH\ CHNH_2$ COOH) are 6.2×10^{-3} and 7.1×10^{-10}, state which ionic form will predominate at the following pH values: 2.0, 5.0, 8.0, and 11.0. What is the isoionic point?

7.11 An 0.1 M solution of sodium palmitate, $C_{15}H_{31}COONa$, is separated from an 0.2 M solution of sodium chloride by a membrane which is permeable to Na^+ and Cl^- but not to palmitate ions. Calculate the concentrations of Na^+ and Cl^- ions on the two sides of the membrane after equilibrium has become established.

7.12 Employ equation (7.35) to make plots of log $y_\pm$ against $\sqrt{I}$ for a uni-univalent electrolyte in water at 25° C, with $B = 0.51$ mol^{-1} dm$^{3/2}$ and $B' = 0.33 \times 10^{10}$ mol^{-1} dm$^{3/2}$ m^{-1}, and for the following values of the interionic distance a: 0, 0.1, 0.2, 0.4, and 0.8 nanometres.

7.13 Calculate the pH of a solution which is 0.20 M with respect to acetic acid and 0.15 M with respect to sodium acetate (K_a for acetic acid is 1.80×10^{-5}).

7.14 Calculate the concentration of acetate ions in a solution which is 0.01 M in acetic acid and 0.02 M in hydrochloric acid (K_a for acetic acid is 1.80×10^{-5}).

7.15 Calculate the pH of a solution of ammonia ($K_a = 5.6 \times 10^{-10}$) in which the degree of ionization is 0.03.

7.16 The degree of ionization of 0.10 M HCN is 8.5×10^{-5} at a certain temperature; calculate K_a at that temperature.

7.17 How many moles of sodium acetate must be added to 1 dm^3 of 0.2 M acetic acid to make a buffer solution of pH 5?

7.18 Calculate the pH of an 0.1 M solution of NH_4Cl, if K_a for NH_4^+ is 5.6×10^{-10}.

7.19 If 20 cm^3 of 0.2 M NaOH is added to 50 cm^3 of 0.1 M acetic acid, what will be the pH?

7.20 The solubility product of Ag_2CrO_4 is 9.0×10^{-12} mol^3 dm^{-9}; calculate the solutibility in (1) 0.1 M $AgNO_3$, (2) 0.2 M K_2KrO_4.

7.21 Calculate the concentration of carbonate ion, CO_3^{2-}, in a solution of 0.01 M sodium bicarbonate, $NaHCO_3$, at pH 8.0; K_a for HCO_3^- is 5.6×10^{-11}.

7.22 100 cm^3 of 0.1 M KH_2PO_4 is mixed with 50 cm^3 of 0.1 M NaOH and the resulting mixture diluted to 1 dm^3. If pK_a for $H_2PO_4^-$ is 6.86, calculate the pH and the ionic strength.

7.23 The acidity of orange juice can be regarded as due to 0.1 M citric acid, $HOC(CH_2COOH)_2COOH$ ($K_1 = 8.4 \times 10^{-4}$). Calculate its pH. If the stomach has a pH of 1.0, will drinking orange juice initially increase or decrease its pH?

7.24 Estimate the change in Gibbs energy ΔG when 1 mol of K^+ ions (radius 0.133 nm) is transported from aqueous solution ($\varepsilon = 78$) to the lipid environment of a cell membrane ($\varepsilon = 4$).

ESSAY QUESTIONS

7.25 Describe briefly the type of hydration found with the following ions in aqueous solution: Li^+, Br^-, H^+, OH^-.

7.26 What modifications to the Debye-Hückel limiting law are required to explain the influence of ionic strength on solubilities?

7.27 Give an account of the behavior of polyelectrolytes in aqueous solution, with special reference to

(a) The existence of zwitterions.
(b) The isoionic and isoelectric points.

SUGGESTED READING

Eisenberg, D., and Kauzmann, W. J. *The Structure and Properties of Water.* Oxford: Clarendon Press, 1969.

Glasstone, S. *Introduction to Electrochemistry.* New York: Van Nostrand, 1942.

Kavanau, J. L. *Water and Solute-Water Interactions.* San Francisco: Holden-Day, 1964.

Moore, W. J. *Physical Chemistry*, 4th ed. Englewood Cliffs, N.J.: Prentice-Hall, 1972.

Robbins, J. *Ions in Solution.* Oxford: Clarendon Press, 1972.

8

ELECTROCHEMICAL CELLS

A very important branch of electrochemistry is concerned with the electromotive force (emf) or voltage developed in electrochemical cells. Investigations on electrochemical cells provide valuable information of various kinds. For example, they lead to thermodynamic quantities such as enthalpies and Gibbs energies for a variety of chemical reactions, including many reactions important in biology. Also, they allow us to obtain activity coefficients for ions in solution. This chapter deals with the general principles of electrochemical cells and with some of their more important applications.

8.1   THE DANIELL CELL

A simple example of an electrochemical cell is the Daniell cell, named after the English chemist John Frederic Daniell (1790–1845). This apparatus, illustrated in Figure 8.1, consists of a zinc electrode immersed in a zinc sulfate solution and a copper electrode immersed in a cupric sulfate solution. The two solutions are separated by a porous partition which is of such a nature that it prevents the bulk mixing of the solutions but allows ions to pass through as the cell operates. The emf developed in the Daniell cell depends upon the concentrations of Zn^{2+} and Cu^{2+} ions in the two solutions. If the concentrations of the two solutions are both one molal (1 m), the cell is called a *standard* cell.

If such a cell is set up, there is a flow of electrons from the zinc to the copper electrode in the outer circuit. This means that a positive current must be moving from left to right in the cell itself. By convention, an emf corresponding to an external flow of electrons from the left-hand electrode to the right-hand electrode is said to be a *positive* emf. The magnitude of the emf developed (with 1 m solutions) is approximately 1.1 volts (V). This positive value is said to be the *standard* emf of the cell, the word "standard" referring to the fact that the concentrations are 1 m. We shall employ the symbol ΔE° for such standard voltages.

The processes which occur when this cell operates are shown in the figure. Since positive electricity moves from left to right within the cell, zinc metal must be dissolving to form Zn^{2+} ions,

$$Zn \rightarrow Zn^{2+} + 2e^-$$

Some of these zinc ions pass through the membrane into the right-hand solution, and at the right-hand electrode, cupric ions interact with electrons to form metallic copper:

$$Cu^{2+} + 2e^- \rightarrow Cu$$

Every time a zinc atom dissolves and a copper atom is deposited, two electrons travel around the outer circuit.

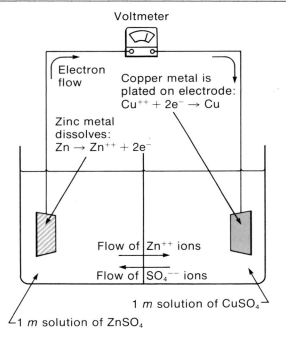

Voltmeter

Electron flow

Copper metal is plated on electrode:
$Cu^{++} + 2e^- \rightarrow Cu$

Zinc metal dissolves:
$Zn \rightarrow Zn^{++} + 2e^-$

Flow of Zn^{++} ions

Flow of SO_4^{--} ions

1 m solution of $CuSO_4$

1 m solution of $ZnSO_4$

Figure 8.1

The standard Daniell cell.

The Daniell cell and many other cells behave in a reversible fashion. It is possible to balance their emf by an external emf, for example by using a potentiometer wire. If the counter-emf is exactly equal to the emf developed by the cell, no current passes. If the counter-emf is slightly less than the emf of the cell, there is a small electron flow from left to right in the outer circuit. If the counter-emf is adjusted to be slightly greater than the cell emf, the cell is forced to operate in reverse; zinc is deposited at the left-hand electrode

$$Zn^{2+} + 2e^- \rightarrow Zn$$

and copper dissolves at the right

$$Cu \rightarrow Cu^{2+} + 2e^-$$

To measure a reversible emf, a variable counter-emf is applied by the use of a potentiometer. The current flow is detected by means of a galvanometer, and the counter-emf is adjusted until the current is zero.

The Daniell cell is a typical example of what is known as a *galvanic, voltaic,* or *electrochemical* cell. The fact that the electrons flow from the zinc to the copper electrode indicates that the tendency for $Zn \rightarrow Zn^{2+} + 2e^-$ to occur is greater than for the reaction $Cu \rightarrow Cu^{2+} + 2e^-$, which is forced to occur in the reverse direction. The magnitude of the emf developed is a measure of the relative tendencies of the two processes. The emf varies with the concentrations of the Zn^{2+} and Cu^{2+} ions in the two solutions. Thus the tendency for $Zn \rightarrow Zn^{2+} + 2e^-$ to occur is smaller when the concentration of Zn^{2+} is large, while the tendency for $2e^- + Cu^{2+} \rightarrow Cu$ to occur increases when the concentration of Cu^{2+} is increased. The precise relationship between emf and the concentrations will be considered later.

8.2   STANDARD ELECTRODE POTENTIALS

It would be very convenient if we could measure the potential of a single electrode, such as the right-hand electrode in Figure 8.1, which we will write as

$$Cu^{2+} \mid Cu$$

The potential of such an electrode would be a measure of the tendency of the process

$$Cu^{2+} + 2e^- \rightarrow Cu$$

to occur. However, there is no way to measure the emf of a single electrode, since in order to obtain an emf there must be two electrodes, with an emf associated with each one. A convenient procedure is to choose one electrode as a standard and measure emf values of other electrodes with reference to that standard. The hydrogen electrode, illustrated in Figure 8.2, has been chosen as the standard. It consists of a platinum electrode immersed in a 1 m solution of hydrogen ions maintained at 25°C and 760 mm Hg pressure. Hydrogen gas is bubbled over the electrode and passes into solution, forming hydrogen ions and electrons:

$$H_2 \rightarrow 2H^+ + 2e^-$$

The emf corresponding to this electrode is arbitrarily assigned to have the value of zero, so that this electrode can be used as a standard for other electrodes.

There are two conventions in common use, and the student should be aware of both methods of procedure. The standard hydrogen electrode can be either the left-hand or the right-hand electrode. In the convention adopted by the International Union of Pure and Applied Chemistry (I.U.P.A.C.) the hydrogen electrode is placed on the *left-hand side*, and the emf of the other electrode is taken to be that of the cell. Such emf values, under standard conditions, are known as *standard electrode potentials* or *standard reduction potentials*, and are given the symbol $E°$. Alternatively, the standard hydrogen electrode may be placed on the right-hand side; the potential so obtained is known as the *standard oxidation potential*. The latter potentials are the standard electrode potentials with the signs reversed, the only difference being that the cells have been turned around.

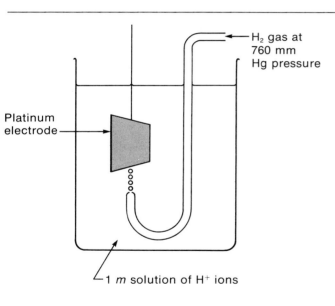

H₂ gas at 760 mm Hg pressure

Platinum electrode

1 *m* solution of H⁺ ions

Figure 8.2

The standard hydrogen electrode.

To illustrate the *standard electrode (reduction) potentials* of the I.U.P.A.C. convention, consider the voltaic cell shown in Figure 8.3. The left-hand electrode is the standard hydrogen electrode, with a hydrogen gas pressure of 1 atm and the acid solution 1 *m* in H⁺ ions. The right-hand electrode is the Cu^{2+} | Cu electrode, with the concentration of Cu^{2+} ions being 1 *m*. The two solutions are connected by a "salt bridge" such as a potassium chloride solution, which conducts electricity but does not allow bulk mixing of the two solutions. Alternatively, an agar gel containing KCl is commonly used. This procedure is somewhat more reliable than separating the two solutions by a porous partition, which itself sets up a small emf. The "salt bridge" minimizes this effect.

The voltaic cell shown in Figure 8.3 can be represented as follows:

$$Pt, \ H_2 \ | \ H^+(1m) \ \| \ Cu^{2+}(1m) \ | \ Cu$$

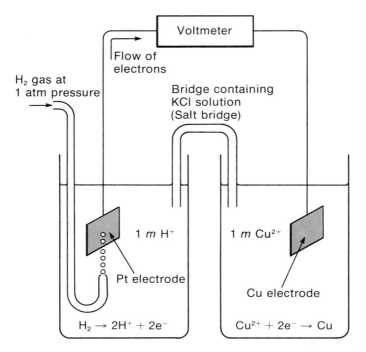

Figure 8.3

A voltaic cell in which a standard hydrogen electrode has been combined with a copper electrode immersed in a $1m$ solution of copper ions, the two solutions being connected by a potassium chloride bridge.

in which the double vertical lines represent the salt bridge. The observed emf is $+0.34$ V. The sign is positive by convention since electrons flow from left to right in the outer circuit. There is therefore a greater tendency for the process

$$Cu^{2+} + 2e^- \rightarrow Cu$$

to occur than for

$$2H^+ + 2e^- \rightarrow H_2$$

to occur; the latter process is forced to go in the reverse direction. By the I.U.P.A.C. convention, the $Cu^{2+} \mid Cu$ electrode being on the right, the standard electrode potential E° of this electrode is $+0.34$ V. Since this is a measure of the tendency for the cupric ions to be *reduced* by the process

$$Cu^{2+} + 2e^- \rightarrow Cu$$

these electrode potentials are also known as *standard reduction potentials*.

The opposite convention is to set up the cell in the reverse manner to that depicted in Figure 8.3; such a cell could be represented as

$$Cu \mid Cu^{2+} \parallel H^+ \mid Pt, H_2$$

where again the double vertical lines represent the salt bridge. The emf of this cell, under standard conditions ($[Cu^{2+}] = [H^+] = 1\ m$; hydrogen gas pressure $= 1$ atm), is obviously -0.34 V. We can write the electrode reactions in such a way as to correspond to the flow of electrons in the conventional direction:

$$Cu \rightarrow Cu^{2+} + 2e^-$$
$$2H^+ + 2e^- \rightarrow H_2$$

The value of -0.34 V is therefore a measure of the tendency of the reaction

$$Cu \rightarrow Cu^{2+} + 2e^-$$

to occur. Since it is an oxidation process, -0.34 V is therefore the *standard oxidation potential* of the $Cu^{2+} \mid Cu$ electrode.

In this book standard electrode (reduction) potentials will be employed, as recommended by the I.U.P.A.C. Table 8.1[†] gives a list of such potentials. The reactions are written as reduction processes, e.g., $Cu^{2+} + 2e^- \rightarrow Cu$. A table of standard oxidation potentials would simply reverse all of the signs and the corresponding reactions would be oxidations, e.g., $Cu \rightarrow Cu^{2+} + 2e^-$.

By combining the standard electrode potentials for two electrodes we can deduce the emf of a cell involving the two electrodes, neglecting the hydrogen electrode. Consider, for example, the following items in Table 8.1:

$$Cu^{2+} + 2e^- \rightarrow Cu \qquad E^\circ = 0.34\ V$$
$$Zn^{2+} + 2e^- \rightarrow Zn \qquad E^\circ = -0.76\ V$$

These values are the emf values for the following cells, in which the electrode processes are shown:

$$Pt, H_2 \mid H^+(1m) \parallel Cu^{2+}(1m) \mid Cu \qquad E^\circ = 0.34\ V$$
$$H_2 \rightarrow 2H^+ + 2e^- \qquad Cu^{2+} + 2e^- \rightarrow Cu$$
$$Pt, H_2 \mid H^+(1m) \parallel Zn^{2+}(1m) \mid Zn \qquad E^\circ = -0.76\ V$$
$$H_2 \rightarrow 2H^+ + 2e^- \qquad Zn^{2+} + 2e^- \rightarrow Zn$$

[†] Some of the values in this table were determined indirectly from other experimental results, since some of the hypothetical electrodes, e.g., Li $\mid$ Li$^+$, are impossible to set up.

Table 8.1   Standard Electrode (Reduction) Potentials[†]

Half Reaction	Standard Electrode Potential, E° (volts)
$F_2 + 2e^- \rightarrow 2F^-$	2.87
$H_2O_2 + 2H^+ + 2e^- \rightarrow 2H_2O$	1.77
$Au^+ + e^- \rightarrow Au$	1.68
$MnO_4^- + 8H^+ + 5e^- \rightarrow Mn^{2+} + 4H_2O$	1.52
$Cl_2 + 2e^- \rightarrow 2Cl^-$	1.36
$Cr_2O_7^{2-} + 14H^+ + 6e^- - 2Cr^{3+} + 7H_2O$	1.33
$MnO_2 + 4H^+ + 2e^- \rightarrow Mn^{2+} + 2H_2O$	1.23
$O_2 + 4H^+ + 4e^- \rightarrow 2H_2O$	1.23
$Pt^{2+} + 2e^- \rightarrow Pt$	1.20
$Br_2 + 2e^- \rightarrow 2Br^-$	1.06
$Hg^+ + 2e^- \rightarrow Hg$	0.85
$Ag^+ + e^- \rightarrow Ag$	0.80
$Hg_2^{2+} + 2e^- \rightarrow 2Hg$	0.79
$Fe^{3+} + e^- \rightarrow Fe^{2+}$	0.77
$O_2 + 2H^+ + 2e^- \rightarrow H_2O_2$	0.68
$I_2 + 2e^- \rightarrow 2I^-$	0.54
$Cu^{2+} + 2e^- \rightarrow Cu$	0.34
$Hg_2Cl_2 + 2e^- \rightarrow 2Hg + 2Cl^-$	0.3338
$AgCl(s) + e^- \rightarrow Ag + Cl^-$	0.2224
$Sn^{4+} + 2e^- \rightarrow Sn^{2+}$	0.15
$2H^+ + 2e^- \rightarrow H_2$	0.00 (by definition)
$Pb^{2+} + 2e^- \rightarrow Pb$	-0.13
$Sn^{2+} + 2e^- \rightarrow Sn$	-0.14
$Ni^{2+} + 2e^- \rightarrow Ni$	-0.25
$Co^{2+} + 2e^- \rightarrow Co$	-0.28
$Fe^{2+} + 2e^- \rightarrow Fe$	-0.44
$Cr^{3+} + 3e^- \rightarrow Cr$	-0.74
$Zn^{2+} + 2e^- \rightarrow Zn$	-0.76
$Al^{3+} + 3e^- \rightarrow Al$	-1.67
$Mg^{2+} + 2e^- \rightarrow Mg$	-2.34
$Na^+ + e^- \rightarrow Na$	-2.71
$Ca^{2+} + 2e^- \rightarrow Ce$	-2.87
$K^+ + e^- \rightarrow K$	-2.92
$Li^+ + e^- \rightarrow Li$	-3.04

[†] The standard oxidation potentials are the negatives of the values given above, the reactions being written in the opposite direction.

We could connect the two cells together as follows

$$\text{Zn} \mid \text{Zn}^{2+}(1m) \parallel \text{H}^+(1m) \mid \text{H}_2, \text{Pt} \quad\text{—}\quad \text{Pt}, \text{H}_2 \mid \text{H}^+(1m) \parallel \text{Cu}^{2+}(1m) \mid \text{Cu}$$

and the emf would then be $\Delta E^\circ = 0.34 - (-0.76) = 1.1$ V. We could also eliminate the hydrogen electrodes altogether, and set up the cell

$$\text{Zn} \mid \text{Zn}^{2+}(1m) \parallel \text{Cu}^{2+}(1m) \mid \text{Cu}$$

The emf of this would be the same, 1.1 V, since we have merely eliminated two identical hydrogen electrodes working in opposition to each other. This last cell is the standard Daniell cell (see Figure 8.1).

Note that in writing down the individual cell reactions it makes no difference whether they are written with one electron or more than one electron. Thus the hydrogen electrode reaction can be written either as

$$2\text{H}^+ + 2\text{e}^- \rightarrow \text{H}_2$$

or as

$$\text{H}^+ + \text{e}^- \rightarrow \tfrac{1}{2}\text{H}_2$$

However, in considering the overall process we must obviously balance the electrons. Thus for the cell

$$\text{Cu} \mid \text{Cu}^{2+} \parallel \text{H}^+ \mid \text{Pt H}_2$$

the individual reactions can be written as

$$\text{Cu} \rightarrow \text{Cu}^{2+} + 2\text{e}^-$$

and

$$2\text{H}^+ + 2\text{e}^- \rightarrow \text{H}_2$$

and the overall process is thus

$$\text{Cu} + 2\text{H}^+ \rightarrow \text{Cu}^{2+} + \text{H}_2$$

This process is accompanied by the passage of *two* electrons around the outer circuit. We could equally well write the reactions as

$$\tfrac{1}{2}\text{Cu} \rightarrow \tfrac{1}{2}\text{Cu}^{2+} + \text{e}^-$$

and

$$\text{H}^+ + \text{e}^- \rightarrow \tfrac{1}{2}\text{H}_2$$

so that the overall process is

$$\tfrac{1}{2}\text{Cu} + \text{H}^+ \rightarrow \tfrac{1}{2}\text{Cu}^{2+} + \tfrac{1}{2}\text{H}_2$$

This tells us that every time 0.5 mol of Cu disappears and 0.5 mol of Cu^{2+} appears, 1 mol of electrons passes from the left-hand electrode to the right-hand electrode.

8.3   OTHER STANDARD ELECTRODES

The standard hydrogen electrode is not the most convenient of electrodes to set up, because of the necessity of bubbling hydrogen over the platinum electrode. Several other electrodes are commonly used as secondary standard electrodes. One of these is the standard silver-silver chloride electrode, in which a silver electrode is in contact with solid silver chloride, which is a highly insoluble salt. The whole is immersed in potassium chloride solution in which the chloride ion concentration is 1 m. This electrode can be represented as

$$\text{Ag, AgCl} \mid \text{Cl}^-(1m)$$

We can set up a cell involving this electrode and the hydrogen electrode,

$$\text{Pt, H}_2 \mid \text{H}^+(1m) \parallel \text{Cl}^-(1m) \mid \text{AgCl, Ag}$$

with a salt bridge connecting the two solutions. The emf is found to be 0.2224 V. The individual reactions are

$$\tfrac{1}{2}\text{H}_2 \rightarrow \text{H}^+ + \text{e}^-$$
$$\text{e}^- + \text{AgCl} \rightarrow \text{Ag} + \text{Cl}^-$$

and the overall process is

$$\tfrac{1}{2}\text{H}_2 + \text{AgCl} \rightarrow \text{H}^+ + \text{Cl}^- + \text{Ag}$$

The standard electrode potential for the silver-silver chloride electrode is thus 0.2224 V.

Another commonly used electrode is the calomel electrode, illustrated in Figure 8.4a. In this, mercury is in contact with mercurous chloride (calomel, Hg_2Cl_2) immersed either in a 1 m solution of potassium chloride or in a saturated solution of potassium chloride. If the cell

$$\text{Pt, H}_2 \mid \text{H}^+(1m) \parallel \text{Cl}^-(1m) \mid \text{Hg}_2\text{Cl}_2, \text{Hg}$$

is set up, the individual reactions are

$$\tfrac{1}{2}\text{H}_2 \rightarrow \text{H}^+ + \text{e}^-$$
$$\text{e}^- + \tfrac{1}{2}\text{Hg}_2\text{Cl}_2 \rightarrow \text{Hg} + \text{Cl}^-$$

and the overall process is

$$\tfrac{1}{2}\text{H}_2 + \tfrac{1}{2}\text{Hg}_2\text{Cl}_2 \rightarrow \text{H}^+ + \text{Cl}^- + \text{Hg}$$

The emf at 25° C is 0.3338 V, which is thus the standard electrode potential E°. If a saturated solution of KCl is used with the calomel electrode, the standard electrode potential is 0.2415 V.

Another electrode commonly used as a secondary standard is the *glass electrode*, illustrated in Figure 8.4*b*. In its simplest form this consists of a tube terminating in a thin-walled glass bulb, the glass being reasonably permeable to ions. The glass bulb contains an 0.1 *m* hydrochloric acid solution and a tiny silver-silver chloride electrode. The theory of the glass electrode is somewhat complicated, but when the bulb is inserted into an acid solution, it behaves like a hydrogen electrode. This electrode is particularly convenient for making pH determinations.

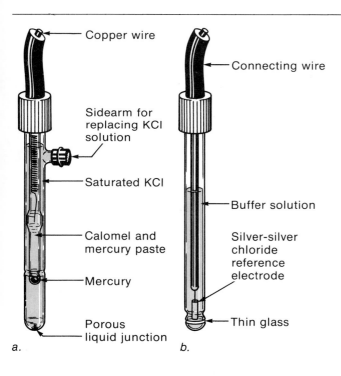

Copper wire

Connecting wire

Sidearm for replacing KCl solution

Saturated KCl

Buffer solution

Calomel and mercury paste

Silver-silver chloride reference electrode

Mercury

Porous liquid junction

Thin glass

a.

b.

Figure 8.4

A calomel electrode (*a*), and a glass electrode (*b*). The pH meter commonly used in chemical and biological laboratories often employs a glass electrode which is immersed in the unknown solution, and is used with a reference calomel electrode.

8.4 THERMODYNAMICS OF ELECTROCHEMICAL CELLS

Consider the standard electrochemical cell

$$\text{Pt, } H_2 \mid H^+(1m) \parallel Cu^{2+}(1m) \mid Cu$$

for which the emf (E°) is $+0.34$ V. We can write the overall reaction as

$$H_2 + Cu^{2+} \rightarrow 2H^+ + Cu$$

Every time one mole of H_2 reacts with one mole of Cu^{2+}, two moles of electrons pass through the outer circuit. According to Faraday's laws, this means the transfer of $2 \times 96\,500$ coulombs of electricity. The emf developed is $+0.34$ V, and the passage of $2 \times 96\,500$ coulombs across this potential drop means that $2 \times 96\,500 \times 0.34$ volt-coulomb of work has been carried out by the system. A volt-coulomb is one joule (J), and therefore the cell has performed

$$2 \times 96\,500 \times 0.34 \text{ J}$$
$$= 65\,620 \text{ J}$$
$$= 15.7 \text{ kcal}$$

of work each time 1 mol of H_2 reacts with 1 mol of Cu^{2+}. Since this emf of 0.34 V corresponds to reversible operation of the cell, this work is the maximum work.

We have seen in equation (5.100) that the amount of reversible non-PV work performed by a system at constant temperature and pressure is the decrease in Gibbs energy of the system. Therefore we may write for this cell process

$$\Delta G^\circ = -2 \times 96\,500 \times 0.34 \text{ J}$$
$$= -65\,620 \text{ J} = -15.7 \text{ kcal}$$

In general, for any standard-cell reaction associated with the passage of n electrons and an emf of ΔE°, the change in Gibbs energy is

$$\Delta G^\circ = -nF\Delta E^\circ \qquad (8.1)$$

where F is the faraday. Since this Gibbs-energy change is calculated from the ΔE° value, which relates to a cell in which the molalities are unity, it is a *standard* Gibbs-energy change, as indicated by the superscript $^\circ$.

The same argument applies to a cell in which concentrations are other than unity. Thus for any cell of emf ΔE

$$\Delta G = -nF\Delta E \qquad (8.2)$$

the ΔG being the increase in Gibbs energy when the reaction occurs with the concentrations having the values employed in the cell. Note that if ΔE is positive, ΔG is negative. A positive ΔE means that the cell is operating spontaneously, with the reactions occurring in the forward direction (e.g., $H_2 + Cu^{2+} \rightarrow 2H^+ + Cu$); this requires ΔG to be negative.

For any reaction

$$aA + bB + \cdots \rightarrow xX + yY + \cdots$$

the Gibbs-energy change which occurs when a moles of A at concentration $[A]$ react with b moles of B at concentration $[B]$, etc., is given by

$$\Delta G = -RT\left[\ln K - \ln \frac{[X]^x[Y]^y\cdots}{[A]^a[B]^b\cdots}\right] \qquad (8.3)$$

(see equation (5.117)).† If the initial and final concentrations are unity, this equation reduces to

$$\Delta G^\circ = -RT \ln K \tag{8.4}$$

where ΔG° is the standard Gibbs-energy change. This is the situation which exists with a cell involving standard electrodes, such as

$$\text{Pt, } H_2 \mid H^+(1m) \parallel Cu^{2+}(1m) \mid Cu$$

Since for such a standard cell, $\Delta G^\circ = -nFE^\circ$, it follows that

$$-nFE^\circ = -RT \ln K \tag{8.5}$$

Thus

$$E^\circ = \frac{RT}{nF} \ln K \tag{8.6}$$

$$= \frac{0.059}{n} \log_{10} K \qquad \text{at } 25^\circ C \tag{8.7}$$

The same type of relationship is obeyed if one of the electrodes is not the standard hydrogen electrode. Consider the standard Daniell cell, for example:

$$\text{Zn} \mid Zn^{2+}(1m) \parallel Cu^{2+}(1m) \mid Cu$$

The emf, now written as ΔE°, since it is the difference between the standard electrode potentials for the two electrodes, is related to the equilibrium constant K for the overall reaction

$$\text{Zn} + Cu^{2+} \rightleftharpoons Zn^{2+} + Cu$$

by the equation

$$\Delta E^\circ = \frac{RT}{2F} \ln K \tag{8.8a}$$

This becomes, at $25^\circ C$

$$\Delta E^\circ = \frac{0.059}{2} \log_{10} K \tag{8.8b}$$

---

† This is the approximate relationship, in which concentrations are used, rather than activities.

Since ΔE° for this cell is 1.10 V

$$\log_{10} K = \frac{2 \times 1.10}{0.059} = 37.2$$

and thus

$$K = 1.5 \times 10^{37}$$

This is a very important method for calculating equilibrium constants and Gibbs-energy changes. Its extension to the calculation of ΔH° and ΔS° values is considered on p. 360.

EXAMPLE

Using the data in Table 8.1, calculate the equilibrium constant K for the reaction

$$Sn + Fe^{2+} \rightleftharpoons Sn^{2} + Fe$$

SOLUTION

From Table 8.1, the standard electrode potentials are

$$Sn^{2+} + 2e^{-} \rightarrow Sn \qquad E^{\circ} = -0.14 \text{ V}$$
$$Fe^{2+} + 2e^{-} \rightarrow Fe \qquad E^{\circ} = -0.44 \text{ V}$$

The ΔE° value for the process

$$Sn + Fe^{2+} \rightarrow Sn^{2+} + Fe$$

is therefore $-0.44 - (-0.14) = -0.30$ V, and the reaction corresponds to the passage of two electrons ($n = 2$). Thus we have from equation (8.8b),

$$-0.30 = \frac{0.059}{2} \log_{10} K$$

or

$$\log_{10} K = -\frac{0.60}{0.059} = -10.2$$

Thus

$$K = 6.3 \times 10^{-11}$$

EXAMPLE

Using the data in Table 8.1, calculate the equilibrium constant for the reaction

$$H_2 + 2Fe^{3+} \rightleftharpoons 2H^+ + 2Fe^{2+}$$

SOLUTION

From Table 8.1, the standard electrode potentials are

$$2H^+ + 2e^- \rightarrow H_2 \qquad E^\circ = 0$$
$$Fe^{3+} + e^- \rightarrow Fe^{2+} \qquad E^\circ = 0.77 \text{ V}$$

The ΔE° value for the process,

$$H_2 + 2Fe^{3+} \rightarrow 2H^+ + 2Fe^{2+}$$

for which $n = 2$, is thus $0.77 - (0) = 0.77$ V. Therefore

$$0.77 = \frac{0.059}{2} \log_{10} K$$

or

$$\log_{10} K = 26.1$$

Thus

$$K = 1.26 \times 10^{26}$$

It should be emphasized that in making this calculation we must *not* multiply the value of 0.77 V by two. The emf of 0.77 V applies equally well to the process

$$2Fe^{3+} + 2e^- \rightarrow 2Fe^{2+}$$

If the problem had been to calculate the equilibrium constant K' for the process

$$\tfrac{1}{2}H_2 + Fe^{3+} \rightleftharpoons H^+ + Fe^{2+}$$

the reactions would have been written as

$$H^+ + e^- \rightarrow \tfrac{1}{2}H_2 \qquad E^\circ = 0$$
$$Fe^{3+} + e^- \rightarrow Fe^{2+} \qquad E^\circ = 0.77 \text{ V}$$

and again $\Delta E^\circ = 0.77$ V. In this case $n = 1$ so that

$$0.77 = \frac{0.059}{1} \log_{10} K'$$

or

$$\log_{10} K' = \frac{0.77}{0.059} = 13.05$$

and therefore

$$K' = 1.12 \times 10^{13}$$

K' is, of course, the square root of K.

EXAMPLE

Calculate the E° value for the process

$$Cu^+ + e^- \rightarrow Cu$$

making use of the following E° values:

(1) $Cu^{2+} + e^- \rightarrow Cu^+$ $E_1^\circ = 0.15$ V

(2) $Cu^{2+} + 2e^- \rightarrow Cu$ $E_2^\circ = 0.34$ V

SOLUTION

The ΔG° values for the latter two reactions are:

$$Cu^{2+} + e^- \rightarrow Cu^+$$
$$\Delta G_1^\circ = -nE_1^\circ F = -1 \times 0.15 \times 96\,500 \text{ J mol}^{-1}$$
$$Cu^{2+} + 2e^- \rightarrow Cu$$
$$\Delta G_2^\circ = -nE_2^\circ F = 2 \times 0.34 \times 96\,500 \text{ J mol}^{-1}$$

The reaction $Cu^+ + e^- \rightarrow Cu$ is obtained by subtracting reaction (1) from reaction (2) and the ΔG° value for $Cu^+ + e^- \rightarrow Cu$ is therefore obtained by subtracting ΔG_1° from ΔG_2°:

$$\Delta G^\circ = -2 \times 0.34 \times 96\,500 - (-1 \times 0.15 \times 96\,500) \text{ J mol}^{-1}$$
$$= (0.15 - 0.68)\,96\,500 \text{ J mol}^{-1}$$
$$= 0.53 \times 96\,500 \text{ J mol}^{-1}$$

Since, for $Cu^+ + e^- \rightarrow Cu$, $n = 1$, it follows that

$$E^\circ = 0.53 \text{ V}$$

In working this problem it is incorrect to directly combine the E° values; they must first be converted into ΔG° values.

In view of this, the student may wonder why then it is legitimate to calculate ΔE° values for overall cell reactions by simply combining the E° values for the individual electrodes. Consider, for example, the following E° values:

$$Fe^{3+} + e^- \rightarrow Fe^{2+} \qquad E^\circ = 0.77 \text{ V}$$
$$I_2 + 2e^- \rightarrow 2I^- \qquad E^\circ = 0.54 \text{ V}$$

In preceding examples we combined two E° values to obtain ΔE° values:

$$2Fe^{3+} + 2I^- \rightarrow 2Fe^{2+} + I_2 \qquad \Delta E^\circ = 0.77 - 0.54$$
$$= 0.23 \text{ V}.$$

The fact that this is justified can be seen by writing down the ΔG° values:

$$Fe^{3+} + e^- \rightarrow Fe^{2+} \qquad \Delta G^\circ = -1 \times 0.77 \times 96\,500 \text{ J}$$
$$I_2 + 2e^- \rightarrow 2I^- \qquad \Delta G^\circ = -2 \times 0.54 \times 96\,500 \text{ J}$$

We combine the two equations by multiplying the first by 2 and subtracting the second:

$$2Fe^{3+} + 2I^- \rightarrow Fe^{2+} + 2I^-$$
$$\Delta G^\circ = 2(-1 \times 0.77 \times 96\,500) - (-2 \times 0.54 \times 96\,500)$$
$$= -2 \times 0.23 \times 96\,500 \text{ J}$$

so that $\Delta E^\circ = 0.23$ V, which is simply $0.77 - 0.54$. Thus, we are justified in simply subtracting E° to find ΔE° for an overall reaction in which there are no electrons left over. However, to obtain E° for a half-reaction (as in the last example above) in general we *cannot* simply combine E° values, but must calculate the ΔG° values individually, as was done above.

8.5 THE NERNST EQUATION

Thus far we have limited our discussion to standard electrode potentials, E°, and to ΔE° values for cells in which the active species are present at 1 m concentrations. The corresponding standard Gibbs energies have been written as ΔG°.

Let us now remove this restriction and consider cells in which the concentrations are other than unity. Consider, for example, the cell

$$\text{Pt, H}_2 \mid \text{H}^+(1m) \parallel \text{Cu}^{2+} \mid \text{Cu}$$

in which the standard hydrogen electrode has been combined with a copper electrode immersed in a Cu^{2+} solution, the concentration of which is other than unity. The overall cell reaction is

$$\text{H}_2 + \text{Cu}^{2+} \rightarrow 2\text{H}^+ + \text{Cu}$$

and the Gibbs-energy change [see equation (8.3)] is

$$\Delta G = -RT\left[\ln K - \ln \frac{[\text{H}^+]^2}{[\text{Cu}^{2+}]}\right] \tag{8.9}$$

$$= -RT\left[\ln K - \ln \frac{1}{[\text{Cu}^{2+}]}\right] \tag{8.10}$$

since $[\text{H}^+] = 1m$. However, $\Delta G^\circ = -RT \ln K = -nE^\circ F$, so that

$$\Delta G = -nE^\circ F + RT \ln \frac{1}{[\text{Cu}^{2+}]} \tag{8.11}$$

Since $\Delta G = -nEF$, where E is the emf of this cell, we obtain

$$-nEF = -nE^\circ F + RT \ln \frac{1}{[\text{Cu}^{2+}]} \tag{8.12}$$

and thus

$$E = E^\circ - \frac{RT}{nF} \ln \frac{1}{[\text{Cu}^{2+}]} \tag{8.13}$$

In general, we may consider any cell for which the overall reaction has the general form

$$a\text{A} + b\text{B} + \cdots \rightarrow x\text{X} + y\text{Y} + \cdots$$

ΔG is given by equation (8.3) and $\Delta G^\circ = -RT \ln K = -n\Delta E^\circ F$, whence

$$G = -n\Delta E^\circ F + RT \ln \frac{[\text{X}]^x[\text{Y}]^y \cdots}{[\text{A}]^a[\text{B}]^b \cdots} \tag{8.14}$$

Note that, as in an equilibrium constant, products are in the numerator and reactants in the denominator. Since $\Delta G = -n\Delta EF$ this leads to the following expression for the emf ΔE:

$$\Delta E = \Delta E^\circ - \frac{RT}{nF} \ln \frac{[X]^x[Y]^y}{[A]^a[B]^b} \tag{8.15}$$

This general relationship was first given in 1889 by the German physical chemist Walter H. Nernst (1864–1941), and is known as the *Nernst equation*. Equation (8.13) is a simple form of it, applicable to a cell in which one electrode is the standard hydrogen electrode.

As an example, suppose that we apply the Nernst equation to the cell

$$\text{Zn} \mid \text{Zn}^{2+} \parallel \text{Ni}^{2+} \mid \text{Ni}$$

for which the overall reaction is

$$\text{Zn} + \text{Ni}^{2+} \rightleftharpoons \text{Zn}^{2+} + \text{Ni}$$

The standard electrode potentials (see Table 8.1) are

$$\text{Ni}^{2+} + 2\text{e}^- \rightarrow \text{Ni} \qquad E^\circ = -0.25 \text{ V}$$
$$\text{Zn}^{2+} + 2\text{e}^- \rightarrow \text{Zn} \qquad E^\circ = -0.76 \text{ V}$$

and ΔE° for the overall process is therefore $-0.25 - (-0.76) = 0.51$ V. The Nernst equation is thus

$$\Delta E = 0.51 - \frac{RT}{nF} \ln \frac{[\text{Zn}^{2+}]}{[\text{Ni}^{2+}]} \tag{8.16}$$

As always the concentrations of solid species such as Zn and Ni are incorporated into the equilibrium constant, and are therefore not included explicitly in the equation. For water at 25° C this equation becomes

$$\Delta E = 0.51 - \frac{0.059}{2} \log_{10} \frac{[\text{Zn}^{2+}]}{[\text{Ni}^{2+}]} \tag{8.17}$$

since $n = 2$. We can see from this equation that increasing the ratio $[\text{Zn}^{2+}]/[\text{Ni}^{2+}]$ decreases the cell emf; this is understandable in view of the fact that a positive emf means that the cell is producing Zn^{2+}, and that Ni^{2+} ions are being removed.

EXAMPLE
Calculate the emf of the cell

$$\text{Co} \mid \text{Co}^{2+} \parallel \text{Ni}^{2+} \mid \text{Ni}$$

if the concentrations are

(1) $[Ni^{2+}] = 1\ m$ and $[Co^{2+}] = 0.1\ m$

(2) $[Ni^{2+}] = 0.01\ m$ and $[Co^{2+}] = 1.0\ m$

SOLUTION

The cell reaction is

$$Co + Ni^{2+} \rightarrow Co^{2+} + Ni$$

and from Table 8.1 the standard electrode potentials are

$$Ni^{2+} + 2e^- \rightarrow Ni \qquad E^o = -0.25\ V$$

$$Co^{2+} + 2e^- \rightarrow Co \qquad E^o = -0.28\ V$$

The cell emf, ΔE^o, is thus $-0.25 - (-0.28) = 0.03$ V, and $n = 2$.
The cell emf at the concentrations specified in (1) is

$$\Delta E = 0.03 - \frac{0.059}{2} \log_{10} \frac{[Co^{2+}]}{[Ni^{2+}]}$$

$$= 0.03 - \frac{0.059}{2} \log_{10} 0.1$$

$$= 0.03 + 0.03 = 0.06\ V.$$

In case (2),

$$\Delta E = 0.03 - \frac{0.059}{2} \log_{10} \frac{1}{0.01}$$

$$= 0.03 - 0.059 = -0.029\ V$$

We see that the cell operates in opposite directions in the two
cases.

8.6   CONCENTRATION CELLS

The cells considered thus far are ones in which there has been a net chemical change
during the production of emf. It is also possible to set up a cell in which there is
no net chemical change, the net effect being the transfer of a species from one solution
to another, in which it is at a different concentration. There is a natural tendency
for a species to pass from a solution in which it is at a high concentration to one
in which it is at a lower concentration. It experiences a decrease in Gibbs energy

in doing so, and the emf results from this Gibbs-energy change, its value being equal to $-\Delta G/nF$. Cells which generate an emf by virtue of a simple change in concentration of a species are known as *concentration cells*.

A simple example of a concentration cell is obtained by connecting two hydrogen electrodes by means of a salt bridge

$$\text{Pt, H}_2 \mid \text{HCl}\,(c_1) \parallel \text{HCl}\,(c_2) \mid \text{H}_2\text{, Pt}$$

The salt bridge could be a tube containing saturated potassium chloride solution, the effect of which is to eliminate any potential at the boundary between the two solutions. The reaction at the left-hand electrode is

$$\tfrac{1}{2}\text{H}_2 \rightarrow \text{H}^+\,(c_1) + \text{e}^-$$

while that at the right-hand electrode is

$$\text{H}^+\,(c_2) + \text{e}^- \rightarrow \tfrac{1}{2}\text{H}_2$$

The net process is therefore

$$\text{H}^+\,(c_2) \rightarrow \text{H}^+\,(c_1)$$

and is simply the transfer of hydrogen ions from a solution of concentration c_2 to one of concentration c_1. If c_2 is greater than c_1 the process will actually occur in this direction and a positive emf will be produced; if c_2 is less than c_1 the emf is negative, and electrons will flow from the right-hand to the left-hand electrode.

The Gibbs-energy change associated with the transfer of H^+ ions from a concentration c_2 to a concentration c_1 is simply[†]

$$\Delta G = RT \ln \frac{c_1}{c_2} \tag{8.18}$$

Since $n = 1$, the emf produced is

$$\Delta E = \frac{RT}{F} \ln \frac{c_2}{c_1} \tag{8.19}$$

This gives a positive emf when $c_2 > c_1$.

---

† This equation follows at once from equation (5.105) on p. 218. Thus $G_1 = G^\circ + RT \ln c_1$ and $G_2 = G^\circ + RT \ln c_2$, whence $\Delta G = G_1 - G_2 = RT \ln (c_1/c_2)$. Note also that equation (8.18) can be obtained from equation (5.116) on p. 219, which leads to $\Delta G = -RT \ln K + RT \ln (c_1/c_2)$. There is no chemical change, so that $K = 1$.

EXAMPLE

Calculate the emf of a cell of the above type in which $c_1 = 0.2\ m$ and $c_2 = 3\ m$; $T = 25°\,C$.

SOLUTION

The emf is given by

$$\Delta E = \frac{RT}{F}\ln\frac{c_2}{c_1}$$

$$= 0.059\ \log_{10}\frac{3.0}{0.2} = 0.0069\ V.$$

8.7   ACTIVITY COEFFICIENTS

So far the equations for Gibbs-energy changes and emf values have been expressed in terms of concentrations. This is an approximation and the errors become serious as concentrations are increased. For a correct formulation *activities* must be employed, the activity being the concentration multiplied by an *activity coefficient*. We will do this for a simple cell and will find that the emf measurements over a range of concentrations lead to values for the activity coefficients.

Consider the cell

$$\text{Pt, H}_2\ |\ \text{HCl}\ |\ \text{AgCl, Ag}$$

The overall process is

$$\tfrac{1}{2}\text{H}_2 + \text{AgCl} \rightarrow \text{Ag} + \text{H}^+ + \text{Cl}^- \qquad (n = 1)$$

and the Gibbs-energy change is

$$\Delta G = \Delta G° + RT\ln a_+ a_- \tag{8.20}$$

where a_+ and a_- are the activities of the H^+ and Cl^- ions, and $\Delta G°$ is the standard Gibbs-energy change when the activities are unity. The emf is

$$E = E° - \frac{RT}{F}\ln a_+ a_- \tag{8.21}$$

Since it is impossible to determine individual activities and activity coefficients, it is convenient to define a mean activity $a_\pm$ as the geometric mean of a_+ and a_-:

$$a_\pm^2 = a_+ a_- \tag{8.22}$$

Thus we can write the emf as

$$E = E^\circ - \frac{RT}{F} \ln a_\pm^2 \tag{8.23}$$

$$= E^\circ - \frac{2RT}{F} \ln m - \frac{2RT}{F} \ln \gamma_\pm \tag{8.24}$$

since $a_\pm$ is the molality m multiplied by the mean activity coefficient $\gamma_\pm$.

This equation provides a means of determining E° for this cell, and values of $\gamma_\pm$ at various concentrations of HCl. The equation can be written as

$$E + \frac{2RT}{F} \ln m = E^\circ - \frac{2RT}{F} \ln \gamma_\pm \tag{8.25}$$

The emf E can be measured at various molalities m of HCl, and the quantity on the left-hand side can therefore be calculated at various molalities. If this quantity is plotted against m, as shown schematically in Figure 8.5, the value extrapolated to zero m gives E°, since at zero m the activity coefficient $\gamma_\pm$ is unity, and the final term vanishes. At any molality the ordinate minus E° then yields

$$\frac{-2RT}{F} \ln \gamma_\pm$$

from which the activity coefficient $\gamma_\pm$ can be calculated (see Fig. 8.5.).

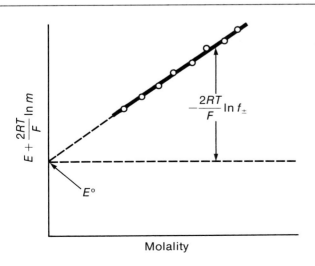

Figure 8.5

A plot which will provide the value of E° for a cell, and the mean activity coefficients at various concentrations.

8.8   REDOX SYSTEMS

All of the electrode processes discussed previously in this chapter involve the occurrence of oxidations or reductions at electrodes, since electrons are given up or accepted. In most of the processes thus far considered, the electrode system (e.g., the H_2 gas in the hydrogen electrode) has participated directly in the oxidation or reduction process. There is also an important class of oxidation-reduction systems, known as *redox* systems, in which both the oxidized and reduced species are in solution, their interconversion being effected by an inert electrode such as one of platinum. Consider, for example, the cell

$$Pt, H_2 \mid H^+ \ (1m) \parallel Fe^{2+}, Fe^{3+} \mid Pt$$

The left-hand electrode is the standard hydrogen electrode and the right-hand electrode consists of a platinum electrode immersed in a solution containing both Fe^{2+} and Fe^{3+} ions. The platinum electrode is able to catalyze the interconversion of these ions, and the reaction at this electrode is

$$e^- + Fe^{3+} \rightarrow Fe^{2+}$$

Since the reaction at the hydrogen electrode is

$$\tfrac{1}{2}H_2 \rightarrow H^+ + e^-$$

the overall process is

$$Fe^{3+} + \tfrac{1}{2}H_2 \rightarrow Fe^{2+} + H^+$$

The emf of this cell represents the ease with which the Fe^{3+} is reduced to Fe^{2+}. The emf of the cell is

$$E = E^\circ - \frac{RT}{F} \ln \frac{[Fe^{2+}]}{[Fe^{3+}]} \tag{8.26}$$

(Note that the E° relates to $p_{H_2} = 1$ atm and $[H^+] = [Fe^{2+}] = [Fe^{3+}] = 1$ m). The E° for this system is $+0.77$ volts (Table 8.1). A similar system, important in biology, involves the oxidized and reduced forms of cytochrome c. Since the former contains iron in its ferric (Fe^{3+}) form and the latter in its ferrous (Fe^{2+}) form the reaction may be written as

$$cytochrome \ c \ (Fe^{3+}) + e^- \rightarrow cytochrome \ c \ (Fe^{2+})$$

REDOX SYSTEMS INVOLVING HYDROGEN IONS

In biological systems, the interconversion of oxidized and reduced forms frequently involves the participation of hydrogen ions. Thus, the half reaction for the reduction of fumarate ions to succinate ions is

$$\begin{array}{c}\text{CHCOO}^- \\ \| \\ \text{CHCOO}^-\end{array} + 2\text{H}^+ + 2\text{e}^- \rightarrow \begin{array}{c}\text{CH}_2\text{COO}^- \\ | \\ \text{CH}_2\text{COO}^-\end{array}$$

$$\qquad\qquad \text{fumarate} \qquad\qquad\qquad\qquad \text{succinate}$$

If we wished to study this system we could set up the following cell:

$$\text{Pt, H}_2 \mid \text{H}^+ (1m) \parallel \text{F}^{2-}, \text{S}^{2-}, \text{H}^+([\text{H}^+]_r) \mid \text{Pt}$$

where F^{2-} and S^{2-} represent fumarate and succinate. The hydrogen ion concentration in the right-hand solution is not necessarily $1\ m$; it will here be denoted as $[\text{H}^+]_r$. If we combine the above equation for the fumarate-succinate half reaction with that for the standard hydrogen electrode

$$\text{H}_2 \rightarrow 2\text{H}^+ (1m) + 2\text{e}^-$$

we obtain, for the overall cell reaction,

$$\text{F}^{2-} + 2\text{H}^+([\text{H}^+]_r) + \text{H}_2 \rightarrow \text{S}^{2-} + 2\text{H}^+ (1m)$$

The equation for the emf is thus

$$E = E^\circ - \frac{RT}{2F} \ln \frac{[\text{S}^{2-}]}{[\text{F}^{2-}][\text{H}^+]_r^2} \qquad (8.27)$$

The E° for this system is related to a standard Gibbs-energy change ΔG° by the usual equation

$$\Delta G^\circ = -nFE^\circ \qquad (8.28)$$

This standard Gibbs-energy change corresponds to the equilibrium constant

$$K = \frac{[\text{S}^{2-}]}{[\text{F}^{2-}][\text{H}^+]_r^2} \qquad (8.29)$$

However, in biological studies it is frequently convenient to deal with the modified equilibrium constant

$$K' = \frac{[\text{S}^{2-}]}{[\text{F}^{2-}]} \qquad (8.30)$$

at some specified hydrogen ion concentration. Often this standard concentration is taken to be 10^{-7} M, corresponding to a pH of 7. In that case

$$K' = (10^{-7})^2 K = 10^{-14} K \qquad (8.31)$$

Table 8.2 Standard Electrode Potentials for Reactions of Biological Interest, at pH 7†

Half Reaction	Standard Electrode Potential at pH 7 and 25° C, $E^{o'}$ (volts)
$NO_3^- + 2H^+ + e^- \rightarrow NO_2^- + H_2O$	0.421
Cytochrome a $(Fe^{3+}) + e^- \rightarrow$ cytochrome a (Fe^{2+})	0.290
Cytochrome c $(Fe^{3+}) + e^- \rightarrow$ cytochrome c (Fe^{2+})	0.254
Butyryl-coenzyme-A $+ 2H^+ + 2e^- \rightarrow$ crotonyl-CoA	0.190
Hemoglobin $(Fe^{3+}) + e^- \rightarrow$ hemoglobin (Fe^{2+})	0.170
Cytochrome b $(Fe^{3+}) + e^- \rightarrow$ cytochrome b (Fe^{2+})	0.077
$\begin{matrix} CHCOO^- \\ \| \\ CHCOO^- \end{matrix} + 2H^+ + 2e^- \rightarrow \begin{matrix} CH_2COO^- \\ \| \\ CH_2COO^- \end{matrix}$ fumarate $\qquad$ succinate	0.031
$HCOCOO^- + 2H^+ + 2e^- \rightarrow HOCH_2COO^-$ glyoxylate $\qquad$ glycollate	-0.090
$\begin{matrix} COCOO^- \\ \| \\ CH_2COO^- \end{matrix} + 2H^+ + 2e^- \rightarrow \begin{matrix} CHOHCOO^- \\ \| \\ CH_2COO^- \end{matrix}$ oxaloacetate $\qquad$ malate	-0.166
$CH_3COCOO^- + 2H^+ + 2e^- \rightarrow CH_3CHOHCOO^-$ pyruvate $\qquad$ lactate	-0.185
$CH_3CHO + 2H^+ + 2e^- \rightarrow C_2H_5OH$ acetaldehyde $\qquad$ ethanol	-0.197
$NAD^+ + H^+ + 2e^- \rightarrow NADH$	-0.320
$CO_2 + H^+ + 2e^- \rightarrow HCOO^-$ formate	-0.420
$2H^+ + 2e^- \rightarrow H_2$	-0.421
$CH_3COO^- + 2H^+ + e^- \rightarrow CH_3CHO + H_2O$ acetate $\qquad$ acetaldehyde	-0.581

† These values have been obtained from W. M. Clark, *Oxidation-Reduction Potentials of Organic Systems* (Baltimore: Williams and Wilkins Co., 1960), and P. A. Loach, "Oxidation-Reduction Potentials," in *Handbook of Biochemistry* (Cleveland, Ohio: Chemical Rubber Co., 1970, pp. J–33). It should be noted that there are rather large discrepancies between $E^{o'}$ values given by different workers. For reliable estimates of $\Delta G^{o'}$ values, $E^{o'}$ values should be known to the third decimal place. Very few reliable values of $\partial \Delta E^{o'}/\partial T$ are available.

In many cases the K' value corresponds to a fairly well-balanced equilibrium at pH 7, whereas K will be larger by the factor 10^{14}; the K' value and the corresponding $\Delta G^{o'}$ at pH 7 therefore give a clearer indication of the situation at that pH.

Equation (8.27) for the emf can be written as

$$E = E^o - \frac{RT}{2F} \ln \frac{[S^{2-}]}{[F^{2-}]} + \frac{2.303RT}{F} \log_{10}[H^+]_r \qquad (8.32)$$

$$= E^o - \frac{RT}{2F} \ln \frac{[S^{2-}]}{[F^{2-}]} + 0.059 \log_{10}[H^+]_r \qquad \text{(at } 25°\text{ C)} \quad (8.33)$$

$$= E^o - \frac{RT}{2F} \ln \frac{[S^{2-}]}{[F^{2-}]} - 0.059 \text{ pH}_r \qquad (8.34)$$

where pH_r is the pH of the solution in which the $S^{2-} : F^{2-}$ system is maintained. Then, if we define a modified standard potential $E^{o'}$ by

$$E = E^{o'} - \frac{RT}{2F} \ln \frac{[S^{2-}]}{[F^2]} \qquad (8.35)$$

it follows that†

$$E^{o'} = E^o - 0.059 \text{ pH}_r \qquad (8.36)$$

This modified standard electrode potential then relates to the equilibrium constant K' at the particular pH of the solution; thus

$$\Delta G^{o'} = -nFE^{o'} \qquad (8.37)$$

and

$$K' = e^{-\Delta G^{o'}/RT} \qquad (8.38)$$

Some values of $E^{o'}$ at pH 7 are given in Table 8.2.

An exactly analogous procedure can be adopted if we combine two oxidation-reduction systems together instead of combining each with the standard hydrogen electrode. Thus, consider the reaction

$$\text{malate}^{2-} + \text{NAD}^+ \rightleftharpoons \text{oxaloacetate}^{2-} + \text{NADH} + \text{H}^+$$

where NAD^+ is the oxidized form and NADH the reduced form of nicotinamide

---

† It must be emphasized that equation (8.36) is not general but applicable only to equilibria of this particular type. Each case must be worked out separately (see Appendix B, p. 563).

adenine dinucleotide. The two half reactions, written as reductions, are

(1)
$$\begin{array}{c}\text{COCOO}^- \\ | \\ \text{CH}_2\text{COO}^- \\ \text{oxaloacetate}\end{array} + 2\text{H}^+ + 2\text{e}^- \rightarrow \begin{array}{c}\text{CHOHCOO}^- \\ | \\ \text{CH}_2\text{COO}^- \\ \text{malate}\end{array}$$

and

(2) $NAD^+ + H^+ + 2e^- \rightarrow NADH$

Each of these reactions has an E^o value, and each has a corresponding $E^{o'}$ value at pH 7 (see Table 8.2). The $\Delta E^{o'}$ value at pH 7 for the reaction with which we are concerned is $E_2^{o'} - E_1^{o'}$. The corresponding $\Delta G^{o'}$ and K' values for the process at pH 7 are thus

$$\Delta G^{o'} = -nF\Delta E^{o'} = -nF(E_2^{o'} - E_1^{o'}) \tag{8.39}$$

and

$$K' = e^{-\Delta G^{o'}/RT} = e^{nF\Delta E^{o'}/RT} \tag{8.40}$$

or

$$\log_{10} K' = \frac{n\Delta E^{o'}}{0.059} \qquad \text{(at 25° C)} \quad (8.41)$$

The K' value corresponds to the equilibrium ratio

$$\frac{[\text{oxaloacetate}^{2-}][\text{NADH}]}{[\text{malate}^{2-}][\text{NAD}^+]}$$

at pH 7. Use of the values in Table 8.2 leads to $K' = 1.73 \times 10^{-6}$.

EXAMPLE
Calculate $\Delta G^{o'}$ at 25° C for the reaction

$$\text{pyruvate}^- + \text{NADH} + \text{H}^+ \rightarrow \text{lactate}^- + \text{NAD}^+$$

using the standard electrode potentials given in Table 8.2. Also, calculate the equilibrium ratio

$$\frac{[\text{lactate}^-][\text{NAD}^+]}{[\text{pyruvate}^-][\text{NAD}]}$$

(a) at pH 7.0, and (b) at pH 8.0.

SOLUTION

The $E^{o'}$ values for the half reactions (Table 8.2) are

(1)   pyruvate$^-$ + 2H$^+$ + 2e$^-$ → lactate$^-$ $E^{o'} = -0.19$ V

(2)   NAD$^+$ + H$^+$ + 2e$^-$ → NADH $E^{o'} = -0.34$ V

The required reaction is obtained by subtracting (b) from (a).
Hence

$$\Delta E^{o'} = -0.19 - (-0.34) = 0.15 \text{ V}$$

Since $n = 2$, the corresponding $\Delta G^{o'}$ is

$$\Delta G^{o'} = -2 \times 96\,500 \times 0.15 \text{ J}$$
$$= -28\,950 \text{ J}$$
$$= -6.92 \text{ kcal.}$$

(a) The equilibrium ratio at pH 7 can be calculated from the relationship

$$\Delta G^{o'} = -4.57 \times T \log_{10} K'$$
$$-6920 = -4.57 \times 298 \log_{10} K'$$
$$\log_{10} K' = 5.08$$
$$K' = 1.21 \times 10^5$$

Alternatively, K' could have been calculated directly from $\Delta E^{o'}$, using equation (8.8):

$$0.15 = \frac{0.059}{2} \log_{10} K'$$
$$\log_{10} K' = 5.08$$
$$K' = 1.21 \times 10^5$$

This K' at pH 7.0 is related to the true (pH-independent) K by the equation

$$K = \frac{[\text{lactate}^-][\text{NAD}^+]}{[\text{pyruvate}^-][\text{NADH}][\text{H}^+]} = \frac{K'}{[\text{H}^+]} = \frac{K'}{10^{-7}}$$

(b) Similarly the K'' at pH 8.0 is related to K by

$$K = \frac{K''}{10^{-8}}$$

Thus

$$\frac{K''}{10^{-8}} = \frac{K'}{10^{-7}}$$

and

$$K'' = 1.21 \times 10^5 \times \frac{10^{-8}}{10^{-7}}$$

$$= 1.21 \times 10^4$$

8.9 TEMPERATURE COEFFICIENTS OF CELL POTENTIALS: ENTHALPY AND ENTROPY CHANGES

We have seen that the standard emf of a reversible cell is related to the standard Gibbs-energy change by equation (8.1). If Gibbs-energy changes are measured over a range of temperature, the $\Delta S°$ and $\Delta H°$ values can be calculated.

The basic relationship is equation (5.154) on p. 232:

$$\Delta S = -\left(\frac{\partial \Delta G}{\partial T}\right)_P \tag{8.42}$$

If the reactants and products of a reaction are in their standard states

$$\Delta S° = -\left(\frac{\partial \Delta G°}{\partial T}\right)_P \tag{8.43}$$

The standard Gibbs-energy change is related to the standard emf of a cell by

$$\Delta G° = -nF\Delta E° \tag{8.44}$$

and therefore

$$\Delta S° = nF\left(\frac{\partial \Delta E°}{\partial T}\right)_P \tag{8.45}$$

For a system at constant pressure, the standard enthalpy change is thus

$$\Delta H° = \Delta G° + T\Delta S° \tag{8.46}$$

$$= -nF\left(\Delta E° - T\frac{\partial \Delta E°}{\partial T}\right) \tag{8.47}$$

The measurement of emf values at various temperatures provides a very convenient method of obtaining thermodynamic values for chemical reactions and has been employed for some reactions of biological interest.

EXAMPLE

Calculate from the values in Table 8.2 the standard cell potential, ΔE°, for the reaction

$$\text{fumarate}^{2-} + \text{lactate}^- \rightarrow \text{succinate}^{2-} + \text{pyruvate}^-$$

The temperature coefficient of the standard potential, $\partial \Delta E^\circ / \partial T$, is found to be 2.18×10^{-5} V K^{-1}. Calculate ΔG°, ΔH°, and ΔS° at 25° C.

SOLUTION

From Table 8.2:

(1) $\text{fumarate}^{2-} + 2\text{H}^+ + 2\text{e}^- \rightarrow \text{succinate}^{2-}$ $E^{\circ\prime} = 0.031$ V

(2) $\text{pyruvate}^- + 2\text{H}^+ + 2\text{e}^- \rightarrow \text{lactate}^-$ $E^{\circ\prime} = -0.185$ V

Subtraction of (2) from (1) gives

$$\text{fumarate}^{2-} + \text{lactate}^- \rightarrow \text{succinate}^{2-} + \text{pyruvate}^-$$
$$\Delta E^{\circ\prime} = 0.216 \text{ V}$$

Note that this is also ΔE°, the hydrogen ions having cancelled out.
  The Gibbs energy change is

$$\Delta G^\circ = -nF\Delta E^\circ = -2 \times 96\,500 \times 0.216$$
$$= -41\,690 \text{ J} = -9.96 \text{ kcal}$$

The entropy change is obtained by use of equation (8.45):

$$\Delta S^\circ = 2 \times 96\,500 \times 2.18 \times 10^{-5} \text{ J K}^{-1}$$
$$= 4.207 \text{ J K}^{-1} = 1.01 \text{ cal K}^{-1}$$

The enthalpy change can be calculated by use of equation (8.47), or more easily from the ΔG° and ΔS° values:

$$\Delta H^\circ = \Delta G^\circ + T\Delta S^\circ$$
$$= -9960 + (298.15 \times 1.01) = -9660 \text{ cal} = -9.66 \text{ kcal}$$

PROBLEMS

8.1 Write down the reaction occurring at the individual electrodes, the overall reaction, and the expression for the emf, for the following reversible cells:

(a) H_2 (g) | HCl | Cl_2 (g)

(b) Hg, Hg_2Cl_2 (s) | HCl | H_2 (g)

(c) Ag, AgCl (s) | KCl ‖ Hg_2Cl_2 (s), Hg

8.2 Calculate the equilibrium constant at $25°$ C for the reaction

$$2Fe^{3+} + 2I^- \rightleftharpoons 2Fe^{2+} + I_2$$

if the following are the standard potentials:

$$e^- + Fe^{3+} \rightarrow Fe^{2+} \qquad E° = 0.77 \text{ V}$$

$$2e^- + I_2 \rightarrow 2I^- \qquad E° = 0.536 \text{ V}$$

8.3 Calculate the emf for the following cell at $25°$ C:

Pt, H_2 | HCl (0.5 m) ‖ HCl (1.0 m) | Pt, H_2

8.4 The standard potential for the process

gluconolacetone $+ 2H^+ + 2e^- \rightarrow$ D-glucose

is 0.77 V. What would be the emf of the cell

Pt, H_2 | H^+ (1 m) ‖ gluconolacetone, D-glucose, H^+ (pH 7) | Pt

if the gluconolacetone and D-glucose are at equal concentrations? Write down the cell reaction and calculate the corresponding equilibrium constant at $25°$ C.

8.5 The standard potential for

cytochrome c $(Fe^{3+}) + e^- \rightarrow$ cytochrome c (Fe^{2+})

is 0.25. Calculate $\Delta G°$ for the process

$\frac{1}{2}H_2 +$ cytochrome c $(Fe^{3+}) \rightarrow H^+ +$ cytochrome c (Fe^{2+})

8.6 Lactate is oxidized by cytochrome c (Fe^{3+}) to yield pyruvate plus cytochrome c (Fe^{2+}):

lactate$^-$ + 2 cytochrome c $(Fe^{3+}) \rightleftharpoons$

pyruvate$^-$ + 2 cytochrome c $(Fe^{2+}) + H^+$

Making use of the standard potentials given in Table 8.2, calculate the value of $\Delta E°'$ for the oxidation reaction at $25°$ C and pH 7. Also, calculate the equilibrium ratio

$$\frac{[\text{cytochrome c } (Fe^{2+})]^2[\text{pyruvate}^-]}{[\text{cytochrome c } (Fe^{3+})]^2[\text{lactate}^-]}$$

at pH 7.0. Calculate also the corresponding equilibrium ratio at pH 6.0.

8.7 At 25°C and pH 7 a solution containing compound A and its reduced form AH_2 has a standard potential of -0.06 V. A solution containing B and BH_2 has a standard potential of -0.16 V. If a cell were constructed with these systems as half cells,

    (a) Is AH_2 oxidized by B, or BH_2 oxidized by A, under standard conditions?

    (b) What is the reversible emf of the cell?

    (c) What would be the effect of pH on the equilibrium constant $[B][AH_2]/[A][BH_2]$?

8.8 Calculate the value of $\Delta G^{o'}$ and the equilibrium constant at pH 7.0 and 25°C, for the reaction

$$\text{succinate}^{2-} + \text{crotonyl CoA} \rightleftharpoons \text{fumarate}^{2-} + \text{butyryl CoA}$$

using the values given in Table 8.2.

8.9 Using the values given in Table 8.1, calculate the standard Gibbs-energy change ΔG^o for the reaction

$$H_2 + \tfrac{1}{2}O_2 \rightleftharpoons H_2O$$

8.10 Using the values given in Tables 8.1 and 8.2, calculate the pressure of oxygen that would be required to keep the cytochrome a system 99.9% in the oxidized form at 25°C and pH 7.0.

8.11 The pyruvate-lactate system has an $E^{o'}$ value of -0.185 at 25°C and pH 7.0. What will be the potential of this system if the oxidation has gone 90% to completion?

8.12 Estimate the Gibbs energy of the formation of the fumarate ion, using the $E^{o'}$ values in Table 8.2 and the following values:

    ΔG_f^o (succinate) $= -165.02$ kcal
    ΔG_f^o (acetaldehyde) $= -33.24$ kcal
    ΔG_f^o (ethanol) $= -43.44$ kcal.

If the $\partial\Delta E/\partial T$ value for the process

$$\text{fumarate}^- + \text{ethanol} \rightleftharpoons \text{succinate}^- + \text{acetaldehyde}$$

is -3.46×10^{-5} V K^{-1}, estimate the heat of formation of the fumarate ion from the following values:

    ΔH_f^o (succinate) $= -217.18$ kcal
    ΔH_f^o (acetaldehyde) $= -50.35$ kcal
    ΔH_f^o (ethanol) $= -68.6$ kcal

ESSAY QUESTION

8.13 Explain the significance of the standard electrode potential, with special reference to metabolic reactions.

SUGGESTED READING

Dawes, E. A. *Quantitative Problems in Biochemistry*, Ch. 10. Edinburgh and London: Churchill Livingstone, 1972.

See also the Suggested Readings for Chapters 6 and 7.

9

REACTION
KINETICS

The preceding chapters have dealt mainly with structure and with systems at
equilibrium. In biological work we are also concerned with the rates at which
processes occur, and with their mechanisms. The functioning of a living system
depends critically on the relative rates of the many reactions which take place, and
these rates in turn depend upon the detailed mechanisms of the processes.

The branch of science that is concerned with rates is known as *kinetics*. If we
deal with the rates of chemical (as opposed to physical) processes we speak of
chemical kinetics or *reaction kinetics*. This chapter presents the fundamental principles
relating to this subject. In developing these principles it is best to keep to rather
simple types of reactions. In the next chapter we shall see how the principles can
be applied to understanding the particular kinds of reactions that are important
in living systems.

9.1   RATES OF REACTIONS

The rate of a chemical reaction is usually expressed as the rate of change of the
concentration of a reactant or product. Thus Figure 9.1 shows for a hypothetical
reaction

$$A + B \rightarrow X + Y$$

the variations in concentrations of the reactant A, and of the product X, as functions
of time. At any time t we can draw a tangent to the curve I for the disappearance
of A; the slope of this tangent, *with the sign omitted so that the rate is always
positive*, is the rate of disappearance of A at that time. As a special case we may
draw the tangent at $t = 0$, corresponding to the beginning of the kinetic experiment,

and thus obtain the *initial rate*. Alternatively, we may deal with curve *II*, for the formation of X, and draw tangents. In this case each slope will represent the rate of formation of X at the appropriate time.

In either case, the rate will be concentration divided by time, and is usually expressed as mol dm^{-3} per second; this is conveniently written as mol dm^{-3} s^{-1} or as M s^{-1}, the symbol M meaning mol dm^{-3}

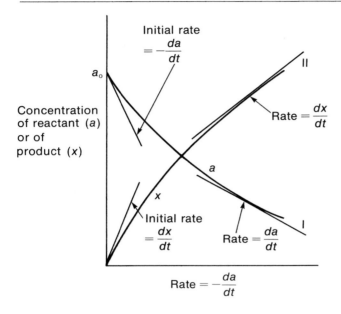

Figure 9.1

The variations with time of the concentrations of a reactant *a* (curve I) and a product *x* (curve II).

Note that the numerical value for the rate may be different according to the substance to which it refers. For example, in the ammonia synthesis

$$N_2 + 3H_2 \rightarrow 2NH_3$$

the rate of appearance of ammonia is twice the rate of disappearance of nitrogen, and the rate of disappearance of hydrogen is three times the rate of disappearance of nitrogen. We should therefore not speak simply of the *rate* of a reaction. Whenever there is ambiguity, we should specify the *reactant* or *product* to which the rate refers.

9.2 ORDER OF A REACTION

The rate, v, of a reaction is sometimes—but by no means always—related to the concentrations of reactants A, B, etc., by an equation of the type

$$v = k[A]^{\alpha}[B]^{\beta} \ldots \tag{9.1}$$

where α, β, etc. are constants. The reaction is then said to be of the αth order with respect to A, the βth order with respect to B, and so on. The *overall order n* is the sum of the various orders α, β, etc.:

$$n = \alpha + \beta + \cdots \tag{9.2}$$

Note that the order of a reaction is a purely experimental quantity.

A very simple case is when the rate is proportional to the first power of the concentration of a single reactant:

$$v = k[A] \tag{9.3}$$

Such a reaction is said to be of the *first order*. An example is the conversion of oxalosuccinate into α-ketoglutarate:

$$
\begin{array}{ccc}
\text{COOH} & & \text{COOH} \\
| & & | \\
\text{C=O} & & \text{C=O} \\
| & & | \\
\text{HOOC}-\text{C}-\text{H} & \rightarrow & \text{CH}_2 + \text{CO}_2 \\
| & & | \\
\text{CH}_2 & & \text{CH}_2 \\
| & & | \\
\text{COOH} & & \text{COOH}
\end{array}
$$

The rate of this reaction is proportional to the first power of the oxalosuccinate concentration. This reaction occurs readily if oxalosuccinate is heated. During metabolism oxalosuccinate is formed from isocitrate but it is not released into the cell; instead it decomposes into α-ketoglutarate.

There are many examples of *second-order* reactions in biological systems. The enzyme-catalyzed reaction between the oxidized form of nicotinamide adenine dinucleotide, NAD^+, with L-lactic acid can be written as follows

$$NAD^+ + \text{L-CH}_3\text{CHOHCOOH} \underset{k_{-1}}{\overset{k_1}{\rightleftharpoons}} NADH + \underset{\text{pyruvic acid}}{CH_3COCOOH} + H^+$$

Under certain conditions the rate equation is

$$v_1 = k_1[NAD^+][\text{L-CH}_3\text{CHOHCOOH}] \tag{9.4}$$

The reaction is first order in NAD^+ and first order in lactic acid, and its overall order is therefore two. The reverse reaction is also second order under certain conditions, the rate of formation of lactic acid from pyruvic acid being, at constant $[H^+]$,

$$v_{-1} = k_{-1}[NADH][CH_3COCOOH] \tag{9.5}$$

However, as we will see in the next chapter, more complicated rate equations are often found for biological reactions.

The rate of a reaction will be proportional to the product of two concentrations [A] and [B] if the reaction simply involves collisions between A and B molecules. Similarly, the kinetics will be third-order if a reaction proceeds in one stage and involves collisions between three molecules A, B, and C. There are a few known reactions of the third order, but reactions of higher order are unknown. The reason for this is that collisions in which three or more molecules all come together at the same time are very unlikely, so that the reaction may well proceed more rapidly by a complex mechanism involving two or more elementary processes each of which is only first or second order.

Note that there is no simple connection between the stoichiometric equation for a reaction and the order of the reaction. A simple example which illustrates this is the decomposition of gaseous acetaldehyde, for which the balanced equation is

$$CH_3CHO \rightarrow CH_4 + CO$$

We might be tempted to think that because there is one molecule on the left-hand side of this equation the reaction should be first order. In fact, it is of three-halves order:

$$v = k[CH_3CHO]^{3/2} \tag{9.6}$$

The reaction occurs by a complex free-radical mechanism of a particular type that leads to three-halves-order behavior.

Not all reactions behave in the manner described by equation (9.1), and the term "order" should not be used for those which do not. For example, as we will discuss in further detail in the next chapter, reactions catalyzed by enzymes frequently follow a law of the form

$$v = \frac{V[S]}{K_m + [S]} \tag{9.7}$$

where V and K_m are constants, and $[S]$ is the concentration of the substance, known as the *substrate*, which is undergoing catalyzed reaction. This equation does not correspond to a simple order, but under two limiting conditions an order may be assigned. Thus if the substrate concentration is sufficiently low that $[S] \ll K_m$, equation (9.7) becomes

$$v = \frac{V}{K_m}[S] \tag{9.8}$$

The reaction is then of the first order with respect to S. Also, when $[S]$ is sufficiently large that $[S] \gg K_m$ the equation reduces to

$$v = V \tag{9.9}$$

The rate is then independent of $[S]$, i.e., is proportional to $[S]^0$, and is said to be of *zero order*.

9.3   RATE CONSTANT

The constant k which appears in the rate equation (9.1) is known as the *rate constant*, the *rate coefficient*, or the *specific rate*; the first expression will be used in this book. The units of the rate constant vary with the order of the reaction. Suppose, for example, that a reaction is of the first order, i.e.,

$$v = k[A] \tag{9.10}$$

If v is in mol dm^{-3} s^{-1} (M s^{-1}), and $[A]$ in mol dm^{-3} (M), the rate constant k is in s^{-1}. For a second-order reaction, for which

$$v = k[A]^2 \qquad \text{or} \qquad k[A][B] \tag{9.11}$$

the units of k will be

$$\frac{\text{mol dm}^{-3}\,\text{s}^{-1}}{(\text{mol dm}^{-3})^2} = \text{dm}^3\,\text{mol}^{-1}\,\text{s}^{-1}\ (\text{M}^{-1}\,\text{s}^{-1})$$

The units corresponding to other orders can easily be worked out.

Note that since the rate in general depends on the reactant or product with which we are concerned, the rate constant also reflects this dependence.

EXAMPLE
Some results for a reaction between two substances A and B are shown below:

[A]	[B]	Rate v
(mol dm^{-3})	(mol dm^{-3})	(mol dm^{-3} s^{-1})
1.2×10^{-1}	4.5×10^{-2}	6.5×10^{-4}
2.4×10^{-1}	9.0×10^{-2}	2.6×10^{-3}
7.2×10^{-1}	9.0×10^{-2}	7.8×10^{-3}

Assuming that the reaction has a simple order, i.e., that the rate equation is of the form

$$v = k[A]^x[B]^y$$

determine the orders x and y and the rate constant k.

SOLUTION
A comparison of rows 2 and 3 shows that, with $[B]$ held constant, increasing $[A]$ by a factor of 3 increases the rate by a factor of 3. The rate is thus proportional to $[A]$; i.e., the order with respect to $[A]$, x, is unity. A comparison of rows 1 and 2 shows that

doubling both [A] and [B] increases v by a factor of 4. Since $x = 1$, doubling [A] alone would double the rate; an additional doubling is thus brought about by doubling [B], so that $y = 1$. The rate law is thus

$$v = k[A][B]$$

The value of k is found by inserting any set of values of v, [A] and [B] into this equation. For example

$$6.5 \times 10^{-4} = k(1.2 \times 10^{-1})(4.5 \times 10^{-2})$$

thus

$$k = \frac{6.5 \times 10^{-4} \text{ mol dm}^{-3} \text{ s}^{-1}}{(1.2 \times 10^{-1})(4.5 \times 10^{-2}) (\text{mol dm}^{-3})^2}$$

$$= 0.120 \text{ dm}^3 \text{ mol}^{-1} \text{ s}^{-1} (\text{M}^{-1} \text{ s}^{-1})$$

The above example is useful in illustrating the principle involved, but orders and rate constants in practice are usually determined in a more systematic fashion, as will now be described.

9.4   ANALYSIS OF KINETIC RESULTS

The first task in the kinetic investigation of a chemical reaction is to measure rates under a variety of experimental conditions and to determine how the rates are affected by the concentrations of reactants, products of reaction, and other substances (e.g., an inhibitor) which may affect the rate.

There are two main methods for dealing with such problems; they are known as

1. The method of integration.
2. The differential method.

In the method of integration we start with a rate equation which we think may be applicable. For example, if the reaction is believed to be a first-order reaction we start with

$$-\frac{dc}{dt} = kc \tag{9.12}$$

where c is the concentration of reactant. By integration we convert this into an equation giving c as a function of t, and we then compare this with the experimental variation of c with t. If there is a good fit we can then, by simple graphical procedures, determine the value of the rate constant. If the fit is not good we must try another rate equation and go through the same procedure until the fit is satisfactory. The method is something of a hit-and-miss one, but is nevertheless very valuable, especially when no special complications arise.

The second method, the differential method, employs the rate equation in its differential, unintegrated, form. Values of dc/dt are obtained from a plot of c against t by taking slopes, and these are directly compared with the rate equation. The main difficulty with this method is that slopes cannot always be obtained very accurately, but in spite of this drawback the method is on the whole the more reliable one and unlike the integration method it does not lead to any particular difficulties when there are complexities in the kinetic behavior.

These two methods will now be considered in further detail.

9.5   METHOD OF INTEGRATION
FIRST-ORDER KINETICS

A first-order reaction may be represented schematically as

$$A \rightarrow X$$

Suppose that at the beginning of the reaction ($t = 0$) the concentration of A is a_o, and that of X is zero. If after time t the concentration of X is x, that of A is $a_o - x$. The rate of formation of X is dx/dt, so that for a first-order reaction

$$\frac{dx}{dt} = k(a_o - x). \tag{9.13}$$

Separation of the variables leads to

$$\frac{dx}{a_o - x} = k \, dt, \tag{9.14}$$

and integration gives

$$-\ln (a_o - x) = kt + k' \tag{9.15}$$

where k' is the constant of integration. This constant may be evaluated using the boundary condition that $x = 0$ when $t = 0$; hence

$$-\ln a_o = k' \tag{9.16}$$

and insertion of this into (9.15) leads to

$$\ln \frac{a_o}{a_o - x} = kt. \tag{9.17}$$

This equation can also be written as

$$x = a_o(1 - e^{-kt}) \tag{9.18}$$

and as

$$a_o - x = a_o e^{-kt} \tag{9.19}$$

This last equation shows that the concentration of reactant, $a_o - x$, decreases exponentially with time, from an initial value of a_o to a final value of zero (see Figure 9.1).

One procedure for testing the applicability of equations (9.17) to (9.19) to a chemical reaction would be to make determinations of x at various times during the course of reaction. At each time a value of k could then be calculated using (9.17), and if k were truly constant during the course of the reaction the conclusion would be that the reaction was of the first order. If the values of k were found to drift, the reaction is not of the first order, and other equations have to be tried.

Alternatively, the first-order equations can be tested and the constant evaluated using a graphical procedure. It follows from (9.17) that a plot of

$$\ln \frac{a_o}{a_o - x} \text{ against } t$$

will give a straight line if the order is the first; this is shown schematically in Figure 9.2a. The rate constant is the slope. We may simply plot $\ln (a_o - x)$ against t, as shown in Figure 9.2b. We may also plot the common logarithms, in which case the slopes are $k/2.303$ or $-k/2.303$.

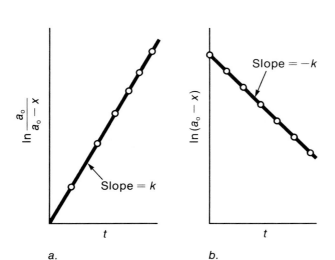

Figure 9.2

Method of integration; analysis of results for a first-order reaction:
a. Plot of $\ln [a_o/(a_o - x)]$ against t.
b. Plot of $\ln (a_o - x)$ against t.

SECOND-ORDER KINETICS

Second-order reactions can be treated in a similar fashion. There are two possibilities for second-order reactions: the rate may be proportional to the product of two equal concentrations, or to the product of two different ones. The first case must occur when a single reactant is involved, in which case the situation may be represented as

$$2A \rightarrow X$$

It may also be found in reactions between two different substances,

$$A + B \rightarrow X$$

provided that their initial concentrations are the same.

In such situations the rate may be expressed as

$$\frac{dx}{dt} = k(a_o - x)^2 \tag{9.20}$$

where x is the amount of A that has reacted in unit volume at time t, and a_o the initial amount of A. Separation of the variables leads to

$$\frac{dx}{(a_o - x)^2} = k\, dt \tag{9.21}$$

which integrates to

$$\frac{1}{a_o - x} = kt + k' \tag{9.22}$$

where k' is the constant of integration. The boundary condition is that $x = 0$ when $t = 0$, so that

$$k' = \frac{1}{a_o} \tag{9.23}$$

Hence

$$\frac{x}{a_o(a_o - x)} = kt \tag{9.24}$$

We see that the variation of x with t is no longer an exponential one.

Graphical methods can again be employed to test this equation and to obtain the rate constant k. A simple procedure is to plot $x/a_o(a_o - x)$ against t ; if the equation is obeyed the points will lie on a straight line passing through the origin (see Figure 9.3), and the slope will be k. Alternatively, we may plot $x/(a_o - x)$ against t, in which case the slope is $a_o k$.

If the rate is proportional to the concentrations of two different reactants, and these concentrations are not initially the same, the integration proceeds differently. Suppose that the initial concentrations of A and B are a_o and b_o, and that in unit volume an amount x of each has reacted. The rate of disappearance of each of the two reactants is then

$$\frac{dx}{dt} = k(a_o - x)(b_o - x) \tag{9.25}$$

The result of the integration, with the boundary condition $t = 0$, $x = 0$, is

$$\frac{1}{a_o - b_o} \ln \frac{b_o(a_o - x)}{a_o(b_o - x)} = kt \tag{9.26}$$

This equation can be tested by plotting the left-hand side against t; the slope of the straight line (if obtained) is k.

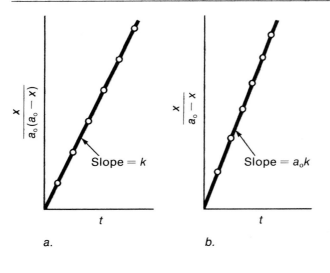

a.   b.

Figure 9.3

Method of integration; analysis of results for a second-order reaction:
a. Plot of $x/a_o(a_o - x)$ against t.
b. Plot of $x/(a_o - x)$ against t.

These are the most common orders; reactions of other orders can be treated in a similar manner.† Table 9.1 summarizes some of the results.

The main disadvantage of the method of integration is that the integrated expressions, giving the variation of x with t, are often quite similar for different types of reactions. The time course of a simple second-order reaction, for example, is closely similar to that of a first-order reaction inhibited by products, and unless the experiments are done very accurately there is danger of confusion. Inhibition by

† Solutions for many systems are given by C. Capellos and B. H. J. Bielski, *Kinetic Systems.* New York: Wiley-Interscience, 1972.

Table 9.1   Rate Equations and Half-Lives

Order	Rate equation Differential form	Rate equation Integrated form	Units of rate constant	Half-life
0	$\dfrac{dx}{dt} = k$	$k = \dfrac{x}{t}$	$M\ s^{-1}$	$\dfrac{a_0}{2k}$
1	$\dfrac{dx}{dt} = k(a_0 - x)$	$k = \dfrac{1}{t} \ln \dfrac{a_0}{a_0 - x}$	s^{-1}	$\dfrac{\ln 2}{k}$
2	$\dfrac{dx}{dt} = k(a_0 - x)^2$	$k = \dfrac{1}{t} \dfrac{x}{a_0(a_0 - x)}$	$M^{-1}\ s^{-1}$	$\dfrac{1}{ka_0}$
2	$\dfrac{dx}{dt} = k(a_0 - x)(b_0 - x)$ (reactants at different concentrations)	$k = \dfrac{1}{t(a_0 - b_0)} \ln \dfrac{b_0(a_0 - x)}{a_0(b_0 - x)}$	$M^{-1}\ s^{-1}$	—
n	$\dfrac{dx}{dt} = k(a_0 - x)^n$	$k = \dfrac{1}{t(n - 1)} \left[\dfrac{1}{(a_0 - x)^{n-1}} - \dfrac{1}{a_0^{n-1}} \right]$	$M^{1-n}\ s^{-1}$	$\dfrac{2^{n-1} - 1}{k(n-1)a_0^{n-1}}$

products can, of course, be tested for directly by measuring the rate after the deliberate introduction of products of reaction.

HALF-LIFE OF A REACTION

The half-life, or half-period, of a reaction is the time that it takes for half of a reacting substance to disappear. The value of the half-life τ is related to the rate constant, the relationship being given by taking the integrated form of the rate equation for a given order and setting t equal to τ and x to $a_0/2$. Thus for a first-order reaction the half-life is given by:

$$k = \frac{1}{\tau} \ln \frac{a_0}{a - (a_0/2)} \tag{9.27}$$

and thus

$$\tau = \frac{1}{k} \ln 2 \tag{9.28}$$

In this case the half-life is independent of the initial concentration. For a second-order reaction the relationship is

$$\tau = \frac{1}{a_0 k} \tag{9.29}$$

and in this case the half-life is inversely proportional to the initial concentration.

In the general case of a reaction of the nth order, the half-period is inversely proportional to a_o^{n-1} (see Table 9.1). These relationships are valid only if the initial concentrations of all reactants are equal.

The order of the reaction can be obtained by determining half-lives at two different initial concentrations a_1 and a_2. The half-lives are related by

$$\frac{\tau_1}{\tau_2} = \left(\frac{a_2}{a_1}\right)^{n-1} \tag{9.30}$$

and thus

$$n = 1 + \frac{\log(\tau_1/\tau_2)}{\log(a_2/a_1)} \tag{9.31}$$

so that n can readily be calculated. This method can give misleading results if the reaction is not of simple order or if there are complications such as inhibition by products.

Since the half-lives of all reactions have the same units, they provide a useful way of comparing the rates of reactions of different orders. Rate constants, as we have seen, have different units for different reaction orders. Thus if two reactions have different orders we cannot at once deduce their relative rates from their rate constants.

9.6   DIFFERENTIAL METHOD

In the differential method, which was first suggested by the Dutch physical chemist Jacobus H. van t' Hoff (1852–1911) in 1884, the procedure is to determine rates directly by measuring tangents to the experimental concentration-time curves, and to introduce these into the equations in their differential forms.

The theory of the method is as follows. The instantaneous rate of a reaction of the nth order involving only one reacting substance is proportional to the nth power of its concentration,

$$v = -\frac{da}{dt} = ka^n \tag{9.32}$$

Taking logarithms of both sides

$$\log_{10} v = \log_{10} k + n \log_{10} a \tag{9.33}$$

A plot of $\log_{10} v$ against $\log_{10} a$ therefore will give a straight line if the reaction is of simple order; the slope will be of the order n.

There are two different ways in which this procedure can be applied. One method is to carry out a single run—that is, to allow the reaction to proceed and determine a at various times. Tangents can then be drawn at different concentrations, as shown schematically in Figure 9.4, and the slopes, da/dt, determined. A plot of $\ln v$ or $\log_{10} v$ is then made against $\ln a$ or $\log_{10} a$, as shown in Figure 9.5. When this procedure is employed there may be interference from the products of reaction.

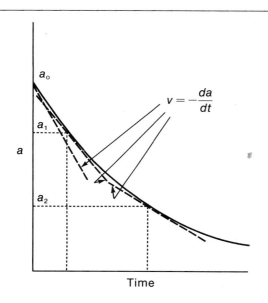

Figure 9.4

Schematic plots of concentration against time, illustrating the use of the differential method. Tangents are shown drawn at the initial concentration a_0 and at the concentrations a_1 and a_2.

In the second method, slopes are measured only at the very beginning of the reaction, and the reaction is run at various initial concentrations. This type of procedure is represented schematically in Figure 9.6. The order, determined by plotting $\log_{10} v$(initial) against $\log_{10} a$, is now known as the *order with respect to concentration*, since it is now the concentration that is varied. This order has been called the *true*

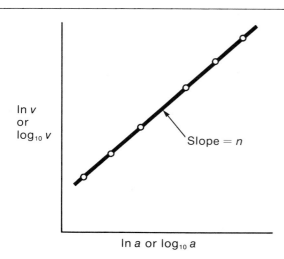

Figure 9.5

A schematic logarithmic plot of v against a; the slope is the order n. Any logarithm can be used (the same for both axes).

order, since it is concerned only with the reacting substances, and not with the products of reaction, which have no effect on the initial rates.

The differential method is a very valuable method, and is particularly useful for enzyme reactions, in which the products frequently interfere and render the method of integration an unreliable procedure. It is often very convenient, in studying enzyme systems, to cause the reaction to occur sufficiently slowly (by reducing the enzyme concentration, for example) that a number of measurements can be made during the very early stages of reaction. In this way it is possible for the initial rates to be determined very accurately. A disadvantage of the differential method is that it is not always easy to obtain accurate values of the slopes of the rate curves; if this difficulty can be overcome, however, the method is more satisfactory than the others.

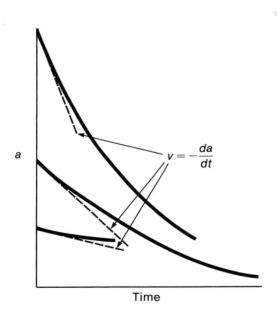

Figure 9.6

Schematic plots of concentration against time, for three initial concentrations. The differential method can be applied to the initial slopes.

In the figure: a, $v = -\dfrac{da}{dt}$, Time

9.7   ISOLATION METHOD

The isolation method is a special way of applying any of the methods already described, and is useful when several reactants are involved. If all of the reactants except one are present in excess, the apparent order of the reaction is the order with respect to the one "isolated" reactant, since the concentrations of those in excess will not change appreciably during the course of reaction. This method is sometimes useful in enzyme systems where there are two substrates.

9.8   REACTIONS HAVING NO SIMPLE ORDER

We have seen that there are many reactions which do not admit of the assignment of an order, and that this is frequently true of enzyme reactions, many of which obey equation (9.7). Although such reactions can be treated by the method of integration, this procedure is seldom entirely satisfactory, owing to the difficulty of distinguishing between various possibilities. In such cases the best method is usually the differential method, rates (slopes) being measured accurately in the initial stages of the reaction, and runs being carried out at a series of initial concentrations. In this way a plot of velocity against concentration can be prepared; an example will be found in Figure 10.1. The dependence of velocity on concentration can then be determined by various methods. One that is frequently used for enzyme reactions will be considered in the next chapter.

9.9   OPPOSING REACTIONS

One complication which frequently exists is that a reaction may proceed to a state of equilibrium which differs appreciably from completion. This situation often arises in biological processes, such as metabolic reactions. For example, the reaction

$$\alpha\text{-ketoglutarate} + \text{alanine} \rightleftharpoons \text{glutamine} + \text{pyruvate}$$

has a ΔG° of 0.06 kcal mol^{-1} at 25° C, which means an equilibrium constant of 0.91. In considering the kinetics of such a reaction in one direction, we would have to take into account the reverse reaction.

The simplest case is when both forward and reverse reactions are of the first order:

$$A \underset{k_{-1}}{\overset{k_1}{\rightleftharpoons}} X$$

If the experiment is started using pure A, of concentration a_0, and if after time t the concentration of X is x, that of A is $a_0 - x$. The rate of $A \rightarrow X$ is then equal to $k_1(a_0 - x)$, while that of $X \rightarrow A$ is $k_{-1}x$. The net rate of production of X is thus

$$\frac{dx}{dt} = k_1(a_0 - x) - k_{-1}x \tag{9.34}$$

If x_e is the concentration of X at equilibrium, when the net rate is zero,

$$k_1(a_0 - x_e) - k_{-1}x_e = 0 \tag{9.35}$$

Elimination of k_{-1} between equations (9.34) and (9.35) gives rise to

$$\frac{dx}{dt} = \frac{k_1 a_0}{x_e}(x_e - x) \tag{9.36}$$

Integration of this, subject to the boundary condition that $x = 0$ when $t = 0$, gives

$$k_1 t = \frac{x_e}{a_o} \ln \frac{x_e}{x_e - x} \qquad (9.37)$$

The amount of x present at equilibrium, x_e, can be measured directly. A procedure for obtaining k_1 is therefore to obtain values of x at various values of t, and then to plot

$$\frac{x_e}{a_o} \ln \frac{x_e}{x_e - x} \text{ against } t$$

The slope of the line then gives the value of k_1. The constant k_{-1} for the reverse reaction can be obtained by use of the fact that the equilibrium constant is k_1/k_{-1}.

Equations for more complicated kinetic situations have also been worked out (see footnote on p. 374).

9.10   THE MEASUREMENT OF REACTION RATES

A variety of experimental procedures can be used in a kinetic investigation. We will mention briefly some conventional methods first, then describe some special techniques.

The older and more conventional methods essentially consist of causing a reaction to occur in a reaction vessel and measuring a concentration as a function of time. There are various ways of starting a reaction. If the reaction is one which takes place between two substances, the reaction can be started by bringing the substances together in the reaction vessel; to avoid errors the time of mixing must be negligible compared with the half-life of the reaction. In the case of a reaction involving only one substance—a decomposition or isomerization—the substance must first be maintained at a temperature at which it is stable. It may then be rapidly heated to a temperature at which reaction occurs at a conveniently measurable rate.

Reaction vessels are usually made of glass, which is unattacked by most chemicals; if it is attacked by the reactants some other material must be chosen. In the case of reactions in solution the reactant and solvent are usually introduced into the vessel by use of conventional volumetric techniques, and the vessel must be stoppered or sealed if there is danger of loss of volatile material. Since the rates of reactions are strongly affected by temperature, reaction vessels must be maintained in a thermostatically-controlled bath, and the temperature should not vary by more than one-tenth of a degree.

A variety of methods may be used for following the course of a chemical reaction. The oldest and simplest procedure involves following the time by means of a stopwatch, and at suitable intervals making a measurement from which the reactant concentration can be deduced. For a reaction in the liquid or solid phases, or in solution, the course of reaction may be followed using analytical methods. Alternatively some physical property, such as light absorption or optical rotation, may be measured during the reaction.

During recent years, following the development of modern electronic techniques, there has been a tendency towards automation in kinetic studies. If a reaction in solution involves a change in absorption at a certain wavelength it is possible to obtain a smooth record of absorption as a function of time.

9.11 TECHNIQUES FOR VERY FAST REACTIONS

The methods so far described for the most part apply to reactions that are fairly slow, having half-lives of half an hour or more. For many reactions it is convenient to choose a temperature range within which they occur sufficiently slowly for them to be studied by conventional means. However, some reactions are so fast that special techniques must be employed. Such techniques are of two main types: the first employ essentially the same principles used for slow reactions, but the methods are modified so as to be suitable for more rapid reactions; the second type are of a different character and employ special principles.

The main reasons why conventional techniques lead to difficulties for very rapid reactions are:

1. The time that it usually takes to mix reactants, or to bring them to a specified temperature, may be significant in comparison with the half-life of the reaction: an appreciable error therefore will be made, since the initial time is not clearly defined.

2. The time that it takes to make a measurement of concentration is significant compared with the half-life.

FLOW TECHNIQUES

The first difficulty can sometimes be surmounted by using special techniques for bringing the reactants very rapidly into the reaction vessel and for mixing them very rapidly. Normally, with the use of conventional techniques, it takes from several seconds to a minute to bring solutions into a reaction vessel, and to have them completely mixed and at the temperature of the surroundings. This time can be reduced greatly by using a rapid flow, and flow techniques are frequently employed for rapid reactions. One particular modification of these methods is the *stopped-flow* technique, shown schematically in Figure 9.7. This particular apparatus is designed for the study of a reaction between two substances in solution. A solution of one of the substances is maintained initially in syringe A, and a solution of the other in syringe B. The plungers of the syringes can be forced down rapidly—perhaps by mechanical means—and a rapid stream of the two solutions passes into the mixing system. This is designed in such a way that jets of the two solutions impinge on one another and give very rapid mixing; with a suitable design of the mixing chamber it is possible for mixing to be essentially complete in one-thousandth of a second. From this mixing chamber the solution passes at once into the reaction cuvette. Alternatively, the two may be combined.

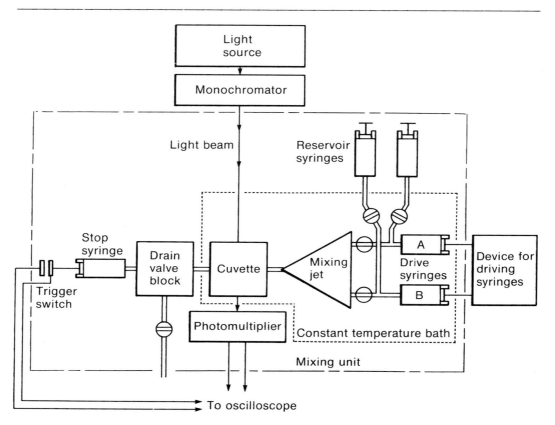

Figure 9.7

Schematic diagram of stopped-flow apparatus. Solutions in the two drive syringes A and B are rapidly forced through a mixing jet into a cuvette. When the flow is stopped, the oscilloscope is triggered and records light absorption as a function of time.

If a reaction is rapid it is not possible to carry out chemical analyses at various stages. The second difficulty referred to above must be resolved by employing techniques which allow properties to be determined instantaneously. For reactions in solution, spectrophotometric methods are commonly employed. If the products absorb differently from the reactants at a particular wavelength, we can pass monochromatic light of this wavelength through the reaction vessel, and by the use of a photoelectric device with suitable electronic circuits display the output on a recorder or oscilloscope. If the reaction is not too fast a pen-and-ink recorder may respond sufficiently rapidly; otherwise an oscilloscope may be used to give a record of absorption against time, and a photograph of the record can be taken. Fluorescence, electrical conductivity and optical rotation are also convenient properties to measure in such high-speed studies.

RELAXATION METHODS

The flow techniques just described are limited by the speed with which it is possible to mix solutions. There is no difficulty, using optical or other techniques, in following the course of a very rapid reaction, but for hydrodynamic reasons it is impossible to mix two solutions in less than about 10^{-3} s. If the half-life is less than this, the reaction will be largely completed by the time that it takes for mixing to be achieved, and any rate measurement made will be of the rate of mixing, not the rate of reaction. The neutralization of an acid by a base, i.e., the reaction

$$H^+ + OH^- \rightarrow H_2O$$

under ordinary conditions has a half-life of 10^{-6} s or less, and its rate therefore cannot be measured by any technique involving the mixing of solutions. The same difficulty applies to numerous biological reactions, including many processes in which enzymes form addition complexes with their substrates (see Chapter 10).

These technical problems were overcome by the development of a group of methods known as *relaxation methods*, the pioneer worker in this field being the German physical chemist Manfred Eigen. These methods differ fundamentally from conventional kinetic methods in that we start with the system at equilibrium under a given set of conditions. We then change these conditions very rapidly; the system is then no longer at equilibrium, and it *relaxes* to a new state of equilibrium. The speed with which it relaxes can be measured, usually by spectrophotometry, and we can then calculate the rate constants.

There are various ways in which the conditions are disturbed. One is by changing the hydrostatic pressure. Another, the most common technique, is to increase the temperature suddenly, usually by the rapid discharge of a capacitor. This method is called the *temperature-jump* or *T-jump* method. It is possible to raise the temperature of a tiny cell containing a reaction mixture by a few degrees in less than 10^{-7} s, which is sufficiently rapid to allow us to study even the fastest chemical processes.

The principle of the method is illustrated in Figure 9.8. Suppose that the system is of the simple type

$$A \underset{k_{-1}}{\overset{k_1}{\rightleftharpoons}} X$$

At the initial state of equilibrium, the product X will be at a certain concentration, and it will stay at this concentration until the temperature jump occurs, when the concentration will change to another value which will be higher or lower than the initial value according to the sign of ΔH° for the reaction. For this simple type of reaction, the shape of the curve during the relaxation phase can be shown to be related to the sum of the rate constants, $k_1 + k_{-1}$, and this sum can therefore be obtained from an analysis of the record. The ratio of the rate constants, k_1/k_{-1}, is the equilibrium constant and can be determined easily. We can then, by simple algebra, calculate the individual rate constants. Similar analysis can be made for other types of reactions, e.g., $A + B \rightleftharpoons X$. Note that the rate constants obtained relate to the second temperature, after the T-jump has occurred. Thus, if we want rate constants at $25°\,C$, and the T-jump is $7°\,C$, we should start at $18°\,C$.

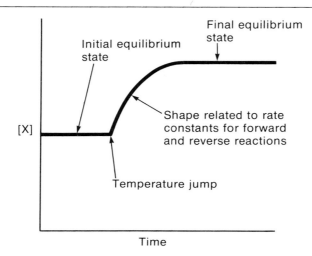

Figure 9.8

Principle of the temperature-jump technique.

(figure labels: Initial equilibrium state; Final equilibrium state; Shape related to rate constants for forward and reverse reactions; Temperature jump; [X]; Time)

During recent years a considerable number of investigations have been made using this technique. An important reaction studied in this way is the dissociation of water and combination of hydrogen and hydroxide ions:

$$2H_2O \rightleftharpoons H_3O^+ + OH^-$$

The reaction was followed by measuring conductivity. After the T-jump the conductivity increases with time, with a relaxation half-life of 3.7×10^{-5} s at $23°\,C$. From this value it can be calculated that the second-order rate constant for the combination of H_3O^+ and OH^- ions is $1.3 \times 10^{11}\ M^{-1}\,s^{-1}$, which is a remarkably high value. The rate constant for the reverse dissociation, which is very small, can be calculated from this value and the equilibrium constant. A considerable number of other hydrogen-ion transfer processes have also been studied using the same technique.

The T-jump method and other relaxation methods have been applied to a number of enzyme systems. For example, it has proved possible to measure rates of interactions between enzymes and substrates† and also the rates of certain conformational changes which occur in enzyme molecules during the course of their reactions.

---

† This term refers to the substance upon which the enzyme acts catalytically.

9.12   MOLECULAR KINETICS

There are several aspects to the kinetic study of a chemical reaction. One of these is concerned with the phenomenological, or empirical, rate laws which are obeyed. The way in which rates depend upon the concentrations of reactants and products is part of this type of investigation, and has been considered earlier in this chapter.

A second important type of kinetic study relates to the way in which rates depend on temperature. The most satisfactory way of dealing with this problem is to investigate how rate constants—or in the case of complex rate equations, the constants appearing in the empirical rate equations—depend upon temperature. Such studies have been of great importance in chemical kinetics, because the temperature dependence leads to a theoretical interpretation of reaction rates that is of very great significance. The temperature dependence is related to molecular properties of the reaction system, and for this reason the subject is frequently referred to as *molecular kinetics*.

A third important aspect of kinetics, to be dealt with later, is concerned with the elucidation of complex reaction mechanisms on the basis of kinetic and other studies. These various types of kinetic investigation are closely related to one another, and molecular-kinetic studies are very valuable in arriving at conclusions about reaction mechanisms.

9.13   ELEMENTARY REACTIONS

Few overall chemical reactions occur in a single stage; when they do they are said to be *elementary*. An example of an elementary reaction is the hydrolysis of an alkyl halide, RX, by hydroxide ions in aqueous solution:

$$R - X + OH^- \rightarrow R - OH + X^-$$

In reactions of this kind the OH^- ion makes an attack on the carbon atom which is bonded to the halogen, somewhat as represented below:

$$HO^- + \overset{\diagdown}{\underset{\diagup}{C}}{-}X \rightarrow HO \cdots \overset{\cdots \diagup}{\underset{|}{C}} \cdots X \rightarrow HO{-}\overset{\diagup}{\underset{\diagdown}{C}} \cdots + X^-$$

Intermediate
complex

On the other hand, when a hydrolysis is catalyzed by an enzyme the evidence, as will be discussed in Section 10.1, is that the process occurs in several stages. One of these is the formation of an addition complex between the enzyme and the substrate, and this complex then undergoes one or more subsequent reactions.

An important problem is how the overall kinetic behavior—i.e., the dependence of rate on the various concentrations—is related to the rate constants of the individual steps. This will be dealt with later in this chapter. First, we will be concerned only with the characteristics of the elementary processes themselves.

MOLECULARITY AND ORDER

Once a process has been identified as an elementary one, an important question which arises is: How many molecules enter into reaction? This number is referred to as the *molecularity* of the reaction. We saw earlier in this chapter that from the variation of rate with concentration we can frequently determine a reaction order. This number, a purely experimental one, should be sharply distinguished from molecularity, which represents a deduction as to the number of molecules. It is permissible to speak of the order of a complex reaction, provided that the rate is proportional simply to concentrations raised to certain powers. It is meaningless, on the other hand, to speak of the molecularity if the mechanism is a complex one.

With certain exceptions, discussed below, we can legitimately assume that the order of an elementary reaction indicates the number of molecules which enter into reaction, i.e., that the order and the molecularity are the same. For example, if an elementary reaction is of the first order with respect to a reactant A and of the first order with respect to another substance B, the conclusion is that the reaction is bimolecular, a molecule of A and a molecule of B entering into reaction.

Sometimes, however, this procedure may lead to incorrect conclusions. Suppose, for example, that one reactant is present in large excess, so that its concentration does not change appreciably as the reaction proceeds; moreover (for example, if it is the solvent) its concentration may be the same in different kinetic runs. If this is so the kinetic investigation will not reveal any dependence of the rate on the concentration of this substance, which would therefore not be considered to be entering into reaction. This situation is frequently found in reactions in solution where the solvent may be a reactant. For example, in hydrolysis reactions in aqueous solution, a water molecule may undergo reaction with a solute molecule. Unless special procedures are employed the kinetic results will not reveal the participation of the solvent. However, its participation is indicated if it appears in the stoichiometric equation.

Another case in which the kinetic study may not reveal that a substance enters into reaction is when a *catalyst* is involved. A catalyst by definition is a substance which influences the rate of reaction without itself being used up; it may be regarded as a substance which is both a reactant and a product of reaction. The concentration of a catalyst therefore remains constant during reaction, and the kinetic analysis of a single run will not reveal the participation of the catalyst in the reaction. However, the fact that it does enter into reaction may be shown by measuring the rate at a variety of catalyst concentrations; generally a linear dependence is found.

It will be clear from what has been said that the decision about the molecularity of an elementary reaction must involve not only a careful kinetic study in which as many factors as possible are varied, but also a consideration of other aspects of the reaction, including the nature of the products.

9.14   THE ARRHENIUS LAW

One of the most important relationships in chemical kinetics, and one which provides much information as to mechanism, is the equation which connects the rate constant of a reaction with the temperature. Many years ago it was discovered empirically that the rate constant k is related to the absolute temperature T by the equation

$$k = Ae^{-B/T} \tag{9.38}$$

where A and B are constants. This relationship was expressed by Jacobus Hendricus van 't Hoff and by Svante A. Arrhenius in the form

$$k = Ae^{-E/RT} \tag{9.39}$$

where R is the gas constant, equal to 1.987 calories per degree per mole. The equation was arrived at in 1887 by van 't Hoff, who argued on the basis of the variation of the equilibrium constant with the temperature, and pointed out that a similar relationship should hold for the rate constant of a reaction. This idea was extended by Arrhenius, who successfully applied it to a large number of reactions, and as a result (9.39) is generally referred to as the Arrhenius law.

The arguments of van 't Hoff are briefly as follows. The variation of the equilibrium constant obeys the law (see (5.165))

$$\frac{d \ln K}{dT} = \frac{\Delta H^{\circ}}{RT^2} \tag{9.40}$$

where K is the equilibrium constant expressed in terms of concentrations, and ΔH° is the standard enthalpy change in the reaction. If a reaction

$$A + B \underset{k_{-1}}{\overset{k_1}{\rightleftharpoons}} X + Y$$

is considered, the equilibrium constant K is equal to the ratio of the rate constants k_1 and k_{-1}:

$$K = \frac{k_1}{k_{-1}} \tag{9.41}$$

Equation (9.40) can therefore be written as

$$\frac{d \ln k_1}{dT} - \frac{d \ln k_{-1}}{dT} = \frac{\Delta H^{\circ}}{RT^2} \tag{9.42}$$

and this may be split into the two equations

$$\frac{d \ln k_1}{dT} = \text{Const.} + \frac{E_1}{RT^2} \tag{9.43}$$

$$\frac{d \ln k_{-1}}{dT} = \text{Const.} + \frac{E_{-1}}{RT^2} \tag{9.44}$$

where $E_1 - E_{-1}$ is equal to $\Delta H°$. Experimentally it is found that the constants appearing in (9.43) and (9.44) can be set equal to zero, and integration of (9.43) and (9.44) then gives rise to

$$k_1 = A_1 e^{-E_1/RT} \tag{9.45}$$

$$k_{-1} = A_{-1} e^{-E_{-1}/RT} \tag{9.46}$$

The quantities A_1 and A_{-1} appearing in these equations are generally known as the *frequency factors* of the reactions, and E_1 and E_{-1} are known as the *activation energies*, or *energies of activation*.

Arrhenius's approach to the law was a little different from that of van 't Hoff. He pointed out that for ordinary chemical reactions the majority of collisions between the reactant molecules are ineffective, the energy being insufficient. In a small fraction of the collisions, however, the energy is great enough to allow reaction to occur. According to the Boltzmann Principle (see pp. 250–252) the fraction of collisions in which the energy is in excess of a particular value E is

$$e^{-E/RT}$$

This fraction is greater the higher the temperature T and the lower the energy E. The rate constant should be proportional to this fraction.

In order to test the Arrhenius equation, equation (9.39), we first take logarithms of both sides:

$$\ln k = \ln A - \frac{E}{RT} \tag{9.47}$$

It follows that, if the Arrhenius law applies, a plot of $\ln k$ against $1/T$ will be a straight line, and the slope will be $-E/R$. Alternatively, we can take common logarithms:

$$\log_{10} k = \log_{10} A - \frac{E}{2.303RT} \tag{9.48}$$

and the slope of a plot of $\log_{10} k$ against $1/T$ is $-E/2.303R$, or

$$-\frac{E(\text{calories})}{4.57}$$

An example of such an Arrhenius plot is shown in Figure 9.9. It follows from equation (9.39) that the frequency factor has the same units as the rate constant itself, e.g., s^{-1} for a first-order reaction, $dm^3\ mol^{-1}\ s^{-1}$ for a second-order reaction. The activation energy is generally expressed as $cal\ mol^{-1}$, or as $kcal\ mol^{-1}$—the

SI recommendation to use joules mol^{-1} ($J\ mol^{-1}$) has not yet been adopted to any extent.

The Arrhenius law has a surprisingly wide applicability. It is obeyed not only by the rate constants of elementary reactions but frequently also by the rates of much more complicated processes. For example, the law is obeyed by the chirping

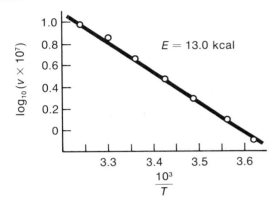

Figure 9.9

Arrhenius plot for the myosin-catalyzed hydrolysis of adenosine triphosphate (ATP). Under the conditions of the experiments the rate v is proportional to a rate constant k; we may therefore plot $\log_{10} v$, instead of $\log_{10} k$, against $1/T$.

of crickets (see Figure 9.10), the creeping of ants, the flashing of fireflies, the α-brain-wave rhythm, and even by psychological processes such as the rates of counting and forgetting.† The reason for the applicability of the law to such varied processes is that they are all controlled by chemical reactions.

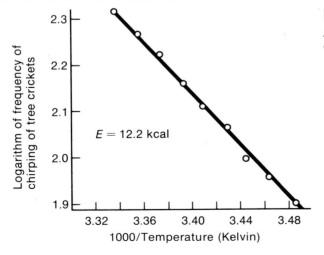

Figure 9.10

Arrhenius plot for the chirping of crickets.

---

† For details, with Arrhenius plots, of these processes see K. J. Laidler, *J. Chem. Education,* **49** (1972) 343.

EXAMPLE

A second-order reaction solution has a rate constant of 5.7×10^{-5} $M^{-1} s^{-1}$ at 25.0° C, and of 1.64×10^{-4} $M^{-1} s^{-1}$ at 40.0° C. Calculate the activation energy and the frequency factor, assuming the Arrhenius law to apply.

SOLUTION

To solve this type of problem it is convenient (but not necessary) to sketch an Arrhenius plot; this is shown in Figure 9.11. It is to be emphasized that in taking the logarithms of the rate constants, and the reciprocals of the rates, it is necessary to use enough significant figures, as we are dealing with relatively small differences between the values.

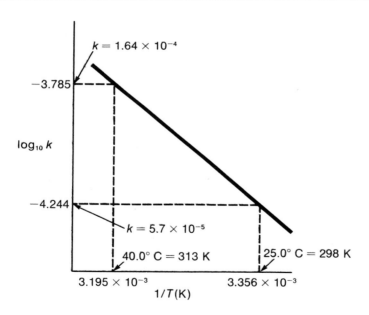

Figure 9.11

Schematic Arrhenius plot for this example.

The slope of the line in Figure 9.11 is

$$\frac{-4.244 - (3.785)}{(3.356 - 3.195) \times 10^{-3}}$$

$$= -2.85 \times 10^3$$

The energy of activation, E, in calories, is the slope multiplied by -4.57:

$$E = 4.57 \times 2.85 \times 10^3 \text{ cal}$$

$$= 13.02 \text{ kcal}$$

The frequency factor A can be obtained from (9.48). The rate constant at either temperature can be used to evaluate A. If we use $25.0°\text{C}$ (see Figure 9.11) we obtain

$$-4.244 = \log_{10} A - \frac{13\,020}{4.57 \times 298}$$

or

$$-4.244 = \log_{10} A - 9.560$$

and thus

$$\log_{10} A = 5.316$$

$$A = 2.07 \times 10^5 \text{ M}^{-1}\text{ s}^{-1}$$

The units of A are the same as those of the rate constant.

ACTIVATION ENERGY

We have seen that the activation energy was looked upon in somewhat different ways in the theories of van 't Hoff and Arrhenius. The interest of van 't Hoff was mainly in thermodynamics, and he placed emphasis on the energy levels of the reactants and products, and of species occurring during the course of reactions. Figure 9.12 shows a potential-energy diagram for a reaction

$$A + B \rightleftharpoons X + Y$$

The internal energy for the products $X + Y$ is larger than that for the reactants $A + B$ by an amount ΔU. In general, there is not much difference between energy and enthalpy changes, so that ΔH will also be positive—that is, the reaction is endothermic.

During the course of the reaction between a molecule of A and a molecule of B the potential energy very often passes through a maximum, as shown in Figure 9.12. The height of this maximum plays a very important role in all theories of the rates of reactions. We call the molecular species having the maximum energy value the *activated complex*, or the *transition state*, and usually denote it by the symbol $\ddagger$. The energy of this complex with respect to the energy of $A + B$, E_1, is the activation energy for the reaction in the left-to-right direction. The energy E_{-1} of the activated complex with respect to X + Y is the activation energy for the reverse reaction. We see from Figure 9.12 that

$$E_1 - E_{-1} = \Delta U \qquad (9.49)$$

This equation is very useful. Sometimes we can measure both E_1 and E_{-1} and then have a value for ΔU, from which ΔH can easily be obtained by making a small correction (see equation (4.28)). More often we know ΔH and hence ΔU, and can measure one activation energy conveniently; the other is then obtained from equation (9.49).

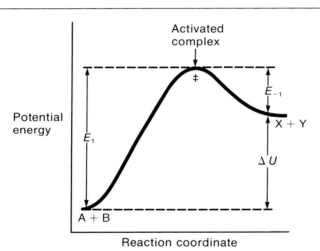

Potential energy

Reaction coordinate

Figure 9.12

Potential-energy diagram for a chemical reaction A + B ⇌ X + Y.

This concept of an energy barrier for a reaction is a very useful and important one. It follows that for an endothermic reaction the activation energy must be at least equal to the endothermicity; for our example, E_1 must be at least equal to ΔU, since E_{-1} cannot be negative. Arrhenius's discussion of the activation energy is closely related to this concept; for reaction between A and B to occur, the molecules must come together with at least the energy E_1, in order to surmount the barrier.

Why is there usually an energy barrier to reaction? Why does not the curve in Figure 9.12 rise smoothly from one level to the other, without passing through a maximum? An important clue is provided by the fact that certain special types of reactions do occur with zero activation energy. There are reactions in which there is simply a pairing of electrons, without the breaking of any chemical bond. A free methyl radical, for example, has an odd electron, its Lewis structure being

$$
\begin{array}{c}
\text{H} \\
\text{..} \\
\text{H : C }\cdot \\
\text{..} \\
\text{H}
\end{array}
$$

When methyl radicals combine to form ethane,

$$2CH_3 \rightarrow C_2H_6$$

they do so with zero activation energy, so that the energy diagram is as shown in Figure 9.13. The activation energy for the reverse reaction, the dissociation of C_2H_6, is obviously equal to the dissociation energy of the bond.

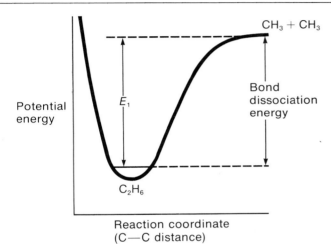

Figure 9.13
Potential-energy diagram for the $2CH_3 \rightarrow C_2H_6$ system.

A zero activation energy, however, only seems to occur in reactions where there is no breaking of a chemical bond. In a few reactions, such as certain processes involving the cytochromes, there is simply the transfer of an electron from one molecule to another, and such reactions are rapid, with a zero or small activation energy. However in most of the reactions with which chemists and biologists are concerned, at least one bond is broken, and at least one bond is formed. For example, in the hydrolysis reaction $C_2H_5X + H_2O \rightarrow C_2H_5OH + HX$, the C—X bond is being broken during reaction, at the same time that the O—C bond is being formed. Even though the overall process is exothermic, the energy released in the formation of the O—C bond is not transferred completely efficiently into the energy required to break the C—X bond, and there is an energy barrier.

The problem of calculating the activation energy of a chemical reaction is a difficult one, even for the simplest of systems. A very important advance was made in 1929 by the American physical chemist Henry Eyring and the Hungarian-British physical chemist (later sociologist) Michael Polanyi (1891–1976). Their contribution was to develop the method of *potential-energy surfaces*, which are like maps of the reaction system. A schematic surface is shown in Figure 9.14. However, even with the aid of the fastest computers now available, it is still not possible to calculate reliable surfaces for most of the reactions in which we are interested. For the time being, therefore, activation energies will have to be obtained experimentally, from the variation of the rate with the temperature.

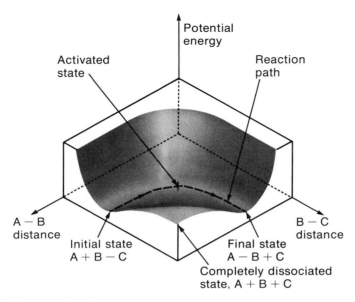

Figure 9.14

A schematic potential-energy surface for a reaction
A + B − C → A − B + C. Any point on the surface represents
the potential energy for particular A − B and B − C distances.
The course of the reaction corresponds to a passage up
the left-hand valley, over the saddle-point, and down into
the right-hand valley.

FREQUENCY FACTOR

The calculation of the frequency factor A is also a matter of some difficulty, although in some cases fairly reliable estimates can be made. We have seen that according to Arrhenius the rate of a reaction is the rate with which collisions occur between the reactant molecules, multiplied by the Boltzmann factor

$$e^{-E/RT}$$

Therefore, if we can calculate the rate of collisions, we can readily obtain the frequency factor.

The first attempt to do this was based on assuming the reacting molecules to be hard spheres, then applying the simple kinetic theory of gases. This works quite well for a few gas reactions, but unfortunately is unsatisfactory for many other reactions, including almost all reactions in solution. It is not difficult to see why this should be so. For two molecules to undergo a chemical reaction they must not merely collide with sufficient mutual energy, but they must come together with such an orientation with respect to each other that the required bonds can be broken and made. It is much too simple to regard the molecules as hard spheres. Other complications arise for reactions in solution, particularly when ions or dipoles are involved. In such cases the reaction may involve changes in electric potential, and these will have important effects on the surrounding solvent molecules.

Complications of these kinds are very difficult to treat by a simple modification of the hard-sphere theory of collisions, and an alternative treatment is necessary. Such a theory was developed in 1935 by Henry Eyring, and is referred to as *activated-complex theory*, or as *transition-state theory*. This theory focuses attention on the activated complex, and is very much concerned with the Gibbs-energy level of the activated complex with reference to the reactants. Figure 9.15 shows a Gibbs-energy diagram and we see that the Gibbs-energy barrier for the reaction from left to right is $\Delta G_1^{\ddagger}$. This diagram is analogous to the potential-energy diagram shown in Figure 9.12, and we have the relationship

$$\Delta G_1^{\ddagger} - \Delta G_{-1}^{\ddagger} = \Delta G^{\circ} \tag{9.50}$$

which resembles (9.49).

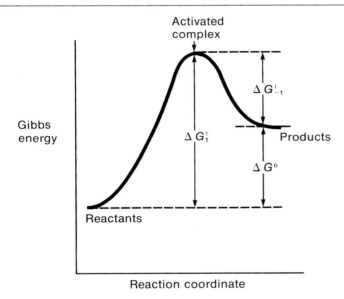

Figure 9.15

Gibbs-energy diagram for a reaction.

According to the Eyring formulation, the rate constant for a reaction involves the very important factor

$$\frac{\mathbf{k}T}{h}$$

where $\mathbf{k}$ is the Boltzmann constant, T the absolute temperature and h is Planck's constant. At 25° C this factor has the magnitude

$$\frac{(1.381 \times 10^{-23} \text{ J K}^{-1})(298 \text{ K})}{6.626 \times 10^{-34} \text{ J s}} = 6.21 \times 10^{12} \text{ s}^{-1}$$

Note that the units are reciprocal seconds, and that the magnitude, roughly 10^{13} s^{-1}, is comparable to the frequency of a molecular vibration. The Eyring theory leads to the result that the rate constant of any reaction is equal to this frequency $\mathbf{k}T/h$ multiplied by the equilibrium constant $K^{\ddagger}$ for the equilibrium between the activated complexes and the reactants:

$$k = \frac{\mathbf{k}T}{h} K^{\ddagger} \qquad (9.51)$$

According to thermodynamics (see equation (5.115)) the standard Gibbs-energy change, $\Delta G^{\ddagger}$, for the formation of activated complexes from reactants is related to this equilibrium constant by

$$\Delta G^{\ddagger} = -RT \ln K^{\ddagger} \qquad (9.52)$$

or

$$K^{\ddagger} = e^{-\Delta G^{\ddagger}/RT} \qquad (9.53)$$

Equation (9.51) therefore takes the very convenient form

$$k = \frac{\mathbf{k}T}{h} e^{-\Delta G^{\ddagger}/RT} \qquad (9.54)$$

It follows that if we have a rate constant at a given temperature we can use (9.54) to calculate from it, at that temperature, the quantity $\Delta G^{\ddagger}$, which is known as the *Gibbs energy of activation* (formerly as the *free energy of activation*).

   Another formulation of the rate equation is also useful and is very commonly employed. The Gibbs energy of activation $\Delta G^{\ddagger}$ is related to the corresponding standard enthalpy and entropy changes by the equation (see equation (5.76))

$$\Delta G^{\ddagger} = \Delta H^{\ddagger} - T\Delta S^{\ddagger} \qquad (9.55)$$

$\Delta H^{\ddagger}$ is known as the *enthalpy of activation* or the *heat of activation*; $\Delta S^{\ddagger}$ as the *entropy of activation*. Combination of (9.54) and (9.55) leads to

$$k = \frac{\mathbf{k}T}{h} e^{\Delta S^{\ddagger}/R} e^{-\Delta H^{\ddagger}/RT} \qquad (9.56)$$

The quantity $\Delta H^{\ddagger}$ differs slightly from the experimental energy of activation E, obtained as previously described (p. 388) from an application of the Arrhenius law; the relationship between the two quantities is, for a reaction in solution,

$$E = \Delta H^{\ddagger} + RT \qquad (9.57)$$

The rate equation (9.56) expressed in terms of E instead of $\Delta H^{\ddagger}$ is therefore

$$k = e \frac{\mathbf{k}T}{h} e^{\Delta S^{\ddagger}/R} e^{-E/RT} \tag{9.58}$$

Comparison of this equation with the Arrhenius equation (9.39) then shows that the frequency factor A is given by

$$A = e \frac{\mathbf{k}T}{h} e^{\Delta S^{\ddagger}/R} \tag{9.59}$$

Therefore, if the frequency factor of a reaction has been determined from an Arrhenius plot, the entropy of activation at a particular temperature can be easily calculated. In work on reactions in solution it is common to quote entropies of activation instead of (or as well as) frequency factors.

   If a reaction involves a considerable loss of entropy when the activated complex is formed from the reactants (i.e., $\Delta S^{\ddagger}$ is negative) the factor $e^{\Delta S^{\ddagger}/R}$ will be a small fraction and the frequency factor will be correspondingly small. A positive $\Delta S^{\ddagger}$ will be associated with a larger frequency factor. If $\Delta S^{\ddagger}$ is zero, the frequency factor is simply $e\mathbf{k}T/h$, and it is of interest to note that the magnitude of this is roughly that given by the simple hard-sphere theory of collisions. We shall later consider some of the factors which influence the magnitudes of entropies of activation, and hence of frequency factors.

EXAMPLE

From the data given in the problem on p. 390, calculate the Gibbs energy of activation at 25.0° C, the entropy of activation, and the enthalpy of activation.

SOLUTION

The Gibbs energy of activation is related to the rate constant by (9.54), which in logarithmic form is

$$\log_{10} k = \log_{10} \frac{\mathbf{k}T}{h} - \frac{\Delta G^{\ddagger}}{4.57T} \tag{9.60}$$

At 25.0° C the value of $\mathbf{k}T/h$ is 6.21×10^{12} s^{-1}. Then, inserting values into (9.60)

$$-4.244 = \log_{10}(6.21 \times 10^{12}) - \frac{\Delta G^{\ddagger}}{4.57 \times 298}$$

or

$$-4.244 = 12.79 - \frac{\Delta G^{\ddagger}}{1326}$$

and thus

$$\Delta G^{\ddagger} = 23\ 195 \text{ cal}$$
$$= 23.2 \text{ kcal}$$

The entropy of activation can be calculated from the frequency factor using (9.59), which in logarithmic form is

$$\log_{10} A = \log_{10} e + \log_{10} \frac{\mathbf{k}T}{h} + \frac{\Delta S^{\ddagger}}{4.57} \qquad (9.61)$$

The value of $\log_{10} A$ was calculated to be 5.316; thus

$$5.316 = \log_{10} 2.718 + 12.79 + \Delta S^{\ddagger}/4.57$$
$$= 0.434 + 12.79 + \Delta S^{\ddagger}/4.57$$

and thus

$$\Delta S^{\ddagger} = -36.1 \text{ cal K}^{-1}$$

From (9.57),

$$\Delta H^{\ddagger} = E - RT = 13\ 020 - 1.986 \times 298$$
$$= 12\ 430 \text{ cal} = 12.4 \text{ kcal}$$

Alternatively, $\Delta H^{\ddagger}$ could have been calculated from $\Delta G^{\ddagger}$ and $\Delta S^{\ddagger}$:

$$\Delta H^{\ddagger} = \Delta G^{\ddagger} + T\Delta S^{\ddagger}$$
$$= 23\ 195 - 36.1 \times 298$$
$$= 12\ 440 \text{ cal}$$
$$= 12.4 \text{ kcal}$$

9.15   REACTIONS IN SOLUTION

In biological systems we are chiefly concerned with reactions occurring in aqueous solution. As we have seen, water is a rather complicated liquid, and reactions in aqueous solution are therefore more difficult to understand in detail than are reactions occurring in the gas phase or in simpler solvents.

We have seen that the theoretical estimation of the magnitude of an activation energy is, for any but the simplest gas reactions, a hopeless task; we must rely on experimental determinations. This is unfortunate, because the activation energy E affects the rate very strongly, through the factor $e^{-E/RT}$.

However, entropies of activation, and the related frequency factors, are much better understood, and it is possible to make reasonably good estimates of their magnitudes for reactions of known types. Conversely, if we do not know much about how a reaction occurs we can, from the experimental frequency factors and entropies of activation, draw some useful inferences about the nature of the reaction. Such inferences are particularly valuable in work with enzyme reactions, where one often does not know the nature of the active region of the enzyme, or what exactly occurs during the course of reaction.

The simplest type of solution reaction to understand is one occurring between two reactant molecules which are relatively nonpolar. The solvent then plays a relatively unimportant role, and there will be only a small change in entropy as the activated complex is formed from the reactants. Theory and experiment indicate that in such cases the second-order rate constant is usually of the order of 10^9 to 10^{11} M^{-1} s^{-1}, which means that there is a small negative entropy of activation. Frequency factors of this magnitude are often referred to as "normal."

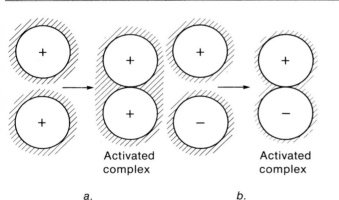

Activated complex   Activated complex

a.   b.

Figure 9.16

An interpretation of entropies of activation in terms of the electrostriction of solvent molecules. In (a) the ions are of the same sign and there is more electrostriction in the activated complex. Therefore there is a decrease in entropy and volume. In (b) there is less electrostriction in the activated complex.

The situation is quite different for reactions between ions. Suppose, for example, that in a reaction in aqueous solution there is an approach of two ions of the same sign, as shown in Figure 9.16a. We have seen on p. 298 that when an ion is present in aqueous solution the surrounding water molecules are bound to it by ion-dipole forces, and are somewhat immobilized. This effect is referred to as *electrostriction*, and leads to loss of entropy. When two ions of the same sign come together, as in Figure 9.16a, there is a much greater loss in entropy, since the activated complex has a larger charge than the reacting ions. There is therefore a loss of entropy when the activated complex is formed; that is, the entropy of activation is negative. A very simple but useful theoretical treatment leads to the result that the loss in entropy will be approximately

$$10 z_A z_B$$

where z_A and z_B are the numbers of unit charges on the reactant molecules. Thus, if the reactants are univalent ions (both positive or both negative), the loss in entropy due to electrostriction will be about 10 cal K^{-1} mol^{-1}. Since the frequency factor is proportional to $e^{\Delta S^{\ddagger}/R}$, which equals $10^{\Delta S^{\ddagger}/4.57}$, it follows that the electrostriction effect in this case will lead to a lowering of the frequency factor by a factor of about $10^{10/4.57}$, i.e., roughly by a hundredfold. In fact, there will be a lowering by a factor of about 100 for each unit of $z_A z_B$.

When reaction involves the approach of opposite charges, the result is a positive contribution to the entropy of activation, and an abnormally high frequency factor. The situation is represented in Figure 9.16b. The activated complex now has a lower charge than either of the reactants and there is less electrostriction in the activated state. There is some release of bound water molecules and a resulting increase in entropy. This simple theory leads to the result that this gain in entropy will be approximately

$$10 \; |z_A| \; |z_B|$$

where $|z_A|$ and $|z_B|$ are the positive values of the charges. Reaction between a unit positive charge and a unit negative charge therefore will lead to a contribution of about 10 cal K^{-1} mol^{-1} to the entropy of activation, and to an enhancement of the frequency factor by a factor of about 100.

These conclusions are very useful in the interpretation of more complex reactions. For example, when an enzyme molecule undergoes interaction with a substrate molecule, often there is an approach of like or opposite charges, with a resulting effect on the entropy of activation. Abnormally large or small entropies of activation therefore provide evidence about such effects. An example will be found on p. 448 for the reaction between the positively-charged enzyme myosin, and the negatively-charged substrate ATP.

INFLUENCE OF IONIC STRENGTH

The ionic strength of a solution can have an important effect on the rate of a reaction occurring in the solution. This is a matter of considerable importance with enzyme reactions, and it is therefore necessary to control the ionic strength. In some cases the effect of changing the ionic strength provides important information about the nature of the reaction.

The theory is very simple for reactions between ions. Consider a reaction of the general type

$$A^{z_A} + B^{z_B} \rightarrow (X^{\ddagger})^{z_A + z_B} \rightarrow \text{products}$$

where z_A and z_B (which may be positive or negative integers) are the values of the charges on the reactants; that on the activated complex is the algebraic sum, $z_A + z_B$. The theory was first worked out in 1922 by J. N. Brønsted. A more modern version of the treatment will now be given, in terms of the concept of the activated complex.

The basis of the treatment is that the rate of a reaction will be proportional to the concentration of the activated complexes $X^{\ddagger}$, rather than to their activity. The rate is therefore given by

$$v = k'[X^{\ddagger}] \tag{9.62}$$

The equilibrium between the activated complexes and the reactants A and B may be expressed as

$$K^{\ddagger} = \frac{a_{\ddagger}}{a_A a_B} = \frac{[X^{\ddagger}]}{[A][B]} \frac{y_{\ddagger}}{y_A y_B} \tag{9.63}$$

where the a's are the activities, and the y's the activity coefficients. Introduction into (9.62) of the expression for $[X^{\ddagger}]$ given by (9.63) gives

$$v = k'K^{\ddagger}[A][B] \frac{y_A y_B}{y_{\ddagger}} = k_o[A][B] \frac{y_A y_B}{y_{\ddagger}} \tag{9.64}$$

where k_o has been written for $k'K^{\ddagger}$. The second-order rate constant k is equal to $v/[A][B]$, so that

$$k = k_o \frac{y_A y_B}{y_{\ddagger}} \tag{9.65}$$

Taking logarithms,

$$\log_{10} k = \log_{10} k_o + \log_{10} \frac{y_A y_B}{y_{\ddagger}} \tag{9.66}$$

According to the Debye-Hückel limiting law (equation 7.24) the activity coefficient of an ion is related to its valency z and the ionic strength I by the equation

$$\log_{10} y = -Bz^2 \sqrt{I} \tag{9.67}$$

The ionic strength is defined by the equation

$$I = \tfrac{1}{2} \sum_i z_i^2 c_i \tag{9.68}$$

where z_i is the valency of the ion and c_i its concentration, the summation being taken of all of the ions in the solution. Introduction of (9.67) into (9.66) the rate equation (9.65) gives us

$$\log_{10} k = \log_{10} k_o + \log_{10} y_A + \log_{10} y_B - \log_{10} y_{\ddagger} \tag{9.69}$$
$$= \log_{10} k_o - B[z_A^2 + z_B^2 - (z_A + z_B)^2]\sqrt{I} \tag{9.70}$$
$$= \log_{10} k_o + 2Bz_A z_A \sqrt{I} \tag{9.71}$$

The value of B is approximately 0.51 for aqueous solutions at 25° C, and (9.71) may thus be written as

$$\log_{10} k = \log_{10} k_o + 1.02 z_A z_B \sqrt{I} \qquad (9.72)$$

This equation has been tested a considerable number of times. The procedure usually has been to measure the rates of ionic reactions in media of varying ionic strength.

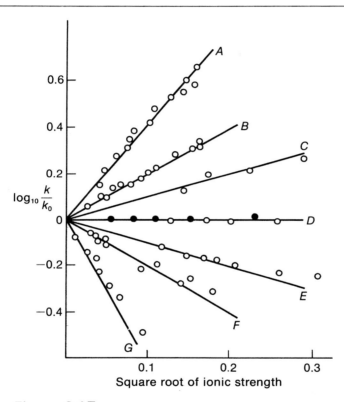

Figure 9.17

Plots of $\log_{10} (k/k_o)$ against the square root of the ionic strength, for ionic reactions of various types. The lines are drawn with slopes equal to $z_A z_B$. The reactions are:

A: $Co(NH_3)_5Br^{2+} + Hg^{2+}$ $\qquad\qquad\qquad$ ($z_A z_B = 4$)
B: $S_2O_8^{2-} + I^-$ $\qquad\qquad\qquad\qquad\qquad\quad$ ($z_A z_B = 2$)
C: $CO(OC_2H_5)N:NO_2 + OH$ $\qquad\qquad\quad$ ($z_A z_B = 1$)
D: $(Cr\,(urea)_6)^{3+} + H_2O$ (open circles) $\qquad$ ($z_A z_B = 0$)
$\quad\;\;$ $CH_3COOC_2H_5 + OH^-$ (closed circles) $\quad$ ($z_A z_B = 0$)
E: $H^+ + Br^- + H_2O_2$ $\qquad\qquad\qquad\qquad\;\;$ ($z_A z_B = -1$)
F: $Co(NH_3)_5Br^{2+} + OH^-$ $\qquad\qquad\qquad\;$ ($z_A z_B = -2$)
G: $Fe^{2+} + Co(C_2O_4)_3^{3-}$ $\qquad\qquad\qquad\quad\;$ ($z_A z_B = -6$)

According to (9.72), a plot of $\log_{10} k$ against $\sqrt{I}$ will give a straight line of slope $1.02 z_A z_B$. Figure 9.17 shows a plot of results for reactions of various types. The lines drawn are those with theoretical slopes, and the points lie very close to them. If one of the reactants is a neutral molecule, $z_A z_B$ is zero, and the rate constant is expected to be independent of the ionic strength. This is true, for example, for the base-catalyzed hydrolysis of ethyl acetate, shown in the figure.

These investigations have provided valuable support for the applicability of the Debye-Hückel limiting law, and for the validity of the assumption that rates are proportional to concentrations of activated complexes.

DIFFUSION-CONTROLLED REACTIONS

When a reaction occurs with a very small, or zero, activation energy, the rate is very large. Such reactions were formerly referred to as "instantaneous" but, as we have seen, their rate constants can now be measured by one of the relaxation techniques (pp. 383–384). If a reaction occurs very rapidly in solution, the rate with which the chemical interaction occurs is often greater than the rate with which the reacting molecules can reach each other by diffusing through the liquid. When this is the case, the rate we measure is not the rate of the chemical interaction, but is the rate of diffusion. Such reactions are referred to as *diffusion controlled*.

We see on p. 473 that the rate constant for the reaction between H_3O^+ and OH^- ions has the very high value of $1.3 \times 10^{11} \, M^{-1} \, s^{-1}$. The actual chemical interaction between a hydrogen and a hydroxide ion, when they come together, involves little or no activation energy; the rate which is measured is the rate with which the ions diffuse past the surrounding water molecules. Diffusion control is often found with proton-transfer and electron-transfer processes, which are very rapid. Such processes commonly occur in biological systems, where diffusion control is therefore very important.

Suppose that we have in solution two uncharged reactant molecules A and B, of radii r_A and r_B cm, and having diffusion coefficients (see p. 000) D_A and D_B cm^2 s^{-1}. It can be shown that the number of encounters v' per cubic centimetre per second between A and B is

$$v' = 4\pi(D_A + D_B)(r_A + r_B)n_A n_B \text{ cm}^{-3} \text{ s}^{-1} \qquad (9.73)$$

where n_A and n_B are the numbers of A and B molecules per cm^3. If the chemical process is so rapid that reaction occurs at each encounter, expression (9.73) is also the rate of reaction in molecules cm^{-3} s^{-1}. The rate in mol cm^{-3} s^{-1} is v'/N_A, where N_A is the Avogadro number, and this is equal to the rate constant k_D (cm^3 mol^{-1} s^{-1}) for the diffusion-controlled reaction multiplied by the product of the concentrations of the reactants in mol cm^{-3}:

$$\frac{v'}{N_A} = k_D \frac{n_A}{N_A} \cdot \frac{n_B}{N_A} \qquad (9.74)$$

or

$$v' = \frac{k_D n_A n_B}{N_A} \tag{9.75}$$

From (9.73) and (9.75) we then find that the rate constant k_D is given by

$$k_D = 4\pi N_A (D_A + D_B)(r_A + r_B) \text{ cm}^3 \text{ mol}^{-1} \text{ s}^{-1} \tag{9.76}$$

The diffusion coefficients may be eliminated from this equation by making use of Stokes's law, which will be considered on p. 481. According to this law, for large spherical particles the individual diffusion coefficients are given by (see (11.66)):

$$D_A = \frac{\mathbf{k}T}{6\pi\eta r_A} \qquad \text{and} \qquad D_B = \frac{\mathbf{k}T}{6\pi\eta r_B} \tag{9.77}$$

where η is the viscosity of the solvent. Introduction of these expressions into (9.76) leads to

$$k_D = \frac{2N_A \mathbf{k}T}{3\eta} \frac{(r_A + r_B)^2}{r_A r_B} \text{ cm}^3 \text{ mol}^{-1} \text{ s}^{-1} \tag{9.78}$$

This equation predicts that, at a given temperature, k_D varies inversely with the viscosity η of the solvent. This prediction has been confirmed experimentally for a number of reactions. The size of the molecules does not have much effect on k_D because $(r_A + r_B)^2/r_A r_B$, equal to

$$2 + \frac{r_A}{r_B} + \frac{r_B}{r_A}$$

does not vary much from one reaction system to another. The reason for this insensitivity of rate to molecular size is that larger molecules will move more slowly than smaller ones, but will present a larger target for encounter with another solute molecule. The two effects largely compensate.

If the molecular radii are assumed equal the expression (9.78) reduces to the simple and useful form

$$k_D = \frac{8RT}{3\eta} \text{ cm}^3 \text{ mol}^{-1} \text{ s}^{-1} \tag{9.79}$$

where $N_A\mathbf{k}$ has been written as R, the gas constant. In the more usual units of $\text{dm}^3 \text{ mol}^{-1} \text{ s}^{-1}$ the expression is

$$k_D = \frac{8RT}{3000\eta} \text{ dm}^3 \text{ mol}^{-1} \text{ s}^{-1} \text{ (M}^{-1} \text{ s}^{-1}) \tag{9.80}$$

For water at 25° C this expression leads to a predicted rate constant of 7.0×10^9 M^{-1} s^{-1}. We therefore expect that reactions between neutral molecules in water should never proceed faster than this, a result which has been confirmed experimentally.

We saw on p. 384 that the observed rate constant for the reaction between hydrogen and hydroxide ions in aqueous solution is 1.3×10^{11} M^{-1} s^{-1}. This very large value is explained partly by the fact that the ions are oppositely charged, so that the diffusion rates are enhanced by the electrostatic attraction. In addition, the hydrogen and hydroxide ions move by the Grotthuss mechanism described on p. 286, which enables them to move at abnormally high speeds.

The theory of diffusion-controlled reactions has also been worked out for charged reactant molecules. If the charges are opposite the rates are enhanced, as we have just seen; if they are the same, abnormally low rates are observed.

Diffusion control is frequently found in enzyme-catalyzed reactions. For example, the addition reactions between enzymes and substrates, with the formation of the enzyme-substrate complex, are sometimes diffusion controlled. We shall see in Section 10.7 that diffusion control becomes particularly important when an enzyme is immobilized in a solid support such as a gel.

9.16   COMPLEX REACTIONS

So far we have been concerned with *elementary* reactions, which occur in a single stage. The reactant molecules come together and form an activated complex, which passes directly into products. However, the majority of chemical reactions in which chemists, biochemists, and biologists are interested are not elementary; instead they involve two or more elementary steps, and then are said to be *complex*. The rest of this chapter is concerned with the characteristics of complex reaction mechanisms.

There are various types of complex reactions. In some of them, relatively stable molecules occur as intermediates. A simple example is the reaction between hydrogen and iodine monochloride, the stoichiometric equation for which is

$$H_2 + 2ICl = I_2 + 2HCl$$

If this reaction went in a single stage it would be third-order—first-order in hydrogen and second-order in ICl, since a molecule of hydrogen and two molecules of ICl would come together to form an activated complex. In fact, the reaction is second-order, being first-order in hydrogen and first-order in iodine monochloride:

$$v = k[H_2][ICl] \tag{9.81}$$

This can be explained if there is initially a slow reaction between one molecule of H_2 and one of ICl,

$$H_2 + ICl \overset{slow}{\rightarrow} HI + HCl$$

This is followed by a rapid reaction between the HI formed in this step and an additional molecule of ICl,

$$\text{HI} + \text{ICl} \xrightarrow{\text{rapid}} \text{HCl} + \text{I}_2$$

Addition of these two reactions gives the overall equation. The HI produced in the first reaction is removed as rapidly as it is formed. The rate of the second process has no effect on the overall rate, which is therefore that of the first step, so that the kinetic behavior is explained. In this scheme the first step is said to be the *rate-determining, rate-controlling,* or *rate-limiting* step, and more is said of it later (p. 408).

Another reaction which involves a fairly stable intermediate, and which has a rate-determining step, is the oxidation of bromide ions by hydrogen peroxide in aqueous acid solution. The stoichiometric equation is

$$2\text{Br}^- + \text{H}_2\text{O}_2 + 2\text{H}^+ \rightarrow \text{Br}_2 + 2\text{H}_2\text{O}$$

and the rate is given by the expression

$$v = k[\text{H}_2\text{O}_2][\text{H}^+][\text{Br}^-] \tag{9.82}$$

This reaction occurs in the following two stages:

(1) $\text{H}^+ + \text{Br}^- + \text{H}_2\text{O}_2 \xrightarrow{\text{slow}} \text{HOBr} + \text{H}_2\text{O}$

(2) $\text{HOBr} + \text{H}^+ + \text{Br}^- \xrightarrow{\text{fast}} \text{Br}_2 + \text{H}_2\text{O}$

The fact that the rate is proportional to $[\text{H}_2\text{O}_2][\text{H}^+][\text{Br}^-]$ suggests that reaction (1) is the rate-determining step. It is reasonable to conclude that HOBr is formed in this reaction, since HOBr is a known chemical substance, although it is not very stable. The slow step (1) of the process can be represented as

activated complex

The activated complex can be formed quite readily as far as the molecular geometry is concerned; thus there is no reason to doubt that reaction (1) may occur. The hypothesis that the reaction does occur by reactions (1) and (2) is supported by the fact that if solutions of HOBr and Br$^-$ are mixed, and the solution is acidified, bromine is produced very rapidly. On the basis of arguments of this kind, we draw conclusions about the elementary reactions that occur in complex mechanisms.

EVIDENCE FOR COMPLEXITY

There are various indications that a reaction is occurring by a complex mechanism. An obvious piece of evidence is when the kinetic law is more complex than would be consistent with the occurrence of the reaction in a single step; two examples of this have already been noted. Enzyme-catalyzed reactions provide additional examples. Thus, if an enzyme reaction involving a single substrate occurred by a simple bimolecular reaction between enzyme and substrate, the kinetics would be first-order with respect to substrate. However, the behavior as far as the substrate is concerned is rarely so simple, and it can therefore be concluded that the reaction occurs in more than one stage.

Another indication of complexity is provided when intermediates can be detected by chemical or other means during the course of reaction. When this can be done a kinetic scheme must be developed which will account for the existence of these intermediates. Sometimes these intermediates are relatively stable substances, while in other cases they are labile substances such as atoms and free radicals. Enzyme-substrate complexes are of the latter class, in that they usually cannot be isolated and preserved and can be detected only by special methods such as spectroscopic ones. Free radicals can sometimes be observed by spectroscopic methods, and evidence for their existence may be obtained by causing them to undergo certain specific reactions which less active substances cannot bring about.

When the nature of the reaction intermediates has been determined by methods such as those outlined above, the next step is to devise a reaction scheme which will involve these intermediates and account for the kinetic features of the reaction. If such a scheme fits the data satisfactorily it can be tentatively assumed that the mechanism is the correct one. It should be emphasized, however, that additional kinetic work frequently leads to the overthrow of schemes which had previously been supposed to be firmly established.

9.17   STEADY-STATE TREATMENT

In order to obtain the overall kinetic equation for a complex reaction scheme it is necessary to write down the differential rate equations for the reactants, products, and intermediates, and to integrate them in order to obtain an expression for the concentrations of reactants and products at various times. However, in many cases there are no mathematical methods available for the solution of the differential equations. When this happens, numerical solutions may be obtained using a computer. It is sometimes possible to avoid this procedure by using an approximate method, known as the *steady-state* method, to obtain the required equation.

The steady-state treatment is based on the fact that the concentrations of certain reaction intermediates, such as free radicals and enzyme-substrate complexes, do not change rapidly during the course of reaction. For example, if $[X]$ is the concentration of such an intermediate, the approximation is made that

$$\frac{d[X]}{dt} = 0 \qquad (9.83)$$

Such relationships lead to a considerable simplification in the equations, and allow a solution to be obtained readily. The steady-state equation (9.83) must be applied only to an intermediate X which is present at minute concentrations compared to the reactants.

Suppose, for example, that the reaction scheme under consideration is

$$A + B \underset{k_{-1}}{\overset{k_1}{\rightleftharpoons}} X$$

$$X \overset{k_2}{\rightarrow} Y$$

The differential rate equations which apply to this set of reactions are:

$$-\frac{d[A]}{dt} = -\frac{d[B]}{dt} = k_1[A][B] - k_{-1}[X] \tag{9.84}$$

$$\frac{d[X]}{dt} = k_1[A][B] - k_{-1}[X] - k_2[X] \tag{9.85}$$

$$\frac{d[Y]}{dt} = k_2[X] \tag{9.86}$$

In order to treat this problem exactly it would be necessary to eliminate [X] and to solve the resulting differential equation to get [Y] as a function of t. Unfortunately, however, in spite of the simplicity of the kinetic scheme, it is not possible to obtain an explicit solution.

The steady-state treatment involves using equation (9.83) so that, from (9.85),

$$k_1[A][B] - k_{-1}[X] - k_2[X] = 0 \tag{9.87}$$

The concentration of [X] is therefore given by

$$[X] = \frac{k_1[A][B]}{k_{-1} + k_2} \tag{9.88}$$

and insertion of this in equation (9.86) gives

$$v = \frac{d[Y]}{dt} = \frac{k_1 k_2 [A][B]}{k_{-1} + k_2} \tag{9.89}$$

Several examples of the application of the steady-state method to enzyme and other systems will be given later.

THE RATE-DETERMINING STEP

We have seen that in certain cases the rate of one step in a reaction mechanism controls the overall rate of the reaction. This step is then known as the *rate-*

determining, *rate-controlling*, or *rate-limiting* step. In biological work the expression *master reaction* has also been used to describe such a step.

Suppose, for example, that in the reaction scheme just considered the intermediate X is converted very rapidly into Y, much more rapidly than it can go back into A + B. In that case the rate of the reaction will be the rate of formation of X from A + B, i.e.,

$$v = k_1[A][B] \tag{9.90}$$

since as soon as X is formed it is transformed into Y. The formation of X is therefore the rate-determining step. The exact condition is

$$k_{-1} \ll k_2$$

and we see that the steady-state rate equation (9.89) becomes (9.90) if this inequality is satisfied.

Alternatively, suppose that the second reaction, $X \overset{k_2}{\rightarrow} Y$, is very slow compared to the reverse of the first reaction, i.e.,

$$k_2 \ll k_{-1}$$

Reaction 2 is then the rate-determining step and the rate is

$$v = k_2[X] \tag{9.91}$$

Reaction 2 is too slow to disturb the equilibrium $A + B \underset{k_{-1}}{\overset{k_1}{\rightleftharpoons}} X$, so that

$$[X] = \frac{k_1}{k_{-1}}[A][B] \tag{9.92}$$

and insertion of this expression into (9.91) gives

$$v = \frac{k_1 k_2}{k_{-1}}[A][B] \tag{9.93}$$

Again, this is the expression to which (9.89) reduces if this latter inequality $(k_2 \ll k_{-1})$ is satisfied.

It is important to work out each kinetic scheme separately, as there are some pitfalls. Note in particular that when there is a chain reaction (see Section 9.19) there is no rate-determining step, the rate *not* being equal to that of any particular step.

9.18   RATE CONSTANTS AND EQUILIBRIUM CONSTANTS

For an elementary reaction it is easy to show that the equilibrium constant must be the ratio of the rate constants in forward and reverse directions. Thus, consider the elementary process

$$A + B \underset{k_{-1}}{\overset{k_1}{\rightleftharpoons}} X + Y$$

The rates in forward and reverse directions are

$$v_1 = k_1[A][B] \tag{9.94}$$

$$v_{-1} = k_{-1}[X][Y] \tag{9.95}$$

If the system is at equilibrium these rates are equal, and hence

$$\frac{k_1}{k_{-1}} = \left(\frac{[X][Y]}{[A][B]}\right)_{eq} = K \tag{9.96}$$

This argument can be extended to a reaction which occurs in two or more stages. Consider, for example, the oxidation of lactate to pyruvate by the oxidized form of nicotinamide adenine dinucleotide, NAD^+, which is reduced to $NADH$:

$$NAD^+ + CH_3CHOHCOO^- \rightleftharpoons NADH + H^+ + CH_3COCOO^-$$

This reaction is catalyzed by the enzyme lactate dehydrogenase, LDH, and the mechanism may be represented in a somewhat simplified form as follows:

(1) $LDH + NAD^+ + CH_3CHOHCOO^- \underset{k_{-1}}{\overset{k_1}{\rightleftharpoons}} LDH\overset{\displaystyle NAD^+}{\underset{\displaystyle CH_3CHOHCOO^-}{<}}$

(2) $LDH\overset{\displaystyle NAD^+}{\underset{\displaystyle CH_3CHOHCOO^-}{<}} \underset{k_{-2}}{\overset{k_2}{\rightleftharpoons}} LDH + NADH + H^+ + CH_3COCOO^-$

If the reaction is allowed to proceed to equilibrium, the rates of each elementary reaction and its reverse must be the same. At equilibrium, therefore, for the first reaction

$$k_1[LDH][NAD^+][CH_3CHOHCOO^-] = k_{-1}[X] \tag{9.97}$$

where X has been written for the ternary complex. This equation can be written as

$$\left(\frac{[X]}{[LDH][NAD^+][CH_3CHOHCOO^-]}\right)_{eq} = \frac{k_1}{k_{-1}} = K_1 \qquad (9.98)$$

where K_1 is the equilibrium constant for the first reaction. Similarly, for the second reaction,

$$k_2[X] = k_{-2}[LDH][NADH][H^+][CH_3COCOO^-] \qquad (9.99)$$

or

$$\left(\frac{[LDH][NADH][H^+][CH_3COCOO^-]}{[X]}\right)_{eq} = \frac{k_2}{k_{-2}} = K_2 \qquad (9.100)$$

where K_2 is the equilibrium constant for the second reaction. Multiplication of (9.98) and (9.100) gives

$$\left(\frac{[NADH][H^+][CH_3COCOO^-]}{[NAD^+][CH_3CHOHCOO^-]}\right)_{eq} = K_1K_2 = K \qquad (9.101)$$

which is the equilibrium expression for the overall reaction; the enzyme concentration has cancelled out. Note that this overall equilibrium constant K is the product of the equilibrium constants for the elementary reactions, and that it is also equal to the product of the rate constants for the reactions in the forward direction divided by the product of those for the reverse reactions:

$$K = K_1K_2 = \frac{k_1k_2}{k_{-1}k_{-2}} \qquad (9.102)$$

This result is easily extended to mechanisms involving a larger number of elementary reactions.

The principle that at equilibrium each elementary process is exactly balanced by its reverse reaction is known as the *principle of microscopic reversibility* or the *principle of detailed balancing*.

9.19 FREE-RADICAL REACTIONS

Reactions frequently occur by a series of reactions in which free radicals play a part. The important distinction between a free radical and an ion may be illustrated by comparison of the hydroxyl radical and the hydroxide ion. Imagine that an oxygen-hydrogen bond in the water molecule is split homolytically, i.e., one of the electrons goes with one fragment and the other with the other:

In this process, two free radicals are produced, an atom being a species of radical. Both radicals are neutral. The hydrogen atom consists of a proton and an electron, and the hydroxyl radical consists of nine protons (one in the nucleus of the hydrogen atom and eight protons in the oxygen nucleus) and nine electrons (seven valence electrons plus two $1s$ electrons). The hydrogen atom and the hydroxyl radical are both one electron short of the inert-gas structures, and therefore are very reactive species. Radicals combine with one another with very low or zero activation energies, and they undergo reactions with stable molecules with quite low activation energies.

In the ionization of water, a bond is split heterolytically, the electron pair remaining with the oxygen atom:

The hydroxide ion is negatively charged, having nine protons and ten electrons. It has the same electronic configuration as neon and therefore is chemically stable and unreactive. Whereas hydroxyl radicals cannot be stored, solutions of hydroxide ions will remain intact for long periods of time.

Ions play little part in gas-phase reactions, owing to the difficulty with which they are formed in the absence of an ionizing solvent. Atoms and free radicals are produced more easily in the gaseous phase and, because they enter readily into further reaction, they are important intermediates in reactions. For example, consider the reaction between hydrogen and bromide. If the product of reaction, HBr, is removed as it is formed, so as to eliminate its effect on the behavior, the rate equation is

$$v = k[H_2][Br_2]^{\frac{1}{2}} \tag{9.103}$$

This behavior can be explained by the mechanism

(1) $Br_2 \overset{k_1}{\to} 2Br$ Initiation

(2) $Br + H_2 \overset{k_2}{\to} HBr + H$ ⎫
(3) $H + Br_2 \overset{k_3}{\to} HBr + Br$ ⎭   Chain propagation

(4) $2Br \overset{k_4}{\to} Br_2$ Termination

The first reaction, the production of bromine atoms from a bromine molecule, is known as the *initiation reaction*, since it starts the whole process. Reactions (2) and (3), the so-called *chain-propagation steps*, play a very important role in reactions of this type. Bromine atoms disappear in reaction (2) and reappear in reaction (3); hydrogen atoms disappear in (3) and come back again in (2). Because of this feature, a small number of Br atoms, produced in reaction (1), can bring about a considerable amount of reaction, since after producing two molecules of hydrogen bromide, one in (2) and one in (3), a bromine atom is regenerated. If it were not for reaction (4), a single pair of bromine atoms could bring about the reaction of all of the H_2 and Br_2 present. Because of the termination reaction (4), however, only a limited amount of reaction is brought about each time a pair of bromine atoms is produced. Bromine atoms are continuously formed by reaction (1), and this keeps the reaction going.

   A reaction of this type is known as a *chain reaction*. The essential feature of a chain reaction is that there must be at least two chain-propagating steps, each regenerating the species removed in the other. Although not all free-radical reactions are chain reactions, many are.

   The way in which the chain-reaction mechanism explains the experimental rate equation (9.103) can be shown as follows. The rate of disappearance of hydrogen is equal to

$$v = k_2[Br][H_2] \tag{9.104}$$

The concentration of bromine atoms can be obtained by use of the steady-state method, which must now be applied to the two unstable intermediates H and Br. The steady state equation for H is

$$\frac{d[H]}{dt} = k_2[Br][H_2] - k_3[H][Br_2] = 0 \tag{9.105}$$

and that for Br is

$$\frac{d[Br]}{dt} = k_1[Br_2] - k_2[Br][H_2] + k_3[H][Br_2] - k_4[Br]^2 = 0 \tag{9.106}$$

We thus have two equations in the two unknowns [H] and [Br], and thus can solve for both of these concentrations. A solution for [Br] is quickly obtained if we add equations (9.105) and (9.106):

$$k_1[Br_2] - k_4[Br]^2 = 0 \tag{9.107}$$

and thus

$$[Br] = \left(\frac{k_1}{k_4}\right)^{\frac{1}{2}} [Br_2]^{\frac{1}{2}} \tag{9.108}$$

Insertion of this expression for [Br] into (9.104) gives

$$v = k_2 \left(\frac{k_1}{k_4}\right)^{\frac{1}{2}} [H_2][Br_2]^{\frac{1}{2}} \qquad (9.109)$$

which is of the same form as (9.103). Notice that the rate constant for the overall reaction, $k_2(k_1/k_4)^{\frac{1}{2}}$, is not that for any elementary reaction, but is a composite constant.

A typical organic free-radical chain reaction is the decomposition of ethane,

$$C_2H_6 = C_2H_4 + H_2$$

Under most conditions this is a simple first-order reaction, and originally it was thought that it occurred in one stage by a molecular reaction; i.e., that a small fraction of the molecules have sufficient energy for two C—H bonds to be ruptured, so that a hydrogen molecule is liberated:

However, the most recent evidence indicates that such a mechanism plays an unimportant role, and that practically all of the decomposition occurs by the following chain mechanism:

(1) $\qquad\qquad C_2H_6 \rightarrow 2CH_3$ ⎫
                                                    ⎬ Initiation
(2) $\qquad CH_3 + C_2H_6 \rightarrow CH_4 + C_2H_5$ ⎭

(3) $\qquad\qquad C_2H_5 \rightarrow C_2H_4 + H$ ⎫
                                                    ⎬ Chain propagation
(4) $\qquad H + C_2H_6 \rightarrow H_2 + C_2H_5$ ⎭

(5) $\qquad\qquad 2C_2H_5 \rightarrow C_4H_{10}$ Termination

The initiation process involves the breaking of a C—C bond, which is the weakest bond in the molecule. Reaction (2) is in a sense part of the initiation reaction, since it converts CH_3 into a radical C_2H_5, which can be involved in propagation; reaction (2) is not a propagation reaction since CH_3 is not regenerated from C_2H_5. Reactions (3) and (4) are chain-propagating steps: C_2H_5 disappears in (3) and appears in (4), while H disappears in (4) and appears in (3). The main products of the reaction, C_2H_4 and H_2, are formed in these propagation steps. The termination step (5) forms butane, C_4H_{10}, which can be detected as a very minor product of the reaction. Methane, formed in reaction (2), has also been observed as a minor product of the reaction.

9.20   IONIC REACTIONS

Some solvents, particularly water, favor the formation of ions, so that reactions in solution frequently involve ions as intermediates. Indeed, the reactants and products in such reactions are often ions. A simple two-stage mechanism involving the transient formation of ions is involved in the hydrolysis of certain organic chlorides, which may be written simply as RCl; the hydrolysis by hydroxide ions follows the stoichiometric equation

$$RCl + OH^- \rightarrow ROH + Cl^-$$

In certain cases, the reaction occurs by an initial slow ionization process,

$$RCl \rightarrow R^+ + Cl^-$$

followed by the rapid reaction between R^+ and the OH^- ion,

$$R^+ + OH^- \rightarrow ROH$$

As soon as an R^+ ion is produced it reacts with OH^-. The overall rate of the reaction thus is equal to the rate of the ionization process. In other words, the ionization process is the rate-determining step. Consequently, the rate does not depend on the concentration of the OH^- ions, and the reaction is kinetically of the first order.

   Ionic mechanisms are very important in organic reactions in solution, including reactions catalyzed by enzymes. In almost all such reactions there is either a total or a partial ionization. Inorganic reactions in aqueous solution also usually involve steps in which ions enter into reaction.

9.21   CATALYSIS†

The rates of many reactions are influenced by the presence of a substance which remains unchanged at the end of the process. Examples are: the conversion of starch into sugars, the rate of which is influenced by acids; the decomposition of hydrogen peroxide, influenced by ferric ions; and the formation of ammonia in the presence of spongy platinum. In 1836 such reactions were classified by the Swedish chemist Jons Jakob Berzelius (1779–1848) under the title of *catalyzed processes*. Substances which decrease the rate of reaction are referred to as *negative catalysts*, or better, as *inhibitors*. It is convenient to classify catalyzed reactions according to whether they occur homogeneously (in a single phase) or heterogeneously (at an interface between two phases).

---

† This word comes from the Greek *kata*, wholly, *lyein*, to loosen.

Various definitions of catalysis have been proposed. An early definition, suggested in 1895 by Friedrick Wilhelm Ostwald, was that a catalyst is "any substance that alters the velocity of a chemical reaction without modification of the energy factors of the reaction." Later, in 1902, Ostwald proposed an alternative definition: "A catalyst is any substance that alters the velocity of a chemical reaction without appearing in the end product of the reaction." A slightly different definition is that "a catalyst alters the velocity of a chemical reaction and is both a reactant and a product of the reaction." These definitions were intended to exclude from the category of catalysts substances that accelerated the rate of a reaction by entering into reaction thus disturbing the position of equilibrium; such substances are reactants in the ordinary sense. In these definitions of catalysis there is no reference to the fact that a small amount of a catalyst has a large effect on the rate; this is frequently the case, but is not an essential characteristic of a catalyst.

Although by definition the amount of a catalyst should be unchanged at the end of the reaction, the catalyst is invariably involved in the chemical process. In the case of one reacting substance, a complex may be formed between this reactant (the *substrate*) and the catalyst. If there is more than one substrate the complex may involve one or more molecules of the substrate combined with the catalyst. These complexes are formed only as intermediates and decompose to give the products of the reaction, with the regeneration of the catalyst molecule. For example, when a reaction is catalyzed by hydrogen ions, an intermediate complex involving the substrate and a hydrogen ion is formed, and this later reacts further with the liberation of the ion and the formation of the products of the reaction.

The catalyst is unchanged at the end of the reaction, and therefore gives no energy to the system, so that according to thermodynamics it can have no influence on the position of equilibrium. It follows that since the equilibrium constant K is the ratio of the rate constants in the forward and reverse directions, i.e., $K = k_1/k_{-1}$, a catalyst must influence the forward and reverse rates in the same proportion. This conclusion has been verified experimentally in a number of instances.

An extremely small amount of a catalyst frequently causes a considerable increase in the rate of a reaction. Colloidal palladium at a concentration of 1 mol in 10^8 dm^3 will cause hydrogen peroxide to decompose at a measurable rate. The effectiveness of a catalyst is sometimes expressed in terms of its *turnover number*, which is the number of molecules of substrate decomposed per minute by one molecule of the catalyst. For example, the enzyme catalase has, under certain conditions, a turnover number of 5 000 000 for the decomposition of hydrogen peroxide ($2H_2O_2 = 2H_2O + O_2$). Since the turnover number generally varies with the temperature and with the concentration of substrate, it is not a particularly useful quantity in kinetic work, in which case the appropriate rate constant is used.

The rate of a catalyzed reaction is often proportional to the concentration of the catalyst

$$v = k[C] \tag{9.110}$$

where $[C]$ represents the concentration of the catalyst, and k is a function of the concentration of the substrate. If (9.110) were obeyed exactly, the rate of reaction in the absence of the catalyst would be zero. Many examples are known for which it is necessary to introduce an additional term which is independent of the catalyst concentration

$$v = k[C] + v_o \qquad (9.111)$$

At zero concentration of the catalyst the reaction occurs with the velocity v_o.

It is usually found that the activation energy of a catalyzed reaction is lower than that of the same reaction when it is uncatalyzed. In other words, catalysts generally work by permitting the reaction to occur by another reaction path which has a lower energy barrier. This is shown schematically in Figure 9.18. It is important to note that inhibitors do *not* work by introducing a higher reaction path; this would not reduce the rate, since the reaction would continue to occur by the alternative mechanism. Inhibitors act either by destroying catalysts already present, or by removing reaction intermediates such as free radicals.

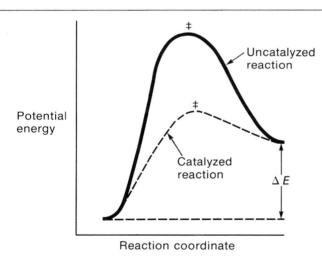

Figure 9.18

The lowering of the energy barrier brought about by a catalyst. Since the catalyst is not used up during the reaction, ΔE is the same for the catalyzed reaction as for the uncatalyzed, but the activation energies in both directions are lower for the catalyzed reaction.

ACID-BASE CATALYSIS

The study of catalysis by acids and bases played a very important part in the development of chemical kinetics, since many of the reactions studied in the early days of the subject were of this type. The early investigations of the kinetics of reactions catalyzed by acids and bases were carried out at the same time that the electrolytic dissociation theory was being developed, and the kinetic studies contributed considerably to the development of that theory. The reactions considered from this point of view were chiefly the inversion of cane sugar and the hydrolysis of esters. It was first realized in 1884 by Ostwald and later by Arrhenius that the ability of an acid to catalyze these reactions is independent of the nature of the anion but is approximately proportional to its electrical conductivity. According to them, the conductivity of an acid is a measure of its strength, i.e., of the concentration of hydrogen ions, and the hydrogen ions were assumed to be the sole effective acid catalyst. Similarly, in catalysis by alkalis the rate is proportional to the concentration of the alkali but independent of the nature of the cation, suggesting that the active species is the hydroxide ion.

The idea that the only catalyzing species in reactions of this type are the hydrogen and hydroxide ions has been found to require modification in a number of instances. Many reactions do exist, however, for which only these two ions are effective catalysts. Examples are the hydrolyses of esters, which have not been shown to be catalyzed by anything (apart from enzymes) except these two ions.

If such reactions are carried out in a sufficiently strong acid solution, the concentration of hydroxide ions may be reduced to such an extent that these ions do not have any appreciable catalytic action. The hydrogen ions are then the only effective catalyst, and the rate of reaction (at least at concentrations of catalyst and substrate which are not too high) would be given by an expression of the type

$$v = k_{H^+}[H^+][S] \tag{9.112}$$

where k_{H^+} is the rate constant for the hydrogen-ion-catalyzed reaction. Such a reaction would be of the second overall order with respect to concentration. If catalysis is effected simultaneously by hydrogen and hydroxide ions and reaction may also occur spontaneously, i.e., without a catalyst, the rate of reaction is

$$v = k_o[S] + k_{H^+}[H^+][S] + k_{OH^-}[OH^-][S] \tag{9.113}$$

The first-order rate constant is therefore given by

$$k = k_o + k_{H^+}[H^+] + k_{OH^-}[OH^-] \tag{9.114}$$

In these equations k_o is the rate constant of the spontaneous reaction, and k_{H^+} and k_{OH^-} are known as the *catalytic constants* for H^+ and OH^-, respectively.

The rate equation may be expressed in a different form by making use of the fact that $[H^+][OH^-] = K_w$, where K_w is the ionic product of water. Elimination of $[OH^-]$ gives

$$k = k_o + k_{H^+}[H^+] + \frac{k_{OH^-}K_w}{[H^+]} \tag{9.115}$$

while elimination of $[H^+]$ gives

$$k = k_o + \frac{k_{H^+}K_w}{[OH^-]} + k_{OH^-}[OH^-] \tag{9.116}$$

In many cases one of these terms containing concentration is negligibly small compared with the other. If work is carried out with 0.1 M hydrochloric acid, for example, the second term in (9.115) is $k_{H^+} \times 10^{-1}$ while the third term is $k_{OH^-} \times 10^{-13}$ (since $K_w = 10^{-14}$). Consequently, unless k_{OH^-} is at least 10^9 greater than k_{H^+}, the third term will be negligible compared with the second so that, at this acid concentration, catalysis by hydroxide ions will be negligible compared with that by hydrogen ions. Similarly, in 0.1 M sodium hydroxide solution, catalysis by hydrogen ions will usually be unimportant compared with that by hydroxide ions. In general, there will be an upper range of hydrogen-ion concentrations at which catalysis by hydroxide ions will be unimportant, and a lower range at which catalysis by hydroxide ions will predominate and catalysis by hydrogen ions will be unimportant. Within these ranges the rate will be a linear function of $[H^+]$ and of $[OH^-]$, respectively. In the former range the value of the catalytic constant k_{H^+} can readily be determined from the experimental data; in the lower range k_{OH^-} can be so determined. The constants for the hydrolysis of ethyl acetate were measured in this manner by J. J. A. Wijs in 1893, and he also obtained a value for K_w by making use of the fact that the velocity is a minimum when the second and third terms in (9.115) are equal; this gives

$$[H^+]_{min} = \left(\frac{k_{OH^-}K_w}{k_{H^+}}\right)^{\frac{1}{2}} \qquad (9.117)$$

so that from the values of $[H^+]_{min}$, k_{H^+}, and k_{OH^-}, the ionic product K_w can be obtained.

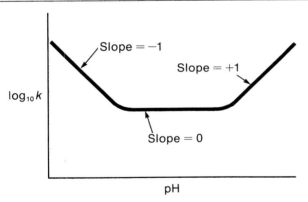

Figure 9.19

A schematic plot of $\log_{10} k$ against pH for a reaction which is catalyzed by both hydrogen and hydroxide ions, and for which the uncatalyzed reaction occurs at an appreciable rate.

A plot of the logarithm of the rate constant against the pH of the solution is shown in Figure 9.19. There are regions of catalysis by hydrogen and hydroxide ions, separated by a region in which the amount of catalysis is unimportant in comparison with the spontaneous reaction. When the catalysis is largely by hydrogen ions, $k = k_{H^+}[H^+]$, so that

$$\log k = \log k_{H^+} + \log [H^+] \qquad (9.118)$$

or

$$\log k = \log k_{H^+} - pH \qquad (9.119)$$

The slope is therefore -1, which is the slope of the left-hand limb. The slope of the right-hand limb is similarly $+1$, and the velocity in the intermediate region is equal to $k_o[S]$, and k_o can thus be determined directly from the rate in this region. If the rate of the spontaneous reaction is sufficiently small, the horizontal part of the curve is not found and two limbs intersect fairly sharply. If either k_{H^+} or k_{OH^-} is negligibly small, the corresponding sloping limb of the curve is not found.

The evidence is that acid and base catalysis involves the transfer of a proton to or from the substrate molecule. It is therefore to be expected that catalysis may be effected by acids and bases other than H^+ and OH^-. This has been found in a number of instances, and *general acid-base catalysis* is then said to occur. For other reactions, catalysis by species other than H^+ and OH^- has not been detected experimentally, and the catalysis is then said to be *specific* with respect to these two ions.

An example of acid-base catalysis is the reaction between acetone and iodine in aqueous solution:

$$CH_3COCH_3 + I_2 = CH_3COCH_2I + HI$$

The rate of this reaction is linear in the acetone concentration and in any acid species present in solution, but is independent of the concentration of iodine; indeed, the corresponding bromination reaction proceeds at the same rate if the iodine is replaced by bromine. This suggests that the iodine or bromine are involved in a rapid step which has no effect on the overall reaction rate. The evidence is that the rate-determining step is the conversion of the ordinary keto form of acetone into its enol form:

The enol form is then rapidly iodinated or brominated:

The way in which acids catalyze the conversion of the keto form into the enol form is as follows. First, the acidic species HA transfers a proton to the oxygen atom on the acetone molecule:

$$\text{HA} + \text{CH}_3-\overset{\overset{\displaystyle O}{\|}}{\text{C}}-\text{CH}_3 \;\rightarrow\; \text{CH}_3-\overset{\overset{\displaystyle \overset{+}{O}-H}{\|}}{\text{C}}-\text{CH}_3 + \text{A}^-$$

The transferred proton is bound to the oxygen atom by one of its lone pairs of electrons. The protonated acetone then gives up one of its other hydrogen atoms to some base B present in solution (which may be water), at the same time forming the enol form of acetone.

$$\text{CH}_3-\overset{\overset{\displaystyle \overset{+}{O}-H}{\|}}{\text{C}}-\text{CH}_3 + \text{B} \;\rightarrow\; \text{CH}_3-\overset{\overset{\displaystyle O-H}{|}}{\text{C}}=\text{CH}_2 + \text{BH}^+$$

It has been demonstrated that this process is catalyzed not only by H^+ ions but by other species present in solution; i.e., there is general acid catalysis. There is also general basic catalysis of this reaction.

Since acid-base catalysis always involves the transfer of a proton from the acid catalyst or to the basic catalyst, it is natural to seek a correlation between the effectiveness of a catalyst and its strength as an acid or base, since this strength is a measure of the ease with which the catalyst transfers a proton to or from a water molecule. The most satisfactory relationship between the rate constant k_a and the acid dissociation constant K_a of a monobasic acid is the equation

$$k_a = G_a K_a^\alpha \tag{9.120}$$

which was proposed in 1924 by J. N. Brønsted and K. J. Pederson. G_a and α are constants, the latter being less than unity. The analogous equation for basic catalysis is

$$k_b = G_b K_b^\beta \tag{9.121}$$

Similarly, the relationship between the catalytic constant of a base and the acid strength of the conjugate acid may be expressed as

$$k_b = G'_b \left(\frac{1}{K_a}\right)^\beta \tag{9.122}$$

β is again always less than unity. Equations (9.120), (9.121), and (9.122) are commonly spoken of as *Brønsted relationships*.

These equations require a modification if they are to be applied to an acid which has more than one ionizable proton, or to a base which can accept more than one proton. The conclusions may be generalized by means of the following relationships, given by Brønsted:

$$\frac{k_a}{p} = G_a \left(\frac{qK_a}{p} \right)^\alpha \tag{9.123}$$

and

$$\frac{k_b}{q} = G_b \left(\frac{p}{qK_a} \right)^\beta \tag{9.124}$$

In (9.123) p is the number of dissociable protons bound equally strongly in the acid, while q is the number of equivalent positions in the conjugate base to which a proton may be attached. Similarly, in (9.124) q is the number of positions in the catalyzing base to which a proton may be attached, while p is the number of equivalent dissociable protons in the conjugate acid. Very satisfactory agreement has been obtained in all the cases to which these equations have been applied.

PROBLEMS

9.1   (a) The rate constant of a first-order reaction is $2.5 \times 10^{-6} \, s^{-1}$, and the initial concentration is $0.1 \, mol \, dm^{-3}$. What is the initial rate in mol $dm^{-3} \, s^{-1}$, in mol $cm^{-3} \, s^{-1}$, and in mol $cm^{-3} \, min^{-1}$?

(b) The initial rate of a second-order reaction is $5.0 \times 10^{-7} \, mol \, dm^{-3} \, s^{-1}$, and the initial concentrations of the two reacting substances are $0.2 \, mol \, dm^{-3}$. What is the rate constant in $dm^3 \, mol^{-1} \, s^{-1}$ and in $cm^3 \, mol^{-1} \, s^{-1}$?

9.2 The stoichiometric equation for the oxidation of bromide ions by hydrogen peroxide in acid solution is

$$2Br^- + H_2O_2 + 2H^+ \rightarrow Br_2 + 2H_2O$$

Since the reaction does not occur in one stage, the rate equation does not correspond to this stoichiometric equation, but is

$$v = k[H_2O_2][H^+][Br^-]$$

(a) If the concentration of H_2O_2 is increased by a factor of 3, by what factor is the rate of disappearance of Br^- ions increased?

(b) If, under certain conditions, the rate of disappearance of Br^- ions is $7.2 \times 10^{-3} \, mol \, dm^{-3} \, s^{-1}$, what is the rate of disappearance of hydrogen peroxide? What is the rate of appearance of bromine?

(c) What is the effect on the rate constant k of increasing the concentration of bromide ions?

(d) If by the addition of water to the reaction mixture the total volume were doubled, what would be the effect on the rate of disappearance of Br^-? What would be the effect on the rate constant k?

9.3 A reaction obeys the stoichiometric equation

$$A + 2B \rightarrow 2C$$

Rates of formation of C at various concentrations of A and B are given in the table below:

$[A]$/mol dm^{-3}	$[B]$/mol dm^{-3}	Rate/mol dm^{-3} s^{-1}
3.5×10^{-2}	2.3×10^{-2}	5.0×10^{-7}
7.0×10^{-2}	4.6×10^{-2}	2.0×10^{-6}
7.0×10^{-2}	9.2×10^{-2}	4.0×10^{-6}

What are x and y in the rate equation

$$v = k[A]^x[B]^y$$

and what is the rate constant k?

9.4 Some results for the rate of a reaction between two substances A and B are shown below. Deduce the order x with respect to A, the order y with respect to B, and the rate constant.

$[A]$/mol dm^{-3}	$[B]$/mol dm^{-3}	Rate/mol dm^{-3} s^{-1}
1.4×10^{-2}	2.3×10^{-2}	7.4×10^{-9}
2.8×10^{-2}	4.6×10^{-2}	5.92×10^{-8}
2.8×10^{-1}	4.6×10^{-2}	5.92×10^{-6}

9.5 A reaction is endothermic by 15.0 kcal. What is the minimum value its activation energy can have?

9.6 The rate constant for a reaction at 30° C is found to be exactly twice the value at 20° C. Calculate the activation energy.

9.7 The rate constant for a reaction at 230° C is found to be exactly twice the value at 220° C. Calculate the activation energy.

9.8 An unstable metabolite is found to disappear according to first-order kinetics, the rate constants at various temperatures being as follows:

Temperature (°C)	Rate Constant/s^{-1}
15.0	4.18×10^{-6}
20.0	7.62×10^{-6}
25.0	1.37×10^{-5}
30.0	2.41×10^{-5}
37.0	5.15×10^{-5}

Plot $\log_{10} k$ against $1/T(K)$ and determine the activation energy. Calculate also, at 25.0° C, the enthalpy of activation, the Gibbs energy of activation, the frequency factor, and the entropy of activation.

9.9 The following data have been obtained for the non-enzymic hydrolysis of adenosine triphosphate, catalyzed by hydrogen ions:

Temperature (°C)	k/s^{-1}
39.9	4.67×10^{-6}
43.8	7.22×10^{-6}
47.1	10.0×10^{-6}
50.2	13.9×10^{-6}

Calculate, at 40.0° C, the Gibbs energy of activation, the activation energy, the enthalpy of activation, the frequency factor, and the entropy of activation.

9.10 The half-life of the thermal denaturation of hemoglobin, a first-order process, has been found to be 3460 seconds at 60.0° C and 530 seconds at 65.0° C. Calculate the heat of activation and entropy of activation at 60.0° C, assuming the Arrhenius law to apply.

9.11 A reaction of stoichiometry A + 2B = Y + Z occurs by the mechanism

$$A + B \xrightarrow{k_1} X \qquad \text{(slow)}$$

$$X + B \xrightarrow{k_2} Y + Z \qquad \text{(very fast)}$$

where X is an intermediate. Write down the expression for the rate of formation of Y in terms of [A], [B], and the rate constants.

9.12 A reaction of stoichiometry A + B = Y + Z is catalyzed by C and occurs by the mechanism

$$A + C \xrightarrow{k_1} X \qquad \text{(slow)}$$

$$X + B \xrightarrow{k_2} Y + Z + C \qquad \text{(very fast)}$$

Write down the expression for the rate of formation of Y.

9.13 A reaction A + 2B = 2Y + 2Z occurs according to the mechanism

$$A \underset{k_{-1}}{\overset{k_1}{\rightleftharpoons}} 2X \qquad \text{(rapid equilibrium)}$$

$$X + B \xrightarrow{k_2} Y + Z \qquad \text{(slow)}$$

Obtain an expression for the rate of formation of the product Y.

9.14 A reaction A + B = Y + Z occurs according to the mechanism

$$A \underset{k_{-1}}{\overset{k_1}{\rightleftharpoons}} X$$

$$X + B \xrightarrow{k_2} Y$$

Apply the steady-state treatment and obtain an expression for the rate. To what expressions does the general rate equation reduce if

(a) Reaction 2 is slow, the initial equilibrium being established rapidly,

(b) Reaction 2 is very rapid compared with the first reaction in either direction?

9.15 A reaction of stoichiometry

$$A + B = Y + Z$$

is found to be second order in A and zero order in B. Suggest a mechanism consistent with this behavior.

ESSAY QUESTIONS

9.16 Give a brief account of the experimental methods which might be used to study the kinetics of (a) a reaction having a half-life of $\sim 10^{-1}$ s; (b) a reaction having a half-life of $\sim 10^{-4}$ s.

9.17 Explain the difference between the *order* and the *molecularity* of a reaction.

9.18 The rate of formation of the product of a reaction is found to give a nonlinear Arrhenius plot, the line being convex to the $1/T$ axis (i.e., the activation energy is higher at higher temperatures). Suggest a reason for this type of behavior.

9.19 An Arrhenius plot is concave to the $1/T$ axis (i.e., a lower activation energy at higher temperatures). Suggest a reason for this type of behavior.

9.20 Predict the effects of (a) increasing the dielectric constant, (b) increasing the ionic strength, on reactions of the following types:

$$A^{2+} + B^- \rightarrow X^+$$
$$A^+ + B^{2+} \rightarrow X^{3+}$$
$$A + B \rightarrow A^+ B^-$$

9.21 Explain the essential features of a free-radical mechanism.

9.22 Give an account of catalysis by acids and bases, distinguishing between (a) specific catalysis, and (b) general catalysis.

SUGGESTED READING

Caldin, E. F. *Fast Reactions in Solution*. Oxford: Blackwell Scientific Publications, 1964.

Campbell, J. A. *Why Do Chemical Reactions Occur?* Englewood Cliffs, N.J.: Prentice-Hall, 1965.

Frost, A. A., and Pearson, R. G. *Kinetics and Mechanism*, 2nd ed. New York: John Wiley, 1961.

Hinshelwood, C. N. *Kinetics of Chemical Change*. Oxford: Clarendon Press, 1945.

King, E. L. *How Chemical Reactions Occur*. New York: Benjamin, 1963.

Laidler, K. J. *Reaction Kinetics*, Vols. I and II. Oxford: Pergamon Press, 1963.

Laidler, K. J. *Chemical Kinetics*, 2nd ed. New York: McGraw-Hill, 1965.

Stevens, B. *Chemical Kinetics*. London: Chapman and Hall, 1961.

10

ENZYME
KINETICS

The enzymes are the biological catalysts. Their action shows some resemblance to the catalytic action of acids and bases, but is considerably more complicated. The details of the mechanisms of enzyme action are still being worked out, and much research remains to be done. The present chapter can give only a brief introduction to the subject, with emphasis on the kinetic effects of concentration, pH and temperature, and on some special aspects of enzyme behavior.

One way in which enzymes differ from acids and bases is that they show a marked degree of *catalytic specificity*. Some enzymes act upon only one substrate, and are said to show *absolute specificity*. A lower degree of specificity is shown by the proteolytic enzymes, which catalyze the hydrolysis of the peptide linkage, provided that certain chemical groups are present in the neighborhood of the linkage. This is known as *group specificity*. Many enzymes exhibit *stereochemical specificity*, in that they catalyze the reactions of one stereochemical form and not the other. The proteolytic enzymes, for example, catalyze only the hydrolysis of peptides made up from amino acids in the L configuration.

Basically the enzymes are all proteins, but they may be associated with nonprotein substances, known as *coenzymes* or *prosthetic groups*, which are essential to the action of the enzyme. Some enzymes are catalytically inactive in the absence of certain metal ions. For a number of enzymes the catalytic activity is due to a relatively small region of the protein molecule, referred to as the *active center*.

In studying the kinetics of an enzyme-catalyzed reaction the most reliable procedure is first to make rate measurements in the early stages of the reaction. If we can follow, by spectrophotometry or in any other way, the change in concentration of a substrate or a product, the initial slope of a concentration-time curve leads at once to the initial rate of the reaction (see Figure 9.4). The variation of this initial rate with substrate concentration can then be investigated. This is a much more reliable procedure than to allow the reaction to proceed to a greater extent and to analyze the concentration-time curves by the method of integration. The difficulty with this latter procedure is that products frequently have an important effect on the rates of enzyme reactions. The concentration-time curves will be influenced by such effects and sometimes will be misleading.

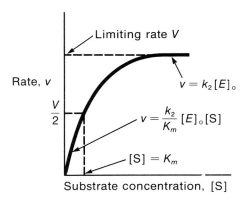

Figure 10.1

The variation of rate with substrate concentration for an enzyme-catalyzed reaction obeying the Michaelis-Menten equation.

10.1   INFLUENCE OF CONCENTRATION
SINGLE-SUBSTRATE REACTIONS

We will consider first an enzyme-catalyzed reaction where there is a single substrate. An example is the hydrolysis of an ester. The dependence on substrate concentration in such cases is frequently as shown in Figure 10.1. The rate varies linearly with the substrate concentration at low concentrations (first-order kinetics), and becomes independent of substrate concentration at high concentrations (zero-order kinetics). This type of behavior was first explained in 1913 by the German-American chemist Leonor Michaelis (1875–1949) and his Canadian assistant Mary L. Menten in terms of the mechanism

$$E + S \underset{k_{-1}}{\overset{k_1}{\rightleftharpoons}} ES$$

$$ES \overset{k_2}{\rightarrow} E + X$$

Here E and S are the enzyme and substrate, X is the product, and ES is an addition complex. Michaelis and Menten arrived at a rate equation by assuming that there is equilibrium between $E + S$ and ES, the concentration of ES not being disturbed by the subsequent reaction. However, this assumption cannot be justified in general, and it is better to make use of the steady-state treatment, as was done in 1925 by the British plant physiologist George Edward Briggs and the British geneticist John Burdon Sanderson Haldane (1892–1964). The steady-state rate of change in the concentration of ES is equal to the rate of its formation, $k_1[E][S]$, minus the net rate of its disappearance, $k_{-1}[ES] + k_2[ES]$:

$$\frac{d[ES]}{dt} = k_1[E][S] - k_{-1}[ES] - k_2[ES] = 0 \tag{10.1}$$

In studies of enzyme reactions the molar concentration of substrate is usually very much greater than that of the enzyme, so that very little of the substrate is bound to the enzyme. The total concentration of enzyme, $[E]_o$, is equal to the concentration of free enzyme, $[E]$, plus the concentration of complex, $[ES]$:

$$[E]_o = [E] + [ES] \tag{10.2}$$

Elimination of $[E]$ between these two equations gives

$$k_1([E]_o - [ES])[S] - (k_{-1} + k_2)[ES] = 0 \tag{10.3}$$

and thus

$$[ES] = \frac{k_1[E]_o[S]}{k_{-1} + k_2 + k_1[S]} \tag{10.4}$$

The rate of reaction is therefore

$$v = k_2[ES] = \frac{k_1 k_2 [E]_o[S]}{k_{-1} + k_2 + k_1[S]} \tag{10.5}$$

$$= \frac{k_2[E]_o[S]}{\dfrac{k_{-1} + k_2}{k_1} + [S]} \tag{10.6}$$

$$= \frac{k_2[E]_o[S]}{K_m + [S]} \tag{10.7}$$

In this equation K_m, equal to $(k_{-1} + k_2)/k_1$, is usually known as the *Michaelis constant*.

When $[S]$ is sufficiently small, it may be neglected in the denominator in comparison with K_m. Under these conditions

$$v = \frac{k_2}{K_m}[E]_o[S] \tag{10.8}$$

and the kinetics are first order in substrate concentration. When, on the other hand, $[S] \gg K_m$,

$$v = k_2[E]_o \tag{10.9}$$

and the kinetics are zero-order. The enzyme is then *saturated* with substrate, and a further increase in $[S]$ has no effect on the rate. Equation (10.7) thus is consistent with the behavior shown in Figure 10.1.

We can conveniently write (10.7) in the form

$$v = \frac{V[S]}{K_m + [S]} \tag{10.10}$$

where V, equal to $k_2[E]_o$, is the limiting rate at high substrate concentrations. We see that when $[S]$ is equal to K_m, (10.10) becomes

$$v = \frac{V[S]}{[S] + [S]} = \frac{V}{2} \tag{10.11}$$

This relationship is illustrated in Figure 10.1. The Michaelis constant K_m thus can be determined from a plot of v against $[S]$, by finding the concentration of substrate that gives one-half of the limiting rate. However, this procedure does not provide a very reliable value.

In order to see whether experimental data are consistent with (10.10) we can recast the equation into a form which will give a linear plot. If we take reciprocals of (10.10),

$$\frac{1}{v} = \frac{K_m}{V[S]} + \frac{1}{V} \tag{10.12}$$

and a plot of $1/v$ against $1/[S]$ should give a straight line. This type of plot, suggested by the American physical chemists Hans Lineweaver and Dean Burk, is shown schematically in Figure 10.2, which gives the intercepts on the axes and the slope. The parameters V and K_m thus can be derived from such a plot. If the enzyme concentration $[E]_o$ is known, in mol dm^{-3}, k_2 can also be calculated, since according to this simple mechanism $V = k_2[E]_o$. However, the individual constants k_1 and k_{-1} cannot be obtained from studies of rate as a function of substrate concentration.

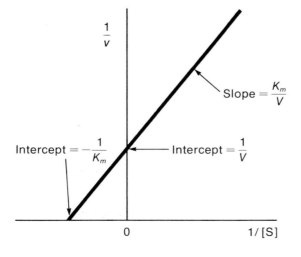

Figure 10.2

A schematic Lineweaver-Burk plot.

Alternatively, the Michaelis equation (10.10) can be put into the form

$$vK_m + v[S] = V[S] \tag{10.13}$$

or

$$\frac{v}{[S]} K_m + v = V \tag{10.14}$$

A plot can therefore be made of $v/[S]$ against v. Figure 10.3 shows the slopes and intercepts. A plot of this kind was first suggested in 1942 by G. S. Eadie. It has the advantage that it tends to spread out the points to a greater extent than the Lineweaver-Burk plot.

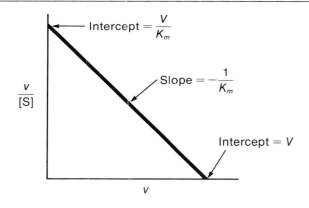

Intercept $= \dfrac{V}{K_m}$

Slope $= -\dfrac{1}{K_m}$

$\dfrac{v}{[S]}$

Intercept $= V$

v

Figure 10.3

A schematic Eadie plot.

Many reactions obey the Michaelis-Menten law, and a considerable number of Michaelis constants have been determined. However, note that adherence to the empirical law does not establish the simple mechanism, since more complicated mechanisms can give exactly the same behavior. An example is the mechanism

$$E + S \underset{k_{-1}}{\overset{k_1}{\rightleftharpoons}} ES \underset{X}{\overset{k_2}{\rightarrow}} ES' \overset{k_3}{\rightarrow} E + Y$$

for which there is considerable evidence for a number of enzyme systems. Here, there are two products of reaction. The first product, X, is formed by the breakdown of the Michaelis complex ES, but this breakdown gives rise to a second intermediate, ES'. This second intermediate breaks down in a subsequent stage into enzyme plus the second product Y.

Application of the steady-state treatment to this mechanism proceeds as follows. The steady-state equation for ES is

$$k_1[E][S] - (k_{-1} + k_2)[ES] = 0 \tag{10.15}$$

while that for ES' is

$$k_2[ES] - k_3[ES'] = 0 \tag{10.16}$$

The total enzyme concentration, $[E]_o$, is the sum of the concentrations of E, ES and ES':

$$[E]_o = [E] + [ES] + [ES'] \tag{10.17}$$

Use of equations (10.15) and (10.16) leads to

$$[E]_o = [ES]\left[\frac{k_{-1} + k_2}{k_1[S]} + 1 + \frac{k_2}{k_3}\right] \tag{10.18}$$

The rate of formation of the product X is†

$$v = k_2[ES] \tag{10.19}$$

$$= \frac{k_2[E]_o}{\dfrac{k_{-1} + k_2}{k_1[S]} + 1 + \dfrac{k_2}{k_3}} \tag{10.20}$$

Equation (10.20) rearranges to

$$v = \frac{\dfrac{k_2 k_3}{k_2 + k_3}[E]_o[S]}{\dfrac{k_{-1} + k_2}{k_1} \cdot \dfrac{k_3}{k_2 + k_3} + [S]} \tag{10.21}$$

This is of the same form as (10.10) with

$$V = \frac{k_2 k_3}{k_2 + k_3}[E]_o \tag{10.22}$$

and

$$K_m = \frac{k_{-1} + k_2}{k_1} \cdot \frac{k_3}{k_2 + k_3} \tag{10.23}$$

In the event that k_3 is very much greater than k_2, (10.21) reduces to

$$v = \frac{k_2[E]_o[S]}{\dfrac{k_{-1} + k_2}{k_1} + [S]} \tag{10.24}$$

---

† Note that this is also the rate of formation of Y, by (10.16).

which is equation (10.6). If $k_3 \gg k_2$ the intermediate ES' exists only at very minute concentrations and can be neglected; the case thus reduces to the simple one-intermediate mechanism.

Because the same type of behavior can arise from a number of different mechanisms it is customary to write the rate equation as

$$v = \frac{k_c[E]_o[S]}{K_m + [S]} \qquad (10.25)$$

where k_c (also often written as k_{cat}) is known as the *catalytic constant*. If the one-intermediate mechanism applies, k_c is k_2. If there are two intermediates k_c is $k_2 k_3/(k_2 + k_3)$. The use of k_c in the equation therefore implies no assumption about the nature of the mechanism, the equation applying to mechanisms having any number of sequential intermediates ES $\to$ ES' $\to$ ES'' $\to$ ES''' $\to$ etc.

TWO-SUBSTRATE REACTIONS

The systems we have just considered involve a single substrate. It is true that hydrolyses involve a water molecule in addition to the molecule undergoing hydrolysis. However, from the point of view of the steady-state equations, the step involving the water molecule is treated by incorporating the water concentration into the rate constant.

When reaction occurs between two solute species, the steady-state equations take a more complicated form than those for the single-substrate reactions. A detailed treatment is outside the scope of this book; here we will consider a few of the possible mechanisms and the corresponding rate equations.

One particular type of reaction to which a treatment of two-substrate systems applies comprises reactions catalyzed by enzymes with which coenzymes are associated. For example, when lactate dehydrogenase catalyzes the conversion of lactate into pyruvate,

$$CH_3CHOHCOOH \to CH_3COCOOH$$

one of the two hydrogen atoms is transferred as H^- to the coenzyme nicotinamide adenine dinucleotide (NAD^+), the other going into solution as H^+. The overall reaction is

$$CH_3CHOHCOOH + NAD^+ \to CH_3COCOOH + NADH + H^+$$

Thus this reaction is a two-substrate reaction.

One mechanism which may apply to a two-substrate reaction is the so-called *random ternary-complex mechanism*. In this mechanism the enzyme E can form binary complexes EA and EB with the two substrates A and B. It can also form the ternary complex EAB, with no restriction on the order in which A and B are attached. In Figure 10.4 this mechanism is represented in two different ways. The lower one is a shorthand notation introduced by the American biophysical chemist

William Wallace Cleland. Unfortunately, the steady-state treatment of this mechanism leads to an extremely complicated expression. Considerable simplification results if the assumption is made that the breakdown of EAB is sufficiently slow that we can regard EA, EB, and EAB as at equilibrium. The rate equation then reduces to the form

$$v = \frac{V[A][B]}{k_A'K_B + K_B[A] + K_A[B] + [A][B]} \tag{10.26}$$

where K_A, K_B, and K_A' are constants. Various graphical procedures are available for determining these constants from the experimental data.

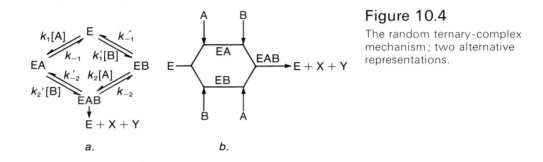

Figure 10.4

The random ternary-complex mechanism; two alternative representations.

If the concentration of B is held constant, the variation of v with $[A]$ is of the Michaelis-Menten form. This may be shown for the case in which we have an excess of B. Thus, if $[B]$ is sufficiently large, we may neglect the first two terms in the denominator of (10.26), and the result is

$$v = \frac{V[A]}{K_A + [A]} \tag{10.27}$$

which is of the Michaelis-Menten form. Thus, K_A is the Michaelis constant for A in the presence of excess of B. Similarly, if we have excess of A,

$$v = \frac{V[B]}{K_B + [B]} \tag{10.28}$$

Another mechanism that may apply to a two-substrate reaction is the *ordered ternary-complex mechanism*. For example, the complex EAB can be formed from EA by addition of B, but not from EB by addition of A; the substrates must become attached in a particular order. This mechanism is represented in Figure 10.5, the second representation being that of Cleland. This mechanism also leads to a rate equation of the form of (10.26), but the significance of the constants is different.

A third possible mechanism for reaction between two substrates is shown in Figure 10.6, in two alternative representations. In this mechanism, one substrate, A, first adds on to the enzyme to form a complex EA, which then eliminates the first product X *before* there is any reaction with B. B then reacts with the complex EA' to form the second product, Y. Experimentally, this mechanism can sometimes be distinguished from the others by observing the production of X from A in the complete absence of the other substrate B. Cleland's name for this mechanism is *ping pong bi bi*, the first "bi" indicating that there are two reactants for the reaction from left to right. The second "bi" means that there are two products.

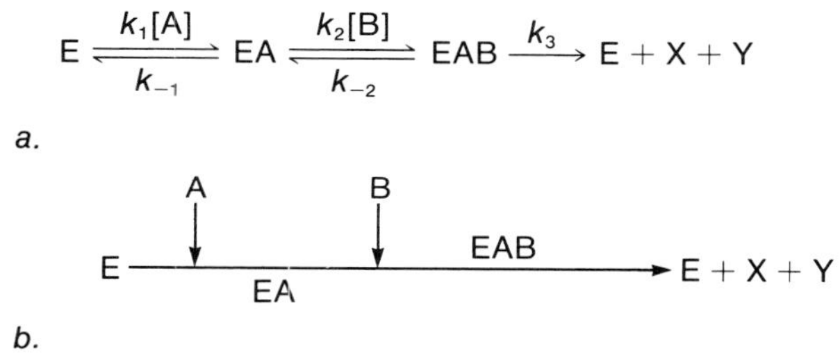

$$E \underset{k_{-1}}{\overset{k_1[A]}{\rightleftharpoons}} EA \underset{k_{-2}}{\overset{k_2[B]}{\rightleftharpoons}} EAB \xrightarrow{k_3} E + X + Y$$

a.

Figure 10.5

The ordered ternary-complex mechanism. The ternary complex EAB can only be formed from EA.

(diagram: A, B, EA, EAB → E + X + Y)

b.

Application of the steady-state treatment to this mechanism leads to an equation of the form

$$v = \frac{V[A][B]}{K_B[A] + K_A[B] + [A][B]}$$

(10.29)

This is of the same form as (10.26) except that the first term in the denominator is missing.

Discrimination between the various possible mechanisms is a matter of some difficulty. If the rate equation is of the form of (10.29) the ping-pong mechanism is suggested, but when it is of the form of (10.26) we cannot tell whether the binding is random or ordered. However, by making carefully planned studies with inhibitors (sometimes the reaction products) it has been found possible in many cases to decide about the mechanism. In addition, isotope studies have been very helpful. For example, in the case of the lactate dehydrogenase reaction, the mechanism is

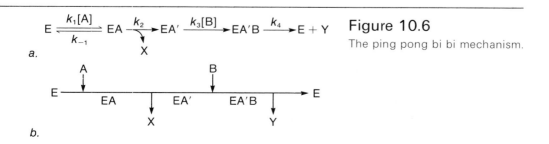

$$E \underset{k_{-1}}{\overset{k_1[A]}{\rightleftharpoons}} EA \xrightarrow{k_2} EA' \xrightarrow{k_3[B]} EA'B \xrightarrow{k_4} E + Y$$

a.

Figure 10.6

The ping pong bi bi mechanism.

(diagram: A, B, EA, EA', EA'B, X, Y)

b.

ordered, with the NAD^+ added first when the reaction goes from left to right, and the NADH added first in the reverse reaction. The complete scheme is as follows:

(a) Addition of NAD^+ to enzyme, with formation of a binary complex,

$$LDH + NAD^+ \rightleftharpoons LDH\!-\!NAD^+$$

(b) Addition of lactate to binary complex, with formation of ternary complex,

$$LDH\!-\!NAD^+ + CH_3CHOHCOO^- \rightleftharpoons LDH\begin{smallmatrix} \diagup NAD^+ \\[2pt] \diagdown CH_3CHOHCOO^- \end{smallmatrix}$$

(c) Oxidation of lactate and reduction of NAD^+, with the splitting off of a proton,

$$LDH\begin{smallmatrix} \diagup NAD^+ \\[2pt] \diagdown CH_3CHOHCOO^- \end{smallmatrix} \rightleftharpoons LDH\begin{smallmatrix} \diagup NADH \\[2pt] \diagdown CH_3COCOO^- \end{smallmatrix} + H^+$$

(d) Release of pyruvate from ternary complex,

$$LDH\begin{smallmatrix} \diagup NADH \\[2pt] \diagdown CH_3COCOO^- \end{smallmatrix} \rightleftharpoons LDH\!-\!NADH + CH_3COCOO^-$$

(e) Release of NADH from binary complex,

$$LDH\!-\!NADH \rightleftharpoons LDH + NADH$$

10.2 INHIBITION

The rates of enzyme-catalyzed reactions are frequently affected by the addition of substances which in general are referred to as *moderators*. When there is a reduction in rate, the substance is known as an *inhibitor*. When a substance increases the rate, it is known as an *activator*. The present section will deal with the effects of inhibitors, which will be denoted by the symbol I.

Enzyme inhibitors are of various types. For example, we can distinguish between irreversible and reversible inhibition. Sometimes an inhibitor reacts so strongly with the active center of an enzyme that the process cannot easily be reversed. Diisopropylfluorophosphate (DFP), for example, undergoes an irreversible reaction with the active centers of certain enzymes, with the liberation of hydrogen fluoride. One of the enzymes with which it reacts is cholinesterase, which is responsible for the functioning of the nerves. As a result DFP is a very powerful nerve gas. Poisons like potassium cyanide exert their action by the irreversible inhibition of enzymes which catalyze oxidative reactions.

On the other hand, reversible inhibitors are those which act in such a way that the inhibition is reduced if the inhibitor is removed from the system; they undergo a simple reversible reaction with the enzyme,

$$E + I \rightleftharpoons EI$$

Many drugs act in this way. For example, eserine combines reversibly with cholinesterase. Substances which are structurally analogous to the natural substrates of enzymes are frequently reversible inhibitors: they can combine with the enzyme at its active center, but the resulting complex does not undergo reaction.

In treating experimental results it is convenient to define a quantity known as the degree of inhibition, i, which is related to the velocity of the uninhibited reaction, v_0, and that of the inhibited reaction, v. It is defined as the reduction in velocity, $v_0 - v$, divided by the velocity of the uninhibited reaction:

$$i = \frac{v_0 - v}{v_0} \tag{10.30}$$

It is convenient to classify inhibitors according to how they bring about various types of behavior:

1. The inhibitor may be such that the degree of inhibition is unaffected by the concentration of substrate, in which case we speak of *pure noncompetitive* inhibition.

2. The degree of inhibition brought about by a given amount of inhibitor may be *reduced* when the substrate concentration is increased, in which case the term *competitive* is used to describe the behavior. This type of behavior *appears* to suggest that the substrate and inhibitor are competing with one another for places on the enzyme, so that addition of substrate displaces the inhibitor. However, the behavior is frequently more complex than this.

3. Sometimes the degree of inhibition is *increased* as the substrate concentration is increased. Here we shall refer to this behavior as *anticompetitive*. (The terms *coupling* and *uncompetitive* are also applied to this type of behavior.)

Now let us consider the simplest enzyme mechanisms which can be put forward to explain these types of behavior. It should be emphasized that enzyme reactions frequently occur in a much more complicated manner and that very detailed studies have to be carried out in order to establish their mechanisms.

The simplest mechanism which explains *competitive* inhibition is one in which there is a single intermediate—the enzyme-substrate complex ES—and in which only the enzyme can combine with the inhibitor to form the complex EI. The mechanism can be written as

$$
\begin{array}{c}
S \\
+ \\
I + E \underset{k_{-i}}{\overset{k_i}{\rightleftharpoons}} EI \\
k_1 \big\Updownarrow k_{-1} \\
ES \\
\Big\downarrow k_2 \\
E + P
\end{array}
$$

This is a *competitive* mechanism in the sense that S and I cannot both be attached to the enzyme—i.e., ESI is not formed. The steady-state equations for ES and EI are

$$k_1[E][S] - (k_{-1} + k_2)[ES] = 0 \tag{10.31}$$

$$k_i[E][I] - k_{-i}[EI] = 0 \tag{10.32}$$

The total enzyme concentration is

$$[E]_o = [E] + [ES] + [EI] \tag{10.33}$$

and use of equations (10.31) and (10.32) leads to

$$[E]_o = [ES]\left[\frac{k_{-1} + k_2}{k_1[S]} + 1 + \frac{k_i[I](k_{-1} + k_2)}{k_{-i}k_1[S]}\right] \tag{10.34}$$

The rate, equal to $k_2[ES]$, is

$$v = \frac{k_2[E]_o}{\dfrac{k_{-1} + k_2}{k_1[S]} + 1 + \dfrac{k_i[I](k_{-1} + k_2)}{k_{-i}k_1[S]}} \tag{10.35}$$

$$= \frac{k_2[E]_o[S]}{\dfrac{k_{-1} + k_2}{k_1} + [S] + \dfrac{k_i(k_{-1} + k_2)}{k_{-i}k_1}[I]} \tag{10.36}$$

which may be written as

$$v = \frac{k_2[E]_o[S]}{K_m\left(1 + \dfrac{[I]}{K_i}\right) + [S]} \tag{10.37}$$

where K_m, equal to $(k_{-1} + k_2)/k_1$, is the Michaelis constant for the reaction occurring in the absence of inhibitor, and K_i, equal to k_{-i}/k_i, is the equilibrium constant for the *dissociation* of the complex EI into E + I.

If equation (10.37) applies, the degree of inhibition is given by

$$i = \frac{v_o - v}{v_o} = 1 - \frac{v}{v_o} \tag{10.38}$$

$$= 1 - \frac{k_2[E]_o[S]}{K_m\left(1 + \dfrac{[I]}{K_i}\right) + [S]} \cdot \frac{K_m + [S]}{k_2[E]_o[S]} \tag{10.39}$$

$$= 1 - \frac{K_m + [S]}{K_m\left(1 + \dfrac{[I]}{K_i}\right) + [S]} \tag{10.40}$$

$$= \frac{\dfrac{K_m}{K_i}[I]}{K_m\left(1 + \dfrac{[I]}{K_i}\right) + [S]} \qquad (10.41)$$

Thus, the degree of inhibition decreases with increasing [S], as is required for competitive behavior.

Noncompetitive inhibition is explained most simply by a mechanism in which *both* the free enzyme and the enzyme-substrate complex can form a complex with the inhibitor:

$$
\begin{array}{c}
\text{S} \\
+ \\
\text{I} + \text{E} \underset{k_{-i}}{\overset{k_i}{\rightleftharpoons}} \text{EI} \\
k_1 \big\downarrow\big\uparrow k_{-1} \quad k_i' \\
\text{I} + \text{ES} \underset{k'_{-i}}{\overset{}{\rightleftharpoons}} \text{ESI} \\
k_2 \big\downarrow \\
\text{E} + \text{P}
\end{array}
$$

We can imagine that there are two separate sites on the enzyme, one of which can bind the substrate and the other the inhibitor molecule. Since the sites are separate it is possible for both S and I to be bound simultaneously; i.e., there is no competition between S and I for a place on the enzyme.

The steady-state equations for EI and ESI are

$$k_i[E][I] - k_{-i}[EI] = 0 \qquad (10.42)$$

$$k_i'[ES][I] - k'_{-i}[ESI] = 0 \qquad (10.43)$$

These are equilibrium equations, and can be written as

$$[E][I] = K_i[EI] \qquad (10.44)$$

and

$$[ES][I] = K_i'[ESI] \qquad (10.45)$$

where $K_i(= k_{-i}/k_i)$ and $K_i'(= k_{-i}'/k_i')$ are the equilibrium constants for the *dissociation* of EI and ESI respectively. The steady-state equation for ES is

$$k_1[E][S] - (k_{-1} + k_2)[ES] - k_i'[I][ES] + k'_{-1}[ESI] = 0 \qquad (10.46)$$

By use of (10.43) this reduces to

$$k_1[E][S] - (k_{-1} + k_2)[ES] = 0 \qquad (10.47)$$

It is a general rule that a "dead-end" complex such as ESI (and EI) can be ignored in steady-state equations.

The total enzyme concentration is

$$[E]_o = [E] + [ES] + [EI] + [ESI] \tag{10.48}$$

and use of equations (10.44), (10.45), and (10.47) leads to

$$[E]_o = [ES]\left[\frac{k_{-1} + k_2}{k_1[S]} + 1 + \frac{[I]}{K_i} \cdot \frac{k_{-1} + k_2}{k_1[S]} + \frac{[I]}{K_i'}\right] \tag{10.49}$$

The rate, equal to $k_2[ES]$, is

$$v = \frac{k_2[E]_o[S]}{\dfrac{k_{-1} + k_2}{k_1}\left(1 + \dfrac{[I]}{K_i}\right) + \left(1 + \dfrac{[I]}{K_i'}\right)[S]} \tag{10.50}$$

$$= \frac{k_2[E]_o[S]}{\dfrac{k_{-1} + k_2}{k_1}\left(1 + \dfrac{[I]}{K_i}\right) + \left(1 + \dfrac{[I]}{K_i'}\right)[S]} \tag{10.51}$$

If the inhibitor is bound equally strongly to ES as to E, $K_i' = K_i$ and (10.51) reduces to

$$v = \frac{k_2[E]_o[S]}{(K_m + [S])\left(1 + \dfrac{[I]}{K_i}\right)} \tag{10.52}$$

where K_m has been written for $(k_{-1} + k_2)/k_1$.

It is easy to show that the degree of inhibition is now independent of [S]. The rate in the absence of inhibitor is

$$v_o = \frac{k_2[E]_o[S]}{K_m + [S]} \tag{10.53}$$

and the degree of inhibition is thus

$$i = \frac{v_o - v}{v_o} = 1 - \frac{v}{v_o} \tag{10.54}$$

$$= 1 - \frac{1}{1 + \dfrac{[I]}{K_i}} \tag{10.55}$$

$$= \frac{[I]}{K_i + [I]} \tag{10.56}$$

The degree of inhibition is thus independent of [S], as is required for noncompetitive behavior.

Anticompetitive inhibition can be simply explained if the inhibitor can become attached to ES but not to E.

10.3   INFLUENCE OF HYDROGEN ION CONCENTRATION

The pH of the solution has a very marked effect on the rates of most enzyme reactions, and the situation is a somewhat complicated one; only a brief account can be given in this book.

In most cases the rates of enzyme reactions pass through a maximum as the pH is varied, as shown in Figure 10.7. The pH corresponding to the maximum rate is known as the *optimum pH*. The optimum pH is sometimes regarded as a characteristic property of an enzyme; however, it varies somewhat with the nature of the substrate and with the substrate concentration.

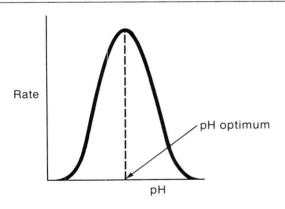

Figure 10.7

The variation with pH of the rate of an enzyme-catalyzed reaction. Contrast this behavior with that found for acid-base catalysis, where rates may pass through a minimum (see Figure 9.19, p. 419).

Two different types of behavior are to be distinguished when the pH is varied: a *reversible* and an *irreversible* effect. If the solution is taken too far to the acid or alkaline side, the enzyme undergoes an irreversible loss of activity, which cannot be restored by changing the pH back to the region of the optimum. This type of behavior is due to an irreversible change in the conformation of the protein, involving the breaking of the hydrogen bonds and hydrophobic bonds which play an important role in maintaining the tertiary structure of the protein. These conformational changes destroy the active center of the enzyme by changing the relative positions of the groups that make up the active center.

Reversible pH changes occur when the pH is not taken too far from the pH optimum: within a certain pH range the pH can be changed back and forth without any permanent effects. This behavior was first explained by Michaelis and his coworkers in terms of the ionizations of groups on the protein. It is necessary to postulate at least two ionizing groups as playing an important role at the active center. If these groups are $-NH_3^+$ and $-COOH$, the ionizations at the active center may be represented as

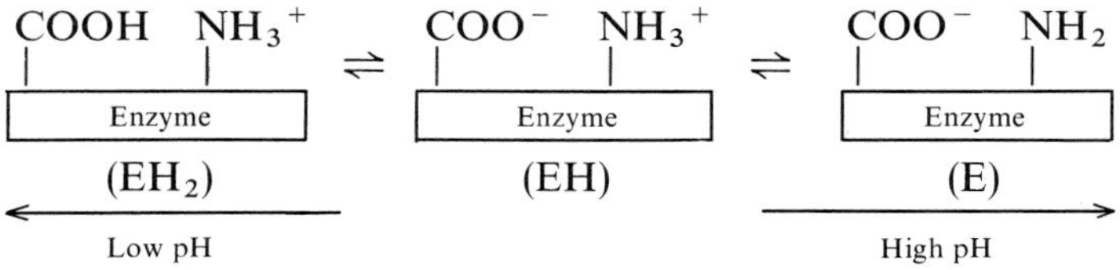

$$
\begin{array}{ccc}
\underset{\text{(EH}_2)}{\boxed{\begin{array}{cc} \overset{|}{\text{COOH}} & \overset{|}{\text{NH}_3^+} \\ \text{Enzyme} \end{array}}} & \rightleftharpoons & \underset{\text{(EH)}}{\boxed{\begin{array}{cc} \overset{|}{\text{COO}^-} & \overset{|}{\text{NH}_3^+} \\ \text{Enzyme} \end{array}}} & \rightleftharpoons & \underset{\text{(E)}}{\boxed{\begin{array}{cc} \overset{|}{\text{COO}^-} & \overset{|}{\text{NH}_2} \\ \text{Enzyme} \end{array}}}
\end{array}
$$

Low pH ← → High pH

The pH behavior now can be explained by postulating that the intermediate (zwitterion) form is enzymically active, but the species to the left and right are inactive. The concentration of the intermediate form goes through a maximum as the pH is varied, and the rate therefore passes through a maximum.

The detailed treatment of pH effects is complicated, since various possibilities have to be taken into account. A very simple situation occurs when the reaction follows the original Michaelis-Menten mechanism, in which the enzyme and substrate form an addition complex which breaks down in a single stage. The enzyme-substrate complex may also exist in three states of ionization, and perhaps only the intermediate form is capable of giving rise to products. This simple case is represented in Figure 10.8. Intuitively we can see that at low substrate concentrations, when the enzyme exists mainly in the free form, the pH behavior will be controlled by the ionization of the free enzyme. Analysis of the experimental pH dependence at low substrate concentrations therefore will allow us to determine the acid dissociation constants K_a and K_b for the free enzyme. On the other hand, if we saturate the enzyme with substrate, analysis of the pH behavior will now give the values of K_a' and K_b', which relate to the ionization of the enzyme-substrate complex.

Application of the steady-state treatment to the mechanism shown in Figure 10.8 leads to the equation

$$
v = \frac{k_2[E]_o[S]}{K_m\left(1 + \dfrac{K_a}{[H^+]} + \dfrac{[H^+]}{K_b}\right) + [S]\left(1 + \dfrac{K_a'}{[H^+]} + \dfrac{[H^+]}{K_b'}\right)} \tag{10.57}
$$

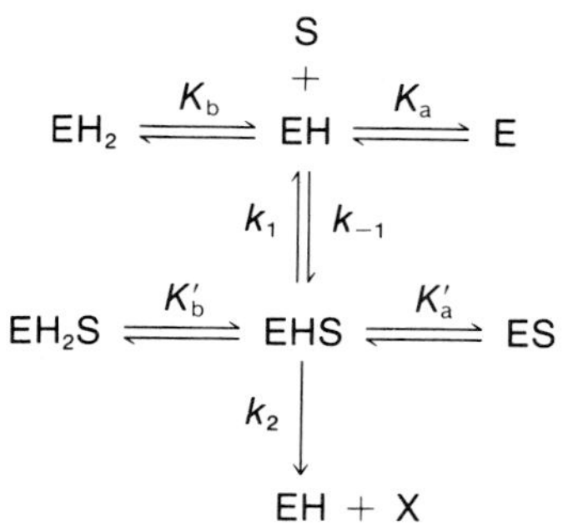

$$
\begin{array}{c}
\text{S} \\
+ \\
\text{EH}_2 \underset{}{\overset{K_b}{\rightleftharpoons}} \text{EH} \overset{K_a}{\rightleftharpoons} \text{E} \\
k_1 \Big\| k_{-1} \\
\text{EH}_2\text{S} \overset{K_b'}{\rightleftharpoons} \text{EHS} \overset{K_a'}{\rightleftharpoons} \text{ES} \\
k_2 \Big\downarrow \\
\text{EH} + \text{X}
\end{array}
$$

Figure 10.8

A simple scheme of reactions, which interprets pH dependence in some systems.

At sufficiently low substrate concentrations

$$v = \frac{k_2[E]_o[S]}{K_m\left(1 + \frac{K_a}{[H^+]} + \frac{[H^+]}{K_b}\right)}$$

(10.58)

while at high concentrations of substrate

$$v = \frac{k_2[E]_o}{1 + \frac{K_a'}{[H^+]} + \frac{[H^+]}{K_b'}}$$

(10.59)

Suppose that the low-concentration rates have been measured over a range of pH values and that we plot the logarithms of these rates against the pH (see Figure 10.9).

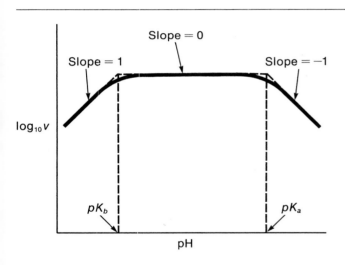

Figure 10.9

Schematic plot of $\log_{10} v$ against pH for systems to which equation (10.58) applies (low substrate concentrations).

In a sufficiently acid solution (low pH) the term $[H^+]/K_b$ predominates in the denominator of (10.58), so that

$$v \approx \frac{k_2[E]_o[S]K_b}{K_m[H^+]}$$

(10.60)

or

$$\log_{10} v = \text{const} - \log_{10}[H^+]$$

(10.61)

$$= \text{const} + \text{pH}$$

(10.62)

The plot of $\log_{10} v$ against pH will therefore have a slope of $+1$ at these low pH values. At intermediate pH values the term unity will predominate in the denominator of (10.58), and thus

$$v = \frac{k_2[E]_o[S]}{K_m} \qquad (10.63)$$

The plot of $\log_{10} v$ against pH will therefore have zero slope in this region. Moreover, we see that these lines, at low and intermediate pH, will intersect when the right-hand sides of (10.60) and (10.63) are equal, and that this occurs when

$$[H^+] = K_b \qquad (10.64)$$

i.e., when

$$pH = pK_b \qquad (10.65)$$

From an experimental plot we can thus determine pK_b; this is shown schematically in Figure 10.9. The lines, of course, do not meet sharply; there is a rounding-off, as shown in the diagram.

In a similar way we can see that the right-hand inflection point corresponds to pK_a. At high pH values the term $K_a/[H^+]$ predominates in the denominator of (10.58), and therefore

$$v = \frac{k_2[E]_o[S][H^+]}{K_m K_a} \qquad (10.66)$$

Thus

$$\log_{10} v = \text{const} + \log_{10}[H^+] \qquad (10.67)$$
$$= \text{const} - pH \qquad (10.68)$$

The slope is now -1 (see Figure 10.9), and by equating the right-hand sides of (10.63) and (10.66) we see that, at the inflection point,

$$pH = pK_a \qquad (10.69)$$

Unfortunately, few enzyme systems conform to this simple mechanism. Many single-substrate systems involve at least one intermediate, which also may exist in various ionization states, and which may influence the behavior. Additional complexities exist for reactions involving more than one substrate. The equations for a large number of cases have been worked out and it has proved possible, by making pH studies over a variety of conditions, to obtain experimental values of dissociation constants for the various groups on the enzyme and the enzyme-substrate complex. The magnitude of these constants provides valuable clues about the nature of the groups. For example, pepsin has a group which has a pK value of 2.2. This must be the carbonyl group, —COOH, which is the only organic group that ionizes so readily. Several enzymes, including trypsin, chymotrypsin, and cholinesterase, have a group of pK about 7. This suggests the ionization of a proton attached to a nitrogen atom in an imidazole ring, which is part of a histidine residue:

$$HN \overset{CH=N^+H}{\underset{C=CH}{\diagup\diagdown}} \rightleftharpoons HN \overset{CH=N}{\underset{C=CH}{\diagup\diagdown}} + H^+$$

Knowledge of the nature of ionizing groups is of great importance in working out the details of the mechanisms of enzyme catalysis.

10.4   INFLUENCE OF TEMPERATURE

Valuable information about mechanisms has been provided by studies of the influence of temperature on rates. We must take account of the fact that the enzymes themselves undergo a deactivation process which has a very high activation energy and also a very high frequency factor. At temperatures of 35° C or higher (depending on the particular enzyme) the enzyme may undergo very rapid deactivation during the course of a kinetic experiment and a low rate of reaction then will be observed. As a result, the rates of enzyme-catalyzed reactions frequently pass through a maximum as the temperature is raised. This is illustrated in Figure 10.10. The temperature at which the rate is a maximum is often referred to as the *optimum temperature*. Its value depends upon the conditions of the experiment. The process of enzyme inactivation, which is due to the denaturation of the protein, has been investigated kinetically in some detail.

By working at sufficiently low temperatures (at which no appreciable inactivation occurs) or by making a correction for the inactivation, it is possible to determine

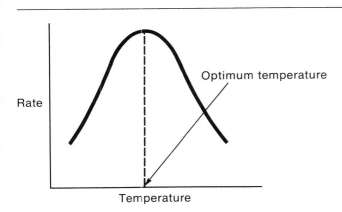

Optimum temperature

Rate

Temperature

Figure 10.10

The variation of rate with temperature for a typical enzyme-catalyzed reaction.

the effect of temperature on the enzyme-catalyzed reaction itself. The analysis of the results must take account of the fact that the rate law may be, for example,

$$v = \frac{k_2[\text{E}]_\text{o}[\text{S}]}{K_\text{m} + [\text{S}]} \qquad (10.70)$$

and a simple dependence of the rate on the temperature is not to be expected. However, at sufficiently high substrate concentrations,

$$v = k_2[\text{E}]_\text{o} \qquad (10.71)$$

so that, since k_2 is expected to vary with temperature according to the Arrhenius law, the same temperature dependence will be found for v, provided that we work at constant enzyme concentration. Under these conditions a plot of $\log_{10} v$ against $1/T$ therefore should give a straight line. This has been found to be the case for a number of enzyme systems.

At sufficiently low substrate concentrations the rate equation becomes

$$v = \frac{k_2}{K_\text{m}}[\text{E}]_\text{o}[\text{S}] \qquad (10.72)$$

Since $K_\text{m} = (k_{-1} + k_2)/k_1$,

$$v = \frac{k_1 k_2}{k_{-1} + k_2}[\text{E}]_\text{o}[\text{S}] \qquad (10.73)$$

In general, a simple dependence of rate on temperature is not to be expected under these conditions of low substrate concentrations. In the special case that k_2 is much greater than k_{-1}, equation (10.73) becomes

$$v = k_1[\text{E}]_\text{o}[\text{S}] \qquad (10.74)$$

A plot of $\log v$ against $1/T$ then should be a straight line, and the activation energy calculated from its slope will correspond to k_1. Therefore, it will apply to the reaction between enzyme and substrate with the formation of the enzyme-substrate complex. If, on the other hand, k_{-1} is much greater than k_2, the rate equation becomes

$$v = \frac{k_1 k_2}{k_{-1}}[\text{E}]_\text{o}[\text{S}] \qquad (10.75)$$

Each of the rate constants k_1, k_2, and k_{-1} will be of the Arrhenius form $Ae^{-E/RT}$, so that (10.75) becomes

$$v = \frac{A_1 e^{-E_1/RT} \, A_2 e^{-E_2/RT}}{A_{-1} e^{-E_{-1}/RT}}[\text{E}]_\text{o}[\text{S}] \qquad (10.76)$$

$$= \frac{A_1 A_2}{A_{-1}} e^{-(E_1 + E_2 - E_{-1})/RT}[\text{E}]_\text{o}[\text{S}] \qquad (10.77)$$

The activation energy obtained from a plot of $\log v$ against $1/T$ now will be equal to $E_1 + E_2 - E_{-1}$, which are the activation energies corresponding to the three elementary reactions.

Table 10.1 Comparison of Kinetic Parameters for Catalyzed Reactions†

Reaction	Catalyst	T, °C	k dm^3 mol^{-1} s^{-1}	A dm^3 mol^{-1} s^{-1}	E kcal mol^{-1}
Hydrolysis of urea	H_3O^+	62.0	7.4×10^{-7}	1.8×10^{10}	24.6
Hydrolysis of urea	Urease	20.8	5.0×10^6	1.7×10^{13}	6.8
Hydrolysis of adenosine triphosphate	H_3O^+	40.0	4.7×10^{-6}	2.4×10^9	21.2
Hydrolysis of adenosine triphosphate	Myosin	25.0	8.2×10^6	1.6×10^{22}	21.1
Decomposition of hydrogen peroxide	None Fe^{2+}	22.0 22.0	10^{-7} 56.0	10^6 1.8×10^9	17–18 10.1
Decomposition of hydrogen peroxide	Catalase	22.0	3.5×10^7	6.4×10^8	1.7

† The catalyzed reactions are all second-order—first order in substrate and first order in catalyst. The uncatalyzed H_2O_2 decomposition is second-order in H_2O_2.

Table 10.1 gives values of rate constants, activation energies, and frequency factors for three enzyme-catalyzed reactions. For comparison, the values for other catalysts are included. Note that molecule for molecule, the enzymes are much more effective catalysts than the nonbiological catalysts. In urease and catalase this higher effectiveness is related to a much smaller activation energy, which is true for a number of other enzyme systems. Enzymes evidently exert their action by allowing the process to occur by a much more favorable reaction path.

The factors involved in the efficient action of enzymes are by no means understood. A considerable amount of work must be done to clarify this problem, and many suggestions have been made. An important clue is provided by the pH dependence of enzyme action, which we have considered earlier in this chapter. We saw that the active form of the enzyme is the intermediate ionic form, which contains a basic group (the —COO$^-$ group in the scheme on p. 442) and an acidic group (the —NH$_3^+$ in the scheme). For example, consider the hydrolysis of an ester,

$$
\begin{array}{c}
\quad\ \overset{\displaystyle O}{\underset{\displaystyle \|}{}} \\
R\!-\!C\!-\!O \\
\qquad\quad \diagdown \\
\qquad\qquad R'
\end{array}
$$

The kinetic evidence indicates that when an ester is hydrolyzed by hydroxide ions, there is a nucleophilic (electron-donating) attack by the OH^- ion, and at the same time a proton is donated by the water molecule:

$$
\begin{array}{ccccc}
\underset{R'}{\overset{\displaystyle O}{\overset{\|}{R-C-O}}} & \longrightarrow & \underset{OH^-}{\overset{\displaystyle O}{\overset{\|}{R-C}}}\cdots O\overset{R'}{\diagup}\ \ H\diagdown O \diagup H & \longrightarrow & \overset{\displaystyle O}{\overset{\|}{R-C}}\diagdown O-H\quad + R'OH \\
+\ OH^- + H_2O & & & & +\ OH^-
\end{array}
$$

Similarly in acid hydrolysis there is addition of a proton to the alcoholic oxygen atom, and at the same time addition of OH^- from a water molecule:

$$
\begin{array}{ccccc}
\overset{\displaystyle O}{\overset{\|}{R-C-O}}\overset{R'}{\diagup} & \longrightarrow & \underset{H^+}{\overset{\displaystyle O}{\overset{\|}{R-C}}}\cdots O\overset{R'}{\diagup}\diagdown H & \longrightarrow & \overset{\displaystyle O}{\overset{\|}{R-C}}-OH + R'OH \\
+\ H_2O + H^+ & \underset{OH}{\ } & & & +\ H^+
\end{array}
$$

In either case, a contribution is made by a water molecule, which is neither a strong acid nor a strong base. In an enzyme, on the other hand, an acidic and a basic group are side by side at the active center of the molecule. There is therefore a *concerted* type of mechanism, which may be more efficient than one in which part of the work has to be done by a water molecule.

This is almost certainly an important factor, but many other aspects have to be considered. The reader is referred to more advanced treatments (see Suggested Reading on p. 461) for a discussion of problems of this kind.

The magnitudes of entropies of activation have provided valuable information regarding the details of the interactions between enzymes and substrates. The process of muscular contraction involves an interaction between the muscle enzyme myosin and adenosine triphosphate (ATP). Myosin is an enzyme which catalyzes the hydrolysis of ATP, a process which we have seen (p. 246) to be more exergonic than is the case for many other phosphates, and this hydrolysis contributes energy for contraction. Because of its catalytic action, myosin is also referred to as adenosine triphosphatase (ATP-ase). When the activated complex is formed from ATP-ase and its substrate ATP, the entropy of activation $\Delta S^\dagger$ is about 41 cal K^{-1} mol^{-1} under approximately normal physiological conditions. We saw on p. 400, on the basis of a very simple electrostatic theory of $\Delta S^\dagger$ values for ionic reactions in aqueous solution, that there will be a positive contribution of about 10 cal K^{-1} mol^{-1} for each unit of the product $|z_A|\,|z_B|$. The long myosin molecules bear a series of positive charges, and under normal physiological conditions ATP bears four negative charges which are fairly close to one another. Therefore it might be supposed, as a very simple way of looking at the problem, that when myosin and ATP come together to form an activated complex, one of the positive charges on a myosin molecule ($|z_A| = 1$) comes close to the four negative charges on the ATP molecule ($|z_B| = 4$), so that $|z_A|\,|z_B|$ is 4. This model leads to a $\Delta S^\dagger$ of about 40 cal K^{-1} mol^{-1}, which is in very close agreement with experiment. As discussed on p. 399, there is electrostriction of water molecules around the ions, and when they come together there is a release of water molecules and a corresponding entropy increase.

When the kinetics of enzyme inactivations are investigated, it is found that the energies of activation are usually very large, with correspondingly large values of the entropies of activation. For example, the inactivation of myosin has under certain conditions an activation energy of ~ 70 kcal, and an entropy of activation of ~ 150 cal K^{-1}. An enzyme inactivation is a special case of a protein denaturation, which is known to involve a very profound change in the conformation of the protein. In particular, there is usually a breaking of hydrogen bonds and other internal bonds, leading to a loosening of the structure, with a corresponding large increase in entropy.

10.5   TRANSIENT-PHASE KINETICS

So far we have been concerned with steady-state kinetics. Most investigations have been concerned with the behavior after the steady state has become established, since the experimental techniques and the analysis of the results are much simpler. However, very valuable information is provided by kinetic studies made prior to the establishment of the steady state. Only a very brief account of this can be given here.

Suppose that a single-substrate system is under investigation, and that reaction occurs by the simple Michaelis-Menten mechanism. If solutions of enzyme and substrate are rapidly brought together there will be, at zero time, no enzyme-substrate complex. Formation of this complex will occur, and its concentration will rise from zero to the steady-state value, as shown in curve a of Figure 10.11. As the complex

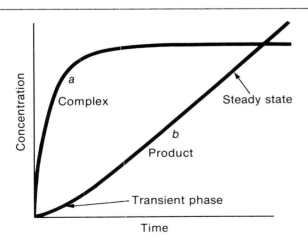

Figure 10.11

The variations with time of the formation of enzyme-substrate complex (curve a) and of the product of reaction (curve b) for a reaction occurring by the simple Michaelis-Menten mechanism: $E + S \rightleftharpoons ES \rightarrow E + X$.

is formed it will become converted into product, the rate of product formation being proportional to the concentration of complex. Thus, as shown in curve *b* of Figure 10.11, the rate of formation of product (the slope of the curve) will be zero at zero time and will rise until it reaches a constant steady-state value.

The time that it takes for the steady state to become established is usually a matter of milliseconds. Therefore it is necessary to use the special high-speed techniques that were referred to in Section 9.11 of the last chapter. The stopped-flow method is frequently used, but the temperature-jump method has also been applied. The procedure is essentially to follow, by spectrophotometric means, the concentration of enzyme-substrate complex or of a reaction product. Then, by applying the theoretical equations to the data it is possible to obtain values for rate constants which could not be obtained from steady-state kinetics. In this way one gains much greater insight into the mechanisms.

10.6   SIGMOID KINETICS AND INTERACTING SUBUNITS

Enzyme systems do not always follow Michaelis-Menten kinetics. Sometimes they show a sigmoidal variation of rate with substrate concentration, as shown schematically in Figure 10.12. There has recently been considerable interest in this kind of behavior, and it turns out that a number of different types of mechanisms can give rise to it. For example, sigmoid kinetics can be found if the enzyme contains an impurity which can combine with the substrate and render it incapable of being acted upon by the enzyme, and also may be found if the enzyme can exist in two forms of differing activities.

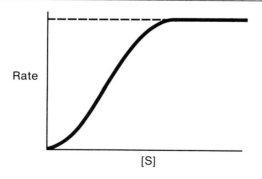

Figure 10.12

Sigmoidal dependence of rate on substrate concentration.

Several theories of sigmoid kinetics are based on the idea that certain enzyme molecules are composed of a number of subunits which interact with each other. One such theory was put forward in 1965 by J. Monod, J. Wyman, and J. P. Changeux. A very simple version of it is as follows. Suppose that the enzyme molecule consists of two subunits, each one of which can occur in two distinct conformations, which we may represent as a circle and a square. Moreover, suppose that there is such a strong interaction between the subunits that in a given molecule both must be in the same conformation. Finally, suppose that the enzyme normally exists with its subunits in the circle form, but that when a substrate molecule becomes attached to a subunit it induces it to change into the square form. The attachment of a substrate molecule to one of the two subunits can therefore be represented as

$$\text{S} \;+\; \text{◯◯} \;\rightleftharpoons\; \boxed{\text{S}\;\;}$$

The other subunit has also been converted into the square form, which can now more readily accept a second substrate molecule. In other words, the equilibrium constant for

$$\text{S} \;+\; \boxed{\text{S}\;\;} \;\rightleftharpoons\; \boxed{\text{S}\;\;\text{S}}$$

is greater than that for the attachment of the first substrate molecule. Thus it follows that after a little substrate has become attached to the enzyme, the rest goes on more readily, and this gives rise to sigmoid behavior. This type of theory has been developed along various lines.

Theories of interacting subunits are often referred to as "allosteric" theories, but the use of this word is unfortunate and should be avoided. The word *allostery* (Greek *allos*, other; *stereos*, solid) refers to the possibility that substances (known as *modifiers* or *effectors*) can be attached at sites other than the site for the attachment of substrate. This is a completely different type of phenomenon and is of great importance in connection with the regulation of metabolic processes, but should be sharply distinguished from subunit interactions. However, in some enzymes the two effects are found together.

10.7 IMMOBILIZED ENZYMES

In biological systems certain enzymes occur in free solution. Examples are the proteolytic enzymes, which bring about hydrolysis in the digestive system. Other enzymes, however, are attached to structural material in the living system and do not have the mobility they would have if they were in free solution. Such enzymes are said to be *immobilized*, or *supported*.

Most kinetic investigations have been carried out with both the enzyme and the substrate present in free solution. Thus it is of great importance, from the standpoint of understanding how enzymes behave in living systems, to investigate the kinetics of immobilized enzymes. During recent years a number of such investigations have been carried out. Besides their significance in leading to an understanding of the functioning of living systems, the results have led to important clinical and technical developments. For example, immobilized enzymes are being used in extracorporeal shunts for the control of certain diseases, and have also been employed in the efficient synthesis of certain chemicals.

The kinetic aspects of immobilized enzymes are rather complicated. A typical situation is when the enzyme is immobilized within some polymeric material, which may be cut into slices and immersed in a suitably buffered solution of the substrate. This is the type of situation that occurs in a biological system, an example being a muscle (in which the enzyme myosin is immobilized) surrounded by a solution of the substrate ATP. For reaction to occur, the substrate has to diffuse through the polymeric material in order to reach the enzyme. Reaction then occurs and the products must diffuse out into the free solution. Since diffusion in polymeric material occurs more slowly than in water, there is now a greater possibility of *diffusion control* (see p. 403): the overall rate of reaction may depend to some extent on the rates with which these diffusion processes occur.

The kinetic investigations so far carried out on immobilized enzymes have indeed provided evidence for various degrees of diffusion control. Enzymes that obey the Michaelis-Menten equation (equation (10.25)) when they are in free solution generally obey it to a good approximation when they are immobilized, but the Michaelis constant is usually significantly different. The rate equation for the immobilized system is

$$v = \frac{k_c[E]_s[S]}{K_m' + [S]} \qquad (10.78)$$

where $[E]_s$ is the concentration of the enzyme in the supporting material. The rate at high substrate concentrations, $k_c[E]_s$, is the same as if the enzyme were in free solution, the value of k_c being unchanged. This is easily understood, because at high substrate concentrations the enzyme is all saturated with substrate, and the rates of diffusion are irrelevant.

On the other hand, at low substrate concentrations the rate becomes

$$v = \frac{k_c[E]_s}{K_m'}[S] \qquad (10.79)$$

This is generally less than the rate in free solution, the value of K_m' being greater than the value K_m in free solution. Under these conditions the rate of the reaction is controlled by the rate with which enzyme and substrate come together, and this in turn is influenced by diffusion rates. Thus there is some measure of diffusion control at lower substrate concentrations, but not at higher ones.

There is obviously less diffusion control if the slices of immobilized enzyme are thin rather than thick, since the substrate then has ready access to the enzyme. Under biological conditions, substrate concentrations are usually substantially less than required to saturate the enzyme. Some diffusion control is therefore to be expected, especially if the macromolecular structural material is fairly thick. It has been estimated that in muscle filaments, of thickness approximately 0.1 micrometres (μm), there is essentially no diffusion control. On the other hand with muscle fibers, of thickness approximately 5 μm, the enzyme reaction is almost completely diffusion controlled. Muscle fibrils, of thickness approximately 2 μm, lie in between, and there is partial diffusion control.

10.8 KINETICS OF BACTERIAL GROWTH

The growth of bacteria contained in a nutrient medium has been studied very extensively, and a number of kinetic aspects are of particular interest. Only a brief account can be given here.

Suppose that, as is usual in an experimental study, bacteria are inoculated into nutrient material contained in a vessel; this is a *batch culture*, growing in a *closed system*. If the logarithm of the number of bacterial cells is plotted against the time, the result is usually as shown in Figure 10.13. The typical growth cycle consists of three phases:

1. An initial phase during which there is no cell division, so that the number of cells remains constant. This is known as the *lag phase*.

2. A subsequent phase during which the logarithm of the number of cells varies linearly with time, which means that the number is increasing exponentially (see below). This is known as the *exponential* or *logarithmic phase*.

3. The *stationary phase*, during which there is no cell division.

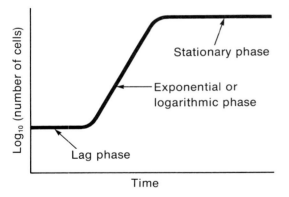

Figure 10.13

A typical bacterial growth cycle for cells growing in a closed system.

Sometimes the transition between phases is very sharp. In other cases there is a considerable rounding off, and one sometimes refers to intermediate phases, i.e., an *acceleration* phase between the lag and the exponential phases, and a *deceleration* phase between the exponential phase and the stationary phase.

THE EXPONENTIAL PHASE

During the exponential phase the logarithm of the number of cells, $\ln n$ or $\log_{10} n$, increases linearly with the time:

$$\ln n = \text{const.} + kt \tag{10.80}$$

where the constant k is known as the *specific growth rate*; its units are reciprocal time (s^{-1}, min^{-1}, etc.). Equation (10.80) can be written as

$$n = n_o e^{kt} \tag{10.81}$$

or as

$$\ln \frac{n}{n_o} = kt \tag{10.82}$$

The significance of n_o, as shown in Figure 10.14, is the value of n extrapolated back to $t = 0$ on this particular plot. Only if there is no lag phase will n_o be the actual number of cells at the start of the experiment.

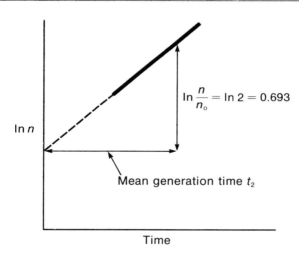

Figure 10.14

Plot of the natural logarithm of the number of cells, against time, for the exponential phase.

It is convenient to define a *mean generation time*, t_2, which is the time it takes for the number of cells to double during the exponential phase. This is the time corresponding to $\ln n/n_o = \ln 2$ so that from (10.82)

$$t_2 = \frac{\ln 2}{k} = \frac{0.693}{k} \tag{10.83}$$

If an individual cell is observed it will be seen to increase in size and eventually divide into two daughter cells, each of which will grow and later divide. The time between one cell division and the next is the *generation time*, and varies very widely for a given type of cell. The *mean generation time* is the mean time for all the cells in a given culture, under the particular conditions of the experiment.

EXAMPLE

A culture medium was inoculated with 4.0×10^6 cells of *Escherichia coli* and incubated at 37° C. There was no lag phase, and after 5 hours there were 8.0×10^9 cells, the system still being in the exponential phase. Calculate the specific growth rate and the mean generation time.

SOLUTION

The specific growth rate k is calculated from equation (10.82),

$$k = \frac{1}{t} \ln \frac{n}{n_o}$$

with $t = 300$ min, $n = 8.0 \times 10^9$, and $n_o = 4.0 \times 10^6$:

$$k = \frac{2.303}{300} \log_{10} 2.0 \times 10^3$$

$$= 0.0253 \text{ min}^{-1}$$

From equation (10.83), the mean generation time is

$$t_2 = \frac{0.693}{0.0253} = 27.4 \text{ minutes}$$

THE LAG PHASE

In the example just considered there was no lag phase, and n_o could therefore be taken to be the number of cells at the beginning of the experiment. If there is a lag phase, the mean generation time must be calculated from data taken within the exponential phase, and the duration of the lag phase can then be calculated. The procedure is best illustrated by an example.

EXAMPLE

A culture medium was inoculated with 5.0×10^5 cells of *Aerobacter aerogenes*. After 100 minutes the system had entered the exponential phase and there were 1.2×10^6 cells. After 300 minutes it was still in the exponential phase and there were 3.5×10^7 cells. Calculate the specific growth rate, the mean generation time within the exponential phase, and the duration of the lag phase.

SOLUTION

We can deal with the exponential phase by shifting the time scale, treating the 100 minute measurement as being at $t = 0$, and the 300 minute measurement as $t = 200$. Then the specific growth rate is

$$k = \frac{2.303}{200} \log_{10} \frac{3.7 \times 10^7}{1.2 \times 10^6}$$

$$= 0.0169 \text{ min}^{-1}$$

The mean generation time is (equation (10.83)):

$$t_2 = \frac{0.693}{0.0169} = 41 \text{ minutes}$$

We can obtain the duration of the lag phase by calculating the time that it would have taken 5.0×10^5 cells to multiply to 3.5×10^7, if there had been no lag phase. This time is given by (equation (10.82)):

$$t = \frac{1}{k} \ln \frac{n}{n_o}$$

$$= \frac{2.303}{0.0169} \log_{10} \frac{3.5 \times 10^7}{5.0 \times 10^5}$$

$$= 251 \text{ minutes}$$

However, the growth actually took 300 minutes. Thus the duration of the lag phase is $300 - 251 = 49$ minutes.

The above calculations are illustrated in Figure 10.15. Note that the calculation of the duration of the lag phase is not dependent on any assumption as to the sharpness of the transition before the lag phase and the exponential phase. The end of the lag phase, in fact, is defined as the point of intersection of the two straight lines.

THE STATIONARY PHASE

The point at which bacteria cease to grow, when they are in a closed system, depends upon a number of factors, an important one being exhaustion of an essential nutrient. In addition, the metabolic processes may cause an adverse pH to develop, or may produce inhibiting substances.

It is often found that over a certain range the total growth (i.e., the total number of bacterial cells produced) varies linearly with the concentration of a given nutrient. This indicates that it is this particular nutrient which is limiting the growth under these conditions.

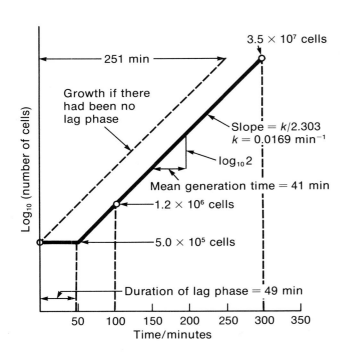

Figure 10.15
Diagram illustrating the calculations on page 456.

PROBLEMS

10.1 The following rates have been obtained for an enzyme-catalyzed reaction at various substrate concentrations:

[S] (mM)	Rate, v (arbitrary units)†
0.4	2.41
0.6	3.33
1.0	4.78
1.5	6.17
2.0	7.41
3.0	8.70
4.0	9.52
5.0	10.5
10.0	12.5

Plot v against $[S]$, $1/v$ against $1/[S]$, and $v/[S]$ against v, and from each plot estimate the Michaelis constant. Which plot appears to give the most reliable value?

---

† Note that the units are irrelevant in determining a Michaelis constant.

10.2 The following data have been obtained for the myosin-catalyzed hydrolysis of ATP, at 25.0° C and pH 7.0:

[ATP]	v
μM	$\mu mol\ dm^{-3}\ s^{-1}$
7.5	0.067
12.5	0.095
20.0	0.119
32.5	0.149
62.5	0.185
155.0	0.191
320.0	0.195

Plot v against [S], $1/v$ against $1/$[S], and $v/$[S] against v, and from each plot calculate the Michaelis constant K_m, and the limiting rate V.

10.3 The following values of V (limiting rate at high substrate concentrations) and K_m have been obtained at various temperatures for the hydrolysis of acetylcholine bromide, catalyzed by acetylcholinesterase:

$T(°C)$	$V \times 10^6$	$K_m \times 10^4$
	$(mol\ dm^{-3}\ s^{-1})$	$(mol\ dm^{-3})$
20.0	1.84	4.03
25.0	1.93	3.75
30.0	2.04	3.35
35.0	2.17	3.05

(a) Assuming the enzyme concentration to be 1.00×10^{-11} M, calculate the energy of activation, the enthalpy of activation, the Gibbs energy of activation and the entropy of activation for the breakdown of the enzyme-substrate complex at 25.0° C.

(b) Assuming K_m to be the *dissociation* constant k_{-1}/k_1 for the enzyme-substrate complex $(ES \underset{k_1}{\overset{k_{-1}}{\rightleftharpoons}} E + S)$, determine the following thermodynamic quantities for the *formation* of the enzyme-substrate complex at 25.0° C: $\Delta G°$, $\Delta H°$, $\Delta S°$.

(c) From the results obtained in (a) and (b) above, sketch a Gibbs-energy diagram and an enthalpy diagram for the reaction.

10.4 In a study of an α-chymotrypsin-catalyzed reaction, the maximum rate of hydrolysis, V, was found to vary with pH as follows:

pH	Maximum rate, V
	(arbitrary units)
5.4	10
5.8	32
6.4	99
6.6	141
6.8	192
7.2	260
7.4	280
7.8	300
8.0	328
8.4	331
9.0	327

Plot $\log_{10} V$ against pH, and from the plot estimate a pK value for the ionization of the enzyme-substrate complex, assuming the simple Michaelis-Menten mechanism to apply.

10.5 The following data relate to an enzyme reaction:

$[S] \times 10^3$	$v \times 10^5$
mol dm^{-3}	mol dm^{-3} s^{-1}
2.0	13
4.0	20
8.0	29
12.0	33
16.0	36
20.0	38

The concentration of the enzyme is 2.0 grams dm^{-3}, and the molecular weight is 50 000. Calculate K_m, the maximum rate V, and k_c.

10.6 The following data have been obtained for the myosin-catalyzed hydrolysis of ATP:

Temperature	$k_c \times 10^6$
(°C)	(s^{-1})
39.9	4.67
43.8	7.22
47.1	10.0
50.2	13.9

Calculate, at 40.0° C, the energy of activation, the enthalpy of activation, the Gibbs energy of activation, and the entropy of activation.

10.7 The following mechanism, known as the Theorell-Chance mechanism, was originally proposed in 1951 by H. Theorell and B. Chance to explain the kinetics of alcohol dehydrogenase:

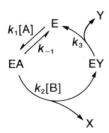

[Note that this is similar to the ordered ternary-complex mechanism shown in Figure 10.5, except that there is no ternary complex EAB. B reacts with EA to form EY + X in one stage.] Apply the steady-state treatment to this mechanism and obtain an expression for the rate of formation of X (equal to the rate of formation of Y) in terms of $[E]_o$, $[A]$, $[B]$, and the rate constants.

10.8 An enzyme-catalyzed reaction proceeds by the simple Michaelis-Menten mechanism:

$$E + S \underset{k_{-1}}{\overset{k_1}{\rightleftharpoons}} ES \overset{k_2}{\rightarrow} E + X$$

The substrate concentration is much higher than that of the enzyme. On the same graph, draw concentration versus time curves for E, S, ES, and X, for the period from instantaneous mixing of substrate and enzyme, to the completion of reaction. Outline briefly how the rate constants k_1, k_{-1} and k_2 could be determined from an analysis of these curves.

10.9 The following data are available for the reaction

$$\text{fumarate} + H_2O \rightleftharpoons \text{malate}$$

catalyzed by the enzyme fumarase:

Overall reaction	$\Delta H° = -3.6 \text{ kcal}$
Fumarate + fumarase $\rightleftharpoons$ enzyme-substrate complex	$\Delta H° = 4.2 \text{ kcal}$
Malate + fumarase $\rightleftharpoons$ enzyme-substrate complex	$\Delta H° = -1.2 \text{ kcal}$
Reaction of fumarate-fumarase complex	$E = 6.1 \text{ kcal}$
Reaction of malate-fumarase complex	$E = 15.1 \text{ kcal}$

Sketch an energy diagram for the course of this reaction.

10.10 A medium was inoculated with 3.0×10^5 cells of *Aerobacter aerogenes*. After 5 hours there were found to be 1.5×10^7 cells, and a lag period of 90 minutes was observed. Calculate the specific growth rate and the mean generation time during the exponential phase.

10.11 A medium was inoculated with 1.00×10^6 bacterial cells and the following counts obtained after various times:

Time (min)	Number of cells
25	1.00×10^6
50	1.00×10^6
75	1.51×10^6
100	2.51×10^6
150	6.61×10^6
200	1.82×10^7
250	4.79×10^7
300	1.32×10^8
350	3.39×10^8
400	3.98×10^8

Plot these data and, from the graph, determine the duration of the lag phase, and the mean generation time during the exponential phase.

ESSAY QUESTIONS

10.12 Give an account of the way in which the rate of an enzyme reaction is affected by the temperature.

10.13 Discuss briefly some of the more important factors that affect the rate of an enzyme reaction.

10.14 Will the rate of an enzyme-catalyzed reaction usually be more sensitive to temperature than that of the same reaction when it is uncatalyzed? Discuss.

10.15 Explain how you would proceed to determine the parameters K_m and k_c for an enzyme system involving a single substrate.

10.16 Male albino mice increase in weight by a factor of about 3.3 from the third to the twentieth week at both $20°$ C and $24°$ C. When maintained at $33°$ C they increase by a factor of only 2.4 during the same period. Discuss this result in the light of your knowledge of the behavior of enzymes.

SUGGESTED READING

Bray, H. G., and White, K. *Kinetics and Thermodynamics in Biochemistry.* London: Churchill, 1967.

Cornish-Bowden, A. *Principles of Enzyme Kinetics.* London: Butterworth, 1976.

Dawes, E. A. *Quantitative Problems in Biochemistry*, Chapter 9. London: Churchill Livingstone, 1967.

Dixon, M., and Webb, E. C. *Enzymes*, 2nd ed., Chapter 4. London: Longmans Green, 1964.

Hinshelwood, C. N. *Chemical Kinetics of the Bacterial Cell*, Chapters 2 and 3. Oxford: Clarendon Press, 1946.

Jencks, W. P. *Catalysis in Chemistry and Enzymology.* New York: McGraw-Hill, 1969.

Johnson, F. H., Eyring, H., and Stover, B. J. *The Theory of Rate Processes in Biology and Medicine.* New York: John Wiley, 1974.

Laidler, K. J., and Bunting, P. S. *The Chemical Kinetics of Enzyme Action*, 2nd ed. Oxford: Clarendon Press, 1973.

Mahler, H. R., and Cordes, E. H. *Biological Chemistry*, Chapter 4. New York: Harper & Row, 1966.

Wesley, J. *Enzymic Catalysis.* New York: Harper & Row, 1969.

Zeffren, E., and Hall, P. L. *The Study of Enzyme Mechanisms.* New York: Wiley-Interscience, 1973.

11

EQUILIBRIUM AND TRANSPORT IN MACROMOLECULAR SYSTEMS

A number of topics are especially applicable to the large molecules that are so important in biology. Some of these topics relate to equilibria, others to the transport properties of macromolecules in solution.

11.1   ADSORPTION ISOTHERMS

Physical chemists have long been interested in the equilibria established at surfaces. The simplest situation is when we have a gas, such as ammonia, in contact with a surface, such as one of copper. Ammonia molecules are rather strongly attached to such a surface, which may become completely covered by a unimolecular layer. A similar, but much more complicated, situation exists if we have an ionic solution in contact with a protein molecule; some of the ions will become attached to the surface of the protein.

It is convenient to have an equation relating the amount of substance attached to the surface, or *adsorbed*, to the concentration of the substance present in the gas phase or in solution. Such equations apply at a fixed temperature, and are known as *adsorption isotherms*. The simplest adsorption isotherm was first obtained by the American physical chemist Irving Langmuir (1881–1957). The basis of the derivation of the Langmuir adsorption isotherm is that all parts of the surface behave in exactly the same way as far as adsorption is concerned. Suppose that, after equilibrium is established, a fraction θ of the surface is covered by adsorbed molecules;

a fraction $1 - \theta$ will not be covered. The rate of adsorption will then be proportional to the concentration [A] of the molecules in the gas phase or in solution, and also to the fraction of the surface that is bare, the assumption being that adsorption can occur only when molecules strike the bare surface. The rate of adsorption is thus

$$v_a = k_a[A](1 - \theta) \tag{11.1}$$

where k_a is a constant relating to the adsorption process. The rate of desorption is proportional only to the number of molecules attached to the surface, which in turn is proportional to the fraction of surface covered:

$$v_d = k_d\theta \tag{11.2}$$

At equilibrium, the rates of adsorption and desorption are the same; thus

$$k_a[A](1 - \theta) = k_d\theta \tag{11.3}$$

or

$$\frac{\theta}{1 - \theta} = \frac{k_a}{k_d}[A] \tag{11.4}$$

The ratio k_a/k_d is an equilibrium constant and can be written as K; then

$$\frac{\theta}{1 - \theta} = K[A] \tag{11.5}$$

or

$$\theta = \frac{K[A]}{1 + K[A]} \tag{11.6}$$

A graph of θ against [A] is shown as Figure 11.1. At sufficiently low concentrations we can neglect $K[A]$ in comparison with unity, and then

$$\theta = K[A] \tag{11.7}$$

At sufficiently high concentrations $K[A] \gg 1$, and (11.6) reduces to

$$\theta = 1 \tag{11.8}$$

which means that the surface is saturated. Note that the Langmuir adsorption isotherm (11.6) is of exactly the same form as the Michaelis equation [see equation (10.10) on p. 430, and compare Figure 11.1 with Figure 10.1 on p. 428].

As we have seen, the Langmuir adsorption isotherm is based on the simplest of assumptions. Systems which obey the equation are often referred to as showing *ideal* adsorption, the Langmuir equation having the same significance in connection with adsorption as has the ideal gas law $PV = nRT$ in connection with the behavior of gases. Systems frequently deviate significantly from the Langmuir equation, and this may occur because the surface is not uniform. In addition, there may be interactions between adsorbed molecules, a phenomenon referred to as *cooperativity*. A molecule attached to a surface may make it more, or less, difficult for another molecule to become attached to a neighboring site, and this will lead to a deviation from the ideal adsorption equation. Nonideal systems can sometimes be fitted to an empirical adsorption isotherm derived by the German physical chemist Herbert Max Finlay Freundlich (1880–1941). According to this isotherm the amount of a substance adsorbed, x, is related to the concentration c by the equation

$$x = kc^n \qquad (11.9)$$

where k and n are empirical constants. This equation does not give saturation of the surface; the amount adsorbed keeps increasing as c increases. If (11.9) applies, a plot of $\log_{10} x$ against $\log_{10} c$ will give a straight line of slope n.

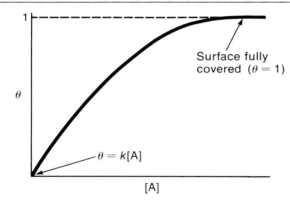

Figure 11.1

A plot of θ (fraction of surface covered) against $[A]$ (concentration of a substance in the gas phase or in solution), for a substance obeying the Langmuir adsorption isotherm [equation (11.6)].

11.2   MULTIPLE EQUILIBRIA

To deal with the binding of solute molecules to macromolecules, we need a more detailed treatment than is provided by these simple isotherms. In biology, we are concerned with such processes as the binding of magnesium ions by DNA and the binding of oxygen by hemoglobin. Substances bound in this way are known as *ligands* (Latin *ligare*, to bind). A number of binding sites are involved, and sometimes they behave identically and act quite independently of each other. This is the simplest situation, and such systems obey an equation of the Langmuir form. However, in a number of important cases the occupation of one site affects the strength of binding at other sites; in other words, the behavior is *cooperative*. Cooperativity is particularly important in biology, in connection with *biological control mechanisms*.

Consider the general case, in which a macromolecule P has n identical binding sites, and reacts with a ligand A, one molecule of which can become attached to each of the sites. Attachment of one molecule of A to P has an *association constant* K_1:†

$$P + A \rightleftharpoons PA \qquad K_1 = \frac{[PA]}{[P][A]} \tag{11.10}$$

Similarly, for subsequent attachments:

$$PA + A \rightleftharpoons PA_2 \qquad K_2 = \frac{[PA_2]}{[PA][A]} \tag{11.11}$$

$$PA_{n-1} + A \rightleftharpoons PA_n \qquad K_n = \frac{[PA_n]}{[PA_{n-1}][A]} \tag{11.12}$$

In the last reaction, all the sites have become occupied. The total concentration of bound A is given by

$$[A]_{bound} = [PA] + 2[PA_2] + 3[PA_3] + \cdots n[PA_n] \tag{11.13}$$

and the total concentration of the macromolecule is

$$[P]_{total} = [P] + [PA] + [PA_2] + \cdots [PA_n] \tag{11.14}$$

If we make use of the equilibrium expressions such as (11.10) through (11.12), we see that

$$[A]_{bound} = [P]\{K_1[A] + 2K_1K_2[A]^2 + \cdots n(K_1K_2\cdots K_n)[A]^n\} \tag{11.15}$$

and

$$[P]_{total} = [P]\{1 + K_1[A] + K_1K_2[A]^2 + \cdots (K_1K_2\cdots K_n)[A]^n\} \tag{11.16}$$

The average number of molecules of A bound per macromolecule is

$$\bar{v} = \frac{[A]_{bound}}{[P]_{total}} \tag{11.17}$$

---

† It is common to work with dissociation constants, which are the reciprocals of the association constants. However, more compact expressions are obtained using association constants.

and is therefore given by

$$\bar{v} = \frac{K_1[A] + 2K_1K_2[A]^2 + \cdots n(K_1K_2\cdots K_n)[A]^n}{1 + K_1[A] + K_1K_2[A]^2 + K_1K_2\cdots K_n[A]^n} \tag{11.18}$$

This equation is sometimes known as the Adair equation, after the British biophysical chemist Gilbert Smithson Adair. Its general form is as shown in Figure 11.2; at high concentrations of ligand, $\bar{v} \to n$, while at low concentrations $\bar{v} \to 0$. However, equation (11.18) is not useful as it stands, since it would be impossible to evaluate the constants from an experimental curve. To make such an analysis of experimental results, some special cases must be considered.

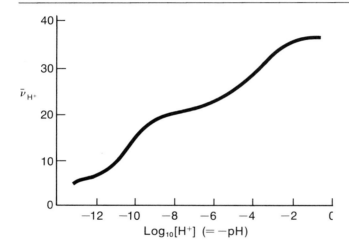

Figure 11.2

A typical plot of $\bar{v}$, the average number of ligand molecules bound per macromolecule against the logarithm of the concentration of ligand for a molecule having several binding sites. This curve was obtained for the binding of H^+ ions to ribonuclease.

IDENTICAL AND INDEPENDENT SITES

The simplest situation is when all of the sites are the same and there is no cooperativity, i.e., binding at any site is independent of the occupation of other sites. We would then expect (11.18) to reduce to the Langmuir form. The reason it does so may be seen as follows.

At first sight, we might think that we should set all the K's equal; i.e., $K_1 = K_2 = \cdots K_n$. However, this is not correct; there are certain statistical factors. This is most readily understood if we simplify the scheme, and consider just four binding sites on the macromolecule, as shown in Figure 11.3. Suppose that K is the equilibrium constant for the binding of A at a *given* site; then K_1 is equal to $4K$, since there are four ways for A to become attached and one way for it to come off. Similarly, K_2 will be $3K/2$; after one molecule of A is attached, there are 3 ways for an additional molecule to go on, and 2 ways for an A to come off. The relationships for K_3 and K_4 are shown in Figure 11.3. For this case of four sites (11.18) reduces to

$$\bar{v} = \frac{K_1[A] + 2K_1K_2[A]^2 + 3K_1K_2K_3[A]^3 + 4K_1K_2K_3K_4[A]^4}{1 + K_1[A] + K_1K_2[A]^2 + K_1K_2K_3[A]^3 + K_1K_2K_3K_4[A]^4}$$

(11.19)

Insertion of the expressions for K_1, K_2, K_3, and K_4 in terms of K then gives

$$\bar{v} = \frac{4K[A] + 12K^2[A]^2 + 12K^3[A]^3 + 4K^4[A]^4}{1 + 4K[A] + 6K^2[A]^2 + 4K^3[A]^3 + K^4[A]^4}$$

(11.20)

$$= \frac{4K[A](1 + 3K[A] + 3K^2[A]^2 + K^3[A]^3)}{1 + 4K[A] + 6K^2[A]^2 + 4K^3[A]^3 + K^4[A]^4}$$

(11.21)

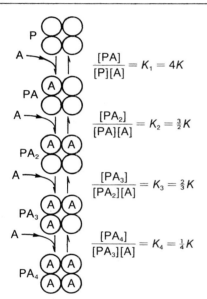

Figure 11.3

The successive binding of a substance A to a macromolecule having four binding sites.

$$\frac{[PA]}{[P][A]} = K_1 = 4K$$

$$\frac{[PA_2]}{[PA][A]} = K_2 = \tfrac{3}{2}K$$

$$\frac{[PA_3]}{[PA_2][A]} = K_3 = \tfrac{2}{3}K$$

$$\frac{[PA_4]}{[PA_3][A]} = K_4 = \tfrac{1}{4}K$$

The coefficients in the numerator and denominator of (11.21) will be recognized as the binomial coefficients; the term in brackets in the numerator is $(1 + K[A])^3$, while the denominator is $(1 + K[A])^4$. Equation (11.21) thus reduces to

$$\bar{v} = \frac{4K[A]}{1 + K[A]}$$

(11.22)

It is easy to show that in general, for a macromolecule having n identical and independent sites,

$$\bar{v} = \frac{nK[A]}{1 + K[A]}$$

(11.23)

The fraction of sites occupied, θ, is $\bar{v}/n$, so that

$$\theta = \frac{K[A]}{1 + K[A]} \tag{11.24}$$

This, as we expected, is the Langmuir isotherm (equation (11.6)).

If data have been obtained for $\bar{v}$ as a function of $[A]$ it is easy to test the applicability of (11.23) and, if the equation does apply, to obtain n. If we take reciprocals of (11.23), we obtain

$$\frac{1}{\bar{v}} = \frac{1}{n} + \frac{1}{nK[A]} \tag{11.25}$$

and a plot of $1/\bar{v}$ against $1/[A]$ will give a straight line. This method—which is equivalent to the Lineweaver-Burk method for rates (p. 430)—is shown schematically in Figure 11.4, and we see that $1/n$ is given by one of the intercepts. Alternatively, we can make a plot equivalent to the Eadie plot (p. 431); equation (11.23) then rearranges to

$$\bar{v} = n - \frac{\bar{v}}{K[A]} \tag{11.26}$$

and a plot of $\bar{v}/[A]$ against $\bar{v}$ will give a straight line, one of the intercepts of which is n (see Figure 11.4b). This type of plot has been applied, for example, to the binding of NADH to the enzyme lactate dehydrogenase. A straight line was obtained and the number of binding sites n found to be 4. Thus it appears that this enzyme has four equivalent binding sites, and exhibits no cooperativity.

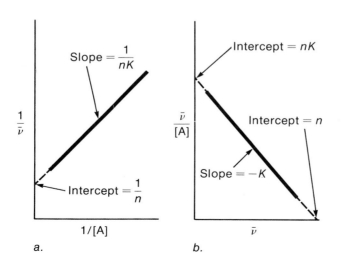

Figure 11.4

Schematic plots of (a) $1/\bar{v}$ against $1/[A]$ and (b) $\bar{v}/[A]$ against $[A]$ for systems that obey equation (11.23).

HIGHLY COOPERATIVE BINDING

The other extreme case is if the binding of the ligand A at one site influences the other sites in such a way that they fill up immediately; now it does not matter whether the sites are identical. In other words, with reference to (11.18), K_n is very much larger than $K_1, K_2, \ldots, K_{n-1}$. Thus, in the numerator of (11.18), we can neglect all terms except the last, all of these terms relating to species other than P and PA_n. Similarly, in the denominator of (11.18) we can neglect all terms except the first and the last. The equation thus reduces to

$$\bar{v} = \frac{nK_1K_2\cdots K_n[A]^n}{1 + K_1K_2\cdots K_n[A]^n} \tag{11.27}$$

We may write the product of the constants, $K_1K_2\cdots K_n$, simply as K, this being the overall equilibrium constant for the binding of n molecules,

$$nA + P \rightleftharpoons PA_n$$

The resulting equation is thus

$$\bar{v} = \frac{nK[A]^n}{1 + K[A]^n} \tag{11.28}$$

The fraction of sites occupied, θ, is equal to $\bar{v}/n$, so that

$$\theta = \frac{K[A]^n}{1 + K[A]^n} \tag{11.29}$$

or

$$\frac{\theta}{1 - \theta} = K[A]^n \tag{11.30}$$

This equation is to be contrasted with the Langmuir equation (11.5), which corresponds to identical sites with no cooperativity. Equation (11.30) can be tested by plotting $\log_{10}\{\theta/(1 - \theta)\}$ against $\log_{10}[A]$. Such plots are known as *Hill plots*, after the British physiologist Archibald Vivian Hill who first employed them in 1910 in his studies of the binding of oxygen to hemoglobin. Figure 11.5 shows schematic Hill plots for two extreme cases. If the sites are identical and independent, equation (11.5) applies and the slope is unity (curve *a*). If there is complete cooperativity, equation (11.30) applies and the slope is n, the number of sites. Curve *b* shows the case of $n = 4$.

INTERMEDIATE CASES

Unfortunately, the analysis of data is much more difficult if the behavior lies in between the two extremes just considered. If we have identical sites equation (11.18) will apply, but it is rarely possible to fit data to the equation in this form. However,

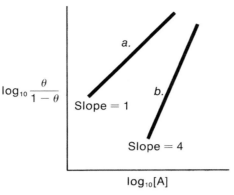

Figure 11.5

Schematic Hill plots. Line (*a*) shows the behavior for identical and independent sites; (*b*) shows the behavior for highly cooperative binding with four sites.

analysis of this equation leads to the result that if a Hill plot is made, the slope will be unity in the limits of low and high concentration, and the maximum slope will be less than *n* and greater than unity. This behavior is illustrated schematically in Figure 11.6. The fact that the slopes are unity at the two limits can be understood as follows. If only a few A molecules have become attached to the macromolecular material, for the most part they will be on different molecules. Therefore, they will not interfere with each other, and cooperativity is unimportant. It can be shown from (11.18) that when [A] is sufficiently large, $\theta/(1 - \theta)$ is proportional to [A], and at large concentrations of A the slope of the Hill plot is again unity. The maximum value of the slope of a Hill plot thus gives a minimum value for the number of binding sites.

So far, we have assumed the binding sites on the macromolecule to be identical. However, many macromolecules possess binding sites of different types. In fact, we have already met this situation in proteins with the binding sites for hydrogen ions; the different ionizing groups have very different dissociation constants. In such a situation, if each site binds only one molecule, we may apply (11.23) to each type of site:

$$\bar{v} = \frac{n_1 K_1 [A]}{1 + K_1 [A]} + \frac{n_2 K_2 [A]}{1 + K_2 [A]} + \cdots \tag{11.31}$$

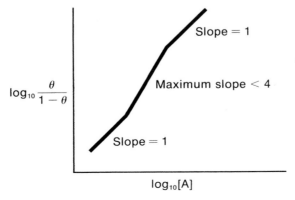

Figure 11.6

Schematic Hill plot for an intermediate situation, with *n* = 4.

There are n_1 sites of type 1 on the molecule, and their binding constant is K_1; the K_2 is for sites of type 2, and so on. The analysis of this type of binding will be difficult unless the values of the binding constants differ greatly. In that case, there will be almost exclusive binding at one type of site over a particular concentration range; over another range of [A] values there will be almost exclusive binding at another type of site.

11.3   CONFORMATIONAL EQUILIBRIA

Another type of equilibrium which plays an important role in biological processes is that occurring between different conformational forms of a macromolecule. For example, proteins in their active forms have a particularly ordered arrangement, sometimes having a considerable degree of α-helix structure, but when they become denatured the structure is much more random. Also, under most circumstances DNA exists as a highly-ordered, double-stranded helix, but when replication occurs, this structure is to some extent broken down.

One property of many such transformations is their *all or none* character. For example, if proteins in solution are slowly heated they often remain in their native states until a certain temperature is reached, when they become denatured rather suddenly. Similarly, if we heat a solution of DNA, and follow the structural changes by spectrophotometry, we find that nothing happens until a temperature of about $80°$ C is reached; then, within a few degrees, the helix "melts" to a form in which the various long-chain molecules are coiled at random. If we write an equilibrium constant as

$$K = \frac{\text{Fraction of residues that are random coils}}{\text{Fraction of residues having helical structure}} \qquad (11.32)$$

we know that

$$K = e^{-\Delta G°/RT} = e^{-\Delta H°/RT}\, e^{\Delta S°/R} \qquad (11.33)$$

For K to change very sharply with temperature it is necessary for $\Delta H°$ to be very large. However, if $\Delta H°$ is very large, K will be abnormally small unless $\Delta S°$ is also very large.

Some thermodynamic values for protein denaturations are given in Table 11.1. These were all determined by measuring the equilibrium constants over a range of temperature. A plot of the logarithm of K against $1/T$ will give a straight line from which the $\Delta H°$ value can be deduced; $\Delta S°$ is then calculated from $\Delta G°$ ($= -RT \ln K$). It will be seen that the $\Delta H°$ and $\Delta S°$ values are very large ones in comparison with those measured for ordinary chemical processes (see Chapter 5). It is similarly found when the *rates* are studied that very large enthalpies and entropies *of activation* are obtained for these denaturation processes. For the inactivation of myosin at pH 7.0, for example, a value of 70 kcal mol^{-1} has been obtained for the energy of activation, and the entropy of activation is 150 cal K^{-1} mol^{-1}.

Table 11.1   Enthalpy and Entropy Changes in Protein Denaturations†

Protein	pH	ΔH° kcal mol^{-1}	ΔS° cal K^{-1} mol^{-1}
Trypsin	2.0	67.6	213
Soybean trypsin inhibitor	3.0	57.3	180
Chymotrypsinogen	2.0	99.6	316
Chymotrypsinogen	3.0	143	432

† These data are further discussed in K. J. Laidler, *The Chemical Kinetics of Enzyme Action*, 1st ed. Oxford: Clarendon Press, 1958, pp. 354–358.

The loosening of a single amino acid or purine unit from the helix structure of a macromolecule involves the breaking of only two or three hydrogen bonds, and the ΔH° value corresponding to this will be about 5 kcal mol^{-1}; the corresponding ΔS° would be about 15 cal K^{-1} mol^{-1}. The larger values given in Table 11.1 show that a number of amino acid or purine units become detached. This detachment cannot occur in successive independent stages, since the overall ΔH° and ΔS° values would then be ~ 5 kcal mol^{-1} and ~ 15 cal K^{-1} mol^{-1} respectively. It follows that the whole process must be cooperative, sections of the structure changing at once from the helix to the random coil.

The reason for the cooperative behavior is that the individual hydrogen-bonded structures are so interdependent. If we examine the α-helix structure of a protein, we see that it is very difficult for one hydrogen bond to break without others breaking at the same time.

11.4   TRANSPORT PROPERTIES

So far we have considered some equilibrium properties of solutions of macromolecules. The remainder of the chapter deals with *transport* or *hydrodynamic* properties, which are concerned with the movements of molecules under the influence of forces of various kinds. For example, if a concentration gradient exists, there is a Gibbs energy difference between various positions in the solution, and this gives rise to a force which produces a molecular motion known as *diffusion*. Molecules subjected to a gravitational field, or to a centrifugal force which produces a larger effective gravitational field, undergo a motion known as *sedimentation*. A shearing force acting on a solution produces a relative motion of different planes in the solution, and the extent of this motion is measured by a property known as the *fluidity*, the reciprocal of which is the *viscosity*.

Motion can also be induced by the application of an electric field. In Chapter 6 we discussed how simple ions move when an electrical potential is applied, this motion in turn contributing to the conductivity of a solution. Larger molecules such as proteins usually contain charged groups, the charges depending on the pH of the solution (see Chapter 7), and such molecules in solution may move in an electric field; this type of motion is referred to as *electrophoresis*. A related type of transport occurs when an electric potential is applied across a charged membrane immersed in a solution; there is a movement of the solvent molecules, and we speak of *electro-endosmosis* or *electro-osmosis*.

We shall now consider the theoretical principles which apply to these various transport properties.

11.5   DIFFUSION

FICK'S LAWS

We have seen in Chapter 3 how measurements of rates of diffusion can provide estimates of molecular weights. In the present section the laws of diffusion of solute species are considered in further detail.

The fundamental law of diffusion is Fick's First Law, according to which the rate of diffusion dn/dt of a solute across an area A is given by

$$\frac{dn}{dt} = -DA \frac{\partial c}{\partial x} \tag{11.34}$$

where $\partial c/\partial x$ is the concentration gradient of the solute; dn is the number of moles of solute crossing the area A in time dt. If the area of cross-section is unity the rate of diffusion is known as the *diffusive flux*, or simply the *flux*, and is usually given the symbol J; thus

$$J = -D \frac{\partial c}{\partial x} \tag{11.35}$$

If SI units are used, dn/dt is in mol s^{-1}, A in m^2 and $\partial c/\partial x$ in mol m^{-4}, so that the diffusion coefficient D is in m^2 s^{-1} and the flux in mol m^{-2} s^{-1}. However, in practice the diffusion coefficient is more commonly expressed as cm^2 s^{-1} and sometimes as cm^2 min^{-1}.

Fick also derived an equation for the rate of change of concentration as a result of diffusion. Figure 11.7 shows a system of unit cross section, having a concentration c at an area at position x and a concentration $c + dc$ at position $x + dx$. Because there is an increase in concentration as x increases, the net diffusion occurs from right to left in the diagram. The flux at x can be written as $J(x)$, and that at $x + dx$ as $J(x + dx)$, which is given by

$$J(x + dx) = J(x) + \frac{\partial J}{\partial x} dx \tag{11.36}$$

The net flux into the region between x and $x + dx$ is thus

$$J_{net} = J(x) - J(x + dx) \tag{11.37}$$

$$= -\frac{\partial J}{\partial x} dx \tag{11.38}$$

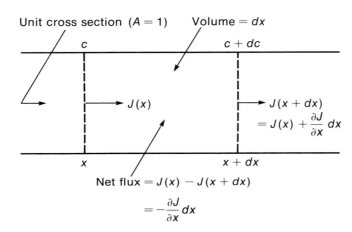

Unit cross section $(A = 1)$ Volume $= dx$

Net flux $= J(x) - J(x + dx)$

$$= -\frac{\partial J}{\partial x} dx$$

Figure 11.7

One-dimensional diffusion illustrating Fick's second law.

The net rate of increase in concentration in this element of volume is the net flux divided by the volume, which is dx since the area is unity:

$$\frac{\partial c}{\partial t} = -\frac{\partial J}{\partial x} \tag{11.39}$$

Introduction of (11.35) then gives

$$\frac{\partial c}{\partial t} = \frac{\partial}{\partial x}\left(D \frac{\partial c}{\partial x}\right) \tag{11.40}$$

If D is independent of the distance x (as is always true to a good approximation) this equation reduces to

$$\frac{\partial c}{\partial t} = D \frac{\partial^2 c}{\partial x^2} \tag{11.41}$$

This is *Fick's second law of diffusion* for the special case of diffusion in one dimension, i.e., along the X axis. In liquids and solutions, which are said to be *isotropic*, D has the same value in all directions. Some solids are *anisotropic*, which means that D is not the same in all directions.

SOLUTIONS OF DIFFUSION EQUATIONS

Equation (11.41) is a second-order, linear, and homogeneous differential equation. Its solution depends upon the nature of the domain through which diffusion is taking place and upon the initial conditions, i.e., the concentrations at various positions at some time which may be taken as $t = 0$. Some initial and boundary conditions present a difficult mathematical problem, and often explicit solutions cannot be obtained.

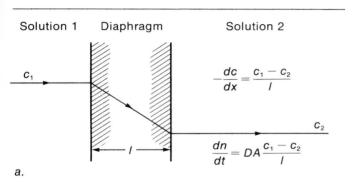

Solution 1   Diaphragm   Solution 2

c_1

$$-\frac{dc}{dx} = \frac{c_1 - c_2}{l}$$

c_2

$$\frac{dn}{dt} = DA\frac{c_1 - c_2}{l}$$

a.

Stirrers

Solution

$l\updownarrow$ Diaphragm

Concentration gradient $= \dfrac{c}{l}$

$$\frac{dn}{dt} = DA\frac{c}{l}$$

Area $= A$

Solvent

b.

Figure 11.8

a. Stirred solutions of concentrations c_1 and c_2 separated by a porous diaphragm of thickness l. In the steady state the concentration within the diaphragm changes linearly from c_1 to c_2.

b. Schematic diagram of simple apparatus for the measurement of diffusion constants.

A particularly simple situation is when solutions at two different concentrations are separated by a porous diaphragm in such a way that the solutions can be stirred and maintained at uniform concentration. A convenient technique for measuring diffusion rates in water is to separate two solutions by a sintered glass membrane of thickness l, as shown schematically in Figure 11.8. Such a membrane contains pores filled with water, and it is assumed that the diffusion through such a membrane occurs at the same rate as in water (a correction can be made for the area of cross-section occupied by the glass). For such a system it is found experimentally that a steady state is soon established; that is, the rate of diffusion does not change with time as long as the concentrations c_1 and c_2 on the two sides remain the same (this will be true to a good approximation if the solution volumes are large). In order for there to be a steady state the concentration gradient must be uniform across the membrane; otherwise the rates of flow would vary across the membrane and there would be accumulation or depletion of material in certain regions of the membrane, and therefore no steady state. In other words, the concentration within the membrane must fall linearly, as shown in Figure 11.8a. The concentration gradient is thus given by

$$-\frac{dc}{dx} = \frac{c_1 - c_2}{l} \qquad (11.42)$$

and by Fick's first law the rate of flow through the diaphragm is

$$\frac{dn}{dt} = DA\frac{c_1 - c_2}{l} \qquad (11.43)$$

The diffusion coefficient D is therefore calculated from the measurement of the rate of flow; the area A, thickness l and concentration difference $c_1 - c_2$ are readily determined. Some diffusion coefficients measured in this and similar ways are given in Table 11.2.

The case just considered is a very simple one in which the concentration has been forced to vary linearly from c_1 to c_2. Another case of interest is when there is an *instantaneous plane source* at a particular plane in a liquid. A simple way of arriving at the equations applicable to this case is to note that a general solution of (11.41) for Fick's second law is

$$c = \alpha t^{-\frac{1}{2}} e^{-x^2/4Dt} \qquad (11.44)$$

Table 11.2   Diffusion Coefficients in Water at 20° C

Substance	Mol. wt.	Diffusion Coefficient, $D/\text{m}^2\,\text{s}^{-1}$
Glucose	180	6.8×10^{-10}
Insulin	41 000	8.2×10^{-11}
Hemoglobin (horse)	68 000	6.3×10^{-11}
Catalase	250 000	4.1×10^{-11}
Urease	470 000	3.5×10^{-11}
Tobacco mosaic virus	31 400 000	5.3×10^{-12}

where α is a constant. That this is a general solution may be verified by substitution into equation (11.41).† When $t \to 0$ this function corresponds to $c = 0$ everywhere except at $x = 0$, where $c \to \infty$. In other words, this case corresponds to solute present at a plane at the origin $x = 0$; since it is present in zero volume the concentration is infinite. The constant α is related to the "strength" of the source, i.e., to the number n_0 of solute molecules initially present at $x = 0$. Since the number of solute molecules remains the same at all times, n is equal to the integral

$$\int_{-\infty}^{\infty} c \, dx$$

at any time t. Introduction of the expression for c (11.44) and integration leads to

$$n_0 = \alpha \int_{-\infty}^{\infty} t^{-\frac{1}{2}} e^{-x^2/4Dt} \, dx \tag{11.45}$$

$$= 2\alpha(\pi D)^{\frac{1}{2}} \tag{11.46}$$

The constant α is thus

$$\alpha = \frac{n_0}{2(\pi D)^{\frac{1}{2}}} \tag{11.47}$$

and equation (11.44) becomes

$$c = \frac{n_0}{2(\pi Dt)^{\frac{1}{2}}} e^{-x^2/4Dt} \tag{11.48}$$

The concentration c in this expression is the number of molecules per unit distance; if the cross-sectional area is unity, c is the number of molecules per unit volume.

---

† From (11.44),

$$\frac{\partial c}{\partial t} = \alpha \left[-\frac{1}{2} t^{-3/2} e^{-x^2/4Dt} + \frac{x^2}{4Dt^{5/2}} e^{-x^2/4Dt} \right]$$

$$\frac{\partial c}{\partial x} = -\frac{\alpha x}{2D} t^{-3/2} e^{-x^2/4Dt}$$

$$\frac{\partial^2 c}{\partial x^2} = -\frac{\alpha t^{-3/2}}{2D} \left[e^{-x^2/4Dt} - \frac{x^2}{2Dt} e^{-x^2/4Dt} \right]$$

and thus

$$\frac{\partial c}{\partial t} = D \frac{\partial^2 c}{\partial x^2}$$

Figure 11.9 shows plots of c/n_o against x, for three different values of Dt. This plot shows how the solute molecules spread out from the instantaneous plane source located at $x = 0$.

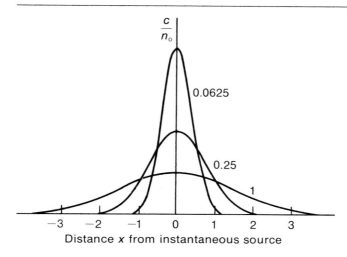

Figure 11.9

Plots of c/n_o against distance from an instantaneous plane source at $x = 0$. At this plane there are initially n_o molecules of solute, at infinite concentration (volume = 0). The numbers on the curves are values of Dt.

An interesting aspect of this problem is to focus attention on an individual solute molecule and to ask, "What is the probability that it will diffuse a distance x in time t?" We must, of course, allow a certain spread of distance, and we may call $p(x)\,dx$ the probability that the molecule has diffused a distance between x and $x + dx$. This probability is the number of solute molecules between x and $x + dx$ divided by the total number in the original source. Thus

$$p(x)\,dx = \frac{c(x)\,dx}{n_o} = \frac{1}{2(\pi Dt)^{\frac{1}{2}}}\,e^{-x^2/4Dt}\,dx \tag{11.49}$$

We now ask, "What is the mean square distance $\overline{x^2}$ traversed by a solute molecule in time t?" (We do not ask what is the mean distance x, since diffusion is equally probable in both directions, so that $\bar{x} = 0$.) The mean square distance is given by

$$\overline{x^2} = \int_{-\infty}^{\infty} x^2 p(x)\,dx \tag{11.50}$$

Substitution of (11.49) into this, and evaluation of the integral, leads to

$$\overline{x^2} = 2Dt \tag{11.51}$$

This simple relation is very useful for providing estimates of mean diffusion distances.

EXAMPLE

A solution of insulin is maintained at a constant temperature of $20°$ C, and without agitation, for one year (3.156×10^7 s). How far would a given insulin molecule be expected to diffuse in that time, if $D = 8.2 \times 10^{-11}$ m^2 s^{-1}?

SOLUTION

From equation (11.51)

$$\overline{x^2} = 2 \times 8.2 \times 10^{-11} \times 3.156 \times 10^7$$

$$= 5.18 \times 10^{-3} \text{ m}^2$$

and thus

$$\left(\overline{x^2}\right)^{\frac{1}{2}} = 0.072 \text{ m} = 7.2 \text{ cm}$$

Under ordinary conditions molecules will move through greater distances, because of agitation of the solution and convection arising from temperature variations.

DRIVING FORCE OF DIFFUSION

Consider the diffusion across a distance dx over which there is a concentration change from c to $c + dc$ (see Figure 11.10). The force which drives the molecule to the more dilute region can be calculated from the difference between the molar Gibbs energy at concentration c and that at concentration $c + dc$. This molar Gibbs energy difference is

$$dG = G_{c+dc} - G_c = RT \ln \frac{c + dc}{c} = RT \ln \left(1 + \frac{dc}{c}\right) \tag{11.52}$$

Since dc/c is very small

$$\ln \left(1 + \frac{dc}{c}\right) \approx \frac{dc}{c} \tag{11.53}$$

so that

$$dG = \frac{RT \, dc}{c} \tag{11.54}$$

This Gibbs energy difference is the work w done on the system in transferring a mole of solute from concentration c to $c + dc$. The work done *by* the system, $-w$, in transferring a *molecule* of solute from $c + dc$ to c is thus

$$-w = -\frac{RT}{N_A}\frac{dc}{c} \tag{11.55}$$

$$= -\frac{kT}{c}dc \tag{11.56}$$

where N_A is the Avogadro number and k is Planck's constant ($k = R/N_A$). This work $-w$ is done over a distance dx, so that if F_d is the driving force leading to diffusion

$$-w = F_d\,dx \tag{11.57}$$

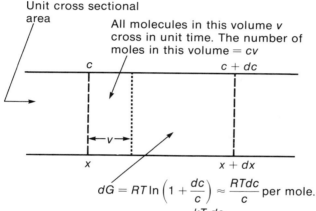

Unit cross sectional area

All molecules in this volume v cross in unit time. The number of moles in this volume $= cv$

c

$c + dc$

x

$x + dx$

$dG = RT\ln\left(1 + \dfrac{dc}{c}\right) \approx \dfrac{RTdc}{c}$ per mole.

Driving force of diffusion, $F_d = -\dfrac{kT}{c}\dfrac{dc}{dx}$ per molecule (equation 11.58).

Frictional force, $F_f = fv$.

Figure 11.10

The driving force of diffusion.

The driving force per molecule is thus given by combining (11.56) and (11.57):

$$F_d = -\frac{kT}{c}\frac{dc}{dx} \tag{11.58}$$

When a driving force is applied to a molecule its velocity increases until the frictional force F_f acting on it is equal to the driving force. The molecule has then attained a *limiting velocity*. Most theories of the frictional force on a molecule (such as that of Stokes, which is considered in the next section) lead to the conclusion that the force is directly proportional to the velocity. Thus

$$F_f = fv \tag{11.59}$$

where f is known as the *frictional coefficient*. The limiting velocity is thus attained when

$$fv = -\frac{kT}{c}\frac{dc}{dx} \tag{11.60}$$

or

$$cv = -\frac{kT}{f}\frac{dc}{dx} \tag{11.61}$$

All molecules within a distance v of a given unit cross-sectional area will cross that area in unit time (see Figure 11.10). The number of molecules in that volume is cv, which is therefore the flux J:

$$J = -\frac{kT}{f}\frac{dc}{dx} \tag{11.62}$$

Comparison of this molecularly-derived diffusion equation with Fick's first law, equation (11.34), shows that the diffusion coefficient is given by

$$D = \frac{kT}{f} \tag{11.63}$$

This equation was first derived by Albert Einstein (1879–1955) in 1905.

STOKES'S LAW

The diffusion coefficient depends upon the ease with which the solute molecules can move. In biology we are mainly concerned with aqueous systems, and the diffusion coefficient of a solute in aqueous solution is a measure of how readily a solute molecule can push aside its neighboring water molecules and move into another position. An important aspect of the theory of diffusion is how the magnitudes of the frictional coefficients f, and hence of the diffusion coefficients D (see (11.63)), depend upon the properties of the solute and solvent molecules.

Examination of the values given in Table 11.2 shows that diffusion coefficients tend to decrease as the molecular size increases. This is easy to understand, since a larger solute molecule has to push aside more water molecules during its progress, and therefore will move more slowly than a smaller molecule. It is difficult to develop a precise theory of diffusion coefficients, but in 1850 the British physicist Sir George Gabriel Stokes (1819–1903) considered a simple situation in which the solute molecules are so much larger than the solvent molecules that the solvent can be regarded as continuous (i.e., as not having molecular character). For such a system Stokes deduced that the frictional force F_f opposing the motion of a large particle of radius r moving at velocity v through a solvent of viscosity η is given by

$$F_f = 6\pi r \eta v \qquad (11.64)$$

The frictional coefficient is therefore

$$f = 6\pi r \eta \qquad (11.65)$$

It then follows from (11.63) that when Stokes's law applies the diffusion coefficient is given by

$$D = \frac{kT}{6\pi r \eta} \qquad (11.66)$$

This is often referred to as the *Stokes-Einstein equation.*

Measurement of D in a solvent of known viscosity therefore permits a value of the radius r to be calculated. However, such a calculation would not be very satisfactory for macromolecules, for several reasons. In the first place, Stokes's law is based on the assumption of very large spherical particles and a continuous solvent, and therefore involves some error even for approximately spherical molecules. Secondly, the macromolecules may not be spherical, and this introduces an additional error. Furthermore, macromolecules are commonly solvated, and in moving through the solution they transport some of their solvation layer. In spite of these drawbacks, equation (11.66) has proved useful in providing approximate values of molecular sizes (see also pp. 99–100).

EXAMPLE

The diffusion coefficient for glucose was found to be 6.81×10^{-10} $m^2\ s^{-1}$ at 25° C. The viscosity of water at 25° C is 0.008937 poise, and the density of glucose is 1.55 g cm^{-3}. Estimate the molecular weight of glucose, assuming that Stokes's law applies and that the molecule is spherical.

SOLUTION

Since Stokes's law applies, D and η are related by (11.66) and we can estimate the radius of the molecule:

$$r = \frac{kT}{6\pi \eta D}$$

One poise is one-tenth of the SI unit (kg $m^{-1}\ s^{-1}$) and thus

$$\eta = 8.937 \times 10^{-4} \text{ kg } m^{-1}\ s^{-1}$$

The radius is thus

$$r = \frac{1.381 \times 10^{-23} \, (\text{J K}^{-1}) \, 298.15 \, (\text{K})}{6 \times 3.1426 \times 8.937 \times 10^{-4} \, (\text{kg m}^{-1} \, \text{s}^{-1}) \times 6.81 \times 10^{-10} \, (\text{m}^2 \, \text{s}^{-1})}$$

$$= 3.59 \times 10^{-10} \, \text{J kg}^{-1} \, \text{m}^{-1} \, \text{s}^2$$

$$= 3.59 \times 10^{-10} \, \text{m} = 0.359 \, \text{nm}$$

since $1 \, \text{J} = 1 \, \text{kg m}^2 \, \text{s}^{-2}$.
  The estimated volume of the molecule is thus

$$\frac{4}{3} \pi (3.59 \times 10^{-10})^3 = 1.94 \times 10^{-28} \, \text{m}^3$$

and its mass is

$$1.94 \times 10^{-28} \, (\text{m}^3) \times 1.55 \times 10^6 \, (\text{g m}^{-3})$$

$$= 3.01 \times 10^{-22} \, \text{g}$$

The molecular weight M is this mass multiplied by the Avogadro number $N_A \, (= 6.022 \times 10^{23} \, \text{mol}^{-1})$:

$$M = 3.01 \times 10^{-22} \times 6.022 \times 10^{23}$$

$$= 181.3 \, \text{g mol}^{-1}$$

This agrees satisfactorily with the true molecular weight of 180.2.

11.6   DIFFUSION THROUGH MEMBRANES

The speed with which molecules and ions can pass through biological membranes is a matter of great importance. In the case of the functioning of nerves, for example, an essential feature is the passage of ions into and out of the fibers. There has been much investigation of these topics, and the problem is by no means fully understood. The present account can only cover the essential physicochemical principles.

  In work with membranes it is convenient to define a quantity known as the *permeability coefficient*, P, which is defined as the rate of flow through unit area (i.e., the flux) when the concentration difference is 1 M. We have previously considered the case of two stirred solutions, of concentrations c_1 and c_2, separated by a sintered glass diaphragm of thickness l. For such a system the rate of permeation is given by (11.43), and the flux is therefore

$$J = D \frac{c_1 - c_2}{l} \tag{11.67}$$

The permeability coefficient P is the flux when $c_1 - c_2 = 1$ M, so that

$$P = \frac{D}{l} \tag{11.68}$$

The SI units for D are $m^2 \, s^{-1}$, so that those for P are $m \, s^{-1}$. However, D is commonly given as $cm^2 \, s^{-1}$ and P as $cm \, s^{-1}$.

It is not necessarily the case that the solute molecules will be just as soluble in the membrane as in the water. For example, a membrane might be composed of a certain amount of lipid material, and a solute molecule might have a certain proportion of nonpolar groups. Such a solute might well be more soluble in the membrane than in the water. This effect can be represented by use of a partition or distribution coefficient K_p, which is an equilibrium constant; it is the ratio of the concentrations of the solute in the membrane and in the solvent. In this situation the concentrations at the surface of the membrane, as shown in Figure 11.11, will be $K_p c_1$ and $K_p c_2$. The gradient is now given by

$$-\frac{dc}{dx} = \frac{K_p c_1 - K_p c_2}{l} = K_p \frac{c_1 - c_2}{l} \tag{11.69}$$

and the rate of diffusion is

$$\frac{dn}{dt} = DAK_p \frac{c_1 - c_2}{l} \tag{11.70}$$

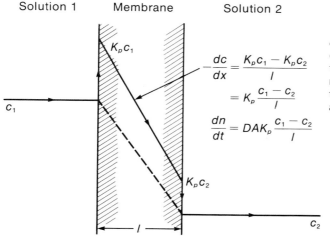

Solution 1   Membrane   Solution 2

$$-\frac{dc}{dx} = \frac{K_p c_1 - K_p c_2}{l}$$

$$= K_p \frac{c_1 - c_2}{l}$$

$$\frac{dn}{dt} = DAK_p \frac{c_1 - c_2}{l}$$

Figure 11.11

Stirred solutions of concentrations c_1 and c_2, separated by a membrane of thickness l. The solubility of the solute is different in the membrane and in the solution. In this diagram the solubility is shown as higher in the membrane.

Thus the rate, the apparent diffusion coefficient (DK_p) and the apparent permeability coefficient (DK_p/l) will all be altered by the factor K_p. Since K_p may be greater or less than unity, there may be an enhancement or a diminution in the rate of permeation of the membrane. Sometimes this partitioning effect leads to rates which are several powers of ten higher than would be obtained in its absence. Conversely, the rate of diffusion through the membrane will be abnormally small if the solute is much less soluble in the membrane than in the solution. For example, a carbohydrate like sucrose will be much less soluble in a fatty membrane than in water, where there is extensive hydrogen bonding; because of this effect it will diffuse slowly through a fatty membrane.

FACILITATED TRANSPORT

Sometimes abnormally high rates of transport of molecules across membranes are associated with other types of behavior, and a special explanation has to be invoked. For example, the rate of permeation of a substance S frequently varies with its concentration according to an equation of the same form as the Michaelis-Menten equation,

$$v = \frac{k[S]}{K + [S]} \tag{11.71}$$

where k and K are constants. This at once suggests that the molecule S becomes attached to a *carrier* molecule C, which may be an enzyme, and that C, which stays in the membrane, carries S from one surface of the membrane to the other, i.e., it is the complex CS which travels. The attachment of S to the carrier C is represented by the equation

$$C + S \rightleftharpoons CS$$

and the *dissociation* constant K is given by

$$K = \frac{[C][S]}{[CS]} \tag{11.72}$$

If the total carrier concentration is $[C]_o$,

$$[C]_o = [CS] + [C] \tag{11.73}$$

$$= [CS]\left(1 + \frac{K}{[S]}\right) \tag{11.74}$$

and thus

$$[CS] = \frac{[C]_o[S]}{K + [S]} \tag{11.75}$$

Therefore, as shown in Figure 11.12, the concentration of the carrier CS will vary within the membrane from

$$\frac{K_p[C]_o[S]_1}{K + [S]_1} \qquad \text{to} \qquad \frac{K_p[C]_o[S]_2}{K + [S]_2}$$

where K_p is the partition coefficient and $[S]_1$ and $[S]_2$ are the solute concentrations in the solution at the two sides of the membrane. In the steady state the concentration of carrier will fall linearly across the membrane, and if l is the thickness the concentration gradient will be

$$-\frac{d[CS]}{dx} = \frac{K_p[C]_o}{l}\left\{\frac{[S]_1}{K + [S]_1} - \frac{[S]_2}{K + [S]_2}\right\} \qquad (11.76)$$

This gradient may be much greater than $([S]_1 - [S]_2)/l$, and thus can lead to enhanced rates. If the second term in (11.76) is much smaller than the first, the rate will approximate to that given by (11.71); otherwise there will be deviations from this equation.

In support of this explanation, certain enzyme inhibitors are also inhibitors for the transport process. For example the glucoside phlorizin, which is a competitive

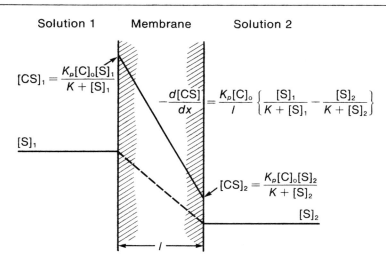

Figure 11.12

Solutions of concentrations $[S]_1$ and $[S]_2$, separated by a membrane of thickness l in which there is a carrier C which transports the solute through the membrane. The carrier molecules may move back and forth across the membrane. Alternatively, they may be supported within the membrane but be capable of undergoing some motion in which one portion of the molecule moves across the membrane and carries S with it.

inhibitor of a number of carbohydrases (which break down carbohydrates), is also a competitive inhibitor for the transport of sugar into and out of cells.

This phenomenon of *facilitated* transport is to be sharply contrasted with *active* transport. In facilitated transport there is not motion *against* the concentration gradient; it is the *rates* of diffusion which are abnormally high. We will now consider active transport.

ACTIVE TRANSPORT

In biological systems there are many instances where a substance passes across a membrane in opposition to the concentration gradient; that is, it moves from a solution of lower concentration to one of higher concentration. For example, amino acids are sometimes found to move from the blood plasma, where they are at relatively low concentrations, into cells where their concentration is higher. Another example is the formation of approximately 0.1 M HCl solution (pH $\sim$ 1) in the gastric juice by the flow of HCl from solution of pH approximately 7 (i.e., $\sim 10^{-7}$ M).

These are both instances of what is known as *active transport*. Care must be taken to distinguish true active transport from certain other effects. For example, the concentration of Mg^{2+} ions is much greater in most cells than in the surrounding fluid. This does not imply active transport, since the Mg^{2+} ions are strongly bound in the cells, a process which reduces the effective concentration of the ions and so disturbs the equilibrium. Another effect which can give a false impression of active transport is the Donnan equilibrium discussed in Section 7.6. We saw there that there can be an abnormal distribution of ions across a membrane because of the presence of large cations or anions to which the membrane is impermeable.

The occurrence of active transport does not, of course, involve a violation of the second law of thermodynamics. Metabolic processes are occurring at the same time; these reactions are spontaneous and thus exergonic, and are coupled with the transport process. Certain enzyme inhibitors, such as CN^- ions and other agents which inhibit ATP formation will cause these chemical processes to cease; then active transport no longer occurs.

Suppose that as a result of active transport the ratio of concentrations of solute species (e.g., of K^+ ions) on opposite sides of a membrane is 10. The Gibbs energy difference per mole of material at 37°C is

$$\Delta G = RT \ln \frac{c_2}{c_1} \tag{11.77}$$

$$= 4.57 \times 310 \log_{10} 10$$

$$= 1.42 \text{ kcal}$$

A reaction which is exergonic by this amount is therefore capable of maintaining such a concentration gradient. The hydrolysis of ATP produces 7.0 kcal mol^{-1}, which is more than enough to provide such a gradient, and the evidence is that ATP does play a key role in active transport. In higher cells the enzyme which is actively concerned with transporting Na^+ and K^+ ions across membranes is itself an ATP-ase, known as NaK-ATP-ase.

A great deal of research work is going on at the present time to establish the mechanisms of active transport. There are several theories, and it appears likely that different mechanisms apply to different biological systems. One possible mechanism, which involves a *carrier*, C, is represented in a simplified and schematic form in Figure 11.13. The carrier is able to travel back and forth across the membrane, and it can change into another form C′ by an endergonic reaction; C′ can spontaneously change back into C. The form C has a high affinity for A, the substance to be transported across the membrane, but C′ has a low affinity for A. The endergonic formation of C′ is driven by the exergonic hydrolysis of ATP, which is present inside the cell. This hydrolysis converts CA into C′A which at once dissociates into C′ + A.

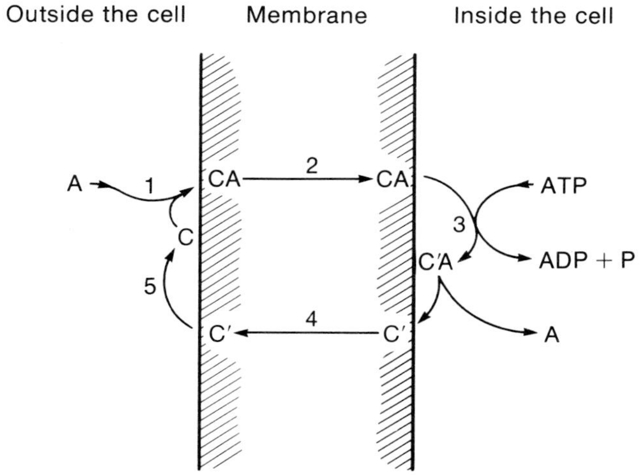

Figure 11.13

A schematic mechanism for active transport, in which a substance A is pumped into a cell against the concentration gradient. The energy for the pumping is provided by the hydrolysis of ATP. In the mechanism shown here the carrier molecule C moves across the membrane. Alternatively, the carrier molecule may be supported in the membrane, the ATP hydrolysis bringing about a conformational change in which one part of the carrier molecule traverses the membrane and carries A with it.

The process occurs in five stages, shown in Figure 11.13. At the outer surface of the membrane, the substance A combines with the carrier C (Stage 1). The species CA then diffuses through the membrane (Stage 2), and at the inside surface comes into contact with ATP and the enzyme. The ATP hydrolysis then converts CA into C′A which at once splits off A since C′ has a low affinity for A (Stage 3). The modified carrier C′ then diffuses into the membrane (Stage 4) and is spontaneously reconverted into C (Stage 5). Stages 4 and 5 may occur simultaneously, but are shown as separate stages in the diagram. The cycle is then ready to begin again with Stage 1. The overall effect is that A is being pumped into the cell, the pump being driven by the exergonic metabolic processes.

In another type of mechanism which has been proposed, the carrier molecule does not move bodily through the membrane, but instead straddles the membrane, part of the molecule being at the outside surface and part at the inside surface of the membrane. There is direct evidence for such an arrangement, particularly for the ATP-splitting molecule which transports K^+ into a cell and Na^+ out of it against the concentration gradients. Some scientists believe that such carrier molecules function by actually transporting ligands across the lipid membrane as a result of conformational changes. Others believe that the binding of ligands causes a conformational rearrangement of the carrier subunits (separate polypeptide chains), resulting in the creation of a "pore" which penetrates the lipid barrier; the ligands then migrate through this pore to the other side of the membrane.

A number of specific carrier molecules have now been identified, but precise mechanisms have not yet been established.

IONIC TRANSPORT THROUGH CELL MEMBRANES

Interesting examples of diffusion and active transport are provided by the distributions of ions across membranes in living systems. For example, K^+ ions show an abnormal distribution across the membranes of nerve fibers, the concentrations being more than ten times as high within the fiber as compared with outside. This behavior cannot be accounted for by the binding of the ions or by the establishment of the Donnan equilibrium; there must be active transport, i.e., a mechanism by which K^+ ions are pumped into the cell and by which Na^+ ions are pumped out.

Some typical data for mammalian muscle cells are illustrated in Figure 11.14. The concentration of K^+ ions is much greater inside the cell, while that of Na^+ ions is much greater outside. We first consider the ordinary diffusion rates, determined by experiments with radioactive isotopes (see pp. 518–523), with inhibition of the enzyme

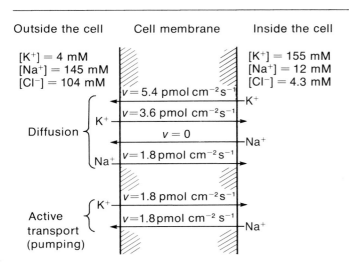

Outside the cell   Cell membrane   Inside the cell

$[K^+] = 4$ mM
$[Na^+] = 145$ mM
$[Cl^-] = 104$ mM

$[K^+] = 155$ mM
$[Na^+] = 12$ mM
$[Cl^-] = 4.3$ mM

Diffusion
K⁺ $v = 5.4$ pmol cm^{-2}s^{-1} —K⁺
     $v = 3.6$ pmol cm^{-2}s^{-1}
     $v = 0$ —Na⁺
Na⁺ $v = 1.8$ pmol cm^{-2}s^{-1}

Active transport (pumping)
K⁺ $v = 1.8$ pmol cm^{-2}s^{-1}
     $v = 1.8$ pmol cm^{-2} s^{-1} —Na⁺

Figure 11.14

Typical distributions of K^+, Na^+ and Cl^- ions across a mammalian muscle cell. Other negative ions are present to balance the charges on each side of the membrane. Rates of diffusion and of active transport are shown.

(NaK-ATP-ase) responsible for the active-transport processes. Such experiments have shown that the K^+ ions readily diffuse through the cell walls, the rate of diffusion into the cell being 3.6 picomol cm^{-2} s^{-1} and the rate of diffusion out being 5.4 picomol cm^{-2} s^{-1}. The K^+ concentration gradient (inside/outside) of $155/4 = 39$ suggests that the rate of diffusion of K^+ out of the cell should be 39 times the rate into the cell. Instead, the ratio is only $5.4/3.6 = 1.5$, which shows that the membrane must be constructed in such a way that the K^+ ions can diffuse into the cell much more rapidly than they can move out. We can visualize the pores in the membrane as funnel shaped, with large areas of cross-section on the outside and small on the inside.

Experiments with radioactive isotopes have shown that Na^+ ions diffuse into the cell at a rate of about 1.8 picomol cm^{-2} s^{-1}, but hardly diffuse out at all. The Na^+ concentration gradient (outside/inside) is $145/12 = 12$ and we should therefore expect that Na^+ ions would diffuse out at a rate of $1.8/12 = 0.15$ picomol cm^{-2} s^{-1}. The fact that the rate is less than this again shows that the pores of the membrane must be constructed in a special way, allowing ions to enter the cell more rapidly than they can leave it. The observation that the Na^+ ions are unable to diffuse out of the cell, whereas the K^+ can do so, may be explained in terms of the greater hydration of the Na^+ ions. We have already seen in Table 6.3 that Na^+ ions have lower conductivities than K^+ because of their greater hydration, which gives them a larger effective radius.

The above considerations have applied to the ordinary diffusion processes. The conditions shown in Figure 11.14 could not apply if there were not active transport in addition. The diffusion rates for K^+ are different in the two directions, and steady concentrations could not be maintained unless K^+ ions were pumped into the cell to make up the difference. The pumping rate for K^+ ions is thus $5.4 - 3.6 = 1.8$ picomol cm^{-2} s^{-1}. Similarly, a mechanism must exist for pumping Na^+ ions out of the cell, and the rate of pumping must be $1.8 - 0 = 1.8$ picomol cm^{-2} s^{-1}. It is of interest that these pumping rates are equal; the details of how the pumping occurs are still being worked out.

It is sometimes convenient to discuss ionic distributions across membranes in terms of the electric potential differences which are established. We saw in Chapter 8 that if the concentrations of an ion are c_1 and c_2 on two sides of a membrane, the emf across the membrane is equal to (see equation (8.19))

$$\frac{RT}{F} \ln \frac{c_1}{c_2}$$

This emf is known as the *Nernst potential* and can be calculated for any ions that pass through the membrane. This applies to the K^+ ions in the example we have considered in Figure 11.14, and if the temperature is $37°$ C (310K) the Nernst potential is calculated to be

$$\frac{8.314 \, (J \, K^{-1} \, mol^{-1}) \, 310 \, (K)}{96 \, 500 \, (C \, mol^{-1})} \ln \frac{4}{155}$$

$$= 0.0267 \ln \frac{4}{155} = -0.0976 \, V = -97.6 \, mV$$

This potential is negative inside the cell and positive outside. We can understand this by noting that because the K^+ concentration is much greater inside the cell than outside there will be a tendency for K^+ ions to diffuse out and thus produce a slight excess of negative charge inside, and hence a negative potential inside.

This potential of -97.6 mV inside the cell is the potential which would exist if the K^+ ions were the only diffusible ions. Chloride ions, however, can also diffuse through the membrane, as can Na^+ ions to a limited extent. As a result of these complications the potential inside the cell is not -97.6 mV, but usually more like -85 mV. Thus the interior of the cell is slighly less negative than is required to maintain the K^+ ratio (inside/outside) at 155/4. Therefore there is a small net diffusion of K^+ ions out of the cell, as we noted in Figure 11.14. This net diffusion rate of $5.4 - 3.6 = 1.8$ pmol cm^{-2} s^{-1} is balanced by the pumping rate into the cell.

Note that the Nernst potential we would calculate for the Na^+ ions is quite different from the true potential which is established; its value is

$$0.0267 \ln \frac{145}{12} = 66.5 \text{ mV}$$

If the membrane were freely permeable to Na^+ ions there would be a *positive* potential inside the cell, and this positive potential would largely neutralize the Nernst potential due to the K^+ ions. However, this occurs only to a minor extent because of the low permeability of the Na^+ ions. On the other hand, the Cl^- ions are freely permeable and they distribute themselves in accordance with the potential established across the membrane. Typical Cl^- concentrations are shown in Figure 11.14, and the corresponding Nernst potential is

$$0.0267 \ln \frac{4.3}{104} = -0.0851 \text{ V} = -85.1 \text{ mV}$$

which is equal to the true potential. This is consistent with the fact that there is no pumping mechanism for Cl^- ions, which are present at equilibrium rather than in a steady state.

Similar ion distributions and potential differences are established across the membranes of resting nerve and muscle cells. However, these membranes have the special property that their permeabilities to K^+ and Na^+ ions are subject to change. When a nerve cell is stimulated, whether electrically, mechanically, or chemically, the cell wall suddenly becomes much more permeable to Na^+ ions, which flow through the membrane into the cell and neutralize the negative charge inside the cell. In fact, for a period of about 0.2 ms, the cell wall is about 100 times as permeable to Na^+ ions as to K^+ ions; the Nernst potential due to Na^+ ions is therefore dominant during this period of time, and the potential inside the cell rapidly changes from -85 mV to $+60$ mV. At the end of this period of stimulation the cell membrane again becomes impermeable to Na^+ ions, and the K^+ ions regain control of the Nernst potential and the potential of -85 mV is reestablished. This sudden rise and fall of the cell potential is referred to as the *action potential*. During the quiescent period which follows the period of action, the excess Na^+ ions which had entered the cell are pumped out again by the active transport mechanism.

11.7   SEDIMENTATION

We have considered how a solute moves as a result of a concentration gradient. Molecules in solution can also be made to move by subjecting them to other forces. A very simple procedure is to allow a solution to stand. If the solute molecules are small there will be no change in the distribution in space, since the thermal motion will counteract the tendency of the molecules to move in the gravitational field. Very large particles, however, will sediment on standing. In order for smaller particles, such as proteins and other molecules, to undergo sedimentation it is necessary to increase the effective gravitational field by subjecting the solution to centrifugal motion. The velocity of sedimentation can lead to values of molecular weights, as will now be explained.

SEDIMENTATION-VELOCITY METHOD

Suppose that a particle of mass m, having a specific volume (volume per unit mass) of v_1, is in a liquid of density ρ (mass per unit volume). The volume of liquid displaced by the particle is $v_1 m$ and the mass displaced is $v_1 m \rho$. The net force F_g acting on the particle as a result of the gravitational field is therefore

$$F_g = mg - v_1 m \rho g = (1 - v_1 \rho)mg \tag{11.78}$$

where g is the acceleration of gravity. The particle will reach a limiting velocity v when this force is equal to the frictional force F_f, which is equal to the frictional coefficient f multiplied by the velocity:

$$F_f = fv = (1 - v_1 \rho)mg \tag{11.79}$$

If Stokes's law applies to the particle, F_f is given by (11.64), so that

$$6\pi r \eta v = (1 - v_1 \rho)mg \tag{11.80}$$

and the limiting velocity is

$$v = \frac{(1 - v_1 \rho)mg}{6\pi r \eta} \tag{11.81}$$

Experiments on sedimentation under the earth's gravitational field were carried out in 1908 by the French physicist Jean Baptiste Perrin (1870–1942), who observed under a microscope the movement of the particles of the pigment gamboge. In order to observe the sedimentation of smaller particles, such as proteins and other macromolecules, it is necessary to employ much higher fields. This is done by means of an ultracentrifuge, in which solutions are rotated at speeds up to 80 000 revolutions per minute, which produces fields up to $3 \times 10^5 \, g$. The development of such ultracentrifugal techniques is largely due to the Swedish physical chemist Theodor

Svedberg, whose work along these lines was started in 1923, and who devoted much study to the characterization of protein molecules and other macromolecules. A schematic diagram of an ultracentrifuge is shown in Figure 11.15a. For a centrifugal field of force we replace g in (11.79) by $\omega^2 x$, where ω is the angular velocity and x is the distance from the center of rotation. Thus

$$v = \frac{(1 - v_1\rho)m\omega^2 x}{f} \qquad (11.82)$$

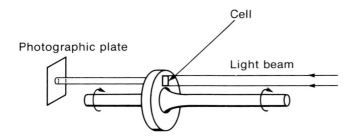

a.

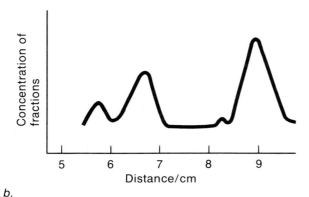

b.

Figure 11.15

a. Schematic diagram of an ultracentrifuge. The absorption of light increases with the concentration, so that the distribution of solute can be measured from the blackening of the photographic plate at various positions. The distribution can be more satisfactorily obtained from the refractive index, which varies approximately linearly with concentration. In the *schlieren* method special illumination is used which leads to darkening of the photographic plate in regions where the refractive index changes rapidly.

b. Record of concentration versus distance for a system of several components, as obtained by schlieren photography.

The quantity

$$s = \frac{v}{\omega^2 x} \tag{11.83}$$

is known as the *sedimentation constant*; it is the sedimentation rate v when the centrifugal acceleration is unity, and the SI unit is the second. For a given molecular species in a given solvent at a given temperature, s is a characteristic constant. It is often expressed in Svedberg units, equal to 10^{-13} seconds.

If Stokes's law applies, the denominator of (11.82) is $6\pi\eta r$, so that

$$v = \frac{(1 - v_1\rho)m\omega^2 x}{6\pi\eta r} \tag{11.84}$$

and the sedimentation constant is

$$s = \frac{(1 - v_1\rho)m}{6\pi\eta r} \tag{11.85}$$

Use of this equation, however, is unreliable, since Stokes's law is valid only for very large spherical particles. It is more satisfactory to express f by the use of (11.63), according to which it is kT/D. Insertion of this expression into (11.82) then gives

$$v = \frac{D(1 - v_1\rho)m\omega^2 x}{kT} \tag{11.86}$$

or

$$s = \frac{D(1 - v_1\rho)m}{kT} \tag{11.87}$$

Multiplication of the molecular mass m by the Avogadro number N_A gives the molecular weight M, and multiplication of k by N_A gives the gas constant R. Thus

$$s = \frac{D(1 - v_1\rho)M}{RT} \tag{11.88}$$

so that the molecular weight M can be calculated from a measurement of the sedimentation constant and the diffusion coefficient:

$$M = \frac{RTs}{D(1 - v_1\rho)} \tag{11.89}$$

This equation, derived by Svedberg in 1929, has been the basis of many measurements of molecular weights. For precise determination, the values of s, D, and v_1 should be extrapolated to infinite dilution.

EXAMPLE

A sample of human hemoglobin had a sedimentation constant of
4.48 Svedbergs in water at 20° C, and a diffusion coefficient of
6.9×10^{-11} m^2 s^{-1}. The specific volume of human hemoglobin
is 0.749 cm^3 g^{-1}, and the density of water at 20° C is 0.998 g cm^{-3}.
Calculate the molecular weight of human hemoglobin.

SOLUTION

The molecular weight is obtained by inserting the following values
into (11.89):

$$R = 8.314 \text{ J K}^{-1} \text{ mol}^{-1}$$
$$T = 293.15 \text{ K}$$
$$s = 4.48 \times 10^{-13} \text{ s}$$
$$D = 6.9 \times 10^{-11} \text{ m}^2 \text{ s}^{-1}$$
$$v_1 = 0.749 \text{ cm}^3 \text{ g}^{-1}$$
$$\rho = 0.998 \text{ g cm}^{-3}$$

Thus we have

$$M = \frac{8.314 \text{ (J K}^{-1} \text{ mol}^{-1}) \, 293.15 \text{ (K)} \, 4.48 \times 10^{-13} \text{ (s)}}{6.9 \times 10^{-11} \text{ (m}^2 \text{ s}^{-1}) \, (1 - 0.749 \times 0.998)}$$

$$= 62.7 \text{ J m}^{-2} \text{ s}^2 \text{ mol}^{-1}$$

To convert this into the usual molecular weight units we note that
$1 \text{ J} = 1 \text{ kg m}^2 \text{ s}^{-2}$; thus

$$M = 62.7 \text{ kg mol}^{-1}$$

$$= 62\,700 \text{ g mol}^{-1}$$

A particular advantage of the sedimentation-velocity technique is that a macromolecular solution containing more than one type of molecule is separated according
to the molecular masses of the components. Figure 11.15b shows the type of
sedimentation diagram obtained for a system containing a number of components.

SEDIMENTATION-EQUILIBRIUM METHOD

An alternative method of using the ultracentrifuge to measure molecular weights is
to allow the distribution of particles to reach equilibrium. As sedimentation occurs
in the ultracentrifuge a concentration gradient is established, and this will cause the
molecules to diffuse in the opposite direction. Eventually the system reaches a state
of equilibrium at which the rate with which the solute is driven outwards by the
centrifugal force just equals the rate with which it diffuses inwards under the influence
of the concentration gradient.

The velocity v with which the particles travel as a result of the centrifugal field is given by (11.82). All particles within a distance v of a given unit cross-sectional area will cross that area in unit time. If the concentration is c there are vc particles which cross; the sedimentation flux is thus vc, and by (11.82) is

$$J(\text{sedimentation}) = vc = \frac{(1 - v_1\rho)m\omega^2 xc}{f} \qquad (11.90)$$

The diffusive flux is

$$J(\text{diffusive}) = -D\frac{dc}{dx} \qquad (11.91)$$

$$= -\frac{kT}{f}\frac{dc}{dx} \qquad (11.92)$$

At equilibrium these two rates are equal, and we obtain (with $kN_A = R$ and $mN_A = M$)

$$\frac{dc}{c} = -\frac{M(1 - v_1\rho)\omega^2 x\,dx}{RT} \qquad (11.93)$$

Integration between two positions x_1 and x_2, at concentrations c_1 and c_2, leads to

$$M = \frac{2RT\ln(c_1/c_2)}{(1 - v_1\rho)\omega^2(x_2{}^2 - x_1{}^2)} \qquad (11.94)$$

Thus, if measurements of the relative concentrations are made at two positions, after equilibrium has been established, this equation can be used to calculate the molecular weight. The value so obtained is $\bar{M}_m$, the mass-average molecular weight.

The sedimentation equilibrium method does not require an independent measurement of the diffusion coefficient, in contrast to the sedimentation-velocity method. However, the time required for complete equilibrium to be established is so long that the method is often inconvenient to use, especially if the molecular weight is greater than 50 000.

APPROACH TO SEDIMENTATION EQUILIBRIUM

In order to overcome this drawback, a modification to the sedimentation-equilibrium method was proposed in 1947 by the Canadian physicist William J. Archibald. At the top meniscus of the cell, and at the bottom of the cell, there can be no net flux, so that the sedimentation-equilibrium equations must hold at these sections at all times. Therefore, shortly after the ultracentrifuge is brought to its top speed, concentrations in these special sections can be determined, and M calculated from (11.89). This modification of the sedimentation-equilibrium method greatly increases its applicability.

DENSITY-GRADIENT ULTRACENTRIFUGATION

If a solution of a substance of low molecular weight is ultracentrifuged, equilibrium is established within a fairly short period of time, and there will be a density gradient in the solution. If a substance of high molecular weight is added to this solution it will float at the position at which its density is equal to the density of the solution. If the macromolecular substance is made up of fractions of different molecular weights it will separate into fractions which will remain at different planes in the cell. This technique, known as density-gradient ultracentrifugation, has proved useful in establishing the different kinds of molecules in a macromolecular sample.

11.8   VISCOSITY

Another transport property which depends on molecular weight is the viscosity of a solution of the substance. However, in contrast to diffusion and sedimentation, viscosity measurements do not lead to absolute values of molecular weights. Comparison must be made with calibration measurements of the viscosities of solutions containing substances whose molecular weights have been determined by other methods.

The viscosity of a fluid was defined on p. 102, where it was seen to be a measure of the resistance to flow of the fluid. According to Newton's law of viscous flow, the frictional force F_f, resisting the relative motion of two adjacent layers in the liquid, is proportional to the area A and to the velocity gradient dv/dx (see Figure 3.1, p. 102):

$$F_f = \eta A \frac{dv}{dx} \tag{11.95}$$

The SI unit for the coefficient of viscosity η is $kg\ m^{-1}\ s^{-1}$ or $N\ s\ m^{-2}$; the commonly used unit, the poise, is one-tenth of the SI unit. The type of flow to which (11.95) applies is called *laminar, streamline,* or *Newtonian* flow. In flow of this kind there is a net component of velocity in the direction of flow superimposed on the random molecular velocities. Streamline flow is observed if the velocity of flow is not too large; with very rapid flow the motion becomes turbulent and (11.95) no longer applies.

MEASUREMENT OF VISCOSITY

Viscosity is usually studied by allowing the liquid to flow through a tube of circular cross section, and measuring the rate of flow. From this rate, and with the knowledge of the pressure acting and the dimensions of the tube, the coefficient of viscosity can be calculated on the basis of a theory developed in 1844 by the French physiologist Jean Leonard Poiseuille (1799–1869). Consider an incompressible liquid flowing through a tube of radius R and length l, with a pressure P_1 at one end and a pressure P_2 at the other (Figure 11.16a, b). The liquid at the walls of the tube is stagnant, the rate of flow increasing to a maximum at the center of the tube. A cylinder of length l and radius r has an area of $2\pi r l$, and according to (11.95) the frictional force is

$$F_f = -\eta \frac{dv}{dr} 2\pi r l \tag{11.96}$$

the velocity gradient dv/dr being a negative quantity. This force is exactly balanced by the force driving the fluid in this cylinder. This force is the pressure difference $P_1 - P_2$ multiplied by the area πr^2 of the cylinder. Thus

$$-\eta \frac{dv}{dr} 2\pi r = \pi r^2 (P_1 - P_2) \tag{11.97}$$

or

$$dv = -\frac{r}{2\eta l} (P_1 - P_2)\, dr \tag{11.98}$$

Integration of this gives

$$v = -\frac{(P_1 - P_2)}{4\eta l} r^2 + \text{const.} \tag{11.99}$$

The velocity v is zero when $r = R$; the constant of integration is thus

$$\text{const.} = \frac{(P_1 - P_2)}{4\eta l} R^2 \tag{11.100}$$

and therefore

$$v = \frac{P_1 - P_2}{4\eta l} (R^2 - r^2) \tag{11.101}$$

The total volume of liquid flowing through the tube per second, dV/dt, is obtained by integrating over each element of cross-sectional area. Each element has an area of $2\pi r\, dr$ (see Figure 11.16b) so that

$$\frac{dV}{dt} = \int_0^R 2\pi r v\, dr \tag{11.102}$$

$$= \frac{(P_1 - P_2)}{2\eta l} \left[R^2 \int_0^R r\, dr - \int_0^R r^3\, dr \right] \tag{11.103}$$

$$= \frac{(P_1 - P_2)\pi R^4}{8\eta l} \tag{11.104}$$

This is the Poiseuille equation, and it enables η to be calculated from measurements of the rate of flow dV/dt in a tube of known dimensions, the pressure difference $P_1 - P_2$ being known.

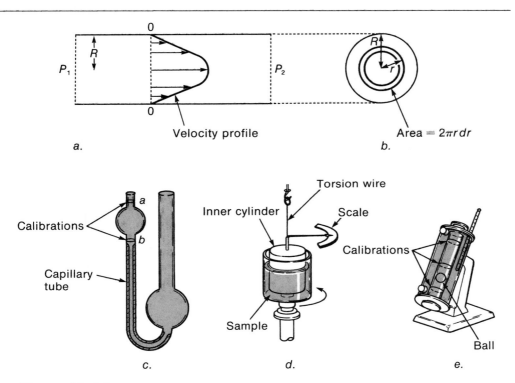

Velocity profile

Area $= 2\pi r\,dr$

a.

b.

Torsion wire

Calibrations

a

b

Capillary tube

Inner cylinder

Scale

Calibrations

Sample

Ball

c.

d.

e.

Figure 11.16

The measurement of viscosity.
a. Flow through a tube
b. Cross-section of tube
c. An Ostwald viscometer
d. Couette rotating-cylinder viscometer
e. Falling-ball viscometer

A commonly employed instrument for measuring the viscosity of a liquid is the *Ostwald viscosimeter* or *viscometer*, illustrated in Figure 11.16c. One measures the time that it takes for a quantity of liquid to pass through the tube, from one position to another, under the force of its own weight. Usually the instrument is calibrated by the use of a liquid of known viscosity.† Another apparatus used for viscosity measurements is the Couette rotating-cylinder viscometer (see Figure 11.16d).

---

† The pressure difference $P_1 - P_2$ varies with the time, but is proportional to the density; the densities of the calibrating liquid and of the sample must therefore be known.

In this instrument a liquid is caused to rotate in an outer cylinder, and it causes a torque to be applied to the torsion wire attached to the inner cylinder. The viscosity is calculated from the torque, the apparatus being calibrated. Another device for measuring viscosity is the falling-ball viscometer (Figure 11.16e). The viscosity is calculated from the time required for the ball to fall from one position to another.

VISCOSITIES OF SOLUTIONS

When macromolecular material is added to a liquid such as water the viscosity is increased. Suppose that the viscosity of a pure liquid is η_o, and that the viscosity of a solution is η. The *specific viscosity* is then defined as

$$\text{specific viscosity} = \frac{\eta - \eta_o}{\eta_o} \tag{11.105}$$

It is the increase in viscosity $\eta - \eta_o$ relative to the viscosity of the pure solvent. Division of the specific viscosity by the concentration c of the solution gives the *reduced specific viscosity*:

$$\text{reduced specific viscosity} = \frac{1}{c} \cdot \frac{\eta - \eta_o}{\eta_o} \tag{11.106}$$

This quantity, however, has a contribution from the intermolecular interactions between the macromolecules. This contribution can be eliminated by extrapolating the reduced specific viscosity to infinite dilution, when we obtain what is known as the *limiting viscosity number* or as the *intrinsic viscosity*, which is given the symbol $[\eta]$:

$$[\eta] = \lim_{c \to o} \left[\frac{1}{c} \frac{\eta - \eta_o}{\eta_o} \right] \tag{11.107}$$

This quantity is the fractional change in the viscosity per unit concentration of macromolecules, at infinite dilution.

Various empirical relationships have been proposed to relate the limiting viscosity number $[\eta]$ to the molecular weight M. The most successful of these is the equation

$$[\eta] = KM^\alpha \tag{11.108}$$

where K and α are constants. This equation was proposed independently by the Austrian-American chemist Herman Francis Mark in 1938 and by R. Houwink in 1941, and is usually called the *Mark-Houwink equation*. If it is obeyed a plot of $\log [\eta]$ against $\log M$ will be a straight line. Figure 11.17 shows some results plotted in this way for various polyisobutenes in diisobutene as solvent. By the use of such calibration curves the molecular weights of unknown samples can be determined.

Attempts have been made to relate the value of α in (11.108) to the shape of the molecule. If the molecules are spherical, the limiting viscosity number $[\eta]$ is independent of the size of the molecules, and α is equal to zero. In agreement with this, all globular proteins, regardless of their size, have essentially the same $[\eta]$. The more elongated a molecule is, the more effective are larger molecules in reducing the viscosity, and the larger is $[\eta]$; values of 1.3 or higher are frequently obtained for molecules which exist in solution as extended chains. Long-chain molecules which are coiled in solution give intermediate values of α, frequently in the range 0.6 to 0.75.

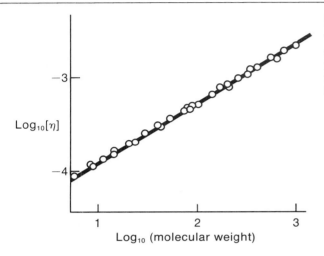

Figure 11.17

The logarithm of intrinsic viscosity of polybutylenes plotted against the logarithm of molecular weight, for solutions in diisobutene at 20° C.

11.9   ELECTROKINETIC EFFECTS

The term *electrokinetic* is applied to a group of effects in which either an electric potential brings about movement, or movement produces an electric potential. For example, if macromolecules are suspended in a liquid, and a potential is applied, the particles often move towards one or other of the electrodes. This phenomenon is called *electrophoresis*. The inverse of it is when the particles undergo sedimentation, in which case a *sedimentation potential* is developed. The occurrence of these electrokinetic effects is due to the existence of potential differences between the solid and liquid phases.

THE ELECTRIC DOUBLE LAYER

Various theories have been suggested to account for these potential differences. The first and simplest theory of the double layer was given in 1879 by the German physiologist and physicist Hermann Ludwig Ferdinand von Helmholtz (1821–1894). According to his model, which is represented in Figure 11.18a, the surface of a solid can be regarded as bearing positive or negative charges. If these are negative, as in the diagram, a unimolecular layer of positive charges will be attracted to the surface from the solution. A *double layer* is therefore formed, and this will correspond to an electric capacitor. If the distance between the layers of positive and negative charges is a, and each layer has a charge density (i.e., a charge per unit area) of σ, it follows from electrostatic theory that the potential difference between the two layers is

$$\phi = \frac{\sigma a}{\varepsilon_0 \varepsilon}$$

(11.109)

where ε is the dielectric constant of the liquid medium.

This simple idea of Helmholtz, that a layer of ions from the solution becomes attached to the surface, was modified in 1910 by the French physicist Georges Gouy (1854–1926) and in 1913 by the British chemist David Leonard Chapman (1869–1958). These workers pointed out that the Helmholtz theory is unsatisfactory in neglecting the Boltzmann distribution of the ions. They suggested that on the

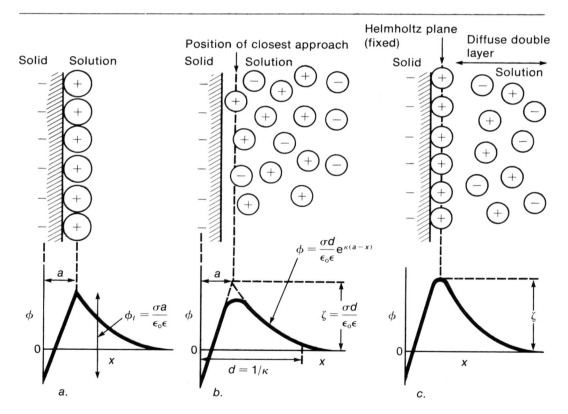

Figure 11.18

Three models for the structure of the electric double layer, showing the variations of electric potential ϕ with distance x from the negative charges on the surface:
a. Helmholtz model of the fixed double layer.
b. Diffuse double-layer model of Gouy and Chapman, showing Chapman's expressions for the variation of ϕ with x, and for the zeta potential ζ.
c. Stern's model, which is a combination of a and b. The theory for this model is more complicated.

solution side of the interface there is not a simple layer of ions, but instead an ionic distribution which extends some distance from the surface; in other words, there is a *diffuse double layer* as shown in Figure 11.18b. Thermal agitation permits the free movement of the ions present in the solution, but the distribution of positive and negative ions is not uniform. In the example shown in the figure, the surface is negative, and there are more positive ions in close proximity to the surface. The idea behind the Gouy-Chapman theory is very similar to that involved in the later Debye-Hückel theory of the ionic atmosphere surrounding an ion (see Section 6.3). Indeed, ten years before the formulation of that theory in 1923 Chapman had worked out an analogous treatment for the distribution of ions around a charged solid.

The Chapman theory is briefly as follows. Suppose that the solid surface has a charge density of σ. A net charge equal in magnitude but opposite in sign will be carried by the ionic atmosphere, which extends into the solution. The value of the charge density ρ per unit volume at any point in the solution portion of the electric double layer can be derived by the method employed in Chapter 6. Provided that $e\phi/kT$ is small in comparison with unity the result is (see equation (6.22))

$$\rho = -\frac{e^2\phi}{kT}\sum_i n_i z_i^2 \tag{11.110}$$

where ϕ is the electric potential at the point under consideration. The charge density ρ and the potential ϕ are related by the Poisson equation, which takes a different form from that used in Chapter 6 (equation (6.23)) since the potential varies only in the direction normal to the plane surface. The appropriate form of the Poisson equation is now

$$\frac{\partial^2\phi}{\partial x^2} = -\frac{\rho}{\varepsilon_0\varepsilon} \tag{11.111}$$

If the value of ρ, given by (11.110), is introduced, the result is

$$\frac{\partial^2\phi}{\partial x^2} = \kappa^2\phi \tag{11.112}$$

where κ is given by

$$\kappa^2 = \frac{e^2}{\varepsilon_0\varepsilon kT}\sum_i n_i z_i^2 \tag{11.113}$$

(this is the same as equation (6.26). The general solution of (11.112) is

$$\phi = A e^{-\kappa x} + A' e^{\kappa x} \tag{11.114}$$

The fact that ϕ must become zero as x becomes large requires that A' is zero, so that

$$\phi = A e^{-\kappa x} \tag{11.115}$$

From equations (11.111) and (11.112) it follows that

$$\rho = -\varepsilon_0 \varepsilon \kappa^2 \phi \tag{11.116}$$

and introduction of (11.115) gives

$$\rho = -A\varepsilon_0 \varepsilon \kappa^2\, e^{-\kappa x} \tag{11.117}$$

The charge on the surface must be equal in magnitude but opposite in sign to that of the solution; thus

$$\sigma = -\int_a^\infty \rho\, dx \tag{11.118}$$

the integration being from a, the distance of closest approach of the ions to the charges on the surface, to infinity. Introduction of (11.117) into (11.118) yields

$$\sigma = A\varepsilon_0 \varepsilon \kappa^2 \int_a^\infty e^{-\kappa x}\, dx \tag{11.119}$$

$$= A\varepsilon\kappa\varepsilon^{-\kappa a} \tag{11.120}$$

and thus

$$A = \frac{\sigma\, e^{\kappa a}}{\varepsilon_0 \varepsilon \kappa} \tag{11.121}$$

Substitution of this expression into (11.115) gives

$$\phi = \frac{\sigma}{\varepsilon_0 \varepsilon \kappa}\, e^{\kappa(a-x)} \tag{11.122}$$

As x becomes very large ϕ approaches zero, while when x is equal to a, the distance of closest approach, ϕ is given by

$$\phi = \frac{\sigma}{\varepsilon_0 \varepsilon \kappa} \tag{11.123}$$

The quantity $1/\kappa$, which has the dimensions of length, can be regarded as the effective thickness d of the double layer, and is to be compared with the effective thickness of the ionic atmosphere in the Debye-Hückel theory (see Section 6.5). Equation (11.123) gives the potential at the distance of closest approach, with respect to the bulk of the solution. If we give this potential the symbol ζ (zeta) we have

$$\zeta = \frac{\sigma d}{\varepsilon_0 \varepsilon} \tag{11.124}$$

This equation is of the same form as (11.109) but this is a coincidence, the significance of the quantities being quite different. Figures 11.18a and 11.18b show the form of the potential changes for the Helmholtz and Gouy-Chapman models.

The Gouy-Chapman theory did not prove entirely satisfactory, and in 1924 a considerable advance was made by the German-American physicist Otto Stern, whose model is shown in Figure 11.18c. Stern combined the fixed double-layer model of Helmholtz with the diffuse double-layer model of Gouy and Chapman. As shown in the figure, there is a fixed layer at the surface, as well as a diffuse layer. On the whole this treatment has proved to be satisfactory, but for certain kinds of investigations it has been found necessary to develop more elaborate models.

The movement induced by a potential difference, as in electrophoresis, depends upon the potential between the outer part of the surface layer, fixed with respect to the solid, and the bulk of the solution. This potential, denoted by the symbol ζ (zeta), is known as the *electrokinetic potential* or the *zeta potential*. For the models shown in Figures 11.18b and 11.18c the zeta potential is the potential drop across the diffuse layer. With reference to the more realistic Stern model, the zeta potential is the difference between the potential at the Helmholtz plane, shown in the Figure 11.18c, and the bulk solution. It is to be emphasized that there is no connection between the zeta potential and the overall potential drop between the solid and the solution; even the signs need not be the same.

ELECTRO-OSMOSIS

If a membrane separates two identical liquids or solutions and a potential difference is applied across the membrane, there results a flow of liquid through the pores of the membrane. This phenomenon is known as *electro-endosmosis*, or as *electro-osmosis*. A simplified version of the theory is as follows.

When a liquid is forced by electro-osmosis through the pores of a membrane, the rate of flow will be determined by two opposing factors: the force of electro-osmosis on the one hand, and the frictional force between the moving liquid layer and the wall on the other. When the two forces are equal there will be a uniform rate of flow. The movement of the liquid will be confined to a distance d across the diffuse double layer; that is, to the distance across which the zeta potential is operating (see Figure 11.19a). On the solid side of the double layer the velocity of flow is zero, whereas on the solution side it has attained the uniform velocity v of the moving liquid. The velocity gradient, assumed to be uniform, is thus v/d. The force due to friction is the product of the velocity gradient and the coefficient of viscosity η of the liquid, and is thus $\eta v/d$. The electric force causing electro-osmosis is equal to the product of the applied electric potential gradient V and the charge density σ at the boundary at which movement occurs. Thus, in the steady state

$$\frac{\eta v}{d} = V\sigma \tag{11.125}$$

The zeta potential ζ is related to the distance d by (11.124), and elimination of d

between equations (11.124) and (11.125) gives

$$v = \frac{\zeta \varepsilon_o \varepsilon}{\eta} V \tag{11.126}$$

If the potential gradient V is unity the uniform velocity obtained, v_o, is known as the *electro-osmotic mobility*, and is given by

$$v_o = \frac{\zeta \varepsilon_o \varepsilon}{\eta} \tag{11.127}$$

If A is the total area of cross section of all of the pores in a membrane, the volume V_1 of liquid transported electro-osmotically per second is equal to $A v_o$, so that from (11.127)

$$V_1 = \frac{\zeta A \varepsilon_o \varepsilon}{\eta} \tag{11.128}$$

Electro-osmosis may alternatively be studied using a single capillary tube instead of a membrane containing many pores. A is then equal to πr^2, where r is the radius of the tube, and (11.128) becomes

$$V_1 = \frac{\pi \zeta r^2 \varepsilon_o \varepsilon}{\eta} V \tag{11.129}$$

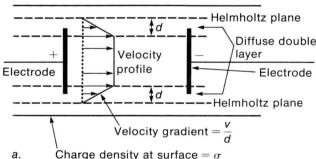

Velocity gradient $= \dfrac{v}{d}$

Charge density at surface $= \sigma$

a.

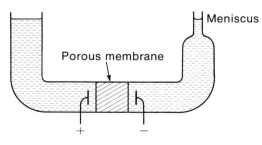

b.

Figure 11.19

a. Electro-osmosis through a capillary tube.
b. Apparatus for studying electro-osmosis. Observations are made of the movement of the meniscus in the capillary tube.

A simple type of apparatus for studying electro-osmosis through a membrane is shown in Figure 11.19*b*.

ELECTROPHORESIS

In the derivation of (11.127) for the electro-osmotic mobility under unit potential gradient, the moving liquid was regarded as a cylinder moving through a capillary tube. The positions of the liquid and wall can be reversed without affecting the argument, and (11.127) will also give the velocity of movement of a solid cylindrical particle through a liquid under the influence of an applied field of unit potential gradient. This quantity is the *electrophoretic mobility*; thus

$$v_{\mathrm{o}} = \frac{\zeta \varepsilon_{\mathrm{o}} \varepsilon}{\eta} \tag{11.130}$$

for a cylindrical particle moving along its axis.

The treatment of particles of different shapes is more complicated, and various theories have been presented. In their theory of electrolytic conductivity (see Section 6.5) Debye and Hückel concluded in 1924 that for a spherical particle equation (11.130) should be replaced by

$$v_{\mathrm{o}} = \frac{2 \zeta \varepsilon_{\mathrm{o}} \varepsilon}{3 \eta} \tag{11.131}$$

This equation is valid only if the thickness of the double layer is large compared with the radius of the particle. The equation may thus be satisfactory for ions, but less so for larger particles, and various improved equations have been suggested. In practice, however, electrophoresis experiments are usually carried out in an empirical manner, without reference to these equations.

The first important experiments on electrophoresis were carried out by the Swedish physical chemist Arne Wilhelm Kaurin Tiselius (1902–1971), who in 1937 made electrophoresis a powerful technique for studying mixtures of proteins. He devised a special type of U-tube, shown schematically in Figure 11.20, along which the protein molecules move under the influence of an electric potential. Different proteins will move at different speeds. The tube consisted of portions fitted together at ground-glass joints, so that one of a mixture of proteins could be isolated in one chamber. Optical methods are used to determine the quantity of each protein present in the mixture. The technique has in particular been applied to the study of proteins in the blood. Careful pH control is required, because the pH has a strong effect on the zeta potential and therefore on the rates of movement. The technique of electrophoresis supplements the ultracentrifuge, which separates according to molecular weight and shape. In the ultracentrifuge, different macromolecules having the same molecular sizes and shapes behave identically, but they may have different electrical properties and hence can be separated by electrophoresis.

A difficulty with the Tiselius tube is that a certain amount of local heating occurs, leading to convection currents which cause some mixing and disturb the separation. This problem is frequently overcome by supporting the solution on a gel, such as a polyacrylamide gel. Another recent development in electrophoresis involves the addition of detergents such as sodium dodecyl sulfate to macromolecular solutions supported on gels. Proteins are then found to have electrophoretic mobilities proportional to their molecular weights, and as a result it is possible to separate proteins which are not easily separable in the absence of the detergent.

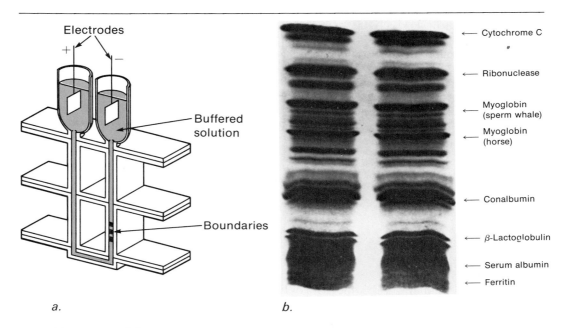

a.   b.

Figure 11.20

a. Cross-section of a Tiselius tube.   *b.* Two records obtained in an isoelectric focusing experiment. A solution containing a mixture of proteins was supported on the surface of a gel, and a pH range of 3 to 10 was maintained over a 40 cm distance. The voltage applied was 400 V for 20 hours followed by 600–800 V for 8 hours. The cathode is at the top. [Courtesy of Bronkman Instruments (Canada) Ltd.]

One special electrophoretic technique is *isoelectric focusing*. In this method a pH gradient is established so that the solution at the cathode is more basic, and that at the anode is more acidic. Molecules will migrate towards the electrode of opposite charge, but as they do so they will be subject to changing pHs which will tend to make them lose their charge. This process will continue until the species is concentrated at the position where the pH of the solution is the isoelectric point of the molecule. If several species are present, each will concentrate at a position corresponding to its isoelectric point. With suitable pH gradients it is possible to separate species with very small differences in isoelectric points. Isoelectric focusing is also frequently carried out with the sample supported on a gel.

REVERSE ELECTROKINETIC EFFECTS

In the two effects just mentioned, electrophoresis and electro-osmosis, the application of an electric field brings about relative motion of two phases. On the other hand, if movement is brought about, the displacement of the charged layers with respect to each other brings about a potential difference between any two points in the direction of motion. For example, if a liquid is forced through the pores of a membrane, or through a capillary tube, a potential difference is observed, its magnitude depending on the zeta potential. This phenomenon, the reverse of electro-osmosis, is known as the *streaming potential.*

The reverse of electrophoresis occurs when small particles are allowed to fall through a liquid under the influence of gravity. A difference of potential is observed between two electrodes placed at different levels, and its magnitude again depends on the magnitude of the zeta potential. This phenomenon is known as the *sedimentation potential* or as the *Dorn effect*, after the German physicist Friedrich Ernst Dorn (1848–1916), who discovered it in 1880.

EXAMPLE

A protein contains the following numbers of ionizing groups, with the average pK_a values shown:

Number	Group	pK_a
11	Carboxyl	4.5
4	Imidazole	7.0
1	α-amino	7.8
6	Phenolic	10.0
10	Side-chain amino	10.2

The protein is subjected to an electrophoresis experiment at pH 7.0. Will the protein move towards the cathode or the anode?

SOLUTION

The various ionizations are as follows:

(1) $-COOH \overset{pK_a = 4.5}{\rightleftharpoons} -COO^- + H^+$

(2)
$$HN\begin{array}{c} CN=N^+H \\ | \\ C=CH \\ | \end{array} \overset{pK_a = 7.0}{\rightleftharpoons} HC\begin{array}{c} CH=N \\ | \\ C=CH \\ | \end{array} + H^+$$

(At pH 7.0, there are equal numbers of the two forms; the average net charge of the 4 groups is thus $+2$).

(3) $-NH_3^+ \overset{pK_a = 7.8}{\rightleftharpoons} -NH_2 + H^+$

(largely $-NH_3^+$ at pH 7.0).

(4) $-OH \underset{}{\overset{pK_a = 10.0}{\rightleftharpoons}} -O^- + H^+$

(largely $-OH$ at pH 7.0).

(5) $-NH_3^+ \underset{}{\overset{pK_a = 10.2}{\rightleftharpoons}} -NH_2 + H^+$

(largely $-NH_3^+$ at pH 7.0).

The charge inventory is thus:

		Charge
11	$-COO^-$	-11
4	imidazole (average charge $= +\frac{1}{2}$)	$+2$
1	$-NH_3^+$	$+1$
6	$-OH$	0
10	$-NH_3^+$	$+10$
	Net charge	$+2$

The protein will therefore move towards the cathode (negative electrode).

In the above it has been assumed that the net charge on the macromolecule is determined entirely by the ionization. This is true to a good approximation. However, as noted on p. 318, complications may arise because of the attachment of other ions to the molecules.

PROBLEMS

11.1 The following data are for the binding of a ligand A to a macromolecule:

$[A] \times 10^3$ mol dm^{-3}	Average number of ligands bound, $\bar{v}$
0.5	1.28
1.0	2.00
2.0	2.56
5.0	3.20
10.0	3.70
20.0	3.85

Are these data consistent with the sites being identical and with no cooperativity? If so, what are n and the binding constant K?

11.2 The following data have been obtained for the binding of glucose-6-phosphate to the corresponding dehydrogenase:

$[A] \times 10^3$	% Saturation
mol dm^{-3}	
0.375	17.5
0.500	32.7
0.625	37.1
0.750	48.0
0.875	52.4
1.000	63.2
1.250	76.9
1.500	80.7
2.000	91.5

Do these data indicate cooperativity? If so, make a Hill plot and determine the minimum number of binding sites.

11.3 The following data relate to the binding of oxygen to hemocyanin:

Pressure of O_2 (mm Hg)	% Saturation
1.13	0.30
7.72	1.92
31.71	8.37
100.5	32.9
136.7	55.7
203.2	73.4
327.0	83.4
452.8	87.5
566.9	89.2
736.7	91.3

Do the data indicate cooperativity? If so, make a Hill plot and determine the maximum number of binding sites.

11.4 Diptheria toxin was found to have, at 20° C, a sedimentation coefficient of 4.60 Svedbergs, and a diffusion coefficient of 5.96×10^{-7} cm^2 s^{-1}. The toxin has a specific volume of 0.736 cm^3 g^{-1}, and the density of water at 20° C is 0.998 g cm^{-3}. Obtain a value for the molecular weight of the toxin.

11.5 Solutions of glucose ($D = 6.8 \times 10^{-10}$ m^2 s^{-1}) and tobacco mosaic virus ($D = 5.3 \times 10^{-12}$ m^2 s^{-1}) were maintained at a constant temperature of 20° C, and without agitation, for 100 days. How far would a given molecule of each be expected to diffuse in that time?

11.6 The diffusion coefficient for horse hemoglobin is 6.3×10^{-11} m^2 s^{-1} at 20° C. The viscosity of water at 20° C is 0.01002 poise, and the specific volume of the protein is 0.75 cm^3 g^{-1}. Assume the hemoglobin molecule to be spherical and to obey Stokes's law, and estimate its radius and the molecular weight.

11.7 The concentration of K^+ ions inside a cell is 155 mM, and that outside is 4 mM; the temperature is 37° C. Calculate the Gibbs energy difference, per mole of K^+ ions. If a reaction A → B is exergonic by 5.0 kcal, what is the minimum amount of A which could maintain this distribution of K^+ ions by a coupling reaction, all other factors being neglected?

11.8 For the protein specified in the Example on p. 509, deduce the direction of movement in an electrophoresis experiment at (a) pH = 6.0, and (b) pH = 7.8. Make a rough estimate of the isoionic point for this protein. (An exact calculation for this system is difficult. A rough estimate is adequate, in view of complications due to binding of other ions, etc.)

11.9 Another protein differs from the one specified in the Example (p. 509) in having 9 instead of 11 carboxyl groups; all other groups are the same. Suggest a pH value at which the two proteins could be separated in an electrophoresis experiment.

11.10 Two proteins have the following specifications:

Group	pK_a	Number of groups	
		Protein A	Protein B
Carboxyl	4.5	12	10
Imidazole	7.0	4	6
α-Amino	7.8	1	1
Phenolic	10.0	6	8
Side-chain amino	10.2	9	10

Estimate the net charges of the two proteins at the following pH values: (a) 6.0, (b) 7.0, (c) 8.6. If an isoelectric focusing experiment is carried out with the pH varying from 5.5 to 9.5, at what pH values will the two proteins congregate?

11.11 In a normal adult at rest the average speed of flow of blood through the aorta is 0.33 m s^{-1}. The radius of the aorta is 9 mm and the viscosity of blood at body temperature, 37° C, is about 4×10^{-3} N s m^{-2}. Calculate the pressure drop along an 0.5 m length of the aorta, in millimetres of mercury.

11.12 A typical capillary is about 1 mm long and has a radius of 2 μm. If the pressure drop along the capillary is 20 mm Hg, calculate (a) the average linear speed of flow of blood of viscosity 4×10^{-3} N s m^{-2}, and (b) the volume of blood passing through each capillary per second. Also, (c) estimate the number of capillaries in the body if they are supplied by the aorta described in Problem 11.11.

11.13 The adult human brain has a power output of about 40 W (see Problem 4.15, p. 415). Suppose that 25 W are utilized to pump Na$^+$ ions out of the nerve cells, the concentration within the cell being 12 mM, that outside being 150 mM. (a) If the pump works with 50% efficiency, calculate the flux of Na$^+$ ions out of the nerve cells ($T = 37°$ C); (b) If there are 10^{10} nerve cells in the brain, and each nerve impulse causes 10^{-11} mol Na$^+$ to leave a cell, what is the average frequency of impulses in a brain cell?

11.14 A cell membrane is permeable to Na$^+$, Cl$^-$ and H$_2$O but not to protein. Suppose that initially there is a NaCl solution of concentration 50 mM inside and outside the cell. Inside the cell there is a protein P at a concentration of 1 mM, and it ionizes completely to give P^{8+} + 8Cl$^-$. Calculate the concentrations of Na$^+$ and Cl$^-$ ions on each side of the membrane after equilibrium has become established, and calculate the Nernst potential ($T = 37°$ C).

11.15 If the diffusion coefficient for insulin is 8.2×10^{-11} m^2 s^{-1} at 20° C, estimate the mean time required for an insulin molecule to diffuse through a distance equal to the diameter of a typical living cell (~ 10 μm).

ESSAY QUESTIONS

11.16 Explain how the rate of diffusion through a membrane depends upon (a) the size of the diffusing substance, and (b) its solubility in the membrane.

11.17 Give a brief explanation of active transport.

11.18 Give an account of the electric double layer, with special reference to the zeta potential.

11.19 Explain what is meant by electrophoresis, and outline some biological applications.

SUGGESTED READING

Stein, W. D. *The Movement of Molecules Across Cell Membranes.* New York: Academic Press, 1967.

Tanford, C. *Physical Chemistry of Macromolecules.* New York: John Wiley, 1961.

Van Holde, K. E. *Physical Biochemistry.* Englewood Cliffs, N.J.: Prentice-Hall, 1971.

12

ISOTOPES IN BIOLOGY

Important advances in the understanding of biological reactions have stemmed from investigations using isotopes. Molecules can be labeled or tagged by the incorporation of isotopes, and by tracing the movement of the labeled atoms we can deduce the subsequent fate of the molecules. Much information about metabolic pathways, not readily ascertained in other ways, has been obtained by such tracer studies. The pioneer of this type of investigation was the German-American biochemist Rudolf Schoenheimer (1898–1941). In 1935, he introduced the use of isotopic tracers into biological research, and was the first to discover that the bodily constituents such as the proteins are not static, but undergo constant interchange with the constituents of the diet.

12.1   NUCLEAR BUILDING BLOCKS

The chemical properties of an atom are determined by its atomic number—that is, by the number of protons in the nucleus, which is equal to the number of orbital electrons in the neutral atom. In considering isotopes, we must concern ourselves with the nucleus, where most of the mass of the atom is concentrated. The most important constituents of the nucleus are the protons and the neutrons, the masses of which are almost identical. The protons bear a unit positive charge, while the neutrons are electrically neutral. The net charge on the nucleus is therefore simply the number of protons, while the *mass number* (approximately equal to the atomic weight) is the sum of the numbers of protons and neutrons. An atom whose nucleus has specified numbers of protons and nuclei is known as a *nuclide*; if it is radioactive it is known as a *radionuclide*.

It follows that the nuclei of a given chemical element, having a specified atomic number (i.e., number of protons), can have different masses, because there can be different numbers of neutrons. For example, the nucleus of the element chlorine must have 17 protons, but different numbers of neutrons are possible. The chlorine nucleus sometimes has 18 neutrons, and sometimes 20. The form that has 18 neutrons has a mass number of $17 + 18 = 35$, and is written as $^{35}_{17}Cl$, the upper number representing the mass number and the lower the atomic number. The nuclide that has 20 neutrons has a mass number of $17 + 20 = 37$, and is represented as $^{37}_{17}Cl$. Atoms of the same element that have different masses are known as *isotopes*. In naturally occurring chlorine there is about three times as much of the $^{35}_{17}Cl$ isotope as of the $^{37}_{17}Cl$ isotope, and the observed atomic weight is about

$$\frac{(3 \times 35) + 37}{4} = 35.5$$

12.2   ISOTOPES

Many elements exist in nature as two or more isotopes, and a number of isotopes have been prepared artificially for all elements. Some elements have more than 20 naturally occurring isotopes. The predominant naturally occurring forms of carbon (at. no. 6) are of mass 12 ($^{12}_6C$) and 13 ($^{13}_6C$) in the proportion of 98.9 to 1.1. Another isotope, $^{14}_6C$, is formed from nitrogen by the action of cosmic rays, and is radioactive; its application in the dating of archeological material is discussed later in this chapter.

The isotopes of hydrogen are of special interest and importance. The most common form of hydrogen has a single proton as its nucleus, and may be written as ^1_1H; it is known as *protium*. There also exist ^2_1H and ^3_1H. In the former the nucleus consists of a proton and a neutron, and in the latter it consists of a proton and two neutrons. The isotope ^2_1H, commonly called *deuterium* and given the symbol D, is now produced industrially in large quantities. Water in which the molecule contains two deuterium atoms, D_2O, is known as *heavy water*, and is frequently employed in biological studies. The isotope ^2_1H, or D, is not radioactive, but ^3_1H, known as *tritium* and often given the symbol T, is radioactive.

Table 12.1   Physical Properties of H_2O and D_2O

	H_2O	D_2O
Molecular weight	18.02	20.03
Density at 0° C, g cm^{-3}	1.000	1.105
Melting point, °C	0.00	3.82
Boiling point, °C	100.00	101.42

Isotopes have almost identical chemical properties, but they and their compounds differ somewhat in physical properties; some properties of H_2O and D_2O are compared in Table 12.1. There are also differences in equilibrium constants and in rate constants. These differences depend for the most part on the ratio of the masses. Thus the two carbon isotopes $^{12}_6C$ and $^{13}_6C$ differ in mass only by the ratio $12/13 = 0.93$, and equilibrium constants and rate constants for reactions involving them differ only by a very small percentage. However, the hydrogen isotopes ^1_1H and ^2_1H differ in mass by a factor of two, and the physical differences are very much more pronounced. One important consequence of this is that it is much easier to separate the isotopes of hydrogen than isotopes of the heavier atoms. Ordinary water contains a small percentage (0.016%) of deuterium; when it is electrolyzed, the H_2 is evolved significantly more readily than HD or D_2, so that the remaining water becomes enriched with heavy water. By successive electrolyses it is possible to produce water having a very high percentage of D_2O. However, the most practical commercial methods for preparing D_2O involve taking advantage of differences in equilibrium constants. Heavy water is made on a large scale in Canada for use as a neutron moderant in the CANDU (Canada Deuterium Uranium) nuclear reactor, and the production process involves the equilibrium

$$D_2S + H_2O \rightleftharpoons H_2S + D_2O$$

At low temperatures this equilibrium lies more to the right than at higher temperatures. The procedure is to bubble hydrogen sulfide through water in a vertical exchange tower, the lower portion of which is hot ($\sim 125°$ C) and the upper portion cold ($\sim 30°$ C). The H_2S gas leaving the lower part of the tower is enriched in deuterium, and it is then passed to the upper part of another tower where it enriches the water with deuterium. A series of such towers is employed, and the final fraction is 99.75% D_2O. Since D_2O has a significantly higher density than H_2O (see Table 12.1), the proportion of D_2O in a sample of water may conveniently be estimated from density measurements.

With the heavier isotopes, separation is also effected by making use of differences in physical properties, but the procedures are more laborious. One usually has to be content with only a partial enrichment, since it is very time-consuming to obtain a sample approaching 100% purity.

If by a suitable enrichment process the proportion of one of the less abundant isotopes is increased, the element is said to be *labeled*; if the element is incorporated into a compound, this is similarly said to be labeled. By suitable organic-chemistry techniques it is frequently possible to enrich the isotope content of one atom in a molecule; for example, citric acid

$$
\begin{array}{c}
CO_2H \\
| \\
H-C-H \\
| \\
HOOC-C-OH \\
| \\
H-C-H \\
| \\
C^*O_2H
\end{array}
$$

has been synthesized with only the starred atom enriched with $^{14}_6C$. Metabolic studies with this compound, in which one traces the fate of the labeled carbon atom, have provided important information about the detailed mechanisms involved.

The *degree of labeling* of a compound is expressed as the *atom percent excess.* Suppose that an element normally contains x atoms of a particular isotope per 100 atoms of the element; the normal abundance would be x atoms percent, and there would be $100 - x$ atoms of other isotopes per 100 atoms. Suppose that the abundance is increased to y atoms percent by the labeling process; the difference, $y - x$, is called the atom percent excess of the particular isotope.

The determination of the amounts of isotopes present can be carried out in a number of ways. If the isotope is radioactive, measurements can be made of the amount of radiation emitted, a matter that is further considered in the next section. The techniques of measuring radiation are relatively simple, and for that reason radioactive isotopes tend to be used whenever possible. However, in the case of nitrogen and oxygen, no suitable radioactive isotopes exist, the stable isotopes $^{15}_7N$ and $^{18}_8O$ being commonly used. There is some advantage to stable isotopes, in that they are permanent and produce no radiation, which in some cases (for example in work with living systems) might have an adverse effect.

The determination of the amounts of stable nuclides can be done using a *mass spectrometer.* This instrument sorts a beam of ionized atoms or molecules according to mass, just as an optical spectrometer resolves a beam of light into its component wavelengths. If a beam of ions of charge e and mass m is passed through a magnetic field, the deflection is independent of the velocity with which they are moving, but is proportional to the ratio e/m. Most of the ions produced in mass spectrometers have a single positive charge, and if this is assumed to be the case the mass m can be determined from the deflection. Some complications arise from the formation of doubly or triply charged positive ions, but enough experience has now accumulated for this difficulty to be easily overcome.

A schematic diagram of a mass spectrometer is shown in Figure 12.1. The substance to be studied is introduced into a chamber where a beam of electrons converts the molecule into positive ions. For example, when a water molecule is struck by an electron of suitable energy, an electron is knocked out of it with the formation of an H_2O^+ ion:

$$H_2O + e^- \rightarrow H_2O^+ + 2e^-$$

In the mass spectrometer the ions so produced are accelerated by an electric field, and the beam is passed through a transverse magnetic field. The deflected beam is focused in such a way that ions of the same e/m ratio arrive together at the detecting device.

Mass spectrometers are calibrated using known species, and in this way unknown samples containing isotopes can be identified and the amount of substance determined. This is the usual procedure for the determination of the amounts of stable isotopes, with the exception of deuterium. For that isotope, since its physical properties and those of its simple compounds differ to an appreciable extent from those of ^1_1H, it is usual to make a measurement of such a property. For example, the hydrogen can be converted into water, and from a density determination the proportion of deuterium can be calculated.

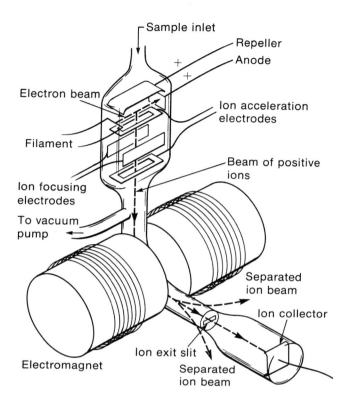

Figure 12.1

Schematic diagram of a mass spectrometer.

RADIOACTIVE ISOTOPES

Radioactive isotopes for tracer studies may be prepared artificially from nonradioactive elements by bombarding them with suitable nuclear particles produced in a cyclotron or a nuclear reactor. The discovery of this effect was made in 1934 by the French physicists Irène Joliot-Curie (1897–1956) and her husband Frédéric Joliot-Curie (1900–1958). They were studying the effect of bombarding light elements such as aluminum with alpha (α) particles, which are beams of helium nuclei, $^4_2$He. They noticed that, after the bombardment had ceased, a new form of radiation continued to be emitted, and they concluded that a new isotope had been formed. In the case of the bombardment of ordinary aluminum, $^{27}_{13}$Al, with α particles, the product is an isotopic form of phosphorus, $^{30}_{15}$P, the most abundant isotope of phosphorus being $^{31}_{15}$P. The process is

$$^{27}_{13}\text{Al} + ^4_2\text{He} \rightarrow ^{30}_{15}\text{P} + ^1_0\text{n}$$

A neutron, $^1_0$n, is produced as well as the radioactive isotope of phosphorus. By methods such as this a very large number of isotopes have been produced, and several of them have been extensively used in biological and medical research.

12.3   NUCLEAR DISINTEGRATION

The disintegration of a radioactive isotope involves the emission of various radiations, of which the most important are the following:

(1) *Alpha (α) particles*, which are helium nuclei, ^4_2He. For example, radium emits α particles with the formation of radon, which is a radioactive gas; the process is

$$^{226}_{88}\text{Ra} \rightarrow ^{222}_{86}\text{Rn} + ^4_2\text{He}$$

However, α radiation is emitted only by nuclei of high mass number, and consequently this type of emission is not of much significance in work with biological materials.

(2) *Beta (β⁻) particles*, which are electrons. These particles are emitted as a result of the transformation, in the nucleus, of a neutron into a proton,

$$^1_0\text{n} \rightarrow ^1_1\text{p} + \text{e}^-$$

For example, tritium, ^3_1H, is a beta emitter. We have seen that the tritium nucleus consists of a proton and two neutrons, and when β emission occurs the product is a nucleus containing two protons and one neutron; this is an isotope of helium:

$$^3_1\text{H} \rightarrow ^3_2\text{He} + \text{e}^-$$

Similarly the radioactive isotope of carbon, $^{14}_6\text{C}$, emits β radiation with the formation of an isotope of nitrogen:

$$^{14}_6\text{C} \rightarrow ^{14}_7\text{N} + \text{e}^-$$

This isotope of nitrogen is the most abundant form in nature.

(3) *Positrons*, which are positive electrons and denoted by the symbol e^+. The emission of positrons accompanies the conversion of a proton to a neutron:

$$^1_1\text{p} \rightarrow ^1_0\text{n} + \text{e}^+$$

For example, the carbon isotope $^{11}_6\text{C}$ is a positron emitter:

$$^{11}_6\text{C} \rightarrow ^{11}_5\text{B} + \text{e}^+$$

The product is an isotope of boron. However, positron emission is not much used in biological work.

(4) *Gamma (γ) ray emission.* Gamma rays are a form of electromagnetic radiation (see Figure 2.1) of low wavelengths and high frequencies and energies; they lie beyond the X rays in the electromagnetic spectrum. Gamma radiation frequently accompanies the other forms of radiation, and is associated with the settling down of the newly formed nuclei into more stable states. In addition, positron emission

always gives rise to γ radiation, because the positrons at once undergo an annihilation reaction when they meet electrons in their path:

$$e^+ + e^- \rightarrow 2 \text{ photons of } \gamma \text{ radiation}$$

The emission of γ rays does not, of course, change the atomic number or atomic mass.

12.4   KINETICS OF NUCLEAR DISINTEGRATION

A radionuclide disintegrates at a rate which is a function only of the constitution of the nucleus; unlike ordinary chemical processes, it cannot be influenced by any chemical or physical means, such as by changing the temperature. The disintegration of a nucleus, with the emission of radiation, is kinetically a first-order process; in other words, the rate of disappearance of the isotope is proportional to the amount n present at any time:

$$-\frac{dn}{dt} = kn \tag{12.1}$$

The proportionality constant k is known as the *decay* or *disintegration constant*. Integration of this equation proceeds as follows:

$$-\int \frac{dn}{n} = k \int dt \tag{12.2}$$

and thus

$$-\ln n = kt + I \tag{12.3}$$

where I is the constant of integration. When $t = 0$, $n = n_o$, where n_o is the amount present at unit time, whence $I = -\ln n_o$ so that

$$\ln \frac{n_o}{n} = kt \tag{12.4}$$

or

$$n = n_o e^{-kt} \tag{12.5}$$

In other words, the nucleus decays exponentially with time. The *half-life* τ is defined as the time that it takes for half of the nuclei to undergo transformation, and is obtained from (12.4) by putting $n = n_o/2$:

$$\ln 2 = k\tau \tag{12.6}$$

and thus

$$\tau = \frac{0.693}{k} \tag{12.7}$$

Some data for radioactive isotopes used in biology are collected in Table 12.2.

Table 12.2   Some Radioactive Isotopes Used in Biological Research

Isotope	Half-life	Type of emission
$^3_1$H (tritium)	12.46 years	Beta
$^{14}_6$C	5730 years	Beta
$^{24}_{11}$Na	15.05 hours	Beta, gamma
$^{32}_{15}$P	14.3 days	Beta
$^{35}_{16}$S	89.0 days	Beta
$^{36}_{17}$Cl	3.1×10^5 years	Beta
$^{40}_{19}$K	1.3×10^9 years	Beta, gamma
$^{42}_{19}$K	12.46 hours	Beta, gamma
$^{99m}_{43}$Tc†	6.02 hours	Gamma
$^{125}_{53}$I	60 days	Gamma
$^{131}_{53}$I	8.05 days	Beta, gamma

† The letter m indicates that this is an isomeric form of the isotope. Another $^{99}_{43}$Tc isomer emits beta radiation and has a half-life of 2.12×10^5 years.

EXAMPLE

The half-life of tritium, $^3_1$H, is 12.4 years; calculate the decay constant in s^{-1}.

SOLUTION

The half-life in seconds is

$$12.4 \times 365.25 \times 24 \times 60 \times 60 = 3.91 \times 10^8 \text{ s}$$

The decay constant is thus

$$\frac{0.693}{3.91 \times 10^8} = 1.77 \times 10^{-9} \text{ s}^{-1}$$

EXAMPLE

The half-life of radium, $^{226}_{88}$Ra, is 1600 years. How many disintegrations per second would be undergone by one gram of radium?

SOLUTION

The half-life in seconds is

$$1600 \times 365.25 \times 24 \times 60 \times 60 = 5.049 \times 10^{10} \text{ s}$$

The decay constant is thus

$$\frac{0.693}{5.049 \times 10^{10}} = 1.372 \times 10^{-11} \text{ s}^{-1}$$

The number of nuclei present in 1 g of radium is the Avogadro number (6.022×10^{23} mol^{-1}) divided by the atomic weight (226); i.e., is

$$\frac{6.022 \times 10^{23}}{226} = 2.665 \times 10^{21}$$

One gram of radium thus undergoes

$$1.37 \times 10^{-11} \times 2.66 \times 10^{21}$$
$$= 3.66 \times 10^{10} \text{ disintegrations per second}$$

UNITS AND DEFINITIONS

A number of special units and technical terms are used in work with radioactive isotopes. The old unit for the amount of an isotope was the *curie* (symbol Ci), which is defined as the amount producing exactly 3.7×10^{10} disintegrations per second. In the preceding example, we have seen that this is approximately the number of disintegrations produced per second by 1 gram of radium. In 1975 the curie was replaced by the *becquerel* (Bq) which is defined as the amount of radioactive substance giving one disintegration per second. Thus 1 Bq = 1 s^{-1} and 1 Ci = 3.7×10^{10} Bq.

The *specific activity* of a preparation can be defined as the number of becquerels or curies of the radioactive isotope divided by the total mass of the element present. For example, if carbon is enriched with the radioactive isotope $^{14}_{6}$C, the specific activity was formerly expressed as

$$\frac{\text{Number of curies of } ^{14}_{6}\text{C}}{\text{Total number of grams of carbon}}$$

and now as, for example,

$$\frac{\text{Number of becquerels of } ^{14}_{6}\text{C}}{\text{Total number of } \mu\text{g of carbon}}$$

The units may be curies per gram (Ci g^{-1}), microcuries per gram (μCi g^{-1}), becquerels per microgram (Bq μg^{-1}), etc. Alternatively, specific activity may be expressed in terms of the compound; thus, we might express the specific activity as becquerels per μg of compound, e.g., as

$$\frac{\text{Number of becquerels of } ^{14}_{6}\text{C}}{\text{Number of micrograms of compound in which it is present}}$$

Specific activity may also relate to the number of moles of the compound; thus, we might have becquerels per micromole (Bq μmol^{-1}).

A simple instrument for measuring radioactivity is the Geiger counter, devised in 1913 by the German physicist Hans Geiger (1882–1945) when he was working in Lord Rutherford's laboratory at Cambridge University. This instrument has a cylinder containing a gas under high electrical potential. If high-energy radiation enters it, ionization occurs and produces an electric current which can be recorded as a "clicking" sound. Modern commercial instruments for measuring radioactivity, such as proportioned counters and scintillation counters, are developments of this basic idea. They record the number of *counts*, i.e., number of times a particle of radiation enters the collector and produces a signal. In practice, radioactivity is usually recorded as *counts per minute* (c.p.m.) for a given weight of material. These are arbitrary units, related to becquerels by a proportionality constant, and usually no effort is made to convert the units into becquerels. A number of corrections have to be made to the counts, the most important being for the background count; the instrument may record counts with no radioactive material inserted, as a result of cosmic rays and random instrumental noise, and this background count must be subtracted.

12.5   STATISTICS OF RADIOACTIVE DISINTEGRATION

Whenever a physical measurement is made, it is important to know the reliability of the quantity determined. This is particularly important with radiochemical assays, for which there may be a wide variation among different measurements. The number of counts measured in a ten-minute interval may be significantly different from that obtained in another ten-minute interval—not because of experimental error but because of statistical fluctuations.

In this type of work we are concerned with how many events occur in a particular continuum, which in this case is a period of time. Precisely the same kind of problem arises if we are observing bacteria through a microscope, and are counting the number present in a square of a particular area; the continuum with which we are then concerned is the area.

The law which applies in problems of this kind is the Poisson distribution law, developed by the French mathematician Siméon Denis Poisson (1781–1840). According to this law, if the mean value is m counts, the probability of finding a value of x counts is

$$p_x = \frac{e^{-m} m^x}{x!} \qquad (12.8)$$

Thus, the probabilities of obtaining counts of 0, 1, 2, and 3 are:

Count of 0: $P_0 = e^{-m}$

Count of 1: $P_1 = e^{-m} m$

Count of 2: $P_2 = e^{-m} m^2 / 2!$

Count of 3: $P_3 = e^{-m} m^3 / 3!$

The total probability, from 0 counts to a count of infinity, is unity:

$$1 = e^{-m} \left(1 + m + \frac{m^2}{2!} + \frac{m^3}{3!} + \frac{m^4}{4!} + \cdots \right) \qquad (12.9)$$

The term in brackets is the expansion of e^m, and this multiplied by e^{-m} is, of course, unity. The form (12.9) is very convenient, successive terms giving us the probability of obtaining 0, 1, 2 … counts.

A plot of P_x against x is shown in Figure 12.2a. If we make a number of counts on a given sample and plot the frequency of occurrence of each value against the count, a histogram such as that shown in Figure 12.2b will be obtained. If the mean value m and the number of determinations are large enough, the histogram will

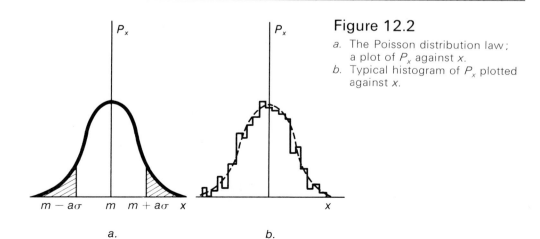

Figure 12.2

a. The Poisson distribution law; a plot of P_x against x.
b. Typical histogram of P_x plotted against x.

approximate closely to the smooth curve shown in Figure 12.2a. The total area under the curve is unity.

The standard deviation in the Poisson distribution is fixed by the value of m; in fact, the standard deviation is equal to $\sqrt{m}$. The standard deviation, a quantity which gives information about the breadth of scatter of observations about the mean, is defined as

$$\sigma = \sqrt{\sum_{x=0}^{\infty} (x - m)^2 P_x} \qquad (12.10)$$

In other words, it is the square root of the mean of the squares of the deviations $x - m$. In distributions like the Gaussian distribution,† the standard deviation is not fixed by the value of the measurement; it varies with the precision with which the experiments are made, being small in high-precision work and large in work of low precision. The Poisson equation, on the other hand, is not concerned with the precision of the experiments; it relates to a situation where there is a natural spread of events.

If equation (12.8) is inserted in (12.10) it is found that

$$\sigma = \sqrt{m} \qquad (12.11)$$

We can then ask what is the fraction of observations that will lie outside the limits $m + a\sigma$ and $m - a\sigma$; this is represented by the shaded area in Figure 12.2a. If we perform the necessary calculations, we find that for $a = 1$, for example, 32% of counts will lie outside the limits. In other words, there is a 68% chance of obtaining counts which lie within the limits $m + \sigma$ and $m - \sigma$ (i.e., $m + \sqrt{m}$ and $m - \sqrt{m}$), and a 32% chance of finding them outside these limits. The error corresponding to the limits $m + \sigma$ and $m - \sigma$ is often known as the *standard error*.

Table 12.3 gives the proportion of counts lying outside the limits $m + a\sigma$ and $m - a\sigma$ for various values of a. The larger is a, the smaller the proportion. The so-called "95% error" is of particular interest. It corresponds to $a = 1.96$, and we see that only 5% of the counts will lie outside the limits $m + 1.96\sigma$ and $m - 1.96\sigma$, 95% lying within. In practice, of course, m is not known; one may measure a count x_s on a sample, and can then say that there is a 95% chance that the correct value is within the limits $x_s + 1.96\sqrt{x_s}$ and $x_s - 1.96\sqrt{x_s}$.

It follows at once from (12.11) for the standard derivation that a count will be more reliable if it is large than if it is small. For example, a count of 100 has a standard error of $\sqrt{100} = 10$, which is 10%; a count of 10 000, however, has a standard

---

† The equation for the Gaussian distribution is

$$P_x = \frac{1}{\sigma\sqrt{2\pi}} e^{-(x-m)^2/2\sigma^2}$$

This equation contains the additional parameter σ, the standard deviation. It was developed by the great German mathematician Johann Karl Friedrich Gauss (1777–1855).

Table 12.3   Error Probabilities

a	Probability that the count lies outside the limits $m + a\sqrt{m}$ and $m - a\sqrt{m}$	Name of error
0.500	0.617	Reliable error
0.675	0.500	Probable error
1.000	0.317	Standard error
1.960	0.050	95% error
2.000	0.0455	—
3.000	0.003	—

error of $\sqrt{10\,000} = 100$, which is only 1%. It is convenient to define a *proportional error* as

$$\frac{a\sqrt{x_s}}{x_s}$$

or as

$$\frac{100a\sqrt{x_s}}{x_s}$$

if expressed as a percentage. Thus the following counts have the standard errors ($a = 1$) indicated:

$$
\begin{array}{lll}
100: & 0.1 & \text{or} \quad 10\% \\
1000: & 0.0316 & \text{or} \quad 3.16\% \\
10\,000: & 0.01 & \text{or} \quad 1\%
\end{array}
$$

95% proportional errors, with $a = 1.96$ (see Table 12.3), are as follows:

$$
\begin{array}{lll}
100: & 0.196 & \text{or} \quad 19.6\% \\
1000: & 0.064 & \text{or} \quad 6.5\% \\
1600: & 0.050 & \text{or} \quad 5.0\% \\
10\,000: & 0.0196 & \text{or} \quad 1.96\%
\end{array}
$$

Thus, a count of 1600 in a single determination will ensure that the true value has a 95% chance of lying within 5% of the observed value. Therefore, it is recommended in radiochemical work that we take a sufficiently long time interval that 1600 counts are made.

It is usually necessary to correct for the background count, and it is therefore important to know how errors are accumulated. If x_1 is an observed count and b

the background count in the same time interval, the 95% error of the net count $x_1 - b$ is $1.96\sqrt{x_1 + b}$, and the percentage error of the net activity is thus

$$E(\%) = \frac{100a\sqrt{x_1 + b}}{x_1 - b} \tag{12.12}$$

EXAMPLE
A sample had an activity of 1600 counts in 10 minutes, with a background count of 900 in the same period. Calculate the 95% error in the net count.

SOLUTION
The 95% error corresponds to $a = 1.96$. The percentage error at this level is given by equation (12.12):

$$E(\%) = \frac{100 \times 1.96 \times \sqrt{1600 + 900}}{1600 - 900}$$

$$= 14\%$$

This means that there is a 5% chance that the result will be in error by 14% or more.

This is a rather large error.† We could improve the reliability of our result by counting the background over a longer period of time. The percentage error in the count is now given by

$$E(\%) = \frac{100a\sqrt{\dfrac{R_s}{t_s} + \dfrac{R_b}{t_b}}}{R_s - R_b} \tag{12.13}$$

where t_s and t_b are the times for the sample and background observations respectively, and R_s and R_b are the rates (counts per minute).

EXAMPLE
A sample had an activity of 1600 counts in 10 minutes, with a background count of 9000 in 100 minutes. Calculate the 95% error in the net count.

---

† The size of this error is worth emphasizing, in view of the fact that conclusions are often drawn from smaller samples by political pollsters, clinical researchers, etc.

SOLUTION

Again, $a = 1.96$. The rates are
$$R_s = 160 \text{ counts per minute}$$
$$R_b = 90 \text{ counts per minute}$$

Equation (12.13) then gives, for the error,

$$E(\%) = \frac{100 \times 1.96 \times \sqrt{\dfrac{160}{10} + \dfrac{90}{100}}}{70}$$

$$= 11.5\%$$

This is less than for the preceding problem, where the counts per minute were the same. However, in the first problem the background time was 10 minutes, and in the second 100 minutes.

12.6   BIOLOGICAL EFFECTS OF RADIATION

High energy radiations, such as the α, β, and γ rays emitted by radioactive substances, can bring about chemical changes in biological material. Living systems contain a great deal of water, and this produces ions when subjected to radiation of sufficiently high energy. An ion that is produced very readily is H_2O^+. When this interacts with a negatively-charged particle such as an electron, there is neutralization with the production of sufficient energy to dissociate the molecule into a hydrogen atom and a hydroxyl radical:

$$H_2O^+ + e^- \rightarrow H_2O \text{ (excited)} \rightarrow H^{\textbf{·}} + .OH$$

This is only one example of the many reactions which occur with water and other molecules when they are subjected to radiation, the end effect being the formation of ions, atoms, and free radicals.

These resulting species are highly reactive and can have a very destructive effect on protein molecules and other molecules present in the organism. For example, free radicals can interact with the hydrogen bonds and sulfur-sulfur linkages important in maintaining the critical conformation of the protein. Such a reaction may bring about protein denaturation, and if the protein is an enzyme it will become catalytically inactive and metabolic processes and cell divisions will be affected. Also, reactions of this kind may make cell membranes more permeable, which will lead to abnormal penetration of substances into and out of the cells, resulting in injury to the organism.

Radiation damage to individuals is classified as being either *somatic* or *genetic*. Somatic (Greek *soma*, body) refers to injury to the body of the irradiated individual. Genetic (Greek *genea*, breed) refers to inheritable changes, or mutations, in the reproductive cells. The effects of exposure to high-energy radiation are well characterized: they include nausea, fatigue, blood and intestinal disorders, damage to the central nervous system, and various forms of cancer such as leukemia.

UNITS FOR DESCRIBING EXPOSURE TO RADIATION

We have seen that the fundamental unit for the amount of a radioactive isotope is the becquerel or the curie, which are measures of the number of disintegrations per second. For the assessment of the biological effects of high-energy radiation, such as that emitted by radioactive substances, we obviously need another kind of unit. There are several in common use.

The *roentgen* (symbol r) was the original unit, introduced to measure the ionization produced in air by radiation. One roentgen is the radiation that produces one electrostatic unit (esu) of pairs of positive and negative ions in 1 cm^3 of air at S.T.P. (standard temperature and pressure: 1 atm and $0°$ C). A singly-charged positive or negative ion has a charge of 1.60×10^{-19} coulombs, which is equal to 4.80×10^{-10} esu. One electrostatic unit therefore corresponds to $1/4.8 \times 10^{-10}$ or 2.08×10^9 ion pairs. Air at S.T.P. has a density of 0.001293 g cm^{-3}, so that

$$1 \text{ r} = 1.6 \times 10^{12} \text{ ion pairs per g of air.}$$

It is found empirically that it requires 5.2×10^{-18} J to create an ion pair in air at S.T.P., from which it follows that

$$1 \text{ r} = 8.3 \times 10^{-6} \text{ J per g of air}$$

Another commonly employed unit is the *rep*, which stands for "roentgen equivalent, physical." This measures the energy absorbed by water or soft tissue. On the average, 1 rep is approximately 9.3×10^{-6} J per g of tissue, which is somewhat more than one roentgen. There is considerable variation of absorption in soft tissue, from 6 to 10×10^{-6} J per g, and bone absorbs about 100×10^{-6} J per g. Therefore, the rep is fairly satisfactory for soft tissue, but not for the whole body.

To circumvent this difficulty of the energy dependence on the medium, the unit *rad* was introduced. Rad stands for "radiation absorbed dose," and is defined as 10^{-2} J kg^{-1}. It relates to the absorption of any kind of energy in any medium. In 1975 the rad was replaced by the *gray* (Gy), defined as 1 J kg^{-1}; 1 rad = 10^{-2} Gy.

Different kinds of radiation have different ionization efficiencies and are absorbed in different ways. In order to compensate for this the unit *rem* ("roentgen equivalent, man") was introduced. It is defined as the quantity of radiation of any type which produces the same biological effect in man as that produced by 1 r of X-ray or γ-ray radiation. However, the biological effect depends upon the part of the body irradiated and on the type of radiation. This problem is dealt with by estimating a *relative biological effectiveness* (RBE), in relation to γ-rays, for various parts of the body. For example, as far as the production of cataracts is concerned, 1 rad of fast neutrons is ten times as effective as 1 rad of γ-rays. Thus if the eyes were irradiated with 0.4 rad of γ-rays and 0.3 rad of fast neutrons, the number of rems is computed as

$$(0.4 \times 1) + (0.3 \times 10) = 3.4 \text{ rem}$$

Table 12.4 shows some typical figures for the yearly exposure to radiation of persons residing in North America. The doses vary considerably according to circumstances. Cosmic radiation gives the largest dosage, and there is nothing that we can do about this. A person residing on the top floor of a tall building will receive 2–3 mrem more cosmic radiation than a person at sea level. We see that, if all goes well, the radiation from nuclear power plants is quite insignificant compared with cosmic radiation.

Table 12.4   Radiation Exposure in North America

Source	Dose per Capita millirem per year
Natural background (cosmic rays)	125
Medical exposure	
Diagnostic	30–90
Therapeutic	5
Persons working at nuclear power stations	$\leqq 2$

COMPARISON OF RADIATIONS

There is considerable variation in the biological effects of different kinds of radiation. The α particles produce many more ion pairs than the β and γ particles. For example, an α particle having 1 million electron volts (1 MeV) of energy produces about 10^5 ion pairs per centimetre in air, in comparison with only about 100 for β particles and 10^4 for γ radiation of the same energy. Because of this high production of ion pairs, the α particles lose their energy quickly as they pass through matter, and do not penetrate far; they may even be stopped by a sheet of paper. For penetration of the skin the α particles must have 7.5 MeV or more, and exposure of the skin to α particles is a negligible hazard. However, ingestion of materials emitting α particles can have a very serious effect.

The β particles are more penetrating. A typical β particle can be stopped by 1 mm thickness of aluminum. Because they penetrate more into the skin, irradiation with β particles is more hazardous than with α particles. Ingested β emitters are more hazardous than they are externally, but less so than α emitters because of the smaller amount of ionization.

The γ rays, like X rays, have a greater destructive effect than the α and β rays. They are much more penetrating; it takes about 5 cm of lead to stop typical γ radiation. Because of their high ionizing power γ rays destroy tissue and inflict serious burns quite rapidly. Both γ and X rays interact with matter in three different ways:

(1) *Photoelectric effect*: they eject electrons particularly from atoms of higher atomic weights, the photons being annihilated.

(2) *The Compton effect*: ejection of an electron without annihilation of the photon. Instead, the scattered photon has lost energy and thus has a higher wavelength.

(3) *Pair production*: the photon is annihilated and produces an electron and a positron.

The electrons and positrons produced in these reactions interact further with the biological material, producing ions, atoms, and free radicals.

12.7 RADIOCARBON DATING

One very important application of radioactive isotopes is the determination of the ages of archaeological remains. The pioneer in this type of investigation is the American chemist Willard F. Libby. The radioactive isotope carbon-14, $^{14}_6C$, was discovered in 1940 by the American biochemists Samuel Ruben and Martin David Kamen, and it has a lifetime of about 5730 years. Libby soon realized that this isotope provides a valuable means of determining the age of carbon-containing materials.

Carbon-14 is continually being produced in the atmosphere from nitrogen, by the action of cosmic rays. These rays generate neutrons, which themselves have a lifetime of only about 12 minutes. However, during their lifetime, the neutrons generate carbon-14 by the reaction

$$^{14}_7N + ^1_0n \rightarrow ^{14}_6C + ^1_1H$$

This $^{14}_6C$ soon finds itself incorporated in the carbon dioxide of the atmosphere, so that the atmospheric carbon dioxide always contains a small but measurable proportion of $^{14}_6C$.

Carbon dioxide is constantly being incorporated into living plant tissues. As a result, plants contain a small proportion of carbon-14, and Libby estimated that the amount corresponds to approximately 15.3 counts per minute per gram of carbon. However, after an organism dies, no more carbon-14 is incorporated into it, and what is already present breaks down at a rate corresponding to a half-life of 5730 years, i.e., to a decay constant of 1.209×10^{-4} year^{-1} (see (12.7)). Thus after this period of time the amount of carbon-14 corresponds to $\frac{1}{2} \times 15.3 = 7.65$ counts per minute per gram of carbon. A measurement of the activity therefore allows us to make an estimate of the time that has elapsed from the death of the plant to the present time; this is the age of the material. This technique is known as *radiocarbon dating*.

For some samples of more recent origin there is good historical or archaeological evidence for the age, and this provides a check on the radiochemical method. For example, the Egyptian king Ptolemy V, who married Cleopatra, is known to have lived from about 203 to 181 B.C., or about 2150 years ago. Articles fabricated at that time would therefore be expected to have a radioactivity of about

$$15.3 \, e^{-1.209 \times 10^{-4} \times 2150}$$

$$= 11.8 \text{ counts per minute per g of carbon}$$

(see (12.5)), which has been found to be the case. Equally good agreement has been obtained for a number of other samples of known age. However, recent research has shown that a correction must be applied when the age is more than about 2000 years, probably because of variations in neutron intensity over long periods of time.

This technique has made very valuable contributions to archaeology. Reliable dates have been assigned to various Indian artifacts, for example, and to the Dead Sea Scrolls (about 1900 years old). Sometimes the results have been surprising: it has been shown that the last retreat of the ice-age glaciers occurred more recently than had been suspected—10 000 years ago rather than 25 000.

Cosmic rays also produce tritium, ^3_1H, and traces of this are therefore present in the atmosphere and in water. Tritium is also radioactive, and its concentration in well water, wine, etc., can be used in dating. Since its half-life (12.46 years) is short, only recent events can be dated.

EXAMPLE
Charcoal from the Lascaux cave in France (famous for its remarkable paintings) was found to have an activity of 2.22 counts per minute per g of carbon. Estimate the age of the sample.

SOLUTION
The age is given by (see (12.4))

$$t = \frac{1}{k} \ln \frac{n_o}{n}$$

with $k = 1.209 \times 10^{-4}$ years^{-1}. Then

$$t = \frac{1}{1.209 \times 10^{-4}} \ln \frac{15.3}{2.22}$$

$$= 16\,000 \text{ years}$$

12.8   APPLICATIONS OF ISOTOPES IN BIOLOGICAL STUDIES

There are a number of different ways in which isotopes can be used in biological investigations. They can be classified under three main headings: *analytical* techniques, *tracer* studies, and *kinetic-isotope* studies. Two important analytical techniques are *isotope dilution* and *radioimmunoassay*, both of which depend upon the fact that small amounts of isotopes are very conveniently determined by measurements of their radioactivity. In a tracer study one labels a compound by making an isotope substitution, and then finds out what happens to the compound by measuring the isotopic composition of it and its reaction products. In a kinetic-isotope study one makes an isotopic substitution into a compound, and then measures the rates of its reactions, in this way reaching conclusions about the reaction mechanism.

Tracer studies can be subdivided as *isotope-incorporation* studies, and *isotope-exchange* studies. In the isotope-incorporation method a compound is labeled and investigations are made of its subsequent reactions, with particular reference to the various possibilities of bond breaking and bond formation, and to the ultimate fate of the substance. One series of investigations of this kind, pioneered by R. Schoenheimer, was concerned with the interchange between the body constituents and the substances present in the diet. Another series has been concerned with establishing the detailed mechanism of the photosynthetic process. By working with isotopes, it is also possible to determine which bonds are broken in molecules when they undergo reaction, and this technique has been applied to a number of biological processes. Other work involving isotope incorporation has led to important conclusions about stereochemical relationships in living systems.

The isotope-exchange technique is essentially different. The isotope-incorporation method can lead to valuable deductions about reaction mechanisms and reaction intermediates, but much more direct information of this kind is provided by isotope-exchange studies. In these, a labeled substance is introduced into a reacting mixture and from the extent to which the isotope undergoes exchange, conclusions can be drawn about the nature of the reaction intermediates, and hence about the overall mechanism.

The kinetic-isotope method is distinctly different from the tracer method. Here we make use of the fact that isotopic substitution leads to changes in rate constants which in some cases can be satisfactorily correlated with the reaction mechanism. The measurement of the rate for a particular system can thus lead to conclusions about the nature of the slow step in the reaction—for example, about whether or not a hydrogen atom is being transferred.

12.9   ANALYTICAL TECHNIQUES

ISOTOPE DILUTION

The importance of the isotope-dilution technique is that it can be used for the quantitative determination of substances present in such small amounts that other methods are difficult to apply. For example, if a protein is hydrolyzed, some of the amino acids are present in very small proportions. In the isotope-dilution technique, we add to such a mixture a pure radioactively-labeled sample of the same compound, and isolate a sample of the substance in pure form regardless of yield. We then measure the specific radioactivity of the product, and from the specific activity of the added material we calculate the amount of unlabeled compound originally present. The method of calculation is best explained by means of an example.

EXAMPLE

A protein was hydrolyzed and it was desired to determine the amount of arginine present. A 5.1 mg sample of $^{14}_{6}C$-labeled arginine giving 2608 counts per minute (corrected for background) was added to the hydrolysis mixture, and then 12.5 mg of arginine was isolated in pure form. The resulting count was 1032 counts per minute (corrected). Calculate the amount of arginine present in the hydrolysate.

SOLUTION

Suppose that there were x mg of arginine present in the hydrolysate. After 5.1 mg of labeled arginine has been added, the total weight is $5.1 + x$ mg and the specific activity is

$$\frac{2608}{5.1 + x} \text{ counts mg}^{-1}$$

After isolation, the specific activity is

$$\frac{1032}{12.5} \text{ counts mg}^{-1}$$

These specific activities are equal:

$$\frac{2608}{5.1 + x} = \frac{1032}{12.5}$$

and thus

$$x = 26.5 \text{ mg}$$

RADIOIMMUNOASSAY

In the isotope-dilution technique it is necessary to isolate, purify, and assay a sample of the substance under study, and this sometimes presents a difficulty. On the other hand, in the radioimmunoassay (RIA) technique it is not necessary to assay the substance, but only to measure the radioactivity.

The principle of RIA is as follows. If we wish to measure the concentration of a substance A we first prepare an antibody to A. With many substances, particularly if the molecules are large, this presents no difficulty. With smaller molecules, such as the steroid hormone estriol, the molecule can be attached to a protein such as albumin. The complex so formed will then act as an antigen, and if it is injected into a rabbit, an antibody will be produced. This antibody will be specific to the estriol.

The next step is to prepare a radioactively labeled molecule, A*. One way of doing this with estriol, for example, is to grow bacteria producing it in a medium labeled with tritium or ^{14}C. Another method is to couple tyrosine chemically to the molecule of interest, and then to iodinate the tyrosine with radioactive ^{125}I.

The assay is carried out by preparing a series of solutions containing various amounts of A* and the antibody B. These form a complex,

$$A^* + B \rightleftharpoons A^*B$$

and the radioactive label is either in the free A* or in the complex A*B. If unlabeled A is added, the equilibria are now

$$
\begin{array}{c}
A \\
+ \\
A^* + B \rightleftharpoons A^*B \\
\downarrow\uparrow \\
AB
\end{array}
$$

The distribution of radioactivity between A* and A*B depends on the amount of A present. If a large amount of A is added, after equilibrium is established there will be little radioactivity in the complex, most of it being in the free antigen; if little A is added, more of the radioactivity is in the antigen-antibody complex. The distribution of radioactivity between the antigen and the complex thus tells us how much A has been added. Measurements can be made of the radioactivity in either the free antigen or the complex, and if known amounts of A are added we can prepare calibration curves which relate [A] to the radioactivity in the antigen or in the complex.

The fact that RIA does not require a chemical assay (once the calibrations have been made), but only measurement of radioactivity, has made the technique of very great value, particularly in clinical biochemistry. The method is exceedingly sensitive, being capable of measuring as little as a femtomole (10^{-15} mol) of material. Automatic systems, known as *gammaflow*, have been designed for use in hospitals.

12.10   ISOTOPE INCORPORATION

THE DYNAMIC STATE OF BODY CONSTITUENTS

The first application of isotopes to biological systems was made in 1923 by the Hungarian-Danish chemist Georg von Hevesy (1885–1966). However, at that time few isotopes were available, and von Hevesy's work was confined to following the absorption by plants of radioactive lead. Lead is not a normal constituent of living systems—indeed it is highly poisonous—and the application of isotopes to biology languished until the discovery of isotopes of elements which are present in the living organism.

In the early 1930s the American chemists Gilbert Newton Lewis and Harold Clayton Urey worked on the preparation of samples containing high proportions of deuterium.

In 1935 Urey was successful in preparing by evaporation a sample of water considerably enriched in D_2O, and he realized the importance of this substance for establishing the details of chemical reactions in living systems. Rudolf Schoenheimer took advantage of this development at once, and until his death in 1941 pioneered an extensive series of biological investigations.

He first worked with lipid molecules into which deuterium atoms had been introduced. These were incorporated into the diet of laboratory animals, which metabolized the deuterated fat much as they did ordinary fat. Up to that time, it had been assumed that the fat stored in an animal was immobile, the molecules remaining there for long periods of time. Under ordinary circumstances the use of fat to supply energy was thought to involve the newly digested fat introduced through the alimentary tract. Schoenheimer's work with deuterated fat showed that on the contrary there is a relatively rapid interchange between ingested fat and stored fat. When he fed rats the deuterated fat he found that at the end of four days the tissue fat contained nearly half of the deuterium that had been fed to the animal.

A rapid turnover was also found to occur with protein molecules. A number of amino acids were labeled with the stable isotope $^{15}_7N$, which had also been concentrated by Urey, and were found to undergo an interchange with the tissue proteins.

PHOTOSYNTHESIS

It has long been known that the growth of green plants involves photosynthesis, in which carbon dioxide and water react together under the influence of light and form starch together with oxygen. This process is the most important of all biological reactions, since all animals (including man) live on plants and breathe oxygen.

The photosynthetic reaction is an enormously complicated one, and involves a number of enzymes and reaction intermediates. Some of the elementary processes which occur are extremely rapid. A great deal of research has been concerned with elucidating the mechanism of the process, the most outstanding contribution being made by the American chemist Melvin Calvin, who made use of radioactive carbon-14, $^{14}_6C$.

Calvin and his coworkers used carbon dioxide enriched in carbon-14, and allowed the photosynthetic process to occur for very short intervals of time. By this means they isolated and identified a number of intermediates containing carbon-14. After much painstaking work they were able to deduce the main details of the mechanism of the photosynthetic process.

The occurrence of photosynthesis provides evidence for the presence of living organisms, and the 1976 Viking I space mission to the planet Mars has applied this test. Samples of Martian soil were exposed to carbon dioxide containing carbon-14 and irradiated with a lamp which simulates sunlight. After a period of incubation the soil was pyrolyzed (decomposed by heating) and tested to see if any of the radioactive carbon had become incorporated into carbon compounds. This test gave positive results. A second experiment was designed to find out if Martian soil contained any component—such as a microorganism—which could assimilate a solution containing organic compounds labeled with carbon-14. Radioactive CO_2 was given

off after such nutrients had been added to a soil sample, which indicates that
something was oxidizing the carbon in the labeled compounds added. The results
of both of these investigations provide some evidence that living organisms may be
present on Mars. However, the results of a third type of experiment cast doubt
on this conclusion. Addition of a nutrient solution to Martian soil was found to
release oxygen. This result suggests that there are some unusual chemical features
of the soil on Mars, and renders the previous results inconclusive. The Viking I
experiments therefore cannot be said to have provided clear evidence of life on Mars,
and further investigation is needed.

POSITION OF BOND BREAKING AND FORMATION

For many reactions the products can be formed in alternative ways, and it is
important to have a method of distinguishing between different mechanisms. For
example, the hydrolysis of any molecule of the type R—O—R' can involve the
breaking of the bond between the oxygen atom and the R group, or the oxygen
and the R' group:

(a)
$$R\!\!\mid\!\!O\!\!-\!\!R'$$
$$H\!\!-\!\!O\!\!\mid\!\!H$$

$$\searrow ROH + R'OH$$

(b)
$$R\!\!-\!\!O\!\!\mid\!\!R'$$
$$H\!\!\mid\!\!O\!\!-\!\!H$$

The products of both modes of reaction are identical. One way of distinguishing
between the two mechanisms is to label the oxygen atom with $^{18}_{8}O$. If the $^{18}_{8}O$ ends
up in the R'OH and not in the ROH, mode (a) is indicated. If the opposite occurs,
mode (b) is indicated.

Work of this kind has been done for nonenzymic reactions, such as the hydrolyses
of esters catalyzed by acids and bases. In most cases the labeled oxygen atom is
found in the resulting alcohol, and not in the acid, indicating that there is breaking
of the bond between the oxygen atom and the carbonyl group:

$$R\!\!-\!\!\overset{\overset{\textstyle O}{\|}}{C}\!\!\mid\!\!{}^{18}O\!\!-\!\!R' \qquad \to R\overset{\overset{\textstyle O}{\|}}{C}OH + R'{}^{18}OH$$
$$H\!\!-\!\!O\!\!\mid\!\!H$$

The same is found when ester hydrolysis is catalysed by enzymes. Thus in the hydrolysis of acetyl choline by acetylcholinesterase (a reaction of great importance for nerve conduction) the labeled atom is found in the choline:

$$
\begin{array}{c}
\overset{\displaystyle O}{\underset{\displaystyle \|}{}} \\
CH_3C\!\!-\!\!^{18}OCH_2CH_2N^+(CH_3)_3 \\
H\!\!-\!\!O\!\!-\!\!H
\end{array}
$$

$$\rightarrow CH_3COOH + H^{18}OCH_2CH_2N^+(CH_3)_3$$

Another reaction to which this method has been applied is the hydrolysis of glucose-1-phosphate catalyzed by the enzyme alkaline phosphatase. In this case the reaction might involve the breaking of either the C—O or the P—O bond. The reaction was caused to occur in water which had been enriched in $H_2^{18}O$. The labeled atom was found in the phosphoric acid and not in the glucose, which proves that the cleavage occurs at the P—O bond:

$$
\begin{array}{c}
CH_2OH \\
\text{(ring structure)} \quad O\!\!-\!\!P\!\!-\!\!OH \rightarrow glucose + H\!\!-\!\!^{18}O\!\!-\!\!P\!\!-\!\!OH
\end{array}
$$

STEREOCHEMISTRY

Isotopic labeling has also led to some very interesting conclusions regarding stereochemistry. It was first pointed out in 1949 by the British biochemist Alexander George Ogston that in order for enzymes to show stereochemical specificity they must become attached to their substrates at three positions at least. This is shown schematically in Figure 12.3. If it is assumed that three groups on an asymmetric substrate can only become attached to the enzyme in the manner shown in Figure 12.3a, with R_2 attached at point a, R_3 at b, and R_4 at c, it follows that the optical enantiomer cannot become properly attached (Figure 12.3b). If there were only two points of attachment, as shown in Figure 12.3c and d, both could become attached to the enzyme molecule.

This ability of enzymes to distinguish between optical enantiomers has an interesting and important extension: *enzymes can distinguish between what would normally be regarded as two identical groups or atoms on a substrate molecule.* This prediction of the Ogston hypothesis has been confirmed by experiments using isotopic tracers. For example, in 1953 the American chemist Frank H. Westheimer and his coworkers carried out a series of investigations on muscle lactate dehydrogenase. This enzyme has associated with it a coenzyme, nicotinamide adenine dinucleotide (NAD$^+$), the structure of which (see p. 138) can be simplified for our present purposes to

The enzyme catalyzes the interconversion of lactate and pyruvate ions, the process being

$$CH_3CHOHCOO^- \,(\text{lactate}) + \text{NAD}^+$$

$$\rightleftharpoons CH_3COCOO^- \,(\text{pyruvate}) + H^+ + \text{NADH}$$

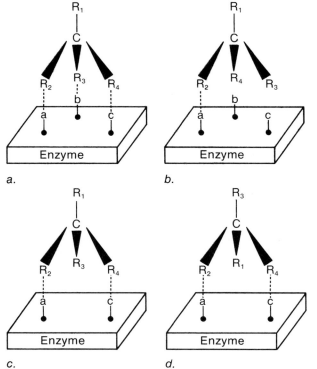

Figure 12.3

Ogston's explanation of stereochemical specificity in terms of three-point attachment between enzyme and substrate.

a.

b.

c.

d.

One of the two hydrogen atoms in position 4 is on one side of the ring, the other is on the other side, as indicated above. In the ordinary chemical sense these two atoms would be regarded as equivalent, but experiments showed that enzymes can discriminate between them. Westheimer and coworkers prepared NADH containing one deuterium atom and caused it to reduce pyruvate in the presence of the enzyme. They found that if the labeled NADH had been prepared by enzymic reduction of NAD^+, by use of alcohol dehydrogenase and CH_3CD_2OH, the reaction with pyruvate produced lactate containing one deuterium atom per molecule, as indicated by the equation

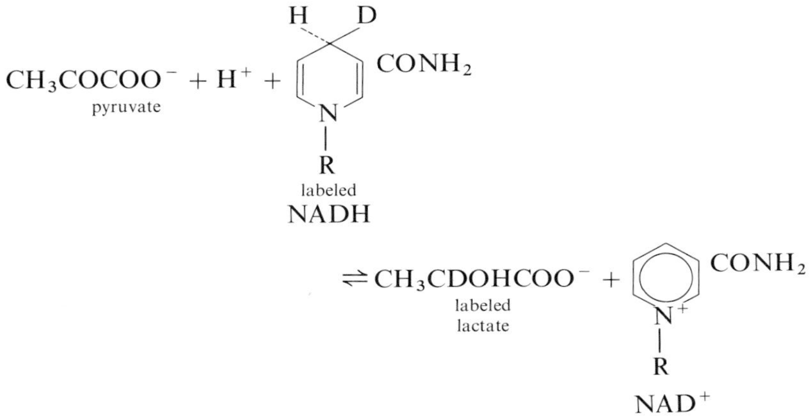

$$CH_3COCOO^- + H^+ +$$

pyruvate

H   D

CONH$_2$

N

R

labeled
NADH

$$\rightleftharpoons CH_3CDOHCOO^- +$$

labeled
lactate

CONH$_2$

N^+

R

NAD$^+$

In other words, only the D atom is transferred, the chemically-equivalent H atom not being transferred. On the other hand, when labeled NADH was prepared by purely chemical means without the use of enzyme, the lactate formed in the above reaction consisted of an equal mixture of $CH_3CDOHCOOH$ and $CH_3CHOHCOOH$. The chemical reduction produces two forms of labeled NADH:

H   D

CONH$_2$

N

R

and

D   H

CONH$_2$

N

R

Because of the stereochemical nature of the attachment of the substrate to the lactate dehydrogenase, the enzyme discriminates between these two forms, transferring only the H or D atom that is on the correct side of the molecule (see Figure 12.4). The alcohol dehydrogenase clearly places the D atom on the side of the ring that is correct for lactate dehydrogenase. That is, both enzymes are stereospecific in the same sense, removing H or D atoms from the same side of the nicotinamide ring.

The way in which an enzyme, by three-point attachment, can discriminate between chemically-equivalent groups is illustrated in Figure 12.5 for the case of aminomalonic acid. This acid has two —COOH groups which are equivalent in the ordinary

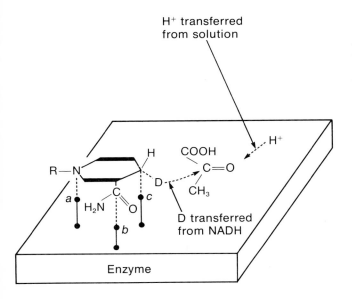

Figure 12.4

A very schematic representation of the reduction of lactic acid by labeled NADH, showing how the enzyme, because of three-point attachment, can discriminate between the H and the D atom.

chemical sense, but are stereochemically different, and since they are attached at different points on the enzyme they will be treated differently by the enzyme. This is of considerable importance as far as reaction intermediates are concerned. For example, serine can be converted into glycine by the action of an enzyme known as a transaminase, or as an aminotransferase:

$$\underset{\text{serine}}{H_2N-\overset{\overset{\displaystyle CH_2OH}{\displaystyle |}}{CH}-COOH} \rightleftharpoons \underset{\text{glycine}}{H_2N-CH_2-COOH}$$

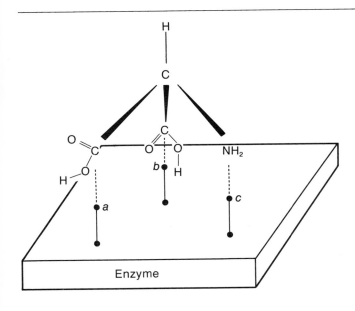

Figure 12.5

The three-point attachment of aminomalonic acid to an enzyme. Since sites a and b on the enzyme are different, the enzyme can discriminate between the two —COOH groups on the substrate.

Serine has been doubly labeled, with ^{15}N on the amino group and ^{13}C in the carboxyl position:

$$CH_2OH$$
$$|$$
$$H_2{}^{15}N—CH—{}^{13}COOH$$

and the glycine produced was found to be purely

$$H_2{}^{15}N—CH_2—{}^{13}COOH$$

At one time it was thought that this result excludes the possibility that a symmetrical substance like aminomalonic acid could be an intermediate in this process. It was thought that if this were the case, there would be an equal possibility of either of the two —COOH groups undergoing reaction,

$$
\begin{array}{c}
{}^{13}COOH \\
\diagup \\
H_2{}^{15}NCH \\
\diagdown \\
COOH
\end{array}
\quad
\begin{array}{c}
\xrightarrow{\;1\;} H_2{}^{15}NCH_2COOH + {}^{13}CO_2 \\
\\
\xrightarrow{\;2\;} H_2{}^{15}NCH_2{}^{13}COOH + CO_2
\end{array}
$$

and that this would disturb the $^{15}N/^{13}C$ ratio in the glycine. However, we can see with reference to Figure 12.5 that an enzyme is able to favor reaction 2 over reaction 1.

There are many other examples of symmetric intermediates leading to asymmetric products. For example, the symmetric molecule citric acid

$$CH_2COOH$$
$$|$$
$$HCCOOH$$
$$|$$
$$CH_2COOH$$

is an intermediate in the tricarboxylic acid cycle, which is important in metabolism.

APPLICATIONS IN CLINICAL MEDICINE

Radioactive isotopes are being used more and more in the diagnosis of disease. Isotopes which emit γ rays or positrons can be administered intravenously, and they become distributed through the organs and tissues of the body according to a pattern which is characteristic of the element and of the chemical form in which it is present. The distribution of the radionuclide is then determined by detection of the radiation which leaves the body. Since the distribution is affected by various pathological conditions, this technique provides valuable diagnostic information.

Various isotopes have been used for this purpose, but at the present time most of the work is done with an artificially-prepared isotope of technetium, $^{99m}_{43}Tc$. The letter m indicates that this isotope is an isomeric form of technetium 99; another form, denoted as $^{99}_{43}$Tc, is produced from $^{99m}_{43}$Tc with emission of γ rays, the half-life being 6.02 hours:

$$^{99m}_{43}\text{Tc} \rightarrow {}^{99}_{43}\text{Tc} + \gamma$$

$^{99m}_{43}$Tc is produced from $^{99}_{42}$Mo by emission of β and γ radiation, this process having a half-life of 67 hours:

$$^{99}_{42}\text{Mo} \rightarrow {}^{99m}_{43}\text{Tc} + \text{e}^- + \gamma$$

Commercially available portable generators are available for producing $^{99m}_{43}$Tc by this process. It is often injected in the form of the ion $^{99m}\text{TcO}_4{}^-$.

This isotope is particularly convenient for clinical work for several reasons. The short half-life means that radiation does not remain in the body for long periods, and millicurie amounts can be administered without giving too high a radiation exposure. At the same time, the half-life is sufficiently long to allow chemical manipulation and localization to be conveniently carried out. The γ rays emitted by this radionuclide have an energy of about 140 keV, which allows a high efficiency of measurement. The most commonly-used instrument for this purpose is the *scintillation camera*, which consists of a large cylindrical crystal of sodium iodide (usually $\frac{1}{2}''$ thick by $11''$ in diameter), backed by an array of photomultipliers. The NaI crystal emits light when a γ photon enters it, and the position of entry of the photon is determined from the amount of light received by each phototube. The camera is thus able to record the distribution of radionuclide throughout the body.

One application of this technique is in making a *brain scan*. When $^{99m}\text{TcO}_4{}^-$ is injected into a normal patient, it is prevented from entering the brain by a physiological "blood brain barrier." However, abnormalities such as tumors and abscesses destroy this barrier, and there is an accumulation of activity in the cerebral tissue. The technique can also be used in the detection of abnormalities in the ureters, kidneys, and bladder, of bone cancer, and of a variety of other pathological conditions.

12.11   ISOTOPE EXCHANGE

When an enzyme catalyzes a reaction between two substances (one of which may be a coenzyme), two different kinds of mechanism are possible. In one of these, the *single-displacement mechanism*, represented in Figure 12.6, the two substrates R—X and Y become attached to the enzyme molecule and undergo reaction with each other to form the products R—Y and X. Alternatively, one of the substrates, R—X, may first become attached to the enzyme molecule and split off the product X before the second substrate Y becomes attached. In a completely separate reaction, Y then reacts with the enzyme intermediate which was formed in the first stage. This is referred to as the *double-displacement mechanism*, and is represented in Figure 12.7. The ping pong mechanism (see Figure 10.6) is of this type.

An important feature of the double-displacement mechanism is that one of the

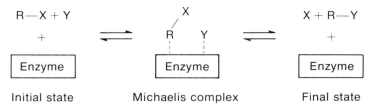

Figure 12.6

The single-displacement mechanism. The reactants RX and Y
become attached side by side on the enzyme and react together
directly. The R—Y bond is formed at the same time that the
R—X bond is broken.

products, X, is formed in the absence of the second substrate. Addition of X will
therefore cause reaction *a* in Figure 12.7 to shift to the left, and if X is isotopically
labeled, the label will appear in R—X. This is impossible in a pure single-displacement
mechanism, since in the absence of the second substrate no X is formed and no
isotopic exchange can occur. The use of isotopes can therefore discriminate between
the mechanisms. Note that a negative result (no isotope exchange) does not prove
anything, since reaction *a* in the double-exchange mechanism may occur essentially
to completion, and the equilibrium may hardly be reversed by addition of X.

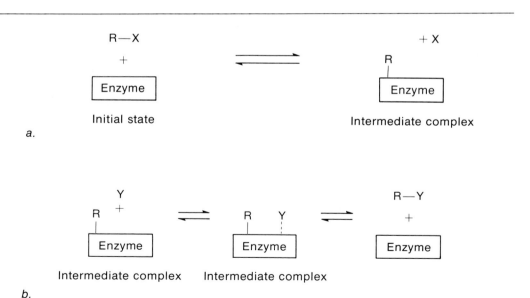

Figure 12.7

The double-displacement mechanism. In *a* an intermediate is formed and
X is liberated, even if there is no Y present. In *b*, Y reacts with the
intermediate to form R—Y. The essential feature of this mechanism is
that the R—X bond is broken before it has any interaction with Y.

One reaction to which this method has been applied is the formation of sucrose from glucose-1-phosphate and fructose. This reaction is catalyzed by the enzyme sucrose phosphorylase, and may be represented in schematic form as follows:

$$G—O—PO_3H^- + \quad F \quad \rightleftharpoons G—O—F + \quad H_2PO_4^-$$

glucose-1-phosphate   fructose   sucrose   phosphoric acid ion

The experiment involved allowing the enzyme to stand in the presence of $G—O—PO_3H^-$ and $H_2^{32}PO_4^-$ ions, but in the absence of fructose. There was an exchange of isotopic phosphorus, which went into the $G—O—PO_3H^-$. This does not occur in the absence of the enzyme, and it is concluded that an equilibrium such as

$$G—O—PO_3H^-$$

is established, thus allowing the exchange to occur.

12.12   KINETIC-ISOTOPE EFFECTS

Studies of kinetic-isotope effects have also provided valuable information about the mechanisms of reactions, and have been very helpful in elucidating biological mechanisms. Unfortunately, the theory is somewhat complicated, and there are a few pitfalls. Only a very general and brief outline can be given here. For further details, with special reference to biological mechanisms, the reader is referred to the book by Laidler and Bunting listed in *Selected Reading* at the end of this chapter.

The most pronounced kinetic-isotope effects are obtained when a hydrogen atom H is replaced by D or T, and when a bond involving this atom is broken or formed during the reaction. Consider, for example, a reaction of the type

$$A—H + B \rightarrow A + H—B$$

in which a hydrogen atom is transferred from a group A to a group B. If the H atom is replaced by D the process is now

$$A—D + B \rightarrow A + D—B$$

The potential-energy surfaces for these two reactions are identical, because the classical potential energy of an atom in a molecule is determined by the electronic configurations, which do not depend on the masses of the nuclei. However, the quantum states for the two systems are different, particularly because the vibrational levels are a function of the vibrational frequencies, which depend upon masses. Figure 12.8 shows the potential-energy curve relating to the A—H and A—D systems, and indicates the zero-point levels. Because of the higher mass of the D atom, the A—D vibrational frequency is lower than that for the A—H vibration. The zero-point energy, equal to $\frac{1}{2}h\nu$, is therefore lower for A—D than for A—H. For a typical A—H or A—D bond occurring in an organic molecule, the difference in zero-point levels is about 1.15 kcal mol^{-1}.

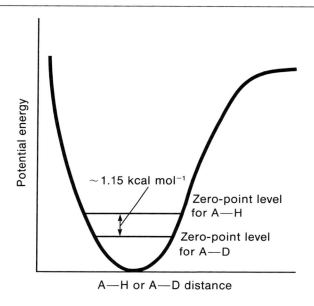

Figure 12.8

Schematic potential-energy curve for the A—H and A—D systems, the group A being treated as if it were an atom. The zero-point levels for A—H and A—D are shown.

This difference has an important effect on the rates of the reactions, as is shown by the potential-energy diagram in Figure 12.9. Calculations for typical organic systems show that the zero-point levels are much the same for the activated complexes in the two reactions. At ordinary temperatures the reactants are for the most part in their zero-point levels, and since the zero-point level for A—H is higher than that for A—D by about 1.15 kcal mol^{-1}, the activation energy for the reaction A—H + B will be less by this amount than that for A—D + B. The frequency factors for the two reactions will be very similar, and the ratio of rate constants is thus

$$\frac{k_H}{k_D} \approx e^{1150/RT} = 6.9 \text{ at } 300K$$

Ratios of this magnitude are frequently observed for reactions in which there is a transfer of an H or D atom. The existence of a kinetic-isotope effect of this magnitude can be used as evidence that there is an H or D transfer in the rate-controlling step of the reactions. A much smaller ratio suggests that the rate-controlling step does not involve such a transfer.

The preceding discussion has been considerably simplified, and a more precise treatment of kinetic-isotope effects requires that other factors be taken into account. One of these is *quantum-mechanical tunneling*. According to classical mechanics, in order for a system to pass from the initial to the final state, for a potential-energy surface such as that shown in Figure 9.14, it must pass over the top of the potential-energy barrier. However, quantum-mechanical theory admits the possibility that a system having less energy than required to surmount the barrier may nevertheless pass from the initial to the final state; it is said to *tunnel*, or *leak*, through the barrier. Tunneling is most important for a particle of small mass, and is particularly important for electrons, H atoms, H^+ ions, and H^- ions. Tunneling is of negligible importance for deuterium and for heavier atoms and ions. As a result of this possibility of tunneling with H but not with D, it follows that the kinetic-isotope ratio can be very much greater than the figure of about 7 that was deduced above; in fact, values of over 20 have been observed, and provide reliable evidence for tunneling.

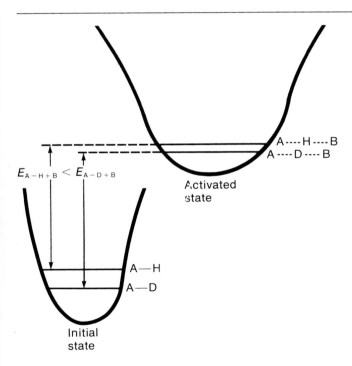

Figure 12.9

Schematic potential-energy diagram for the reactions A—H + B → A + H—B and A—D + B → A + D—B, showing why there is a substantial isotope effect.

The H/T kinetic-isotope effect is larger than that for H/D, because of the large ratio of masses; a value of 17 is typical if there is no tunneling with H. With isotope substitutions involving heavier atoms, the kinetic-isotope effect is much smaller, since the mass ratio is closer to unity. Some examples are given in Table 12.5. The smaller ratios with the heavier atoms are much harder to detect, but there have been a few important investigations with such systems.

Table 12.5   Typical Kinetic-Isotope Ratios at 25°C

Isotopic Forms	Ratio
H, D	7†
H, T	17†
^{12}C, ^{13}C	1.04
^{12}C, ^{14}C	1.08
^{14}N, ^{16}N	1.04
^{16}O, ^{18}O	1.04

† These values may be significantly larger if there is tunneling with H.

EXCHANGE WITH THE SOLVENT

One complication in the study of isotope effects is that there may be isotopic exchange with the solvent. For example, if we replace the alcoholic H atom in an alcohol ROH by D or T, and dissolve the product in ordinary water, there will be very rapid exchange of H and D or T atoms, the equilibrium

$$ROD + H_2O \rightleftharpoons ROH + DOH$$

being established. Since the water will be in excess, practically all of the ROD will be converted into ROH. Therefore, in order to carry out a kinetic-isotope study, we would have to dissolve the ROH in H_2O, and the ROD in D_2O, in which case both alcohols remain intact. This introduces the complication that we are comparing the kinetics in two different solvents, and this factor must be carefully taken into consideration in interpreting the experimental results.

This problem does not arise with kinetic-isotope substitutions involving heavier atoms. For example $^{12}_6C$ can be replaced by $^{13}_6C$ and the reactions of both substances investigated in ordinary water. Thus, although the expected kinetic-isotope ratio $^{12}k/^{13}k$ is 1.04 if a bond involving the C atoms is broken during the reaction, there is no complication arising from the use of different solvents.

EQUILIBRIUM EFFECTS

Kinetic-isotope effects are frequently complicated by the effects of isotope substitutions on equilibrium constants. We have seen in Chapter 9 that reactions frequently occur in more than a single step, and that the overall rate may depend upon the magnitude of an equilibrium constant. A simple case is when there is a rapidly established preequilibrium, as in the mechanism

$$A + B \underset{k_{-1}}{\overset{k_1}{\rightleftharpoons}} X \qquad \text{(rapid)}$$

$$X + B \overset{k_2}{\rightarrow} Y \qquad \text{(slow)}$$

The overall rate is

$$v = k_2 \cdot \frac{k_1}{k_{-1}} [A][B]^2 \tag{12.14}$$

The effect of an isotopic substitution would therefore depend on the effect on the equilibrium constant k_1/k_{-1} as well as on the rate constant k_2.

Acid-catalyzed reactions frequently occur by the initial attachment of a proton to the substrate S, followed by the slow reaction of the protonated substrate:

$$S + H^+ \rightleftharpoons SH^+ \qquad \text{(rapid)}$$

$$SH^+ \rightarrow \text{products} \qquad \text{(slow)}$$

Experimentally, it is often found that the rates of such reactions are 2–3 times *faster* in D_2O than in H_2O. In D_2O the process is

$$S + D^+ \rightleftharpoons SD^+ \qquad \text{(rapid)}$$

$$SD^+ \rightarrow \text{products} \qquad \text{(slow)}$$

We have seen that replacement of H by D generally decreases reaction rates, so that the breakdown of SD^+ is expected to occur more slowly than that of SH^+. The greater rate with the deuterated product must therefore be because the equilibrium

$$S + D^+ \rightleftharpoons SD^+$$

lies more to the right than does

$$S + H^+ \rightleftharpoons SH^+$$

This means that acid dissociation constants are smaller for the deuterated acid than for the undeuterated; i.e., for the dissociations

$$HA \rightleftharpoons H^+ + A^- \qquad K_H = \frac{[H^+][A^-]}{[HA]} \tag{12.15}$$

$$DA \rightleftharpoons D^+ + A^- \qquad K_D = \frac{[D^+][A^-]}{[DA]} \tag{12.16}$$

K_H is greater than K_D. In practice the pK difference, $pK_D - pK_H$, is about 0.6 in many cases. A somewhat oversimplified explanation for this is illustrated in Figure 12.10. Because of the higher zero-point level for the undeuterated molecule, less energy is required to dissociate it than to dissociate the deuterated molecule. In reality, of course, the dissociation of the acids involves the transfer of H^+ to H_2O, or of D^+ to D_2O, and the true explanation for the higher acidity of HA is more complicated.

It is evident that in the analysis of kinetic-isotope effects careful consideration

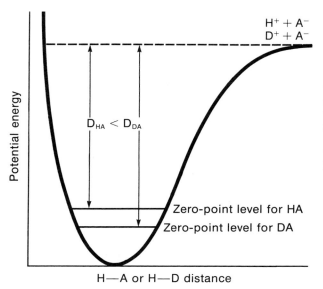

Figure 12.10

Schematic potential-energy diagram for the dissociation of HA into $H^+ + A^-$, and of DA into $D^+ + A^-$. This interpretation of the isotope effect neglects the solvation of the ions; in reality there is a difference between the zero-point levels for solvated H_3O^+ and solvated D_3O^+, but the difference is less than that between HA and DA.

has to be given to the possible complexities of the reaction, particularly to the existence of preequilibria which may have a dominating effect on the overall behavior. An important general principle with regard to equilibria is that *substitution with a heavier atom will always favor the formation of the stronger bond*. This may be understood with respect to the two energy diagrams shown in Figure 12.11. The essential point is that the stronger the bond the greater is the vibrational frequency, so that there is a wider separation between the zero-point levels for the two isotopes, these levels being $\frac{1}{2}h\nu$ higher than the classical ground states. In case *a* shown in Figure 12.11, with the weak bond at a higher energy level than the strong, E_{heavy} is seen to be greater than E_{light}, and the equilibrium in the light system will be more in favor of the weak bond than in the heavy system. In case *b* the weak bond is at a lower energy level than the strong, and now $E_{\text{light}} > E_{\text{heavy}}$ and the weak bond is still more favored by the light system.

SOME BIOLOGICAL EXAMPLES

Studies of kinetic-isotope effects have been carried out for several enzyme-catalyzed reactions, with the object of obtaining information about mechanisms.

Detailed studies on reactions catalyzed by chymotrypsin were made in 1964 by the American biophysical chemist Myron L. Bender and his coworkers. This enzyme brings about reaction by the mechanism

$$E + S \underset{k_{-1}}{\overset{k_1}{\rightleftharpoons}} ES \underset{X}{\overset{k_2}{\rightarrow}} ES' \overset{k_3}{\rightarrow} E + Y$$

where ES is the enzyme-substrate addition complex and ES′ is an acylated enzyme;

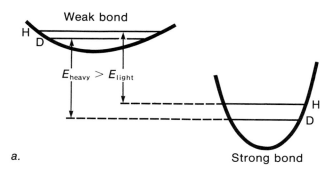

Figure 12.11

Potential-energy diagrams illustrating the general principle that the heavy atom favors the strong bond.

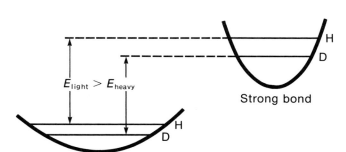

X and Y are the two products of the reaction, the latter being an acid formed by the hydrolysis of the acyl enzyme.

For some substrates at high substrate concentrations the conversion of ES into ES′, the *acylation* reaction, is rate limiting ($k_2 \ll k_3$), while for others the *deacylation* (ES′ → E + Y) is rate limiting ($k_3 \ll k_2$). With substrates of both types, Bender found a k_H/k_D ratio of 2.0 to 2.5 over a range of pH values. The work with the deuterated substrates had to be done in D_2O, because of isotope exchange, and this introduces the usual complications resulting from the change of solvent. However, even when these factors are taken into consideration there appears to be a genuine isotope effect for both types of substrate. It therefore appears that the processes of acylation and deacylation both involve a direct transfer of an H or D atom.

Work with heavier-atom isotopes has been done by M. H. O'Leary and coworkers, who in 1969–1970 studied reactions with glutamic acid decarboxylase, obtained from the bacterium *E. coli*. This enzyme catalyzes the reaction

$$
\begin{array}{ccc}
\text{COOH} & & \text{COOH} \\
| & & | \\
\text{CH}_2 & & \text{CH}_2 \\
| & \rightarrow & | \\
\text{CH}_2 & & \text{CH}_2 \\
| & & | \\
\text{H}_2\text{NCHCOOH} & & \text{H}_2\text{NCH}_2 \\
\text{glutamic acid} & & \gamma\text{-aminobutyric acid}
\end{array}
$$

This reaction appears to occur by the simple Michaelis-Menten mechanism

$$E + S \underset{k_2}{\overset{k_1}{\rightleftharpoons}} ES \overset{k_2}{\rightarrow} E + X$$

The rate equation is

$$v = \frac{k_2[E]_o[S]}{\dfrac{k_{-1} + k_2}{k_1} + [S]} \qquad (12.17)$$

(see equation (10.6)) and at low substrate concentrations

$$v = \frac{k_1 k_2}{k_{-1} + k_2} [E]_o[S] \qquad (12.18)$$

O'Leary and coworkers measured rates at low substrate concentrations with ordinary glutamic acid, and also with glutamic acid labeled with ^{13}C in its carboxyl group. They observed a k^{12}/k^{13} ratio of 1.02, somewhat smaller than the value of 1.04 expected for a process involving the breaking of a bond to the labeled carbon atom, and occurring in a single stage.

Two extreme cases may be considered. If step 2 in the reaction is rate controlling (i.e., if $k_2 \ll k_{-1}$), (12.18) reduces to

$$v = \frac{k_1 k_2}{k_{-1}} [E]_o[S] \qquad (12.19)$$

A kinetic-isotope effect is then expected, since step 2 involves the breaking of a bond to the labeled carbon atom. On the other hand, if step 1 is rate controlling ($k_{-1} \ll k_2$), (12.18) reduces to

$$v = k_1[E]_o[S] \qquad (12.20)$$

We would now expect no isotope effect, since the formation of the addition complex does not involve the breaking of a bond to the labeled atom.

The observation of a kinetic-isotope effect is therefore evidence that step 2 in the reaction is more or less rate determining.

PROBLEMS

12.1 What are the numbers of protons and neutrons in each of the following?

    (a) $^{3}_{1}H$ (d) $^{14}_{6}C$

    (b) $^{4}_{2}He$ (e) $^{131}_{53}I$

    (c) $^{30}_{15}P$ (f) $^{226}_{88}Ra$

12.2 The isotope $^{32}_{15}P$ emits β radiation and has a half-life of 14.3 days. Calculate the decay constant. What percentage of the initial activity remains after (a) 10 days, (b) 20 days, (c) 100 days?

12.3 The following counts per minute were recorded on a counter for the isotope $^{35}_{16}S$, at various times:

Time (days)	Counts per minute (after correction)
0	4280
1	4245
2	4212
3	4179
4	4146
5	4113
10	3952
15	3798

Determine the half-life in days, and the decay constant in s^{-1}. How many counts per minute would be expected after (a) 60 days, (b) 365 days?

12.4 Two isotopes of the same heavy element have very different half-lives. Which weighs more, one microcurie of the long-lived isotope, or one microcurie of the short-lived isotope?

12.5 The iodine isotope $^{131}_{53}I$ has a half-life of 8.1 days. In an experiment in which there was a background count of 15 counts per minute, a sample gave a recorder count of 2470 in 10 minutes. If the sample was assayed 16.2 days after the beginning of the experiment, what was the recorded count (in counts per minute) from the same amount of the isotope as at the beginning of the experiment?

12.6 Methionine was enriched with $^{35}_{16}S$, and an 0.1 μg sample was found to produce 11 100 disintegrations per minute. Calculate the specific activity in μCi per mg.
    1 mg of this enriched methionine was added to a protein hydrolysate, and a pure sample of methionine was isolated. The activity of 0.1 μg of this methionine was found to be 700 disintegrations per minute. Calculate the weight of methionine in the protein hydrolysate.

12.7 A quantity of fat was hydrolyzed to various fatty acids, one of which is palmitic acid. A 0.216 g sample of deuterated palmitic acid was added to the mixture, a sample of pure palmitic acid isolated, and its isotope content determined in the mass spectrometer. If the added palmitic acid was 21.5% deuterated, and the final palmitic acid was 2.75% deuterated, how much palmitic acid was present initially? [The deuterated compound may be assumed to have the same molecular weight as the undeuterated.]

12.8 21.65 g of horse hemoglobin was completely hydrolyzed, and 0.074 g of $^{15}_{7}N$-enriched L-tyrosine was added; the added sample was 6.85% enriched. A pure specimen of L-tyrosine was isolated and found to be 0.70% enriched. Calculate the weight of tyrosine produced in this hydrolysis. If the molecular weight of the protein is 66 700, how many tyrosine residues are present in the molecule?

12.9 A radioactive sample gave 2450 counts in 10 minutes. The background count for 1000 minutes was 12 500 counts. What is the net count over 10 minutes? Calculate the percentage standard error ($a = 1.000$) and the 95% error ($a = 1.960$).

12.10 A radioactive sample gave 24 500 counts in 100 minutes. The background count for 1000 minutes was 12 500 counts. What is the net count rate? Calculate the percentage standard error ($a = 1.000$) and the 95% error ($a = 1.960$) and compare with the results of the previous problem.

12.11 Certain samples of wood from an Egyptian coffin are believed, from archaeological evidence, to date from 3100 BC. Measurements made in 1975 gave values ranging from 8.25 to 8.35 counts per minute per g of carbon. Do these results support the archaeological conclusion?

12.12 The following radiocarbon counts have been measured: Lake mud from Ireland: 3.7 counts per minute per g of carbon. Charcoal sample from Stonehenge: 9.5 counts per minute per g of carbon. Estimate the ages of these samples.

12.13 A patient weighing 60 kg is being treated by exposure to a 1 g $^{60}_{27}$Co source, which emits two γ photons per decay, at an average energy of 1.25 MeV. The patient is irradiated for 3 minutes, and the arrangement is such that 1% of the radiation falls on the patient, and that half of that is absorbed (i.e., 0.5% is absorbed). The half-life of $^{60}_{27}$Co is 5.26 years. Calculate the dose in rads.

12.14 A patient weighing 75 kg having a brain scan is injected with 20 millicuries of $^{99m}_{43}$Tc, which on each disintegration emits one γ photon of energy of 0.143 MeV per photon. The half-life is 6 hours, and it can be assumed that all of the $^{99m}_{43}$Tc decays while still in the body and that all of the radiation is absorbed by the patient. Calculate the dose in rads.

ESSAY QUESTIONS

12.15 Give an account of the preparation and properties of heavy water.

12.16 Describe the nature of α, β, and γ radiations. Explain what nuclear changes accompany the emission of these radiations.

12.17 Give an account of the biological effects of α, β, and γ radiations.

12.18 Explain the method of isotope dilution, with particular reference to the determination of the proportions of the various amino acids in a protein.

12.19 Give an account of tracer techniques in biology, distinguishing clearly between isotope incorporation and isotope exchange.

12.20 Explain the significance of kinetic-isotope studies in connection with the mechanisms of biological reactions.

SUGGESTED READING

Dawes, E. A. *Quantitative Problems in Biochemistry*, Chapter 11. London: Churchill Livingstone, 1972.

Glasstone, S. *Sourcebook on Atomic Energy*, 3rd ed. Princeton, N.J.: Van Nostrand, 1968.

Kamen, M. D. *Isotopic Tracers in Biology*. New York: Academic Press, 1957.

Laidler, K. J., and Bunting, P. S. *The Chemical Kinetics of Enzyme Action*, Chapter 8 (kinetic-isotope effects). Oxford: Clarendon Press, 1973.

Lapp, R. E., ed. *Matter*. New York: Time Incorporated, Life Science Library, 1963.

Libby, W. F. *Radiocarbon Dating*. Chicago: Univ. of Chicago Press, 1952; 2nd edition, 1955.

Schoenheimer, R. *The Dynamic State of Body Constituents*. Cambridge, Mass.: Harvard University Press, 1949. [This small book is based on lectures given at Harvard in 1941, just before Schoenheimer's death in September of that year. Although this book is out of date, it still makes very interesting reading.]

Wernick, R., *et al. The Monument Builders*. New York: Time-Life Books, 1973. (See pp. 27–33 for an interesting account of radiocarbon dating in relation to archaeology.)

Williams, V. R., and Williams, H. B. *Basic Physical Chemistry for the Life Sciences*, Chapter 8. San Francisco: W. H. Freeman, 1973.

APPENDIX A

UNITS AND CONVERSION FACTORS

It is recommended that in scientific work use should be made of a coherent system of units known as the Système International d'Unités, abbreviated as the SI units. A system is said to be coherent when the units for all derived physical quantities are obtained from certain basic units by multiplication or division without the use of any numerical factors. A number of articles have been written about the SI units,[†] which represent an extension of the metric system.

The basic SI physical quantities, units, and their symbols, are as follows:

Basic physical quantity	Symbol for quantity	SI unit	Symbol for unit
Length	l	metre	m
Mass	m	kilogram	kg
Time	t	second	s
Thermodynamic temperature	T	kelvin	K
Electric current	I	ampere	A
Luminous intensity	I_v	candela	cd
Amount of substance	n	mole	mol

Note that the symbol for a physical quantity is always printed in *italics*, while a symbol for a unit is printed in roman type. Luminous intensity is seldom used in physical chemistry, but the other units are of importance and are considered in further detail later.

The following prefixes are used with SI units:

Fraction	Prefix	Symbol	Multiple	Prefix	Symbol
10^{-1}	deci	d	10	deca	da
10^{-2}	centi	c	10^2	hecto	h
10^{-3}	milli	m	10^3	kilo	k
10^{-6}	micro	μ	10^6	mega	M
10^{-9}	nano	n	10^9	giga	G
10^{-12}	pico	p	10^{12}	tera	T
10^{-15}	fento	f			
10^{-18}	atto	a			

LENGTH

The metre† is defined as the length equal to 1 650 763.73 wavelengths in vacuum of the radiation of a spectral line in the orange-red region, corresponding to the transition between certain levels ($2p_{10}$ and $5d_5$) of the krypton-86 atom.

The following SI units of lengths are also commonly used:

$$
\begin{aligned}
1 \text{ decimetre} &= 10^{-1} \text{ metres} & (1 \text{ dm} &= 10^{-1} \text{ m}) \\
1 \text{ centimetre} &= 10^{-2} \text{ metres} & (1 \text{ cm} &= 10^{-2} \text{ m}) \\
1 \text{ millimetre} &= 10^{-3} \text{ metres} & (1 \text{ mm} &= 10^{-3} \text{ m}) \\
1 \text{ micrometre} &= 10^{-6} \text{ metres} & (1 \text{ } \mu\text{m} &= 10^{-6} \text{ m}) \\
1 \text{ nanometre} &= 10^{-9} \text{ metres} & (1 \text{ nm} &= 10^{-9} \text{ m}) \\
1 \text{ picometre} &= 10^{-12} \text{ metres} & (1 \text{ pm} &= 10^{-12} \text{ m}) \\
1 \text{ kilometre} &= 10^{3} \text{ metres} & (1 \text{ km} &= 10^{3} \text{ m})
\end{aligned}
$$

The micrometre was formerly called the micron, and the nanometre was formerly called the millimicron; the use of these older terms is discouraged. The angstrom (Å), a non-SI unit, is still commonly employed:

$$1 \text{ Å} = 10^{-10} \text{ m} = 10^{-8} \text{ cm} = 10^{-1} \text{ nm} = 100 \text{ pm}$$

---

† Note the spelling of metre, even in U.S. scientific usage, metre being an international unit.

Although the angstrom is not recommended, it is still acceptable and is in very common use. Its advantage is that interatomic distances in molecules range from about 1 to 2 Å. Use of the nanometer requires fractions for such distances (0.1–0.2 nm), but some workers are now expressing interatomic distances in picometres, such distances ranging from about 100 to 200 pm.

VOLUME

In SI the basic unit of volume is the cubic metre, m^3. However, this is an inconveniently large unit for chemical and biological work, and the liter† is still commonly used. The liter is now defined as equal to one cubic decimetre (dm^3). The cubic centimetre should be written as cm^3, not as cc or as ml (milliliter).

MASS

The kilogram is the mass of a platinum-iridium block in the custody of the Bureau International des Poids et Mesures at Sèvres, France.

Although the kilogram is the basic unit, its multiples are expressed with reference to the gram (g); it would obviously be absurd to express a gram as a millikilogram! Thus 10^{-6} kg is written as 1 milligram (1 mg).

TIME

The second is the duration of 9 192 631 770 periods of the radiation corresponding to the transition between the two hyperfine levels of the ground state of the cesium-133 atom.

The second should be used as far as possible, but it is sometimes convenient to use larger units:

$$
\begin{aligned}
1 \text{ minute (min)} &= 60 \text{ s} \\
1 \text{ hour} &= 3600 \text{ s} \\
1 \text{ day} &= 86\,400 \text{ s} \\
1 \text{ year} &= 3.156 \times 10^7 \text{ s}
\end{aligned}
$$

THERMODYNAMIC TEMPERATURE

The kelvin is strictly defined as the fraction 1/273.16 of the temperature interval between the absolute zero and the triple point of water. In practice one obtains the number of kelvins by adding 273.15 to the temperature in degrees Celsius;

e.g., $25.0°\,C = 298.15$ K.

The use of the symbols °K and deg is not recommended.

---

† Since this is not an SI unit the spelling liter is permissible and is normally used in the U.S.

ELECTRIC CURRENT

The ampere is that constant current which, if maintained in two straight parallel conductors of infinite length, of negligible cross-section, and placed 1 metre apart in vacuum, would produce between these conductors a force equal to 2×10^{-7} newton per metre of length (1 newton = 1 kg m s^{-2}).

AMOUNT OF SUBSTANCE

The mole is the amount of substance which contains as many elementary units as there are atoms in 12 g of carbon-12, $^{12}_{6}C$. The elementary unit must be specified and may be an atom, a molecule, an ion, a radical, an electron, or any other elementary particle. It may also be any specified group or fraction of such entities, such as a polymer or a fraction of a molecule. For example,

> 1 mol of HgCl has a mass of 236.04 g
> 1 mol of Hg_2Cl_2 has a mass of 472.08 g
> 1 mol of Hg has a mass of 200.59 g
> 1 mol of $\frac{1}{2}CuSO_4$ has a mass of 79.80 g
> 1 mol of $\frac{1}{2}Cu^{2+}$ has a mass of 31.77 g
> 1 mol of electrons has a mass of 5.486×10^{-4} g

CONCENTRATION AND MOLALITY

The word *concentration* means the amount of substance divided by the volume of the solution. The usual unit is moles per liter (mol dm^{-3}) which is conveniently abbreviated as M. The use of equivalents and normalities (N) is not permitted in SI. Using the word *molarity* to mean concentration is not recommended because of the possibility of confusion with molality.

The word *molality* means the amount of solute divided by the mass of the solvent. The usual units are mol kg^{-1}, for which the symbol m is frequently used; to avoid confusion with the symbol for metre, it is best to use italics.

DERIVED SI UNITS

Table A-1 gives some derived SI units, together with some other units which are still commonly employed. It is likely that the calorie will remain in use for some time, since there has accumulated a vast scientific literature which employs calories; for example, practically every table of enthalpies of formation gives the values in calories. In this book we have retained the use of calories, and also of atmospheres and occasionally millimetres of mercury, which are still in common use.

The convention for writing the abbreviations should be noted. Prefixes are not separated from the basic unit (e.g., kg, not k g); units are otherwise separated (e.g., m s^{-1}, not ms^{-1}). The solidus (/) has been avoided in this book (e.g., m s^{-1} and not m/s). The solidus is acceptable but it should never be used more than once in the same expression unless parentheses are used. Thus entropy may be expressed as J K^{-1} mol^{-1} or as J/(K mol) but never as J/K/mol. The solidus is very useful in table headings and in figures (see, e.g., Table 6.4).

Table A-1   Units

Quantity	Symbol for Quantity	SI Unit and Symbol	Other Units
Length	l	metre (m)	1 angstrom (Å) $= 10^{-10}$ m $= 0.1$ nm 1 micron $= 10^{-6}$ m (not recommended)
Volume	v	cubic metre (m^3)	1 liter $= 1$ dm^3 (by definition)
Mass	m	kilogram (kg)	
Time	t	second (s)	1 min $= 60$ s; 1 year $= 3.156 \times 10^7$ s
Frequency	v	hertz (Hz $=$ s^{-1})	
Temperature	T	kelvin (K)	°C $=$ K $- 273.15$
Electric current	I	ampere (A)	
Electric charge	Q	coulomb (C $=$ A s)	1 esu $= 3.336 \times 10^{-10}$ C
Electric potential	V, ϕ	volt (V $=$ kg m^2 s^{-3} A^{-1})	
Electric resistance	R	ohm ($\Omega =$ V A^{-1})	
Electric capacitance	C	farad (F $=$ C V^{-1})	
Force	F	newton (N $=$ kg m s^{-2})	1 dyne $= 10^{-5}$ N
Pressure	P	pascal (Pa $=$ kg m^{-1} s^{-2})	1 atm $= 1.01325 \times 10^5$ Pa 1 bar $= 10^5$ Pa 1 torr $= 1$ mm Hg $= 133.322$ Pa
Energy	E, U	joule (J $=$ kg m^2 s^{-2})	1 erg $= 10^{-7}$ J 1 cal $= 4.184$ J (by definition) 1 electron volt (eV) $= 1.602 \times 10^{-19}$ J 1 eV molecule^{-1} $= 96.47$ kJ mol^{-1} $\quad\quad = 23.06$ kcal mol^{-1} 1 liter-atm $= 101.325$ J $\quad\quad = 24.22$ cal
Power	P	watt ($W =$ J s^{-1})	

ELECTRICAL UNITS

The electrical units require some special discussion. The force of attraction or repulsion between charged bodies was first used by the French engineer Charles Coulomb (1736–1806) as the basis for the definition of the unit of charge. He defined the electrostatic unit (esu) of charge as

that point charge which, when placed 1 cm from a similar point charge in a vacuum, is repelled by a force of 1 dyne.

The force is proportional to the product of the charges and inversely proportional to the distance of separation. Thus, if charges Q_1 and Q_2 are expressed in electrostatic units, and the distance r in centimetres, the force is

$$F(\text{dynes}) = \frac{Q_1 Q_2}{r^2} \tag{A-1}$$

One dyne is the force which gives an acceleration of 1 cm s^{-2} to a mass of one gram, and is therefore 10^{-5} newton (N), the newton being the force which gives an acceleration of 1 m s^{-2} to a mass of 1 kg.

The SI unit of charge is the coulomb (C), and the SI unit of distance is the metre. It can be shown that 1 coulomb is equal to 1 esu unit multiplied by ten times the velocity of light in metres per second:

$$1 \text{ C} = 2.998 \times 10^9 \text{ esu}$$

The force between two like charges of 1 C, separated by 1 m, is thus

$$\frac{(2.998 \times 10^9)^2}{100^2} = 8.99 \times 10^{14} \text{ dynes} = 8.99 \times 10^9 \text{ N}$$

The force between two charges Q_1 and Q_2, expressed in coulombs, separated by a distance of 1 metre in a vacuum, is thus

$$F = \frac{8.99 \times 10^9 Q_1 Q_2}{r^2} \text{ N} \tag{A-2}$$

It has proved convenient to write this equation as

$$F = \frac{Q_1 Q_2}{4\pi\varepsilon_0 r^2} \tag{A-3}$$

where ε_0, known as the *permittivity of a vacuum*, is equal to $1/4\pi \times 8.99 \times 10^9$:

$$\varepsilon_0 = 8.854 \times 10^{-12} \text{ C}^2 \text{ N}^{-1} \text{ m}^{-2}$$

The reason for the introduction of 4π is that certain equations are simplified if this is done. For example, the Poisson equation in polar coordinates takes the form

$$\frac{1}{r^2} \frac{\partial}{\partial r} \left(r^2 \frac{\partial \phi}{\partial r} \right) = -\frac{\rho}{\varepsilon_0} \tag{A-4}$$

whereas it involves 4π if electrostatic units are employed.

In a medium other than a vacuum, ε_0 in all of the equations is replaced by $\varepsilon_0\varepsilon$, where ε is known as the *relative permittivity* or more usually as the *dielectric constant*:

$$F = \frac{Q_1 Q_2}{4\pi\varepsilon_0\varepsilon r^2} \text{ N} \tag{A-5}$$

The corresponding energy of interaction is

$$E = \frac{Q_1 Q_2}{4\pi\varepsilon_0\varepsilon r} \text{ J} \tag{A-6}$$

and the Poisson equation in polar coordinates is now

$$\frac{1}{r^2} \frac{\partial}{\partial r} \left(r^2 \frac{\partial \phi}{\partial r} \right) = -\frac{\rho}{\varepsilon_0\varepsilon} \tag{A-7}$$

APPENDIX B

PHYSICAL CONSTANTS

Constant	SI Symbol	Value in SI and other units
Avogadro's number	N_A	6.022×10^{23} mol^{-1}
Electronic charge	e	1.602×10^{-19} C
Faraday constant	$F = N_A e$	96 485 C mol^{-1}
Gas constant	R	8.314 J K^{-1} mol^{-1}
		1.987 cal K^{-1} mol^{-1}
		0.08205 liter-atm K^{-1} mol^{-1}
Planck's constant	h	6.626×10^{-34} J s
Boltzmann constant	$\mathbf{k} = R/N_A$	1.381×10^{-23} J K^{-1}
Velocity of light	c	2.998×10^8 m s^{-1}
Permittivity of vacuum	ε_o	8.854×10^{-12} C^2 N^{-1} m^{-2}
Ratio of circumference to diameter of circle	π	3.1416
Base of natural logarithms	e	2.7183

APPENDIX C

RELATIONSHIPS BETWEEN $E^{\circ'}$ AND E°

Equation (8.36) on p. 357 was derived for a reaction of the type

$$aA + bB + 2H^+ \rightleftharpoons xX + yY$$

Other expressions are required when H^+ appears in different ways in the equation, and for other values of n; some are listed below

n	Type of equilibrium	Expression
1	$aA + bB + H^+ \rightleftharpoons xX + yY$	$E^{\circ'} = E^{\circ} - 0.059 \text{ pH}$
1	$aA + bB \rightleftharpoons xX + yY + H^+$	$E^{\circ'} = E^{\circ} + 0.059 \text{ pH}$
2	$aA + bB + H^+ \rightleftharpoons xX + yY$	$E^{\circ'} = E^{\circ} - \dfrac{0.059}{2} \text{ pH}$
2	$aA + bB \rightleftharpoons xX + yY + H^+$	$E^{\circ'} = E^{\circ} + \dfrac{0.059}{2} \text{ pH}$
2	$aA + bB + 2H^+ \rightleftharpoons xX + yY$	$E^{\circ'} = E^{\circ} - 0.059 \text{ pH}$
2	$aA + bB \rightleftharpoons xX + yY + 2H^+$	$E^{\circ'} = E^{\circ} + 0.059 \text{ pH}$

In general, for

$$aA + bB + hH^+ \rightleftharpoons xX + yY$$

$$E^{\circ'} = E^{\circ} - \frac{0.059h}{n} \text{ pH}$$

(this applies to positive and negative values of h).

ANSWERS TO NUMERICAL PROBLEMS

Chapter 1

1.1   (a) 1.212×10^{-11} m $= 12.12$ pm
      (b) 2.93×10^{-11} m $= 29.3$ pm
      (c) 6.65×10^{-15} m $= 6.65$ fm
      (d) 2.38×10^{-38} m

1.2   (a) 1.23 nm
      (b) 28.6 pm

1.3   0.1 nm $= 1.0$ Å

1.4   1875 nm

1.5   0.336 nm; -340.5 J mol^{-1}

1.6   (a) 0.47
      (b) 6.0%

1.10 (a) $\dfrac{1}{\sqrt{2}} (\psi_1 + \psi_2)$

     (b) $\dfrac{1}{\sqrt{2}} (\psi_1 - \psi_2)$

     (c) $\dfrac{1}{\sqrt{3}} (\psi_1 + \psi_2 + \psi_3)$

     (d) $\dfrac{1}{2} (\psi_1 - \psi_2 + \psi_3 - \psi_4)$

     (e) $\dfrac{1}{\sqrt{3}} \left(\psi_1 - \dfrac{1}{\sqrt{2}} \psi_2 + \dfrac{\sqrt{3}}{\sqrt{2}} \psi_3 \right)$

Chapter 2

2.1 5890 Å; 1.70×10^4 cm^{-1}; 5.09×10^{14} s^{-1}

2.2   (a) 3.37×10^{-19} J; 48.5 kcal mol^{-1}; visible (yellow)
      (b) 7.82×10^{-19} J; 112.5 kcal mol^{-1}; near ultraviolet

2.3 9.90×10^{13} s^{-1}; in the infrared

2.4 $A = 0.152$; $T = 0.705$; $T\% = 70.5$

2.5 $\varepsilon_{1\,cm}^{1\,M}$ (580 nm) $= 11.9$ dm^3 mol^{-1} cm^{-1} $= 1.19 \times 10^4$ cm^2 mol^{-1} $= 11.9$ cm^2 mmol^{-1}

2.6 2.16×10^{-4} M (216 μM); 13.8 g dm^{-3}

2.7   (a) 238.2 nm;   (b) 242.3 nm; both in the ultraviolet

2.8 3.14×10^{-4} M $= 314$ μM

2.9 56.7 μM

2.10 NAD$^+$, 12.6 μM; NADH, 34.6 μM

2.11 0.373; 42.4%

2.12 0.755 M

Chapter 3

3.1 105; $n = 1$

3.2   Fe: $13\,960$; S: 6680; $M_{min} \approx 13\,500$ with $n = 1$ for Fe and 2 for S

3.3   Fe: $16\,600$; S: 6680; arginine: 4110; $M_{min} \approx 33\,300$ if the molecule has 2 Fe, 5 S, and 8 arginine

3.4 1042

3.5   47 acidic groups and 48 basic groups

3.6 $M_n = 21\,000$; $M_w = 23\,333$

3.7 $M_n = 50\,000$; $M_w = 54\,000$

3.8 0.30 m

3.9 0.197 M

3.10 $31\,000$

3.11 60

3.12 $75\,600$

3.13 $+52.7°$

Chapter 4

4.1   (a) 4 atm
      (b) 44.8 atm dm^3 = 1084 cal
      (c) ~7 cal K^{-1} mol^{-1}

4.2   (a) Zero
      (b) 1000 cal
      (c) 1000 cal
      (d) 5.47 atm
      (e) 1483 cal
      (f) 1400 cal

4.3   (a) 15.3 atm
      (b) 400 cal
      (c) 1400 cal
      (d) 1400 cal
      (e) 1000 cal

4.4   (a) Zero
      (b) 8 atm
      (c) 751 cal
      (d) 751 cal

4.5   (a) −48.6 kcal
      (b) −69.6 kcal

4.6   −115.7 kcal; this differs slightly from −48.6 −69.6 = −118.2. The data in 4.5 are for the liquid, those in 4.6 for aqueous solutions.

4.7   3.74 kcal g^{-1}

4.8   −23.4 kcal

4.9   −0.5 kcal

4.10 −217.2 kcal mol^{-1}

4.11 4.88 kJ = 1.17 kcal

4.12 2.71 kcal = 11.34 kJ

4.13 1.4 kJ

4.14 (a) 73°C;   (b) 4.35 kg

4.15 (a) 10.4 cal;   (b) 43.5 W

Chapter 5

5.1   (a) 0.8 or 80%
      (b) 10 kcal
      (c) 50 cal K^{-1}
      (d) 50 cal K^{-1}

(e) Zero
(f) Zero
(g) -50 kcal

5.2 C_6H_6:20.8 cal K^{-1} mol^{-1}
CHCl$_3$:21.0 cal K^{-1} mol^{-1}
H_2O:26.0 cal K^{-1} mol^{-1}
C_2H_5OH:26.2 cal K^{-1} mol^{-1}
Hydrogen bonding in liquids H_2O and C_2H_5OH

5.3 $\int_{T_1}^{T_2} \dfrac{C_p}{T}\, dT = C_p \ln \dfrac{T_2}{T_1}$ if C_p is constant

5.4   (a) -58.0 cal K^{-1} mol^{-1}
(b) -109.1 cal K^{-1} mol^{-1}

5.5   (a) Positive
(b) Positive
(c) Negative
(d) Positive
(e) Negative

5.6   (a) $\Delta S^\circ = 8.39$ cal K^{-1}
(b) $K = 1.67 \times 10^5$ mol dm^{-3}

5.7   (a) $\Delta H^\circ = 14.44$ kcal
(b) $\Delta G^\circ = 2.94$ kcal
(c) $\Delta S^\circ = 37.9$ cal K^{-1}

5.8 4.94×10^{19} mol^3 dm^{-6}

5.9   (a) $\Delta G^\circ = -108.03$ kcal
(b) $K = 1.9 \times 10^{79}$ mol dm^{-3}

5.10 (a) $\Delta G^\circ = 1.59$ kcal
(b) Equilibrium lies somewhat to the left
(c) $K = 6.81 \times 10^{-2}$
(d) $\Delta G^\circ = -64.3$ kcal
(e) Yes, it will bring about almost complete removal of citrate

5.11 (a) 1.99×10^{50}
(b) Yes

5.12 $\Delta G^\circ = -17.7$ kcal

5.13 $\Delta G^\circ = -1.83$ kcal

5.14 (a) $\Delta G^\circ = 0$
(b) $K = 1$
(c) >1

5.15 (a) No
(b) Yes
(c) Yes

5.16   (a) $K = 19$; $\Delta G^\circ = -1.74$ kcal
         (b) $\Delta G = -4.47$ kcal

5.17   (a) $\Delta G^\circ = -60$ cal
         (b) No
         (c) (i) $\Delta G = 5.39$ kcal;   (ii) No

5.18   4.0 kcal

5.19   (a) Yes
         (b) $\Delta G^\circ = -3.0$ kcal
         (c) $K = 1.31 \times 10^2$

5.20 $\Delta G^\circ = 0$
         $\Delta H^\circ = 66.4$ kcal
         $\Delta S^\circ = 209.5$ cal K^{-1}

5.21   Work $= 731$ cal mol^{-1}; $\Delta U = 8.97$ cal mol^{-1}; $\Delta G = 0$; $\Delta S = 26.0$ cal K^{-1} mol^{-1}

5.22   Mercury: -1.50 cal K^{-1}
         Water and vessel: 1.69 cal K^{-1}
         Net $\Delta S = 0.19$ cal K^{-1}

5.23   48.1 J $= 11.5$ cal

5.24   671.8 J $= 160.6$ cal

5.25   5.32 atm

5.26   8.57 kcal

Chapter 6

6.1   33.77 milliamperes (mA)

6.2   2.38 mA

6.3   Yes; $K = 1.51 \times 10^{-3}$

6.4 9.12×10^{-6} M $= 9.12 \ \mu$M

6.5   (a) 30.5 nm
        (b) 0.673 nm

6.6   129.9 Ω^{-1} cm^2 mol^{-1}

6.7   238.2 Ω^{-1} cm^2 mol^{-1}; $\alpha = 0.0403$

6.8 $t_{Cl^-} = 0.683$; $t_{Li^+} = 0.317$; $\lambda^\circ_{Cl^-} = 78.6 \ \Omega^{-1}$ cm^2 mol^{-1};
        $\lambda^\circ_{Li^+} = 36.4 \ \Omega^{-1}$ cm^2 mol^{-1}

Chapter 7

7.1   NaCl: -97.3 kcal
      CaCl$_2$: -209.8 kcal
      ZnBr$_2$: -94.2 kcal

7.2   H$^+$: -251.3 kcal mol^{-1}
      Na$^+$: -89.0 kcal mol^{-1}
      Mg^{2+}: -437.1 kcal mol^{-1}
      Al^{3+}: -1075.7 kcal mol^{-1}
      Cl$^-$: -85.0 kcal mol^{-1}
      Br$^-$: -81.7 kcal mol^{-1}

7.3 4.23×10^{-4} M

7.4   KNO$_3$:   0.1 M
      K$_2$SO$_4$:   0.3 M
      ZnSO$_4$:   0.4 M
      ZnCl$_2$:   0.3 M
      K$_4$Fe(CN)$_6$:   1.0 M

7.5   0.026

7.6   0.0249 M

7.7   pH $= 7.3$

7.8   (a)

(b) $pI = 6.0$

(c) (1) at pH 2.0;   (2) at pH 3.0 and 9.0;   (3) at pH 9.8;   (4) at pH 10.5

7.9 7.19×10^{-4} M

7.10 (a) $H_3N^+\overset{\displaystyle CH_2OH}{\underset{\displaystyle |}{CH}}COOH$

(b) and (c) $H_3N^+\overset{\displaystyle CH_2OH}{\underset{\displaystyle |}{CH}}COO^-$

(d) $H_2N\overset{\displaystyle CH_2OH}{\underset{\displaystyle |}{CH}}COO^-$

$pI = 5.68$

7.11 On the palmitate side $[Na^+] = 0.18$ M; $[Cl^-] = 0.08$ M; on the other side $[Na^+] = [Cl^-] = 0.12$ M

7.13 $pH = 4.62$

7.14 9.0×10^{-6} M

7.15 $pH = 10.8$

7.16 7.225×10^{-10}

7.17 0.36 mol

7.18 $pH = 5.13$

7.19 $pH = 5.35$

7.20 (a) 9.0×10^{-10} M;   (b) 3.35×10^{-6} M

7.21 5.57×10^{-5} M

7.22 $pH = 6.86$; $I = 0.02$ M

7.23 2.04; increase

7.24 123.8 kJ mol^{-1} $= 29.6$ kcal mol^{-1}

Chapter 8

8.1   (a) $H_2 \rightarrow 2H^+ + 2e^-$

$Cl_2 + 2e^- \rightarrow 2Cl^-$

$H_2 + Cl_2 \rightarrow 2H^+ + 2Cl^-$

$$\Delta E = \Delta E^\circ - \frac{RT}{2F} \ln [H^+]^2[Cl^-]^2$$

(b) $2Hg + 2Cl^- \rightarrow Hg_2Cl_2 + 2e^-$
$2H^+ + 2e^- \rightarrow H_2$
$2Hg + 2H^+ + 2Cl^- \rightarrow Hg_2Cl_2 + H_2$

$$\Delta E = \Delta E^\circ + \frac{RT}{2F} \ln [H^+]^2 [Cl^-]^2$$

(c) $Ag + Cl^- \qquad \rightarrow AgCl + e^-$
$2e^- + Hg_2Cl_2 \rightarrow 2Hg + 2Cl^-$
$2Ag + Hg_2Cl_2 \rightarrow 2AgCl + 2Hg$
$\Delta E = \Delta E^\circ$

8.2 $8.55 \times 10^7 \; M^{-2}$

8.3 $0.0178 \; V$

8.4 $0.357 \; V$; $H_2 + $ gluconolactone $+ 2H^+ \; (10^{-7} \; m) \rightarrow$ D-glucose $+ 2H^+ \; (1 \; m)$
1.26×10^{12} (for $H_2 + $ gluconolactone $\rightarrow$ D-glucose at pH 7)

8.5 $-24.1 \; kJ = -5.76 \; kcal$

8.6 $0.44 \; V$; $\Delta G^{\circ'} = -20.30 \; kcal$
$K = 8.06 \times 10^{14}$ at pH 7; $K = 8.06 \times 10^{12}$ at pH 6

8.7 (a) BH_2 is oxidized by A
(b) $0.1 \; V$
(c) None

8.8 $-30.8 \; kJ = -7.38 \; kcal$
$K' = 2.62 \times 10^5$

8.9 $-237.4 \; kJ = -57.1 \; kcal$

8.10 $1.86 \times 10^{-24} \; atm$

8.11 $-0.147 \; V$

8.12 $\Delta G_f^\circ = -144.3 \; kcal$; $\Delta H_f^\circ = -186.4 \; kcal$

Chapter 9

9.1 (a) $2.5 \times 10^{-7} \; mol \; dm^{-3} \; s^{-1}$
$2.5 \times 10^{-10} \; mol \; cm^{-3} \; s^{-1}$
$1.5 \times 10^{-8} \; mol \; cm^{-3} \; min^{-1}$
(b) $1.25 \times 10^{-5} \; dm^3 \; mol^{-1} \; s^{-1}$
$1.25 \times 10^{-2} \; cm^3 \; mol^{-1} \; s^{-1}$

9.2 (a) 3
(b) Both rates are $3.6 \times 10^{-3} \; mol \; dm^{-3} \; s^{-1}$
(c) No effect
(d) Rate of disappearance of Br^- decreased by a factor of 8; no effect on k.

9.3 $x = 1$; $y = 1$; $k = 6.21 \times 10^{-4}$ dm^3 mol^{-1} s^{-1}

9.4 $x = 2$, $y = 1$, $k = 1.64 \times 10^{-3}$ dm^6 mol^{-2} s^{-1}

9.5 15.0 kcal

9.6 12.2 kcal

9.7 34.1 kcal

9.8 $E = 20.1$ kcal; $\Delta H^{\ddagger} = 19.5$ kcal; $\Delta G^{\ddagger} = 24.0$ kcal
$A = 7.90 \times 10^9$ s^{-1}; $\Delta S^{\ddagger} = -15.2$ cal K^{-1}

9.9 $\Delta G^{\ddagger} = 26.0$ kcal mol^{-1}; $E = 21.5$ kcal mol^{-1};
$\Delta H^{\ddagger} = 20.9$ kcal mol^{-1}; $A = 5.21 \times 10^9$ s^{-1};
$\Delta S^{\ddagger} = -16.1$ cal K^{-1}

9.10 $\Delta H^{\ddagger} = 83.3$ kcal; $\Delta S^{\ddagger} = 174$ cal K^{-1}

9.11 $v = k_1[\text{A}][\text{B}]$

9.12 $v = k_1[\text{A}][\text{C}]$

9.13 $v = k_2\left(\dfrac{k_1}{k_{-1}}\right)^{\frac{1}{2}}[\text{A}]^{\frac{1}{2}}[\text{B}]$

9.14 $v = \dfrac{k_1 k_2[\text{A}][\text{B}]}{k_{-1} + k_2[\text{B}]}$

(a) $v = \dfrac{k_1 k_2}{k_{-1}}[\text{A}][\text{B}]$

(b) $v = k_1[\text{A}]$

9.15 $2\text{A} \rightarrow \text{X}$ (slow)
$\text{X} + 2\text{B} \rightarrow 2\text{Y} + 2\text{Z}$ (fast)

9.18 Two simultaneous reactions

9.19 Two consecutive reactions

Chapter 10

10.1 2.0×10^{-3} M $= 2.0$ mM

10.2 $K_m = 16 \times 10^{-6}$ M $= 16$ μM; $V = 0.22$ μM dm^{-3} s^{-1}

10.3 (a) $E = 1.95$ kcal mol^{-1}; $\Delta H^{\ddagger} = 1.35$ kcal mol^{-1};
  $\Delta G^{\ddagger} = 10.2$ kcal mol^{-1}; $\Delta S^{\ddagger} = -29.7$ cal K^{-1} mol^{-1}
  (b) $\Delta G^{\circ} = -4.7$ kcal mol^{-1}; $\Delta H^{\circ} = 3.4$ kcal mol^{-1};
  $\Delta S^{\circ} = 27.0$ cal K^{-1} mol^{-1}

10.4 pK ≈ 6.8

10.5 $K_m = 5.8 \times 10^{-3}$ M; $V = 5.0 \times 10^{-4}$ mol dm^{-3} s^{-1};
$k_c = 12.5$ s^{-1}

10.6 $E = 21.6$ kcal mol^{-1}; $\Delta H^{\ddagger} = 21.0$ kcal mol^{-1};
$\Delta G^{\ddagger} = 25.9$ kcal mol^{-1}; $\Delta S^{\ddagger} = -15.8$ cal K^{-1} mol^{-1}

10.7 $$v = \frac{k_1 k_2 k_3 [E]_o [A][B]}{k_{-1}k_3 + k_1 k_3 [A] + k_2 k_3 [B] + k_1 k_2 [A][B]}$$

10.10 0.0186 min^{-1}; 37 min

10.11 Lag, 53 min; $t_2 = 34.4$ min

Chapter 11

11.1 Yes; $n = 4$; $K = 940$ M^{-1}

11.2 Yes; $n = 3$

11.3 Yes; $n = 3$

11.4 70 900 g mol^{-1}

11.5 Glucose: 10.8 cm; tobacco mosaic virus: 0.96 cm

11.6 3.4 nm; 132 000 g mol^{-1}

11.7 2.25 kcal mol^{-1}; 0.45 mol

11.8 (a) Net charge $= +4$; movement towards cathode
(b) Net charge $= -\frac{1}{2}$; movement towards anode
    p$I \approx 7.6$

11.9 At pH 7.8 the two proteins would move in opposite directions (net charges $-\frac{1}{2}$ and $+1\frac{1}{2}$)

11.10 Net charges are:

	pH 6.0	pH 7.0	pH 8.6
Protein A	+2	0	−3
Protein B	+7	+4	0

A focuses at pH 7.0, B at pH 8.6

11.11 0.49 mm Hg

11.12 (a) 0.33 mm s^{-1}; (b) 4.19×10^{-15} m^3 s^{-1}; (c) 2×10^{10}

11.13 (a) 1.92×10^{-3} mol s^{-1}; (b) 1.92×10^{-2} impulse s^{-1} cell^{-1}

11.14 Inside: $[Na^+] = 48.1$ mM; $[Cl^-] = 56.1$ mM
Outside: $[Na^+] = 51.9$ mM; $[Cl^-] = 51.9$ mM
2.03 mV

11.15 0.61 s

Chapter 12

12.1 (a) $1, 2$; (b) $2, 2$; (c) $15, 15$
(d) $6, 8$; (c) $53, 78$; (f) $88, 138$

12.2 $k = 0.0485 \text{ days}^{-1} = 5.61 \times 10^{-7} \text{ s}^{-1}$
(a) 61.6%; (b) 37.9%; (c) 0.78%

12.3 $\tau = 87.0 \text{ days}$; $k = 9.22 \times 10^{-8} \text{ s}^{-1}$
(a) 2653; (b) 233

12.5 1033 counts per minute

12.6 $50 \ \mu\text{Ci mg}^{-1}$; 14.9 mg

12.7 1.47 g

12.8 0.65 g; 11.

12.9 Standard error $= 2.13\%$
95% error $= 4.17\%$

12.10 Standard error $= 0.67\%$
95% error $= 1.32\%$

12.11 Radiocarbon dates: 3135 to 3035 B.C.

12.12 Irish mud: 11 700 years old
Stonehenge: 3940 years old

12.13 25 rad

12.14 70.5 rad

INDEX

T

UNITS

Quantity	Symbol for Quantity	SI Unit and Symbol	Other Units
Length	l	metre (m)	1 angstrom (Å) = 10^{-10} m = 0.1 nm 1 micron = 10^{-6} m (not recommended)
Volume	V	cubic metre (m^3)	1 liter = 1 dm^3 (by definition)
Mass	m	kilogram (kg)	
Time	t	second (s)	1 min = 60 s; 1 year = 3.156×10^7 s
Frequency	ν	hertz (Hz = s^{-1})	
Temperature	T	kelvin (K)	°C = K − 273.15
Electric current	I	ampere (A)	
Electric charge	Q	coulomb (C = A s)	1 esu = 3.336×10^{-10} C
Electric potential	V, ϕ	volt (V = kg m^2 s^{-3} A^{-1})	
Electric resistance	R	ohm (Ω = V A^{-1})	
Electric capacitance	C	farad (F = C V^{-1})	
Force	F	newton (N = kg m s^{-2})	1 dyne = 10^{-5} N
Pressure	P	pascal (Pa = kg m^{-1} s^{-2})	1 atm = 1.01325×10^5 Pa 1 bar = 10^5 Pa 1 torr = 1 mm Hg = 133.322 Pa
Energy	E, U	joule (J = kg m^2 s^{-2})	1 erg = 10^{-7} J 1 cal = 4.184 J (by definition) 1 electron volt (eV) = 1.602×10^{-19} J 1 eV $molecule^{-1}$ = 96.47 kJ mol^{-1} $\qquad$ = 23.06 kcal mol^{-1} 1 liter-atm = 101.325 J $\qquad$ = 24.22 cal
Power	P	watt (W = J s^{-1})	

PHYSICAL CONSTANTS

Constant	SI Symbol	Value in SI and other units
Avogadro's number	N_A	6.022×10^{23} mol^{-1}
Electronic charge	e	1.602×10^{-19} C
Faraday constant	$F = N_A e$	96 485 C mol^{-1}
Gas constant	R	8.314 J K^{-1} mol^{-1} 1.987 cal K^{-1} mol^{-1} 0.08205 liter-atm K^{-1} mol^{-1}
Planck's constant	h	6.626×10^{-34} J s
Boltzmann constant	$\mathbf{k} = R/N_A$	1.381×10^{-23} J K^{-1}
Velocity of light	c	2.998×10^8 m s^{-1}
Permittivity of vacuum	ε_o	8.854×10^{-12} C^2 N^{-1} m^{-2}
Ratio of circumference to diameter of circle	π	3.1416
Base of natural logarithms	e	2.7183